P9-BJX-125

The Finite Element Method in Heat Transfer and Fluid Dynamics

Second Edition

J.N. Reddy
D.K. Gartling

CRC Press

Boca Raton   London   New York   Washington, D.C.

Library of Congress Cataloging-in-Publication Data

Reddy, J.N. (Junuthula Narasimha), 1945-
    The finite element method in heat transfer and fluid dynamics / J.N. Reddy, D.K.
Gartling.—2nd ed.
      p. cm.
    Includes bibliographical references and index.
    ISBN 0-8493-2355-X (alk. paper)
    1. Fluid dynamics—Mathematical models. 2. Heat—Transmission—Mathematical
models. 3. Finite element method. I. Gartling, David K. II. Title.

TA357 .R43 2000
620.1′06—dc21
                                                00-048638

This book contains information obtained from authentic and highly regarded sources. Reprinted material is quoted with permission, and sources are indicated. A wide variety of references are listed. Reasonable efforts have been made to publish reliable data and information, but the author and the publisher cannot assume responsibility for the validity of all materials or for the consequences of their use.

Neither this book nor any part may be reproduced or transmitted in any form or by any means, electronic or mechanical, including photocopying, microfilming, and recording, or by any information storage or retrieval system, without prior permission in writing from the publisher.

The consent of CRC Press LLC does not extend to copying for general distribution, for promotion, for creating new works, or for resale. Specific permission must be obtained in writing from CRC Press LLC for such copying.

Direct all inquiries to CRC Press LLC, 2000 N.W. Corporate Blvd., Boca Raton, Florida 33431.

Trademark Notice: Product or corporate names may be trademarks or registered trademarks, and are used only for identification and explanation, without intent to infringe.

© 2001 by CRC Press LLC

No claim to original U.S. Government works
International Standard Book Number 0-8493-2355-X
Library of Congress Card Number 00-048638
Printed in the United States of America 1 2 3 4 5 6 7 8 9 0
Printed on acid-free paper

To our wives

Aruna and Laura

About the Authors

J. N. Reddy earned a Ph.D. in Engineering Mechanics and worked as a Postdoctoral Fellow at the University of Texas at Austin, Research Scientist for Lockheed Missiles and Space Company during 1974-75, and taught at the University of Oklahoma from 1975 to 1980 and Virginia Polytechnic Institute & State University from 1980 to 1992. Currently, he is a Distinguished Professor and the inaugural holder of the *Oscar S. Wyatt Endowed Chair* at Texas A&M University. Dr. Reddy authored over 250 journal papers and 12 text books on the theory and finite element analysis of problems in applied mechanics, laminated composite plates and shells, computational fluid mechanics and heat transfer, and applied mathematics. Other books of Dr. Reddy include *An Introduction to the Finite Element Method*, McGraw-Hill (1984; 1993); *Energy and Variational Methods in Applied Mechanics*, John Wiley (1984); *Applied Functional Analysis and Variational Methods in Engineering*, McGraw-Hill (1986); *Mechanics of Laminated Composite Plates: Theory and Analysis*, CRC Press (1997); *Theory and Analysis of Elastic Plates*, Taylor & Francis (1999). Dr. Reddy is the recipient of the Walter L. Huber Civil Engineering Research Prize of the American Society of Civil Engineers (ASCE), the 1992 Worcester Reed Warner Medal and 1995 Charles Russ Richards Memorial Award of the American Society of Mechanical Engineers (ASME), and the 2000 Excellence in the Field of Composites Award. Dr. Reddy serves on the editorial boards of numerous journals, including *International Journal for Numerical Methods in Engineering* and *International Journal for Numerical Methods in Fluids*. He is the editor of *Mechanics of Composite Materials and Structures* (Taylor & Francis), and *International Journal of Computational Engineering Science* (World Scientific Publishers).

David K. Gartling is a Senior Scientist in the Engineering Sciences Center at Sandia National Laboratories, Albuquerque, New Mexico. He earned his B.S. and M.S. in Aerospace Engineering at the University of Texas at Austin and completed the diploma course at the von Kármán Institute for Fluid Dynamics in Brussels, Belgium. After completion of his Ph.D. in Aerospace Engineering at the University of Texas at Austin, he joined the technical staff at Sandia National Laboratories. Dr. Gartling was a Visiting Associate Professor in the Mechanical Engineering Department at the University of Sydney, Australia under a Fulbright Fellowship, and later he was a Supervisor in the Fluid and Thermal Sciences Department at Sandia National Laboratories. Dr. Gartling published numerous papers dealing with finite element model development and finite element analysis of heat transfer and fluid dynamics problems of practical importance. He is presently a member of several professional societies, and serves on the editorial boards of several archival journals, including *International Journal for Numerical Methods in Fluids*, *Communications in Applied Numerical Methods* and *International Journal of Computational Engineering Science*.

Preface to the Second Edition

In the six years since the first edition of this book appeared some significant changes have occurred in the area of computational mechanics in general and in computational fluid mechanics and heat transfer in particular. Foremost among these changes have been the extraordinary increase in performance in desktop computing platforms and the arrival in significant numbers of parallel computers. This widespread availability of capable computing hardware has predictably lead to the increased demand for computer simulation of products and processes during the engineering design and manufacturing process. Our original thesis that the finite element method was very well suited to general purpose and commercial software continues to hold true, as numerous programs are now available for the simulation of all types of applied mechanics problems. The range of applications of finite element analysis in fluid mechanics and heat transfer has become quite remarkable with complex, realistic simulations being carried out on a routine basis. The combination of hardware performance and reliable finite element algorithms has made these advances possible. Another significant change in computational mechanics is the increase in multidisciplinary (multiphysics) problems and their solution via finite element methods. Again, the increase in hardware performance has contributed to these types of computationally intensive problems. However, the inroads made by the finite element method in all areas of mechanics have also had a positive influence on coupled analyses. The commonality of finite element formulation, approximation and solution among the various boundary value problems in mechanics eases considerably the contemplation of multiphysics solutions and software. A final change in the numerical simulation arena comes from the implementation side of the finite element method. The demand for software capability and reliability has increased in step with the hardware performance. The use of parallel computers has added to the complexity of the implementation. All of these attributes lead to the conclusion that finite element implementation, if done well, will require some significant knowledge from areas in computer science. The time of general purpose codes being developed and maintained by one or two individuals is past and multi-talented teams now provide the most modern software.

Our focus for the present edition of this book remains the same – the education of the individual who is interested in good numerical methods for the study of fluid mechanics and heat transfer phenomena. The text remains practical in scope and content with an emphasis on computational procedures that we have found effective on a wide spectrum of applications. Little, if any, material has been deleted from the first edition. New material has been added in almost all chapters along with some rearrangement of topics to improve overall clarity and maintain the step-wise addition of increasingly complex material. Chapter 1 contains the general description of the boundary value problems of interest and has been augmented with a section on chemically reactive systems and additional discussion of change of phase. Chapter 2 continues with the introduction of the finite element method and is essentially unchanged. The thermal conduction and radiation problem is discussed in great detail in Chapter 3, and it has new sections covering specialized finite elements and advanced topics in thermal analysis. The advanced topics section includes descriptions of difficult boundary conditions, such as multipoint constraints, contact and bulk nodes, material motion and kinematics, and methods for chemically reactive solids. The isothermal, viscous flow problem is the topic of Chapter 4. New sections in this chapter cover stabilized finite element methods and a general discussion of methods for free surface problems; the section on turbulence modeling has also been moved to this chapter. Chapter 5 extends the finite element method to non-isothermal flows and is largely unchanged. Non-Newtonian flow problems, both inelastic and viscoelastic, are now included in a revised and updated Chapter 6. A completely new Chapter 7 is focused on formulations and algorithms for multidisciplinary problems involving fluid mechanics, heat transfer, solid mechanics and electromagnetics. This chapter outlines many of the possible types of coupling, describes the finite element equations for each mechanics area, and presents a number of realistic numerical examples. The last chapter on advanced topics is now devoted exclusively to a discussion of parallel computing including some general discussion of the parallel architecture and sections on parallel implementation of finite element models. Organization of the text, equation numbering, references and symbols retain the same style as used in the first edition.

The second author thanks his numerous present and former colleagues at Sandia National Laboratories who continue to provide a wealth of challenging problems in applied and computational mechanics. Specific acknowledgments must go to Drs. Mike Glass, Rick Givler, Charles Hickox, Roy Hogan and Phil Sackinger for collaboration and assistance in much of the algorithm development and demonstration simulations cited in this work. Portions of this book are adapted from work performed at Sandia National Laboratories under Contract No. DE–AC04–94AL85000 awarded by the U.S. Department of Energy and are used with permission. The authors dedicate this book to their wives, Aruna and Laura, who have graciously tolerated the authors' preoccupation with the writing of the book.

J. N. Reddy
College Station, Texas

D. K. Gartling
Albuquerque, New Mexico

Preface to the First Edition

The numerical simulation of fluid mechanics and heat transfer problems has become a routine part of engineering practice as well as a focus for fundamental and applied research. Though there are still various topical areas where our physical understanding and/or ineffective numerical algorithms limit the investigation, a large number of complex phenomena can now be confidently studied via numerical simulation. Though finite difference methods have and will continue to play a major role in computational fluid dynamics (CFD) and heat transfer, finite element techniques have spurred the explosive development of "general purpose" methods and the growth of commercial software. The inherent strengths of the finite element method such as unstructured meshes, element-by-element formulation and processing, and the simplicity and rigor of boundary condition application are being coupled with modern developments in automatic mesh generation, adaptive meshing, and improved solution techniques to produce accurate and reliable simulation packages that are widely accessible. Improvements in computer hardware and system software (e.g., powerful workstations and window environments) have contributed significantly to streamlining the numerical simulation process. The finite element method in fluid mechanics and heat transfer has rapidly caught up with the well-established solid mechanics community in simulation capabilities.

As in any rapidly developing field, the education of the non-expert user community is of primary importance. The present text is an attempt to fill a need for those interested in using the finite element method in the study of fluid mechanics and heat transfer. It is a pragmatic book that views numerical computation as a means to an end—we do not dwell on theory or proof. Other fundamental and theoretical text books that cover these aspects are available or anticipated. The emphasis here is on presenting a useful methodology for a limited but significant class of problems dealing with heat conduction, incompressible viscous flows, and convection heat transfer.

The text has been developed out of our experience and course notes used in teaching graduate courses and continuing education courses to a wide spectrum of students. To gain the most from the text the student should have a reasonable background in fluid mechanics and heat transfer as would normally be found in most mechanical, aerospace, chemical or engineering mechanics curriculums. An introductory knowledge of finite element techniques would be very helpful but not essential; some familiarity with basic numerical analysis, linear algebra, and numerical integration would also be of assistance.

Our approach to the finite element method for fluid mechanics and heat transfer has been designed as a series of incremental steps of increasing complexity. In Chapter 1, the continuum boundary value problems that form the central focus of the book are described in some detail. We have tried to be as general as possible in describing the varied physical phenomena that may be encountered within the limits of non-isothermal, incompressible, viscous flows. Chapter 2 introduces the finite element method by application to a simplified, two-dimensional heat conduction problem. All of the necessary machinery for constructing weak forms of a partial differential equation and building a finite element model are introduced here and demonstrated by application. Chapter 3 recaps parts of Chapter 2 and extends the finite element method to three dimensions, time dependence, and practical applications in conduction heat transfer. Isothermal viscous fluid mechanics formulations are described in Chapter 4 along with a significant section on the solution of nonlinear equations developed from the flow problem. Chapter 5 extends the viscous flow problem to consider convective heat transfer formulations and applications. Inelastic non-Newtonian flows and free surface problems are covered in Chapter 6. The complex topic of viscoelastic flow simulation is surveyed in Chapter 7. The last chapter concludes the text with a survey of several advanced topics, including turbulence modeling. The coverage of each topic is sufficient to allow the reader to understand the basic methodology, use existing simulation software with confidence, and allow development of some simpler, special purpose computer codes. Example problems ranging from simple benchmarks to practical engineering solutions are included with each topical area. Adequate references to the relevant literature have also been included for those desiring a more encyclopedic coverage of a specific topic.

The text is organized into major sections within each chapter. Equations are numbered consecutively within each major section. Within a section, reference to an equation is by its sequential number; references to equations outside the current section have a full section, equation number citation. Vectors, tensors and matrices are denoted by boldface letters. The vectors of interpolation (shape) functions in this book are denoted by Greek symbols (Ψ, Θ, Φ). Bibliographic information for literature cited in the text is numbered sequentially within each chapter and collected at the end of the chapter.

The first author would like to thank Mr. M. S. Ravisankar for his help with Chapter 8 topics and Mr. Praveen Gramma for reading the manuscript. The second author would like to thank his numerous present and former colleagues at Sandia National Laboratories who have provided a seemingly endless stream of challenging problems and taught him much about the practice of computational mechanics. Specific acknowledgments must go to Drs. Charles Hickox, Rick Givler, Roy Hogan, Phil Sackinger, Randy Schunk, Rekha Rao, and Steve Rottler for their suggestions and comments on early versions of the text. Portions of this book are adapted from work performed at Sandia National Laboratories under Contract No. DE–AC04–76DP00789 awarded by the U.S. Department of Energy and are used with permission.

J. N. Reddy
College Station, Texas

D. K. Gartling
Albuquerque, New Mexico

Contents

Chapter 1

Equations of Heat Transfer and Fluid Mechanics

1.1 Introduction

The disciplines of heat transfer and fluid dynamics are subjects that are closely coupled and form the core of many engineering and scientific studies. Areas as diverse as aerodynamics, meteorology, geology and geophysics, biology, material science and manufacturing all rely heavily on the ability to predict, understand and control complex fluid and thermal systems. In order to properly outline the scope of the present text, it is important to first provide brief definitions of the general areas of heat transfer and fluid dynamics.

1.1.1 Heat Transfer

Heat transfer is a branch of engineering that deals with the transfer of thermal energy from one point to another within a medium or from one medium to another due to the occurrence of a temperature difference. Heat transfer may take place in one or more of its three basic forms: conduction, convection and radiation. The transfer of heat within a medium due to a diffusion process is called conduction heat transfer. The Fourier heat conduction law states that the heat flow is proportional to the temperature gradient. The constant of proportionality depends, among other things, on a material parameter known as the *thermal conductivity* of the material.

Convection heat transfer is usually defined as energy transport effected by the motion of a fluid. The convection heat transfer between two dissimilar media is governed by Newton's law of cooling. It states that the heat flow is proportional to the difference of the temperatures of the two media. The proportionality constant is called the *convection heat transfer coefficient* or *film conductance*.

Thermal radiation is defined as radiant (electromagnetic) energy emitted by a medium and is due solely to the temperature of the medium. Radiant energy exchange between surfaces or between a region and its surroundings is described by the Stefan–Boltzmann law, which states that the radiant energy transmitted is proportional to the difference of the fourth power of the temperatures of the surfaces. The proportionality parameter is known as the *Stefan–Boltzmann constant*.

1.1.2 Fluid Mechanics

Fluid mechanics is one of the oldest branches of physics, and is concerned with the motion of gases and liquids and their interaction with the surroundings. For example, the flight of birds in the air and the motion of fish in the water can be understood by the principles of fluid mechanics. Such understanding helps us design airplanes and ships. The formation of tornadoes, hurricanes, and thunderstorms can also be explained with the help of the equations of fluid mechanics.

A fluid state of matter, as opposed to the solid state of matter, is characterized by the relative mobility of its molecules.   Very strong intermolecular attractive forces exist in solids which are responsible for the property of relative rigidity (or stiffness) in solids. The intermolecular forces are weaker in liquids and extremely small in gases. The stress in a solid body is proportional to the strain (i.e., deformation per unit length), while the stress in a fluid is proportional to the time rate of strain (i.e., rate of deformation). The proportionality parameter in the case of fluids is known as the *viscosity*. It is a measure of the intermolecular forces exerted as layers of fluid attempt to slide past one another. The viscosity of a fluid, in general, is a function of the thermodynamic state of the fluid and in some cases the strain rate.

Fluid mechanics is a very broad area and is traditionally divided into smaller topical areas based on characteristics of the fluid properties or the basic nature of the flow.   An *inviscid fluid* is one where the viscosity is assumed to be zero.   An *incompressible fluid* is one with constant density or an *incompressible flow* is one in which density variations (compared to a reference density) are negligible. An inviscid and incompressible fluid is termed an *ideal* or a *perfect* fluid. A *real fluid* is one with finite viscosity, and it may or may not be incompressible. When the viscosity of a fluid depends only on thermodynamic properties, and the stress is linearly related to the strain rate, the fluid is said to be *Newtonian*. A *non-Newtonian fluid* is one which does not obey the Newtonian (i.e., linear) stress-strain rate relation. A non-Newtonian constitutive relation can be of algebraic (e.g., power-law), differential, or integral type.

The flow of viscous fluids can also be classified into two major types: a smooth, orderly motion is called *laminar flow*, and a random, fluctuating motion is called *turbulent flow*.   The nature of a viscous flow can be characterized by a nondimensional parameter known as the *Reynolds number*, $Re = \rho U L / \mu$, which is defined as the ratio of inertial forces ρU^2 to viscous forces $\mu U / L$. Here ρ denotes the density of the fluid, U the characteristic flow velocity, μ is the fluid viscosity, and L is a characteristic dimension of the flow region. High viscosity fluids and/or small velocities produce relatively small Reynolds numbers and a laminar flow. The flow of less viscous fluids and/or higher velocities lead to higher Reynolds numbers and a turbulent flow. Transitional flows contain regions of both laminar and turbulent flows.

1.2 Present Study

The purpose of the present book is to describe the use of the finite element method for the solution of engineering problems in well-defined areas of heat transfer and fluid mechanics. In particular, we will concentrate on finite element computational procedures for the areas of conduction and convection heat transfer. Since convection relies heavily on fluid dynamics, the study of isothermal flows will play a prominent role in developing the overall subject. To focus the presentation, all fluid mechanics applications here will be limited to the area of viscous, incompressible flows. Radiation heat transfer will be discussed to the extent that it affects thermal boundary conditions and heat transfer in enclosures. In addition, nonisothermal flows in porous media will be considered as an extension of the viscous flow problem. Complexities such as change of phase, free surface boundaries, and non-Newtonian constitutive behavior will also be considered. Also, there is some description of the interaction between fluid and thermal problems and other areas of applied mechanics.

In the remainder of this chapter, we will provide a résumé of the partial differential equations that describe heat transfer and fluid mechanics in the continuum. The boundary conditions for typical problems are also presented, and the special conditions associated with change of phase problems, free surface flows, and enclosure radiation are discussed.

1.3 Governing Equations of a Continuum

1.3.1 Introduction

In the present study we are interested in the combined convective and conductive transport of thermal energy in a material region, Ω. In the general case, the region Ω is made up of subregions containing a moving fluid in Ω_f and a solid body in Ω_s, where $\Omega = \Omega_f \cup \Omega_s$. The motion or equilibrium of a continuous medium is governed by global conservation principles.

There are two alternative descriptions used to express the conservation laws in analytical form. In the first, one considers the motion of all matter passing through a *fixed spatial location*. Here one is interested in various properties (e.g., velocity, pressure, temperature, density, and so on) of the matter that instantly occupies the fixed spatial location. This description is called the *Eulerian description* or *spatial description*. In the second, one focuses attention on a *set of fixed material particles*, irrespective of their spatial locations. The relative displacements of these particles and the stress caused by external forces and temperature are of interest in this case. This description is known as the *Lagrangian description* or *material description*. The Eulerian description is most commonly used to study fluid flows and convective heat transfer, while the Lagrangian description is generally used to study solid body heat conduction and the stress and deformation of solid bodies. Here we present the governing equations of a continuous medium based primarily on the Eulerian description. For a derivation of the equations, the reader may consult the books on continuum mechanics (e.g., Bird, Stewart, Lightfoot [1], Malvern [2], and Reddy and Rasmussen [3]), heat transfer (e.g., see Bejan [4], Holman [5], and Özisik [6,7]), and fluid mechanics [e.g., Batchelor [8] and Schlichting [9]).

1.3.2 Conservation of Mass; the Continuity Equation

The principle of conservation of mass can be stated as *the time rate of change of mass in a fixed volume is equal to the net rate of flow of mass across the surface.* The mathematical statement of the principle results in the following equation, known as the *continuity equation*

$$\frac{\partial \rho}{\partial t} + \nabla \cdot (\rho \mathbf{u}) = 0 \tag{1.3.1}$$

where ρ is the density (kg/m^3) of the medium, $\mathbf{u}$ the velocity vector (m/s), and ∇ is the gradient operator. Introducing the *material derivative* or *Eulerian derivative* operator D/Dt

$$\frac{D}{Dt} = \frac{\partial}{\partial t} + \mathbf{u} \cdot \nabla, \tag{1.3.2}$$

Eq. (1.3.1) can be expressed in the alternative form

$$\frac{D\rho}{Dt} + \rho \nabla \cdot \mathbf{u} = 0 \tag{1.3.3}$$

For steady-state conditions, the continuity equation becomes

$$\nabla \cdot (\rho \mathbf{u}) = 0 \tag{1.3.4}$$

When the density changes following a fluid particle are negligible, the continuum is termed *incompressible* and we have $D\rho/Dt = 0$. The continuity equation (1.3.3) then becomes

$$\nabla \cdot \mathbf{u} = 0 \tag{1.3.5}$$

which is often referred to as the incompressibility condition.

1.3.3 Conservation of Momenta

The principle of conservation of linear momentum (or Newton's Second Law of motion) states that *the time rate of change of linear momentum of a given set of particles is equal to the vector sum of all the external forces acting on the particles of the set, provided Newton's Third Law of action and reaction governs the internal forces.* Newton's Second Law can be written as

$$\rho \frac{D\mathbf{u}}{Dt} = \nabla \cdot \sigma + \rho \mathbf{f} \tag{1.3.6}$$

where σ is the Cauchy stress tensor (N/m^2) and $\mathbf{f}$ is the body force vector, measured per unit mass. Equation (1.3.6) describes the motion of a continuous medium, and in fluid mechanics they are also known as the *Navier equations*.

The principle of conservation of angular momentum can be stated as *the time rate of change of the total moment of momentum of a given set of particles is equal to the vector sum of the moments of the external forces acting on the system*. In the absence of distributed couples, the principle leads to the symmetry of the stress tensor:

$$\sigma = (\sigma)^{\mathrm{T}} \tag{1.3.7}$$

where the superscript T denotes the transpose of the enclosed quantity.

1.3.4 Conservation of Energy

The law of conservation of energy (or the First Law of Thermodynamics) states that *the time rate of change of the total energy is equal to the sum of the rate of work done by applied forces and the change of heat content per unit time*. For an incompressible fluid, the First Law of Thermodynamics can be expressed as

$$\rho C \frac{DT}{Dt} = -\nabla \cdot \mathbf{q} + Q + \Phi \tag{1.3.8}$$

where T denotes the temperature $(^\circ C)$, $\mathbf{q}$ the heat flux vector (W/m^2), Q is the internal heat generation (W/m^3), Φ is the viscous dissipation function

$$\Phi = \tau : \mathbf{D} \tag{1.3.9}$$

and C is the specific heat $[J/(kg\cdot^\circ C)]$. For constant density processes, the specific heat at constant pressure, C_p, and the specific heat at constant volume, C_v, are the same and are labeled C. In Eq. (1.3.9), τ is the viscous part of the stress tensor σ, and $\mathbf{D}$ is the strain rate tensor. Other types of internal heat generation may arise from other physical processes such as chemical reactions and Joule heating. These quantities are defined more completely in the next section and in particular applications throughout the text.

1.3.5 Constitutive Equations

The fluids of interest, for the moment, are assumed to be Newtonian (i.e., the constitutive relations are linear). Non-Newtonian fluids will be considered in a later chapter. Further, the fluids are assumed to be incompressible, and the flow is laminar. For flows involving buoyancy forces, an extended form of the Boussinesq approximation (see Gray and Giorgini [10] and Gartling and Hickox [11]) is invoked, which allows the fluid properties to be expressed as functions of the thermodynamic state (e.g., pressure and temperature) and the density ρ to vary with temperature T according to the relation

$$\rho = \rho_0[1 - \beta(T - T_0)] \tag{1.3.10}$$

where β is the coefficient of thermal expansion (per $°\mathrm{C}$) and the subscript zero indicates a reference condition. The variation of density as given in Eq. (1.3.10) is permitted only in the description of the body force; the density in all other situations is assumed to be that of the reference state, ρ_0.

For viscous incompressible fluids the total stress σ can be decomposed into hydrostatic and viscous parts:

$$\sigma = -P\mathbf{I} + \tau \tag{1.3.11}$$

where P is the hydrostatic pressure, $\mathbf{I}$ is the unit tensor and τ is the viscous stress tensor. For Newtonian fluids, the viscous stress tensor is related to the strain rate tensor $\mathbf{D}$ by

$$\tau = \mathbf{C} : \mathbf{D} \tag{1.3.12}$$

where $\mathbf{C}$ is the fourth-order tensor of fluid properties and $\mathbf{D}$ is the strain rate tensor (or rate of deformation tensor)

$$\mathbf{D} = \frac{1}{2}\left[(\nabla\mathbf{u}) + (\nabla\mathbf{u})^T\right] \tag{1.3.13}$$

For an isotropic fluid (i.e., whose material properties are independent of direction), the fourth-order tensor $\mathbf{C}$ can be expressed in terms of two constants λ and μ, called Lamé constants, and Eq. (1.3.12) takes the form

$$\tau = \lambda(\mathrm{tr}\ \mathbf{D})\mathbf{I} + 2\mu\mathbf{D} \tag{1.3.14a}$$

where $(tr\mathbf{D})$ denotes the *trace* (or sum of the diagonal elements) of the matrix $\mathbf{D}$. For an incompressibe fluid, we have $tr\mathbf{D} = 0$ and Eq. (1.3.14a) becomes

$$\tau = 2\mu\mathbf{D} \tag{1.3.14b}$$

The Fourier heat conduction law states that

$$\mathbf{q} = -\mathbf{k} \cdot \nabla T \tag{1.3.15a}$$

where $\mathbf{k}$ denotes the conductivity tensor of order two. For an isotropic medium, $\mathbf{k}$ is of the form

$$\mathbf{k} = k\mathbf{I} \tag{1.3.15b}$$

where k denotes the thermal conductivity $[\mathrm{W}/(\mathrm{m}\cdot°\mathrm{C})]$ of the medium, and $\mathbf{I}$ is the unit tensor.

The material coefficients, μ, C, β, and $\mathbf{k}$ are generally functions of the fluid temperature; property variations as a function of a second thermodynamic variable (e.g., pressure) are possible but usually are unimportant for an incompressible fluid.   Solid materials may also have specific heats and conductivities that vary with temperature. In addition, the solid conductivity may vary with spatial position and, in general, remains a symmetric second-order tensor (i.e., $\mathbf{k}^{\mathrm{T}} = \mathbf{k}$) for anisotropic materials.   The volumetric heat source for the fluid and/or solid may be a function of temperature, time, and spatial location. In developing the finite element models in the coming chapters, the dependence of the material properties on the spatial location is assumed. The dependence of the viscosity and conductivity on the strain rates and/or temperature are also discussed in the sequel.

1.4 Governing Equations in Terms of Primitive Variables

1.4.1 Vector Form

Equations (1.3.3), (1.3.6), and (1.3.8) can be expressed in terms of the primitive variables $(\mathbf{u}, P, T)$ by means of Eqs. (1.3.12)–(1.3.15). The results are summarized below for isotropic, Newtonian, viscous, incompressible fluids in the presence of buoyancy forces:

$$\nabla \cdot \mathbf{u} = 0 \tag{1.4.1}$$

$$\rho_0 \left(\frac{\partial \mathbf{u}}{\partial t} + \mathbf{u} \cdot \nabla \mathbf{u} \right) = -\nabla P + \nabla \cdot \left\{ \mu \left[(\nabla \mathbf{u}) + (\nabla \mathbf{u})^T \right] \right\}$$
$$+ \rho_0 \mathbf{f} - \rho_0 \mathbf{g} \beta (T - T_0) \tag{1.4.2}$$

$$\rho_0 C \left(\frac{\partial T}{\partial t} + \mathbf{u} \cdot \nabla T \right) = \nabla \cdot (k \nabla T) + Q + \Phi \tag{1.4.3}$$

where $\mathbf{u}$ represents the velocity vector, ρ_0 the density, $\mathbf{g}$ the gravity force vector per unit mass, T the temperature, C the specific heat of the fluid, and Q the rate of heat generation.

The above equations are valid for the fluid region Ω_f. In the solid region Ω_s, the fluid velocity is zero, $\mathbf{u} = \mathbf{0}$, and the only relevant equation is (1.4.3). The energy equation (1.4.3) for the solid region is given by

$$\rho_s C_s \frac{\partial T}{\partial t} = \nabla \cdot (k_s \nabla T) + Q_s \tag{1.4.4}$$

In writing Eq. (1.4.4) it is assumed that the solid is stationary with respect to the coordinate frame such that the advective transport of energy [i.e., the velocity-dependent part of Eq. (1.4.3)] need not be considered.

1.4.2 Cartesian Component Form

The vector form of the equations in (1.4.1)–(1.4.3) allows us to express them in any coordinate system. In the Cartesian coordinate system (x_1, x_2, x_3), the kinematic and constitutive relations become

$$D_{ij} = \frac{1}{2} \left(\frac{\partial u_i}{\partial x_j} + \frac{\partial u_j}{\partial x_i} \right) \tag{1.4.5}$$

$$\sigma_{ij} = -P \delta_{ij} + \tau_{ij} \; ; \quad \tau_{ij} = 2\mu D_{ij} \tag{1.4.6}$$

The conservation equations can be expressed as

$$\frac{\partial u_i}{\partial x_i} = 0 \tag{1.4.7}$$

$$\rho_0 \left(\frac{\partial u_i}{\partial t} + u_j \frac{\partial u_i}{\partial x_j} \right) = \frac{\partial}{\partial x_j} \left[-P \delta_{ij} + \mu \left(\frac{\partial u_i}{\partial x_j} + \frac{\partial u_j}{\partial x_i} \right) \right]$$
$$+ \rho_0 f_i - \rho_0 g_i \beta (T - T_0) \tag{1.4.8}$$

$$\rho_0 C \left(\frac{\partial T}{\partial t} + u_j \frac{\partial T}{\partial x_j} \right) = \frac{\partial}{\partial x_i} \left(k \frac{\partial T}{\partial x_i} \right) + Q + 2\mu D_{ij} D_{ij} \tag{1.4.9}$$

for the fluid region Ω_f and

$$\rho_s C_s \frac{\partial T}{\partial t} = \frac{\partial}{\partial x_i}\left(k_s \frac{\partial T}{\partial x_i}\right) + Q_s \tag{1.4.10}$$

for the solid region Ω_s. Equations (1.4.5)–(1.4.10) are written for a Cartesian geometry in an Eulerian reference frame, with the indices $i, j = 1, 2, 3$ (or $i, j = 1, 2$ for two-dimensional problems); the Einstein summation convention on repeated indices is used (see Reddy and Rasmussen [3], pp. 18–20).

1.4.3 Cylindrical Component Form

Equations (1.4.1)–(1.4.3) can also be expressed in a cylindrical coordinate system, (r, θ, z), by writing all vectors and tensors, including the del operator, in terms of components in a cylindrical coordinate system. For example, the del operator and the material time derivative operators in the cylindrical coordinate system are defined by (see Reddy and Rasmussen [3], p. 94 and p. 128)

$$\nabla = \hat{e}_r \frac{\partial}{\partial r} + \hat{e}_\theta \frac{1}{r}\frac{\partial}{\partial \theta} + \hat{e}_z \frac{\partial}{\partial z} \tag{1.4.11}$$

$$\frac{D}{Dt} = \frac{\partial}{\partial t} + u_r \frac{\partial}{\partial r} + \frac{u_\theta}{r}\frac{\partial}{\partial \theta} + u_z \frac{\partial}{\partial z} \tag{1.4.12}$$

where $(\hat{e}_r, \hat{e}_\theta, \hat{e}_z)$ are the unit basis vectors and (u_r, u_θ, u_z) are the velocity components in the r, θ, and z directions, respectively. Note that the basis vector $\hat{e}_z$ is constant while the vectors $\hat{e}_r$ and $\hat{e}_\theta$ depend on the angular coordinate θ. Thus the derivatives of the basis vectors $\hat{e}_r$ and $\hat{e}_\theta$ with respect to the coordinates r and z are zero, and the derivatives with respect to θ are given by

$$\frac{\partial \hat{e}_r}{\partial \theta} = \hat{e}_\theta; \quad \frac{\partial \hat{e}_\theta}{\partial \theta} = -\hat{e}_r \tag{1.4.13}$$

The kinematic and constitutive relations are

$$\begin{aligned}
D_{rr} &= \frac{\partial u_r}{\partial r}; & D_{\theta\theta} &= \frac{1}{r}\frac{\partial u_\theta}{\partial \theta} + \frac{u_r}{r} \\
D_{zz} &= \frac{\partial u_z}{\partial z}; & 2D_{r\theta} &= \frac{\partial u_\theta}{\partial r} - \frac{u_\theta}{r} + \frac{1}{r}\frac{\partial u_r}{\partial \theta} \\
2D_{\theta z} &= \frac{1}{r}\frac{\partial u_z}{\partial \theta} + \frac{\partial u_\theta}{\partial z}; & 2D_{zr} &= \frac{\partial u_r}{\partial z} + \frac{\partial u_z}{\partial r}
\end{aligned} \tag{1.4.14}$$

$$\sigma_{rr} = -P + 2\mu D_{rr}; \quad \sigma_{\theta\theta} = -P + 2\mu D_{\theta\theta}; \quad \sigma_{zz} = -P + 2\mu D_{zz} \tag{1.4.15a}$$
$$\sigma_{r\theta} = 2\mu D_{r\theta}; \quad \sigma_{\theta z} = 2\mu D_{\theta z}; \quad \sigma_{zr} = 2\mu D_{zr} \tag{1.4.15b}$$

The governing equations are summarized below:

$$\frac{1}{r}\frac{\partial}{\partial r}(r u_r) + \frac{1}{r}\frac{\partial u_\theta}{\partial \theta} + \frac{\partial u_z}{\partial z} = 0 \tag{1.4.16}$$

$$\rho_0 \left(\frac{Du_r}{Dt} - \frac{u_\theta^2}{r} \right) = \rho f_r + \frac{1}{r} \left[\frac{\partial (r\sigma_{rr})}{\partial r} + \frac{\partial \sigma_{r\theta}}{\partial \theta} + \frac{\partial (r\sigma_{zr})}{\partial z} \right] - \frac{\sigma_{\theta\theta}}{r} \qquad (1.4.17a)$$

$$\rho_0 \left(\frac{Du_\theta}{Dt} + \frac{u_r u_\theta}{r} \right) = \rho f_\theta + \frac{1}{r} \left[\frac{\partial (r\sigma_{r\theta})}{\partial r} + \frac{\partial \sigma_{\theta\theta}}{\partial \theta} + \frac{\partial (r\sigma_{\theta z})}{\partial z} \right] + \frac{\sigma_{r\theta}}{r} \qquad (1.4.17b)$$

$$\rho_0 \left(\frac{Du_z}{Dt} \right) = \rho f_z + \frac{1}{r} \left[\frac{\partial (r\sigma_{zr})}{\partial r} + \frac{\partial \sigma_{z\theta}}{\partial \theta} + \frac{\partial (r\sigma_{zz})}{\partial z} \right] \qquad (1.4.17c)$$

$$\rho_0 C \left(\frac{\partial T}{\partial t} + u_r \frac{\partial T}{\partial r} + \frac{u_\theta}{r} \frac{\partial T}{\partial \theta} + u_z \frac{\partial T}{\partial z} \right) = \frac{1}{r} \frac{\partial}{\partial r} \left(r k_{rr} \frac{\partial T}{\partial r} \right) + \frac{1}{r^2} \frac{\partial}{\partial \theta} \left(k_{\theta\theta} \frac{\partial T}{\partial \theta} \right)$$
$$+ \frac{\partial}{\partial z} \left(k_{zz} \frac{\partial T}{\partial z} \right) + \Phi + Q \qquad (1.4.18)$$

where the stress components are known in terms of the velocity components *via* equations (1.4.14) and (1.4.15), and the viscous dissipation Φ is given by

$$\Phi = 2\mu \left[\left(\frac{\partial u_r}{\partial r} \right)^2 + \left(\frac{1}{r} \frac{\partial u_\theta}{\partial \theta} + \frac{u_r}{r} \right)^2 + \left(\frac{\partial u_z}{\partial z} \right)^2 \right] + \mu \left(\frac{\partial u_\theta}{\partial r} - \frac{u_\theta}{r} + \frac{1}{r} \frac{\partial u_r}{\partial \theta} \right)^2$$
$$+ \mu \left(\frac{1}{r} \frac{\partial u_z}{\partial \theta} + \frac{\partial u_\theta}{\partial z} \right)^2 + \mu \left(\frac{\partial u_r}{\partial z} + \frac{\partial u_z}{\partial r} \right)^2 \qquad (1.4.19)$$

Note that the time derivative of a vector (or tensor) in a rotating reference frame is given by (see Reddy and Rasmussen [3], pp. 69–74)

$$\left[\frac{D(\cdot)}{Dt} \right]_{nonrot} = \left[\frac{D(\cdot)}{Dt} \right]_{rot} + \omega \times (\cdot) \qquad (1.4.20)$$

where $\omega = \frac{d\theta}{dt} \hat{e}_z = \frac{u_\theta}{r} \hat{e}_z$ is the angular velocity vector of the rotating frame of reference. Therefore, we have

$$\left(\frac{Du}{Dt} \right)_{nonrot} = \left(\frac{Du}{Dt} \right)_{rot} + (\frac{u_\theta}{r} \hat{e}_z) \times u$$
$$= \hat{e}_r \left(\frac{Du_r}{Dt} - u_\theta \frac{u_\theta}{r} \right) + \hat{e}_\theta \left(\frac{Du_\theta}{Dt} + u_r \frac{u_\theta}{r} \right) - \hat{e}_z \frac{Du_z}{Dt} \qquad (1.4.21)$$

Equation (1.4.21) could be used in conjunction with (1.4.17a–c) to provide a description of fluid motion in a rotating cylindrical coordinate system.

1.4.4 Closure

Equations (1.4.1)–(1.4.5) describe the convective and conductive heat transfer problems of primary interest. Although the velocities, pressure, and temperature were selected as the dependent variables, other choices for the description of the fluid motion are possible. Indeed, equations employing the stream function and vorticity, or the stream function alone, can be derived and used effectively with the energy equation to analyze convection problems. The choice of velocity, pressure, and temperature variables, the so-called "primitive variables", is made here for a number of reasons. The velocity and pressure variables are a natural description for fluid mechanics and permit boundary conditions to be described in a straightforward and intuitive manner. The description of free surface flows, problems in multiply connected domains, and phase change problems is direct and uncomplicated with the primitive variables. Also, the use in two or three dimensions and various coordinate systems is relatively straightforward, as is the inclusion of non-Newtonian and/or porous (Darcy) flow effects. Although none of the reasons cited above is by itself a compelling argument for exclusive use of the primitive variable approach, it is the most commonly used approach. All of the subsequent developments in this book will be carried out using the primitive variable description.

1.5 Porous Flow Equations

Suppose that the fluid domain, Ω_f, is composed of two subregions denoted by Ω_{vf} and Ω_{pf} such that $\Omega_f = \Omega_{vf} \cup \Omega_{pf}$. In the subregion Ω_{vf}, the viscous flow equations (1.4.7)–(1.4.9) are assumed to hold. The subregion Ω_{pf} is assumed to contain a rigid porous medium that is saturated with a viscous, incompressible fluid. The saturating fluid in Ω_{pf} is the same fluid as in Ω_{vf} if the two regions share a common permeable interface; otherwise the fluids in the two regions may be different. If the porous medium is assumed to be homogeneous and isotropic and the fluid and solid are in thermal equilibrium, then the equations describing the fluid motion and energy transport in the region Ω_{pf} can be written for a Cartesian rectangular coordinate system as

$$\frac{\partial u_i}{\partial x_i} = 0 \tag{1.5.1}$$

$$\rho_0 \alpha_{ij} \frac{\partial u_j}{\partial t} + \left(\frac{\rho_0 \hat{c}}{\sqrt{\kappa}} \|u\| + \frac{\mu_e}{\kappa} \right) u_i = \frac{\partial}{\partial x_j} \left[-P\delta_{ij} + \mu_e \left(\frac{\partial u_i}{\partial x_j} + \frac{\partial u_j}{\partial x_i} \right) \right]$$
$$+ \rho_0 f_i - \rho_0 g_i \beta (T - T_0) \tag{1.5.2}$$

$$(\rho C)_e \left(\frac{\partial T}{\partial t} \right) + (\rho_0 C) \left(u_j \frac{\partial T}{\partial x_j} \right) = \frac{\partial}{\partial x_i} \left(k_e \frac{\partial T}{\partial x_i} \right) + Q \tag{1.5.3}$$

Analogous equations are available for other coordinate systems. In Eqs. (1.5.1)–(1.5.3), κ is the permeability, $\hat{c}$ is the inertia coefficient, $\|u\|$ is the magnitude of the velocity vector, and α_{ij} are the components of the acceleration tensor. The porosity, ϕ, of the medium is defined as the fraction of the total volume of material occupied by interconnected pores (voids). The velocity components in (1.5.1)–(1.5.3) represent the volume averaged Darcy velocity (also termed the seepage or filtration velocity) which is related to the interstitial pore level velocity, u_i^p, by the inverse of porosity of the medium, i.e., $u_i = u_i^p / \phi$. The subscript e indicates an effective property; all other symbols are as defined earlier. The effective capacitance and conductivity are generally functions of the fluid and solid matrix properties and the porosity.

For example, the capacitance is usually a porosity weighted function such that $(\rho C)_e = \phi(\rho_0 C) + (1 - \phi)(\rho_s C_s)$. The effective conductivity can also be a function of the fluid velocity when thermal dispersion effects are important. The acceleration tensor is dependent on the geometry of the porous matrix, though in most cases it is proportional to $\frac{1}{\phi}$. Flow transients in a porous material usually decay very rapidly and the time-dependent term in (1.5.2) is often neglected. The effective viscosity is often taken to be $\mu_e = \mu$, although other approximations are possible. Suitable empirical models are available for all these quantities, and they can be found in the books by Kaviany [12], Nield and Joseph [13], Nield [14], and Nield and Bejan [15].

Equations (1.5.1)–(1.5.3) represent a generalization of the standard Darcy equations for nonisothermal flow in a saturated porous medium. This system is sometimes referred to as the *Forchheimer–Brinkman* or *Darcy–Forchheimer–Brinkman model* for porous flow. No attempt will be made here to delineate all the conditions under which such a model is appropriate. However, it should be noted that by selectively including or omitting certain terms, a number of other standard porous flow models can be derived. For example, if $\hat{c} = 0$, a Brinkman model is obtained, while a prescription of $\hat{c} = 0$ and $\mu_e = 0$ produces the standard Darcy's model. For further discussion of these models and their regions of applicability, the reader is referred to Nield [14], Nield and Bejan [15], Beavers and Joseph [16], Bear [17], and Gartling, Hickox, and Givler [18].

The general form of the porous flow equations is very similar to the equations for the viscous flow of a bulk fluid. By simply redefining certain terms and coefficients in equations (1.5.1)–(1.5.3), they can in fact be obtained from (1.4.7)–(1.4.10). As a result of this simple transformation process, it is quite straightforward to include a porous flow model in the general computational framework for nonisothermal flow problems. The region Ω_{pf} may exist in conjunction with Ω_{vf} or in place of Ω_{vf}. The solid region, Ω_s, may or may not be included as the problem dictates. In the following sections specific reference will be made to the porous flow equations only when their treatment differs significantly from the equations for a viscous fluid.

1.6 Auxiliary Transport Equations

The equations governing nonisothermal, viscous flow problems outlined in the previous sections are very general. However, there are still a significant number of flow problems that cannot be described by the stated equations. For flows in which transport processes other than those of a thermal nature are important, additional equations are required. The added equations are of the advection-diffusion type and are generic in the sense that they are not specifically associated with a particular physical process. Possible uses for these additional equations include the description of mass transport in a multicomponent system, the simulation of certain types of chemical reactions, and the prediction of volume fraction or particle orientation for flows containing suspended particles or fibers. Here we give the additional equations describing such flows.

Consider the material region Ω. The transport of a scalar quantity ϕ_i within this region is governed, for fixed i (i.e., no sum on i) by

$$C_i \left(\frac{\partial \phi_i}{\partial t} + u_j \frac{\partial \phi_i}{\partial x_j} \right) = \frac{\partial}{\partial x_j} \left(D_i \frac{\partial \phi_i}{\partial x_j} \right) + Q_i \qquad (1.6.1)$$

In Eq. (1.6.1), C_i is a "capacitance" coefficient [analogous to ρC in Eq. (1.4.10)], D_i is a diffusion coefficient, and Q_i is a volumetric source term. The subscript i runs between 1 and n, and it indicates that up to n transport equations could be added to the general flow problem. Note that Eq. (1.6.1) is valid in the solid regions of Ω if u_j is set to zero; the transport equation can also be used in a porous region if C_i and D_i are interpreted as "effective" properties of the fluid/matrix system.

In order to fully couple the above transport equation to the general nonisothermal flow problem, certain extensions must be made to the previously stated functional form for the fluid density variation. The possibility of buoyancy forces due to variations in the auxiliary variables ϕ_i requires that the equation of state (1.3.10) be modified to

$$\rho = \rho_0 \left[1 - \beta(T - T_0) - \gamma_1(\phi_1 - \phi_{1_0}) \ldots - \gamma_n(\phi_n - \phi_{n_0}) \right] \qquad (1.6.2)$$

where γ_i is an expansion coefficient and the subscript zero refers to a reference condition. As before this variation in density is only permitted in the body force term. The body force term in Eq. (1.4.8) must therefore be altered to include the last n terms in Eq. (1.6.2) whenever auxiliary transport equations of the appropriate type are included in the problem formulation.

1.7 Chemically Reacting Systems

The inclusion of chemical reactions in a fluid or thermal problem adds considerably to the difficulty of the numerical simulation. However, such effects are central to many important technology areas. The majority of these applications involve combustion or other gas phase processes that are described by compressible flows and are therefore outside the scope of this book. Chemical reactions in incompressible flows are often associated with gelation or polymerization reactions; some low-speed gas flows, such as chemical vapor deposition processing, are frequently modeled as incompressible. Thermal diffusion problems coupled with chemical kinetics are important for thermal ignition studies in energetic materials (e.g., pyrotechnics and explosives) and in the simulation of material decomposition or ablation. In this section we will outline the general form of the continuum equations that describe a chemically reactive system and make these equations specific for a few particular applications.

For a chemically reactive fluid, the boundary value problems described in Sections 1.3 and 1.4 are altered by the fact that the incompressible fluid must now be considered a mixture of various chemical species. Conservation principles must be invoked for the individual constituents of the fluid as well as for the fluid mixture as a whole. The overall mass, momentum, and energy balances for the mixture remain as given in Section 1.3 and are rewritten here only for reference.

$$\frac{\partial u_i}{\partial x_i} = 0 \tag{1.7.1}$$

$$\rho\left(\frac{\partial u_i}{\partial t} + u_j\frac{\partial u_i}{\partial x_j}\right) = -\frac{\partial P}{\partial x_i} + \frac{\partial}{\partial x_j}\left[\mu\left(\frac{\partial u_i}{\partial x_j} + \frac{\partial u_j}{\partial x_i}\right)\right] + \rho g_i \tag{1.7.2}$$

$$\rho C\frac{\partial T}{\partial t} + \rho C u_j\frac{\partial T}{\partial x_j} = \frac{\partial}{\partial x_i}\left(k\frac{\partial T}{\partial x_i}\right) + Q_R \tag{1.7.3}$$

In equations (1.7.1)–(1.7.3) the thermophysical properties will normally be strong functions of the chemical composition of the fluid. Assume that the fluid is composed of I individual species and that these species may react according to J reactions. The primary effect of the reaction process on the flow is to alter the internal energy of the system; this change in energy provides the volume source term Q_R in equation (1.7.3).

A mass conservation equation for each of the I species is required and these have the form

$$\frac{\partial[N_i]}{\partial t} + u_j\frac{\partial[N_i]}{\partial x_j} = \frac{\partial}{\partial x_j}\left(D_{jk}\frac{\partial[N_i]}{\partial x_k}\right) + R_i \quad \text{for}\ \ i = 1, 2, \ldots, I \tag{1.7.4}$$

where $[N_i]$ is the concentration variable (mole fraction) for species i, D_{jk} is the diffusion tensor and R_i is the chemical reaction rate. Note that there is no summation on i in Eq. (1.7.4). In writing Eq. (1.7.4), an incompressible flow was assumed and the diffusion process was assumed to follow Fick's law. More general species conservation equations can be derived that account for other types of diffusion processes (see Herschfelder et al. [19] and Oran and Boris [20]), but these need not be considered here.

To describe the details of the chemical reaction, the stoichiometry, reaction kinetics, and material property behavior must be specified. For a material with I species and J reactions, the stoichiometry is provided by

$$\sum_{i=1}^{I}\nu'_{ij}\mathcal{M}_i \quad \longrightarrow \quad \sum_{i=1}^{I}\nu''_{ij}\mathcal{M}_i \quad \text{for}\ \ j = 1, 2, \ldots, J \tag{1.7.5}$$

where ν'_{ij} and ν''_{ij} are stoichiometric coefficients (usually integer values) and $\mathcal{M}_i$ is the chemical symbol for the ith species. Generally, these expressions are given as reversible reactions; however, they are treated here as irreversible, and the reversed reactions are specified as additional reaction steps. To accommodate expressions for global reactions, the stoichiometric coefficients are allowed to be non-integer.

For each step of the reaction, a reaction rate, r_j, is defined in the form:

$$r_j = k_j(T)\prod_{i=1}^{I}[N_i]^{\mu_{ij}} \quad \text{for}\ \ j = 1, 2, \ldots, J \tag{1.7.6}$$

where μ_{ij} are the concentration exponents (usually $\mu_{ij} = \nu'_{ij}$ in kinetic theory, but here they are treated independently) and the symbol Π denotes the product operation. Typically, the expressions for the kinetic coefficients, $k_j(T)$, are given in an Arrhenius form

$$k_j(T) = T^{\beta_j} A_j \ exp(-E_j/RT) \tag{1.7.7}$$

where β_j is the coefficient for a steric factor, A_j is the pre-exponential factor, E_j is the activation energy, and the universal gas constant is R. It is convenient to define $\nu_{ij} = (\nu''_{ij} - \nu'_{ij})$ and thus the rate of change of the species (neglecting diffusion) is given as

$$R_i = \sum_{j=1}^{J} \nu_{ij} r_j \ \text{for} \ i = 1, 2, \ldots, I \tag{1.7.8}$$

The chemical reaction process is coupled directly to the energy equation (1.7.3) by the volumetric source term

$$Q_r = \sum_{j=1}^{J} q_j r_j \tag{1.7.9}$$

where q_j represents the known endothermic or exothermic energy release for reaction step j.

The material properties for the mixture are usually represented as mole fraction weighted averages of the I constituents. That is,

$$(\rho C) = (\rho C)_{mix} = \sum_{i=1}^{I} [N_i](\rho C)_i$$

$$\mu = (\mu)_{mix} = \sum_{i=1}^{I} [N_i](\mu)_i \tag{1.7.10}$$

$$k = (k)_{mix} = \sum_{i=1}^{I} [N_i](k)_i$$

where the constituent properties could still be functions of the temperature or other variable.

The equations described above include a wide spectrum of multistep chemical reactions. For many flow or heat transfer problems such generality is not required and the chemical process can be simplified. For a polymerization or curing reaction, or in fact any binary mixture problem, the set of mass conservation equations for the species reduces to a single equation

$$\frac{\partial \alpha}{\partial t} + u_j \frac{\partial \alpha}{\partial x_j} = \frac{\partial}{\partial x_j}\left(D\frac{\partial \alpha}{\partial x_j}\right) + R \tag{1.7.11}$$

where α is now integrated as a mass fraction or extent of reaction variable that varies from 0 to 1. Equation (1.7.11) describes the simple process of a material being transformed from species A to species B with an accompanying change in energy. The reaction ratio R in (1.7.11) would typically be of the Arrhenius type. In a fluid system the heat release from the reaction could cause a buoyancy-induced flow and thus couple the flow problem to (1.7.11). Such a problem will be described in Chapter 5 in the context of an epoxy solidification example.

In the case of a thermal conduction problem, the elimination of the fluid velocity allows additional simplification of the species equations by elimination of the advection term in (1.7.4). Also, for most solids, the species diffusion is negligibly small and the conservation equations in (1.7.4) can be reduced to a set of ordinary differential equations

$$\frac{d[N_i]}{dt} = R_i = \sum_{j=1}^{J} \nu_{ij} r_j \tag{1.7.12}$$

The equations in (1.7.12) must be solved in conjunction with the standard heat conduction equation and the appropriate definitions for the heat source and kinetic parameters. Algorithms for and numerical studies of reactive systems of this type will be described in Chapter 3.

1.8 Boundary Conditions

To complete the description of the general initial and/or boundary value problem posed in the previous sections, suitable boundary and initial conditions are required. Boundary conditions are most easily understood and described by considering the fluid mechanics separate from other transport processes. In the following sections, boundary conditions associated with the viscous flow, porous flow, and thermal transport are discussed, followed by a discussion of the initial conditions.

1.8.1 Viscous Flow Boundary Conditions

For the fluid dynamic part of the problem, either the velocity components or the total surface stress or traction must be specified on the boundary of the fluid region. In general the boundary conditions can be classified into two types (see Reddy [21], pp. 28–32), as given below.

Dirichlet or essential boundary conditions:

$$u_i = f_i^u(s_k, t) \qquad \text{on } \Gamma_u \tag{1.8.1a}$$

Neumann or natural boundary conditions:

$$T_i = \sigma_{ij}(s_k, t) n_j(s_k) = f_i^T(s_k, t) \qquad \text{on } \Gamma_T \tag{1.8.1b}$$

where s_k are the coordinates along the boundary, t is the time, n_i is the outward unit normal to the boundary, and Γ_f is the total boundary enclosing the fluid domain, Ω_f, with $\Gamma_f = \Gamma_u \cup \Gamma_T$ (see Figure 1.8.1). Note that the conditions written in (1.8.1a,b) are in component form, i.e., there must be a condition on each component of the velocity. The specified functions f_i^u and f_i^T are generally simple expressions for standard situations where the fluid is contained by fixed boundaries ($f_i^u = 0$) or planes/lines of symmetry ($f_i^T = 0$), or enters/leaves the domain Ω_f ($f_i^T = $ constant). The boundary conditions listed in (1.8.1a,b) are adequate for most situations involving a single fluid in an uncomplicated domain.

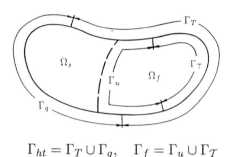

$$\Gamma_{ht} = \Gamma_T \cup \Gamma_q, \quad \Gamma_f = \Gamma_u \cup \Gamma_T$$

Figure 1.8.1. Schematic for boundary condition definitions.

A slight generalization of the previous Dirichlet and Neumann condition leads to the possibility of representing periodic conditions in a flow. In this case the velocity field would be specified by

$$u_i(s_k, t) = u_i(s_k + \Delta s_k, t) \tag{1.8.2a}$$

and the traction condition by

$$T_i(s_k, t) = T_i(s_k + \Delta s_k, t) + g_i^T(s_k + \Delta s_k, t) \tag{1.8.2b}$$

Here the quantity Δs_k indicates a spatial offset from one boundary to another; this offset is usually a simple translation or rotation. The velocity components on the periodic surfaces are equal, but unknown. The tractions on the two surfaces may differ by a function g_i^T. The most common situation is for the normal tractions (essentially the pressure) to have an offset equal to the pressure change over the length Δs_k, while the tangential tractions are equal.

In some specialized applications it may be required to specify that the fluid slips along a solid surface. In this case the tangential Dirichlet conditions on the surface would be replaced by traction conditions of the form

$$\alpha \sigma_{ij} n_j t_i = (u_i - u_i^s) t_i \tag{1.8.3}$$

where α is the slip coefficient, t_i is tangent vector to the surface and u_i^s is the velocity of the surface.

The general boundary or interface conditions between two immiscible fluids are more complex and are described here for several important types of problems. It is assumed that there is no mass transfer across the interface which allows all components of the velocity field to be continuous across the boundary. Note that this condition is often retained even for change of phase problems in order to simplify the formulation. In concert with this velocity specification, a kinematic condition is also imposed that requires the interface to remain an interface, i.e., the interface is a stream surface. This requirement can be represented by the equation

$$\frac{DF}{Dt} = \frac{\partial F}{\partial t} + u_j \frac{\partial F}{\partial x_j} = 0 \qquad (1.8.4)$$

where $F(x_i, t)$ is the equation for the fluid interface. Since the location and shape of the interface between two or more fluids is usually not known a priori, this type of problem is termed a free boundary problem. For time-dependent free surface flows, Eq. (1.8.4) provides an evolution equation for the free surface that must be solved in conjunction with the viscous flow equations; the fluid domain Ω_f is now a function of time, $\Omega_f(t)$. In the case of a time-independent flow the kinematic condition simplifies to

$$u_i n_i = 0 \qquad (1.8.5)$$

which simply states that the free surface is a stream surface. For the steady case there is no explicit equation for the free surface; equation (1.8.5) only provides a constraint condition that must be applied along the unknown interface. The numerical treatment of free surface flows will be considered in Chapter 4.

The final condition at the free surface specifies that the change in total stress (traction) between the two fluids is related to surface tension forces. The interfacial stress condition is expressed as

$$\sigma_{ij}^2 n_j - \sigma_{ij}^1 n_j = \gamma \left(\frac{1}{R_1} + \frac{1}{R_2} \right) n_i - (\delta_{ij} - n_i n_j) \frac{\partial \gamma}{\partial x_j} \qquad (1.8.6)$$

where γ is the surface tension, R_1 and R_2 are the principal radii of curvature of the interface, and the superscripts '2' and '1' on σ_{ij} refer to the two fluids. The last term in (1.8.6) allows variations in the surface tension (e.g., due to temperature) to produce a shear stress in the fluid (Marangoni flows).

The complex conditions expressed in (1.8.6) are necessary when the motion in both fluids is of equal importance (e.g., two immiscible liquids). However, in many cases, one of the fluids is a gas and its motion relative to a liquid may be neglected. For this simplified case the continuity condition on the velocity is removed since the gas motion is ignored, and shear stresses (drag forces) on the interface are neglected. If the gas is assumed to be at a uniform pressure, P_{amb}, then (1.8.6) becomes

$$\sigma_{ij} n_j = \gamma \left(\frac{1}{R_1} + \frac{1}{R_2} \right) n_i - P_{amb} n_i \qquad (1.8.7)$$

which relates the stress normal to the free surface to the surface tension, surface curvature and ambient external pressure.

Other types of computational boundary conditions, such as multipoint constraints, are possible and may be encountered in various modeling or simulation situations. These conditions have their theoretical basis in the conditions outlined here but are best described in detail as part of the numerical algorithm.

1.8.2 Porous Flow Boundary Conditions

The flow conditions that can be applied to the boundary of a fluid-saturated porous medium depend on the specific model used to describe the problem. For the general Forchheimer–Brinkman equation listed in (1.5.2), the admissible boundary conditions are the same as those given for the Navier–Stokes equations, Eq. (1.8.1a,b). However, in cases where the Brinkman terms are excluded from the momentum equation, $(\mu_e = 0)$ the order of the equation is reduced and fewer boundary conditions are needed. A Forchheimer or simple Darcy model allows the following types of boundary conditions:

$$u_i n_i = f^u(s_k, t) \qquad \text{on } \Gamma_u \qquad (1.8.8a)$$

$$T_i n_i = -P(s_k, t)\delta_{ij}n_j(s_k)n_i = -P(s_k, t) = f^T(s_k, t) \qquad \text{on } \Gamma_T \qquad (1.8.8b)$$

In essence these conditions state that the fluid velocity normal to the boundary may be specified (outflow/inflow) or the normal force (pressure) on the boundary may be imposed. These simplified models do not allow a tangential velocity to be imposed (e.g., no-slip walls are not admissible) nor the specification of viscous forces. Also, note that only one type of boundary condition can be specified at any point on the boundary.

For many applications a saturated porous layer will adjoin a clear fluid region and certain continuity conditions will be required at the open interface. The applicable conditions vary with the type of porous media model and in many cases have not been rigorously verified. The major difficulty stems from the heterogeneous nature of the porous medium and the fact that a continuum description such as (1.4.1)–(1.4.3) is derived by an averaging process. Averaging within a distance $\approx \sqrt{\kappa}$, where κ is the permeability, of a boundary or interface is not valid and some loss of information will occur; detailed microscopic analyses or observations must be used to generate valid macroscopic boundary conditions for the continuum models. In both the Darcy and Brinkman formulations the (seepage) velocity normal to the interface is assumed continuous; the Brinkman model assumes all tangential velocity components are also continuous between the porous layer and the clear fluid. Many investigators have assumed that all components of the viscous stress and the pressure are continuous at an interface with a Brinkman formulation, a situation that is very convenient from a computational point of view. However, Nield [14] and Nield and Bejan [15] believe that the stresses and pressure are not continuous, and they advocate the use of a Navier (slip) condition for the tangential stress at the interface. No proposal has been offered for estimating the changes in normal stress or pressure at the interface. The situation for the Darcy model is just as confused since no stress terms are present in this formulation. The standard assumptions in this case are that the pressure is continuous and the shear stress in the bulk fluid is related to the tangential velocity in the porous layer by the Navier type condition proposed by Beavers and Joseph [16]

$$\frac{\partial u_f}{\partial n} = \frac{\alpha}{\sqrt{\kappa}}(u_f - u_p) \tag{1.8.9}$$

Here u_f is the clear fluid velocity, u_p is the velocity in the porous media, and α is a dimensionless (porous) material parameter. Based on the above discussion it is clear that some caution must be used when considering interface and boundary conditions for a saturated porous material. A further discussion and demonstration of interface conditions may be found in Gartling, Hickox, and Givler [18].

1.8.3 Thermal and Transport Boundary Conditions

The thermal part of the boundary value problem for the fluid or solid requires the temperature (Dirichlet or essential condition) or the heat flux (Neumann or natural condition) to be specified on all parts of the boundary enclosing the heat transfer region:

$$T = f^T(s_k, t) \quad \text{on} \ \Gamma_T \tag{1.8.10a}$$

$$-\left(k_{ij}\frac{\partial T}{\partial x_j}\right)n_i \equiv q_i n_i = q_c + q_r + q_a = f^q(s_k, t) \quad \text{on} \ \Gamma_q \tag{1.8.10b}$$

where Γ_{ht} is the total boundary enclosing the heat transfer region, and $\Gamma_{ht} = \Gamma_T \cup \Gamma_q$. Note that the condition on Γ_q for a fluid or an isotropic solid can be simplified, since k_{ij} reduces to a scalar k. Also, q_a refers to a specified flux and q_c and q_r refer to the convective and radiative flux components given by

$$q_c = h_c(s_k, T, t)(T - T_c) \tag{1.8.11a}$$
$$q_r = h_r(s_k, T, t)(T - T_r) \tag{1.8.11b}$$

In Eq. (1.8.11a), h_c is the convective heat transfer coefficient which, in general, depends on the location on the boundary, temperature, and time, and T_c is a reference (or sink) temperature for convective transfer. Also, the effective radiation heat transfer coefficient is $h_r = \mathcal{F}\sigma(T + T_r)(T^2 + T_r^2)$ where σ is the Stefan–Boltzmann constant, $\mathcal{F}$ is a form factor, and T_r the reference temperature for radiative transfer. The form factor $\mathcal{F}$ is related to the boundary emissivity ϵ and the position of the boundary relative to surrounding surfaces (see Arpaci [22]). Note that this type of radiation boundary condition is appropriate when a body or surface radiates to a black body environment that can be characterized by a single temperature. A more complex condition of surface to surface or enclosure radiation is described in Section 1.10. As in the case of the fluid boundary conditions, the specified temperature f^T and total flux f^q are simple functions for the cases normally encountered in practical problems. One of the more complex conditions occurs in change of phase problems and this will be treated separately in the next section.

Another condition that is of concern when considering a material interface between two or more solid regions is the problem of gap or contact resistance. In some situations, the thermal resistance at the interface is attributed to a fictitious material that has a (generally) negligible specific heat and a nonlinear conductivity. The boundary or interface conditions in this situation are the usual continuity conditions on temperature and heat flux since the gap resistance is dictated by property variations.   A more mathematical representation of contact resistance provides that the heat flux across the interface be described by an internal boundary condition of the general form given in (1.8.11):

$$q_{gap} = h_{gap}(s_k, T_{gap}, t)(T_M - T_S) \qquad (1.8.12)$$

where h_{gap} is an effective heat transfer coefficient for the contact surface, and T_{gap} is an average temperature between T_M and T_S. The subscripts M and S designate the "master" and "slave" sides of the contact surface, a distinction that is important in the numerical implementation of this condition. Note that in the limit as h_{gap} becomes large, the temperatures of the two surfaces will become equal, a situation that is useful in some applications.

As a final point on the boundary condition specification note that the boundaries Γ_f and Γ_{ht} need not coincide (see Figure 1.8.1). Indeed, for problems involving regions of both convection and solid body conduction the boundary regions are not the same. When both fluid and solid regions exist in a problem, continuity conditions on the temperature and heat flux must prevail along the interface of the two materials. Examples of this type of conjugate problem will be given in a later section.

Although not specifically written here, boundary and initial conditions for the auxiliary transport variables are required if these equations are included in the formulation. The form of these boundary conditions generally follows the mathematical form of the boundary conditions associated with the energy equation for a fluid region.   This is due to the close similarity between the transport equation (1.6.1) and the energy equation (1.4.10). Thus the boundary conditions on ϕ_i are the same as those given in Eq. (1.8.10) with T replaced by ϕ_i and the radiation condition neglected.

1.8.4 Initial Conditions

For time-dependent problems, a set of initial conditions are necessary for the dependent variables. Very often these conditions consist of a solid body at a uniform temperature and a quiescent fluid at a uniform temperature and hydrostatic pressure. A second common possibility is for the transient motion to be initiated from an established steady-state flow and temperature field. In all cases, the dependent variables must be known for all $\mathbf{x}$ at $t = 0$ and must satisfy the basic conservation equations (e.g., the initial fluid velocity field must be divergence free and have a compatible pressure field).

1.9 Change of Phase

Many problems of importance involving conductive and convective heat transfer have the added complication of a material change of phase. Such problems are part of a general class of boundary value problems termed *free* or *moving boundary problems* (see Ockendon and Hodgkins [23]). The major difficulty in analyzing such problems is the occurrence of a boundary (or interface) surface, across which certain jump conditions must be satisfied and whose location is unknown a priori. Thus, the location of the moving boundary becomes part of the solution. In this section the equations for a limited class of such problems will be described.

The present discussion will be limited to melting/solidification phase transitions where the solid phase is stationary. Also, it is assumed that density changes upon change of phase may be neglected and that the effects of latent heat release may be accounted for through an appropriate modification of the specific heat. The present development also ignores the explicit modeling of the porous layer or mushy zone that may occur between the solidus and liquidus temperatures in some materials. Methods for treating the mushy zone are easily incorporated into a numerical simulation following the works of Voller and Prakash [24] and Prescott, Incropera, and Bennon [25], and primarily involve the use of the porous media equations described previously. With these assumptions, the phase change problem is described by the equations of the previous sections plus a special set of conditions on the melt/solid interface. At the phase change boundary the following conditions are required:

$$u_i|_X = 0 \tag{1.9.1}$$

$$T_f|_X = T_s|_X \tag{1.9.2}$$

$$k_f \frac{\partial T}{\partial n}\bigg|_X - k_s \frac{\partial T}{\partial n}\bigg|_X = \rho L \frac{\partial X}{\partial t} = \rho L u^\star \tag{1.9.3}$$

where L is the latent heat, $X(t)$ is the spatial position of the phase boundary, $u^\star$ is the velocity of the surface and the subscripts f and s refer to fluid and solid, respectively. The function $X(t)$ is unknown and must be determined as part of the solution. Basically, the melt/solid interface is taken to be a no-slip surface [Eq. (1.9.1)], with continuous temperature [Eqs. (1.9.2)] and a jump in the heat flux [Eq. (1.9.3)]. Note that the solid is assumed stationary; a moving solid would require an additional velocity on the right-hand side of Eq. (1.9.3). Following the work of Bonacini et al. [26] and Comini et al. [27], the jump condition in Eq. (1.9.3) may be rewritten using the so-called "enthalpy method".

By observing that the latent heat, L, corresponds to the isothermal change in the enthalpy, H, for a material at the transition temperature, T_t, the following relation can be introduced

$$H(T) = \int_{T_{ref}}^{T} C(\xi)d\xi + L\,\eta(T - T_t) \tag{1.9.4}$$

with

$$\eta(T) = \begin{cases} 1 & \text{if } T \geq 0 \\ 0 & \text{if } T < 0 \end{cases} \tag{1.9.5}$$

where η is the Heaviside step function with argument T. The equivalent specific heat, C^*, is then introduced by

$$C^*(T) = \frac{dH}{dT} = C(T) + L\,\delta(T - T_t) \tag{1.9.6}$$

where δ is the Dirac delta function. Through the use of Eq. (1.9.6) latent heat effects may be included *via* the specific heat function, and the jump in the heat flux [see Eq. (1.9.3)] is eliminated from the problem formulation. This particular approach to the problem has a theoretically sound basis, as outlined by Bonacini et al. [26]. Moreover, it is a computationally effective modification since a two-region problem with a jump condition has been converted to a single-region problem with rapidly varying properties. For use in a finite element model, Eq. (1.9.6) requires additional modification, as explained in a later section.

The enthalpy method described above is the most heavily used formulation for the phase change problem but certainly not the only method available. Another approach for treating the latent heat is the heat source method. The development follows the enthalpy method up to Eq. (1.9.6) but proceeds one step further. When Eq. (1.9.6) is substituted into the energy equation, Eq. (1.4.3), the latent heat term may be viewed as a temperature dependent heat source and moved to the right-hand side of the energy balance. An effective heat source is then

$$Q_{lh} = \rho L \frac{\partial T}{\partial t}\,\delta(T - T_t) \tag{1.9.7}$$

This term is active only when the local temperature is at the transition temperature. For computational purposes, Eq. (1.9.7) requires modification because of the difficulties inherent in the delta function $\delta(\cdot)$.

A final method for resolution of the phase change problem draws on the similarities between a chemical reaction with energy release and the phase transition process. This method requires that the phase change be viewed as a reaction with material A going to material B with an appropriate "kinetic" description. The energy release from this reaction appears as a continuous source term in the energy balance with its magnitude dependent on the reaction rates. This type of formulation avoids many of the computational difficulties associated with the enthalpy and heat source methods though it does introduce the complexity of treating a chemically reactive material. Details of this method are presented in Chapter 3.

The previous phase change discussion has been presented under the rather strong assumption that any density changes involved in the phase transition were negligible. The relaxation of this assumption leads to a significant increase in problem complexity. Though the no-slip velocity condition in Eq. (1.9.1) continues to apply for the velocity components tangent to the interface, a normal mass balance augments Eq. (1.9.1) for the velocity normal to the interface,

$$\rho_f n_j (u_j^f - u_j^*) = \rho_s n_j (u_j^s - u_j^*) \qquad (1.9.8)$$

with the special case of a stationary solid being

$$\rho_f n_j (u_j^f - u_j^*) = -\rho_s n_j u_j^* \qquad (1.9.9)$$

The subscripts and superscripts s and f refer to the fluid and solid. A nonzero essential velocity condition, such as the one in Eq. (1.9.8) or (1.9.9), is difficult to implement in many computational schemes. This velocity condition implies that some specific algorithm is required to explicitly locate, resolve and follow the phase boundary. Though a number of computational methods exist for this type of problem, this boundary condition is sufficiently difficult to keep the constant density assumption as a standard procedure.

1.10 Enclosure Radiation

Radiant energy exchange between neighboring surfaces of a region or between a region and its surroundings can produce large effects in the overall heat transfer problem. Though the radiation effects generally enter the heat transfer problem only through the boundary conditions, the coupling is especially strong due to the nonlinear dependence of the radiation on the surface temperature. The present description considers a restricted class of radiation problems that ignore participation of the fluid phase; the radiative exchange only influences the flow and/or solid body conduction problem through modification of the surface temperature distribution.

Enclosure or surface-to-surface radiation is limited to diffuse, gray, opaque surfaces. This assumption implies that all energy emitted or reflected from a surface is diffuse. Further, surface emissivity ϵ, absorbtivity α, and reflectivity ρ are independent of wavelength and direction so that $\epsilon(T) = \alpha(T) = 1 - \rho(T)$. Each individual area or surface that is considered in the radiation process must be at a uniform temperature; emitted and reflected energy are uniform over each such surface. Note that the definition of a surface is arbitrary and can be based on geometry alone or be defined to specifically satisfy the uniform temperature criterion.

With the above assumptions, the radiation problem can be approached using the net-radiation method (see Siegel and Howell [28]). For purposes of discussion, consider the two-dimensional enclosure made up of N distinct surfaces as shown in Figure 1.10.1. Associated with each surface is a uniform temperature $\overline{T}_j$, an area A_j, and a surface emissivity ϵ_j. An energy balance for each surface in the enclosure leads to the following system of equations

$$\sum_{j=1}^{N}\left[\frac{\delta_{kj}}{\epsilon_j} - F_{k-j}\left(\frac{1-\epsilon_j}{\epsilon_j}\right)\right]\frac{Q_j}{A_j} = \sum_{j=1}^{N}(\delta_{kj} - F_{k-j})\,\sigma\,\overline{T}_j^4 \qquad (1.10.1)$$

Equation (1.10.1) relates the net energy loss, Q_j, from each surface to the temperature of each surface, where δ_{kj} is the unit tensor, σ is the Stefan–Boltzmann constant, and F_{k-j} are radiation view (configuration) factors. The view factor is defined as the fraction of energy leaving a surface that arrives at a second surface. For surfaces with finite areas, the view factors are defined by

$$F_{k-j} = \frac{1}{A_k}\int_{A_k}\int_{A_j}\frac{\cos\theta_k\cos\theta_j}{\pi S^2}\,dA_j\,dA_k \qquad (1.10.2)$$

where S is the distance from a point on surface A_j to a point on surface A_k. The angles θ_j and θ_k are measured between the line S and the normals to the surface as shown in Figure 1.10.1. It is clear from Eq. (1.10.2) that the view factors are purely geometric quantities that can in principle be evaluated for any given distribution of surfaces. Although not explicitly shown in the figure, note that view factors may be modified by the presence of intervening bodies or surfaces that shadow surface A_j from surface A_k. Methods for evaluating F_{k-j} will be outlined later in the book.

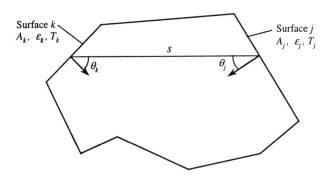

Figure 1.10.1. Nomenclature for enclosure radiation.

For purposes of computation it is convenient to rearrange Eq. (1.10.1) into the following equations:

$$\sum_{j=1}^{N}\left[\delta_{kj} - (1-\epsilon_k)F_{k-j}\right]q_j^o = \epsilon_k\sigma\overline{T}_k^4 \qquad (1.10.3)$$

$$q_k = q_k^o - \sum_{j=1}^{N}F_{k-j}q_j^o \qquad (1.10.4)$$

Equations (1.10.3) and (1.10.4) are expressed in terms of the outgoing radiative flux for each surface, q_j^o, and the net flux from each surface $q_k = Q_k/A_k$. For known surface temperatures $\overline{T}_k$ in the enclosure, Eq. (1.10.3) can be solved for the outgoing radiative flux at each surface. Equation (1.10.4) then allows the net flux at each surface to be evaluated. The actual method of solution using this formulation in a finite element context will be discussed in the later chapters.

1.11 Summary of Equations

All of the governing differential equations, constitutive relations, and boundary conditions of heat transfer and fluid mechanics relevant to the present discussion were presented in the previous sections. It is useful to summarize the complete boundary value problem in one section. Therefore, the complete set of equations are presented here in Cartesian component form for each physical process. Sum on repeated indices is assumed, and free indices have the range of 1, 2, and 3.

Flows of Viscous Incompressible Fluids:

$$\frac{\partial u_i}{\partial x_i} = 0 \tag{1.11.1}$$

$$\rho_0 \left(\frac{\partial u_i}{\partial t} + u_j \frac{\partial u_i}{\partial x_j} \right) = \frac{\partial}{\partial x_j} \left[-P\delta_{ij} + \mu \left(\frac{\partial u_i}{\partial x_j} + \frac{\partial u_j}{\partial x_i} \right) \right] + \rho_0 f_i - \rho_0 \hat{g}_i \tag{1.11.2a}$$

where

$$\hat{g}_i = g_i \left[\beta(T - T_0) + \gamma_1(c_1 - c_{1_0}) + \gamma_2(c_2 - c_{2_0}) + \cdots \right] \tag{1.11.2b}$$

Porous Flow:

$$\frac{\partial u_i}{\partial x_i} = 0 \tag{1.11.3}$$

$$\rho_0 \alpha_{ij} \frac{\partial u_j}{\partial t} + \left(\frac{\rho_0 \hat{c}}{\sqrt{\kappa}} \|\mathbf{u}\| + \frac{\mu}{\kappa} \right) u_i = \frac{\partial}{\partial x_j} \left[-P\delta_{ij} + \mu_e \left(\frac{\partial u_i}{\partial x_j} + \frac{\partial u_j}{\partial x_i} \right) \right] + \rho_0 f_i - \rho_0 \hat{g}_i$$

$$\tag{1.11.4a}$$

where

$$\hat{g}_i = g_i \left[\beta(T - T_0) + \gamma_1(c_1 - c_{1_0}) + \gamma_2(c_2 - c_{2_0}) + \cdots \right] \tag{1.11.4b}$$

Heat Transfer (Convection):

$$\rho_0 C \left(\frac{\partial T}{\partial t} + u_j \frac{\partial T}{\partial x_j} \right) = \frac{\partial}{\partial x_i} \left(k \frac{\partial T}{\partial x_i} \right) + Q + \Phi \tag{1.11.5}$$

Heat Transfer (Conduction):

$$\rho_s C_s \left(\frac{\partial T}{\partial t} \right) = \frac{\partial}{\partial x_i} \left(k_{s_{ij}} \frac{\partial T}{\partial x_j} \right) + Q_s \tag{1.11.6}$$

Auxiliary Transport:

$$C_i \left(\frac{\partial \phi_i}{\partial t} + u_j \frac{\partial \phi_i}{\partial x_j} \right) = \frac{\partial}{\partial x_j} \left(D_i \frac{\partial \phi_i}{\partial x_j} \right) + Q_i \tag{1.11.7}$$

Equations (1.11.1)–(1.11.7) with the noted boundary and initial conditions form a complete set for the determination of the velocity, pressure, temperature, and auxiliary variables in a fluid, porous, or solid region. Embedded in this description are numerous classes of flow problems that can be recovered through the omission or inclusion of specific equations, the neglect of certain terms within an equation, and the variation of material properties.

Problems

1.1 Let $\rho(\mathbf{x}, t)$ denote the mass density of a continuous region, Ω. Then conservation of mass for a material region requires that

$$\frac{D}{Dt} \int_\Omega \rho \, d\Omega = 0 \tag{i}$$

Show that for a fixed region, conservation of mass can also be expressed as

$$\int_\Omega \frac{D\rho}{Dt} \, d\Omega + \int_\Gamma \rho \hat{\mathbf{n}} \cdot \mathbf{u} \, d\Gamma = 0 \tag{ii}$$

Interpret the equation in physical terms.

1.2 Newton's second law of motion (or the principle of conservation of linear momentum) applied to a continuum states that the rate of change of momentum following a material region Ω of fixed mass is equal to the sum of all the forces acting on the region. This can be expressed as

$$\frac{D}{Dt} \int_\Omega \rho \mathbf{u} \, d\Omega = \int_\Gamma \hat{\mathbf{n}} \cdot \sigma \, d\Gamma + \int_\Omega \rho \mathbf{f} \, d\Omega \tag{i}$$

where σ is the total stress tensor, $\mathbf{f}$ is the body force per unit mass, ρ is the mass density, and $\mathbf{u}$ is the material velocity. Since the material particle mass $\rho d\Omega$ is constant with respect to the material time derivative D/Dt, make use of the divergence theorem and obtain the differential form of Newton's second law of motion:

$$\rho \frac{D\mathbf{u}}{Dt} = \nabla \cdot \sigma + \rho \mathbf{f} \tag{ii}$$

1.3 Let ϵ denote the thermodynamic internal energy per unit mass of a material. Then the first law of thermodynamics applied to a material region Ω can be written as

$$\frac{D}{Dt} \int_\Omega \rho(e + \frac{1}{2}\mathbf{u} \cdot \mathbf{u}) \, d\Omega = \int_\Gamma \hat{\mathbf{n}} \cdot \sigma \cdot \mathbf{u} \, d\Gamma + \int_\Omega \rho \mathbf{f} \cdot \mathbf{u} \, d\Omega - \int_\Gamma \mathbf{q} \cdot \hat{\mathbf{n}} \, d\Gamma + \int_\Omega Q \, d\Omega \tag{i}$$

The first two terms on the right-hand side describe the rate of work done on the material region by the surface stresses and the body forces. The third integral describes the net outflow of heat from the region, causing a decrease of energy inside the region, where $\mathbf{q}$ denotes the heat flux vector. The fourth integral describes the internal heat generation in the region, where Q denotes the internal heat generation per unit volume. Show that the differential form of the first law of thermodynamics is given by

$$\rho \frac{De}{Dt} = -\nabla \cdot \mathbf{q} + Q + \sigma : \mathbf{D} \tag{ii}$$

Make use of the momentum equation and the identity

$$\text{div}(\sigma \cdot \mathbf{u}) - \mathbf{u} \cdot \text{div}\sigma = \sigma : \mathbf{D} \tag{iii}$$

For a viscous incompressible fluid, we take $\rho(De/Dt) = \rho C_v(DT/Dt)$.

1.4 Use equations (1.4.11)–(1.4.13) in equation (1.4.4) and derive the energy equation in the cylindrical coordinate system (r, θ, z).

1.5 For axisymmetric flows of viscous, incompressible fluids (i.e., flow field is independent of θ−coordinate), show that the Stokes flow equations (i.e., when the nonlinear terms in the convective parts are neglected) are given by

Conservation of Momentum:

$$\rho \frac{\partial u_r}{\partial t} = \frac{1}{r}\frac{\partial}{\partial r}(r\sigma_{rr}) - \frac{\sigma_{\theta\theta}}{r} + \frac{\partial \sigma_{rz}}{\partial z} + f_r \tag{i}$$

$$\rho \frac{\partial u_z}{\partial t} = \frac{1}{r}\frac{\partial}{\partial r}(r\sigma_{rz}) + \frac{\partial \sigma_{zz}}{\partial z} + f_z \tag{ii}$$

Conservation of Mass:

$$\frac{1}{r}\frac{\partial}{\partial r}(ru_r) + \frac{\partial u_z}{\partial z} = 0 \tag{iii}$$

where

$$\sigma_{rr} = -P + 2\mu\frac{\partial u_r}{\partial r}, \quad \sigma_{\theta\theta} = -P + 2\mu\frac{u_r}{r}$$

$$\sigma_{zz} = -P + 2\mu\frac{\partial u_z}{\partial z}, \quad \sigma_{rz} = \mu(\frac{\partial u_r}{\partial z} + \frac{\partial u_z}{\partial r}) \tag{iv}$$

1.6 Using equations (1.4.11)–(1.4.13) in equations (1.4.1)–(1.4.3), derive equations (1.4.16)–(1.4.18).

1.7 Use the definition of the rate of strain tensor $\mathbf{D}$ in equation (1.3.13) and equations (1.4.11) and (1.4.13) to verify the relations in equation (1.4.14).

References for Additional Reading

1. R. B. Bird, W. E. Stewart, and E. N. Lightfoot, *Transport Phenomena,* John Wiley & Sons, New York (1960).

2. L. E. Malvern, *Introduction to the Mechanics of a Continuous Medium,* Prentice Hall, Englewood Cliffs, New Jersey (1969).

3. J. N. Reddy and M. L. Rasmussen, *Advanced Engineering Analysis,* John Wiley & Sons, New York (1982); reprinted by Krieger Publishing, Melbourne, Florida (1991).

4. A. Bejan, *Convection Heat Transfer,* Second Edition, John Wiley & Sons, New York (1995).

5. J. P. Holman, *Heat Transfer,* Seventh Edition, McGraw–Hill, New York (1990).

6. M. N. Özisik, *Heat Transfer: A Basic Approach,* McGraw–Hill, New York (1985).

7. M. N. Özisik, *Heat Conduction,* Second Edition, John Wiley & Sons, New York (1993).

8. G. K. Batchelor, *An Introduction to Fluid Dynamics,* Cambridge University Press, Cambridge (1967).

9. H. Schlichting, *Boundary–Layer Theory,* (translated by J. Kestin), Seventh Edition, McGraw–Hill, New York (1979).

10. D. D. Gray and A. Giorgini, "The Validity of the Boussinesq Approximation for Liquids and Gases," *International Journal of Heat and Mass Transfer,* **19**, 545–551 (1976).

11. D. K. Gartling and C. E. Hickox, "A Numerical Study of the Applicability of the Boussinesq Approximation for a Fluid-Saturated Porous Medium," *International Journal for Numerical Methods in Fluids,* **5**, 995–1013 (1985).

12. M. Kaviany, *Principles of Heat Transfer in Porous Media,* Springer–Verlag, New York (1991).

13. D. A. Nield and D. D. Joseph, "Effects of Quadratic Drag on Convection in a Saturated Porous Medium," *Physics of Fluids,* **28**, 995–997 (1985).

14. D. A. Nield, "The Limitations of the Brinkman-Forchheimer Equation in Modeling Flow in a Saturated Porous Medium and at an Interface," *International Journal of Heat and Fluid Flow,* **12**, 269–272 (1991).

15. D. A. Nield and A. Bejan, *Convection in Porous Media,* Springer–Verlag, Berlin (1992).

16. G. S. Beavers and D. D. Joseph, "Boundary Conditions at a Naturally Permeable Wall," *Journal of Fluid Mechanics*, **30**, 197–207 (1967).

17. J. Bear, *Dynamics of Fluids in Porous Media*, American Elsevier, New York (1972).

18. D. K. Gartling, C. E. Hickox, and R. C. Givler, "Simulations of Coupled Viscous Porous Flow Problems," *Computational Fluid Dynamics*, **7**, 23–48 (1996).

19. J. O. Herschfelder, C. F. Curtis, and R. B. Bird, *Molecular Theory of Gases and Liquids*, John Wiley & Sons, New York (1954).

20. E. S. Oran and J. P. Boris, *Numerical Simulation of Reactive Flows*, Elsevier, New York (1987).

21. J. N. Reddy, *An Introduction to the Finite Element Method*, Second Edition, McGraw–Hill, New York (1993).

22. V. S. Arpaci, *Conduction Heat Transfer*, Addison–Wesley Publishing, Reading, Massachusetts (1966).

23. J. R. Ockendon and W. R. Hodgkins (Eds.), *Moving Boundary Problems in Heat Flow and Diffusion*, Clarendon Press, Oxford (1975).

24. V. R. Voller and C. Prakash, "A Fixed Grid Numerical Modeling Methodology for Convection Diffusion Mushy Region Phase Change Problems," *International Journal for Numerical Methods in Engineering*, **30**, 1709–1719 (1987).

25. P. J. Prescott, F. P. Incropera, and W. D. Bennon, "Modeling of Dendrite Solidification Systems: Reassessment of the Continuum Momentum Equation," *International Journal of Heat and Mass Transfer*, **34**, 2351–2359 (1991).

26. C. Bonacini, G. Comini, A. Fasano, and M. Primicerio, "Numerical Solution of Phase-Change Problems," *International Journal of Heat and Mass Transfer*, **16**, 1825–1832 (1973).

27. G. Comini, S. DelGuidice, R. Lewis, and O. C. Zienkiewicz, "Finite Element Solution of Non–Linear Heat Conduction Problems with Special Reference to Phase Change," *International Journal for Numerical Methods in Engineering*, **8**, 613–624 (1974).

28. R. Siegel and J. R. Howell, *Thermal Radiation Heat Transfer*, McGraw–Hill, New York (1972).

Chapter 2

The Finite Element Method: An Overview

2.1 Introduction

The finite element method is a powerful computational technique for the solution of differential and integral equations that arise in various fields of engineering and applied science. The method is a generalization of the classical variational (i.e., the Ritz) and weighted-residual (e.g., Galerkin, least-squares, collocation, etc.) methods, which are based on the idea that the solution u of a differential equation can be represented as a linear combination of unknown parameters c_j and appropriately selected functions ϕ_j in the *entire domain* of the problem. The parameters c_j are then determined such that the differential equation is satisfied, often, in a weighted-integral sense. The functions ϕ_j, called the *approximation functions*, are selected such that they satisfy the boundary conditions of the problem. For additional details on the variational and weighted-residual methods, the reader may consult Mikhlin [1,2], Finlayson [3], Oden and Reddy [4], Reddy [5–7], and Reddy and Rasmussen [8].

The traditional variational and weighted-residual methods suffer from one major shortcoming: construction of the approximation functions that satisfy the boundary conditions of the problem to be solved. Most real-world problems are defined on regions that are geometrically complex, and therefore it is difficult to generate approximation functions that satisfy different types of boundary conditions on different portions of the boundary of the complex domain. However, if the domain can be represented as a collection of "simple subdomains" that permit generation of the approximation functions for any arbitrary but physically meaningful boundary conditions, then the traditional variational methods can be used to solve practical engineering problems. The basic idea of the finite element method is to view a given domain as an assemblage of simple geometric shapes, called *finite elements*, for which it is possible to systematically generate the approximation functions needed in the solution of differential equations by any of the variational and weighted-residual methods. The ability to represent domains with irregular geometries by a collection of finite elements makes the method a valuable practical tool for the solution of boundary, initial, and eigenvalue problems arising in various fields of engineering. The approximation functions are often constructed using ideas from interpolation theory, and hence they are also called *interpolation functions*. Thus the finite element method is a piecewise (or elementwise) application of the variational and weighted-residual methods. For a given differential equation, it is possible to develop different finite element approximations (or finite element models), depending on the choice of a particular variational or weighted-residual method.

The primary objective of this chapter is to present an introduction to the finite element method. For elementary introductions to the finite element method, the reader is advised to consult Becker, Carey, and Oden [9], Reddy [6,7], and Burnett [10], among others. The major steps in the finite element analysis of a typical problem are (see Reddy [5–7]):

1. Discretization of the domain into a set of finite elements (*mesh generation*).

2. Weighted-integral or weak formulation of the differential equation to be analyzed.

3. Development of the finite element model of the problem using its weighted-integral or weak form.

4. Assembly of finite elements to obtain the global system of algebraic equations.

5. Imposition of boundary conditions.

6. Solution of equations.

7. Postcomputation of the solution and quantities of interest.

In the following sections, we will discuss the first four steps with the help of a model differential equation, namely, the equation governing conductive heat transfer in a two-dimensional solid medium. Since this is an overview of the method only the major steps are included.

2.2 Model Differential Equation

Consider the problem of finding the steady-state temperature $T(x,y)$ distribution in a two-dimensional orthotropic medium Ω, with boundary Γ. The equation governing the temperature distribution is given by setting the time derivative term to zero and $x_1 = x, x_2 = y$ in Eq. (1.11.6) (also see Carslaw and Jaeger [11], Holman [12], and Myers [13]):

$$-\left[\frac{\partial}{\partial x}\left(k_{xx}\frac{\partial T}{\partial x}\right) + \frac{\partial}{\partial y}\left(k_{yy}\frac{\partial T}{\partial y}\right)\right] = Q \quad \text{in } \Omega \qquad (2.2.1)$$

where k_{xx} and k_{yy} are conductivities in the x and y directions, respectively, and $Q(x,y)$ is the known internal heat generation per unit volume. For a nonhomogeneous conducting medium, the conductivities k_{xx} and k_{yy} are functions of position (x,y). For an isotropic medium, we set $k_{xx} = k_{yy} = k$ in Eq. (2.2.1) and obtain the Poisson equation

$$-\frac{\partial}{\partial x}(k\frac{\partial T}{\partial x}) - \frac{\partial}{\partial y}(k\frac{\partial T}{\partial y}) = Q \quad \text{in} \quad \Omega \qquad (2.2.2)$$

or in vector form

$$-\nabla \cdot (k\nabla T) = Q \quad \text{in} \quad \Omega \tag{2.2.3}$$

where ∇ is the gradient operator

$$\nabla = \hat{e}_x \frac{\partial}{\partial x} + \hat{e}_y \frac{\partial}{\partial y} \tag{2.2.4}$$

and $\hat{e}_x$ and $\hat{e}_y$ denote the unit vectors directed along the x and y axes, respectively.

Equation (2.2.1) must be solved in conjunction with specified boundary conditions of the problem. The following two types of boundary conditions are assumed in the following development [see Eqs. (1.8.10a,b)]:

$$T = f^T(s) \quad \text{on} \quad \Gamma_T \tag{2.2.5a}$$

$$q_n = \left(k_{xx} \frac{\partial T}{\partial x} n_x + k_{yy} \frac{\partial T}{\partial y} n_y \right) + q_c = f^q(s) \quad \text{on} \quad \Gamma_q \tag{2.2.5b}$$

where Γ_T and Γ_q are disjoint portions of the boundary Γ such that $\Gamma = \Gamma_T \cup \Gamma_q$, q_c refers to the convective component of heat flux

$$q_c = h_c(s, T)(T - T_c) \tag{2.2.6}$$

and (n_x, n_y) denote the direction cosines of the unit normal vector on the boundary. In Eq. (2.2.6) h_c denotes the convective heat transfer coefficient (see Section 1.8). The radiative heat transfer boundary condition is not considered in the present development.

2.3 Finite Element Approximation

As stated in the introduction to this chapter, the main idea of the finite element method is to generate the approximation functions required in the solution of differential equations by a variational or weighted-integral method. In the finite element method, this is accomplished by subdividing the given domain $\bar{\Omega} = \Omega \cup \Gamma$ into a set of subdomains, called finite elements (see Figure 2.3.1). The phrase *finite element* often refers to both the geometry of the element and degree (or order) of approximation used for the dependent unknown over the element, both of which would be known in the context of the discussion. Any geometric shape for which the approximation functions can be derived uniquely qualifies as an element. We shall discuss various geometric shapes and orders of approximation shortly. To keep the formulative steps very general (i.e., not confine the formulation to a specific geometric shape or order of element), we denote the domain of a typical element by the symbol Ω^e and its boundary by Γ^e. The element Ω^e can be a triangle or quadrilateral in shape, and the degree of interpolation over it can be linear, quadratic, and so on. The non-overlapping sum (or *assembly*) of all elements used to represent the actual domain will be denoted by Ω^h, and it is called the *finite element mesh* of the domain Ω. In general, Ω^h may not equal the actual domain Ω

because of the geometric complexity of the actual domain. Of course, for simple geometries (e.g., polygonal domains) the finite element mesh exactly duplicates the actual domain.

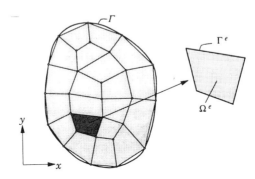

Figure 2.3.1. Finite element discretization of a domain.

For the moment let us assume that it is possible to systematically generate the approximation functions for the element Ω^e. Suppose that the dependent unknown T is approximated over a typical finite element Ω^e by the expression

$$T(x,y) \approx T^e(x,y) = \sum_{j=1}^{n} T_j^e \psi_j^e(x,y) \qquad (2.3.1)$$

where $T^e(x,y)$ represents an approximation of $T(x,y)$ over the element Ω^e, T_j^e denote the values of the function $T^e(x,y)$ at a selected number of points, called *element nodes*, in the element Ω^e, and ψ_j^e are the approximation functions associated with the element. As we shall see shortly, the interpolation functions depend not only on the number of nodes in the element, but also on the shape of the element. The shape of the element must be such that its geometry is uniquely defined by a set of points, which serve as the element nodes in the development of the interpolation functions. A triangle $(n = 3)$ is the simplest two-dimensional geometric shape in two dimensions.

Equation (2.3.1) is of the same form as that used in the traditional variational and weighted-residual methods, except for the fact that the approximation functions used in the finite element method are often polynomials and the undetermined parameters T_j^e denote the values of the function being approximated. The choice of polynomials for approximation functions is dictated largely by two factors: (1) interpolation theory can be readily used to derive them, and (2) integral expressions of polynomials can be evaluated exactly using numerical integration methods. The use of function values as unknown parameters has the primary objective of imposing continuity of the dependent variable across the boundary between elements. This will be clear when assembly of elements is discussed.

2.4 Weighted-Integral Statements and Weak Forms

The n parameters (or nodal values) T_j^e in Eq. (2.3.1) must be determined such that the approximate solution $T^e(x,y)$ satisfies the governing Eq. (2.2.1) and specified boundary conditions of the problem. As in the case of a variational and weighted-residual method, we seek to satisfy the governing differential equation in a weighted-integral sense, as described below. Since we are working with a typical element, we satisfy the differential equation in a weighted-integral sense over the element Ω^e. This process leads to n algebraic equations among the nodal values $(T_1^e, T_2^e, \cdots, T_n^e)$. The set of algebraic equations is termed a *finite element model* of the original differential equation. The type of finite element model depends on the weighted-integral form used to generate the algebraic equations. Thus, if one uses a variational form, also called a *weak form*, the resulting model will be different from those obtained with a weighted-residual statement in which the weight function can be any one of several choices (see Reddy [5–7] and Reddy and Rasmussen [8] for additional details). Throughout the present study, we will be primarily concerned with the weak-form finite element models, in which the weight functions are selected to be the same as the approximation functions (the so-called Ritz–Galerkin models).

The weak form of a differential equation is a weighted-integral statement that is equivalent to both the governing differential equation as well as the associated natural boundary conditions. We shall develop the weak form of Eqs. (2.2.1) and (2.2.5b) over the typical element Ω^e. There are three steps in the development of a weak form. The first step is to take all nonzero expressions in Eq. (2.2.1) to one side of the equality, multiply the resulting equation with a weight function w, and integrate the equation over the element domain Ω^e:

$$0 = \int_{\Omega^e} w \left[-\frac{\partial}{\partial x} \left(k_{xx} \frac{\partial T^e}{\partial x} \right) - \frac{\partial}{\partial y} \left(k_{yy} \frac{\partial T^e}{\partial y} \right) - Q(x,y) \right] dxdy \qquad (2.4.1)$$

The expression in the square brackets of the above equation represents a residual of the approximation of the differential equation (2.2.1), because $T^e(x,y)$ is only an approximation of $T(x,y)$. Therefore, Eq. (2.4.1) is called the *weighted-residual statement* of Eq. (2.2.1). For every choice of the weight function $w(x,y)$, we obtain an algebraic equation from Eq. (2.4.1) among the nodal values T_j^e. For n independent choices of w, we obtain a set of n linearly independent algebraic equations. This set is called a *weighted-residual finite element model*. The drawback of this model for second- and higher-order differential equations is that the approximation functions ψ_j^e should be differentiable as many times as the actual solution T in the differential Eq. (2.2.1). For the present case, the approximation functions should be twice-differentiable with respect to both x and y. This in turn requires that ψ_j^e be a quadratic or higher-order polynomial in the x and y coordinates. In the weak form, as the name suggests, this continuity requirement is reduced (or weakened) by moving some of the differentiation to the weight function.

In the second step, we distribute the differentiation among T and w equally, so that both T and w are required to be differentiable only once with respect to x and y. To achieve this we use integration-by-parts (or the Green–Gauss theorem) on the first two terms in Eq. (2.4.1). First we note the following identities for any differentiable functions $w(x,y)$, $F_1(x,y)$, and $F_2(x,y)$:

$$\frac{\partial}{\partial x}(wF_1) = \frac{\partial w}{\partial x}F_1 + w\frac{\partial F_1}{\partial x} \quad \text{or} \quad -w\frac{\partial F_1}{\partial x} = \frac{\partial w}{\partial x}F_1 - \frac{\partial}{\partial x}(wF_1) \qquad (2.4.2a)$$

$$\frac{\partial}{\partial y}(wF_2) = \frac{\partial w}{\partial y}F_2 + w\frac{\partial F_2}{\partial y} \quad \text{or} \quad -w\frac{\partial F_2}{\partial y} = \frac{\partial w}{\partial y}F_2 - \frac{\partial}{\partial y}(wF_2) \qquad (2.4.2b)$$

Next, we recall the component form of the gradient (or divergence) theorem,

$$\int_{\Omega^e} \frac{\partial}{\partial x}(wF_1)\,dxdy = \oint_{\Gamma^e} (wF_1)n_x\,ds \qquad (2.4.3a)$$

$$\int_{\Omega^e} \frac{\partial}{\partial y}(wF_2)\,dxdy = \oint_{\Gamma^e} (wF_2)n_y\,ds \qquad (2.4.3b)$$

where n_x and n_y are the components (i.e., the direction cosines) of the unit normal vector

$$\hat{\mathbf{n}} = n_x\hat{\mathbf{e}}_x + n_y\hat{\mathbf{e}}_y = \cos\alpha\,\hat{\mathbf{e}}_x + \sin\alpha\,\hat{\mathbf{e}}_y \qquad (2.4.4)$$

on the boundary Γ^e, and ds is the arc length of an infinitesimal line element along the boundary (see Figure 2.4.1). Using Eqs. (2.4.2a,b) and (2.4.3a,b), with

$$F_1 = k_{xx}\frac{\partial T}{\partial x}, \quad F_2 = k_{yy}\frac{\partial T}{\partial y}$$

we obtain

$$0 = \int_{\Omega^e} \left(k_{xx}\frac{\partial w}{\partial x}\frac{\partial T}{\partial x} + k_{yy}\frac{\partial w}{\partial y}\frac{\partial T}{\partial y} - wQ \right) dxdy$$
$$- \oint_{\Gamma^e} w\left(k_{xx}\frac{\partial T}{\partial x}n_x + k_{yy}\frac{\partial T}{\partial y}n_y \right) ds \qquad (2.4.5)$$

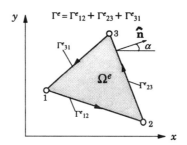

Figure 2.4.1. The unit normal vector on the boundary of a finite element.

From an inspection of the boundary term in Eq. (2.4.5), we note that the specification of T constitutes the essential boundary condition, and T is called the *primary variable*. The specification of the coefficient of the weight function in the boundary expression

$$q_n = k_{xx}\frac{\partial T}{\partial x}n_x + k_{yy}\frac{\partial T}{\partial y}n_y \qquad (2.4.6)$$

constitutes the natural boundary condition [see Eq. (2.2.5b)], and q_n is the *secondary variable* of the formulation. The function $q_n = q_n(s)$ denotes the projection of the vector $\mathbf{k} \cdot \nabla T$ along the unit normal $\mathbf{n}$. By definition q_n is positive outward from the surface as we move counterclockwise along the boundary Γ^e. The secondary variable q_n denotes the heat flux normal to the boundary of the element (into the element).

The third and last step of the formulation is to use the definition (2.4.6) in Eq. (2.4.5) and write it as

$$0 = \int_{\Omega^e}\left(k_{xx}\frac{\partial w}{\partial x}\frac{\partial T}{\partial x} + k_{yy}\frac{\partial w}{\partial y}\frac{\partial T}{\partial y} - wQ\right)dxdy - \oint_{\Gamma^e} wq_n\,ds \qquad (2.4.7)$$

or

$$B(w,T) = \ell(w) \qquad (2.4.8)$$

where the *bilinear form* $B(\cdot,\cdot)$ and *linear form* $\ell(\cdot)$ are defined by

$$B(w,T) = \int_{\Omega^e}\left(k_{xx}\frac{\partial w}{\partial x}\frac{\partial T}{\partial x} + k_{yy}\frac{\partial w}{\partial y}\frac{\partial T}{\partial y}\right)dxdy \qquad (2.4.9a)$$

$$\ell(w) = \int_{\Omega^e} wQ\,dxdy + \oint_{\Gamma^e} wq_n\,ds \qquad (2.4.9b)$$

The word 'bilinear' indicates that $B(w,T)$ is linear in *both* w and T.

This completes the three-step procedure of constructing the weak form of a differential equation. The weak form in Eq. (2.4.7) or Eq. (2.4.8) forms the basis of the finite element model of Eq. (2.2.1). By substituting for T from Eq. (2.3.1) and setting $w = \psi_1, \psi_2, \cdots, \psi_n$ (for the Ritz–Galerkin model), we obtain n algebraic equations from Eq. (2.4.7) or (2.4.8), as shown in the next section.

The statement in Eq. (2.4.8) is called the *variational problem* associated with Eqs. (2.2.1) and (2.2.5b). Whenever the bilinear form is symmetric in its arguments (w,T)

$$B(w,T) = B(T,w),$$

it is possible to construct the associated quadratic functional from the formula (see Oden and Reddy [4])

$$I(T) = \frac{1}{2}B(T,T) - \ell(T) \qquad (2.4.10)$$

However, it is *not necessary* to have a quadratic functional in order to develop a finite element model. What is needed is a weighted-integral statement or a weak form of the differential equation to be solved. It is always possible to construct the weighted-integral statement of *any* differential equation. The weak form exists for any second- and higher-order equations, because for such equations it is possible to trade differentiation from the dependent unknown to the weight function and hence include the natural boundary condition (or the secondary variable) into the weighted-integral statement. These observations hold for linear as well as for nonlinear problems. For example, the Navier-Stokes equations governing the flow of a viscous incompressible fluid do not admit a functional; however, a weak form can be constructed for the equations, as will be shown later in this book.

2.5 Finite Element Model

The weak form in Eq. (2.4.7) requires that the approximation chosen for T should be at least linear in both x and y so that there are no terms in (2.4.7) that become identically zero. Suppose that T is approximated over a typical finite element Ω^e by the expression [cf. Eq. (2.3.1)]

$$T(x,y) \approx T^e(x,y) = \sum_{j=1}^{n} T_j^e \psi_j^e(x,y) \tag{2.5.1}$$

where T_j^e is the value of $T^e(x,y)$ at the jth node with coordinates $(x_j,\ y_j)$ of the element, and ψ_j^e are the interpolation functions, with the property (see Reddy [6])

$$\psi_i^e(x_j, y_j) = \delta_{ij} \tag{2.5.2}$$

The specific form of ψ_i^e will be presented for triangular and rectangular elements in Section 2.6. Additional discussion of the elements is given in Section 2.11 and Chapter 3.

Substituting the finite element approximation (2.5.1) for T into the weak form (2.4.7), we obtain

$$0 = \int_{\Omega^e} \left[\frac{\partial w}{\partial x} \left(k_{xx} \sum_{j=1}^{n} T_j^e \frac{\partial \psi_j^e}{\partial x} \right) + \frac{\partial w}{\partial y} \left(k_{yy} \sum_{j=1}^{n} T_j^e \frac{\partial \psi_j^e}{\partial y} \right) - wQ \right] dx\,dy$$

$$- \oint_{\Gamma^e} w q_n\ ds \tag{2.5.3}$$

This equation must hold for any weight function w.   Since we need n independent algebraic equations to solve for the n unknowns, $T_1^e, T_2^e, ..., T_n^e$, we choose n independent functions for w: $w = \psi_1^e, \psi_2^e, ..., \psi_n^e$. This particular choice of weight functions is a natural one when the weight function is viewed as a virtual variation of the dependent unknown (i.e., $w = \delta T^e = \sum_{i=1}^{n} \delta T_i^e \psi_i^e$). For this choice, the solution of (2.5.3) is termed the "best approximation" (see Mikhlin [1,2] and Reddy [7]) in the variational sense.

For each choice of w we obtain an algebraic relation among $(T_1^e, T_2^e, ..., T_n^e)$. We label the algebraic equation resulting from substitution of ψ_1^e for w into Eq. (2.4.7) as the first algebraic equation. The ith algebraic equation is obtained by substituting $w = \psi_i^e$ into Eq. (2.4.7):

$$\sum_{j=1}^{n} K_{ij}^e T_j^e = Q_i^e + q_i^e \tag{2.5.4}$$

where the coefficients K_{ij}^e, Q_i^e, and q_i^e are defined by

$$K_{ij}^e = \int_{\Omega^e} \left(k_{xx} \frac{\partial \psi_i^e}{\partial x} \frac{\partial \psi_j^e}{\partial x} + k_{yy} \frac{\partial \psi_i^e}{\partial y} \frac{\partial \psi_j^e}{\partial y} \right) dxdy \tag{2.5.5a}$$

$$Q_i^e = \int_{\Omega^e} Q \psi_i^e \, dxdy \tag{2.5.5b}$$

$$q_i^e = \oint_{\Gamma^e} q_n \psi_i^e \, ds \tag{2.5.5c}$$

In matrix notation, Eq. (2.5.4) takes the form

$$[K^e]\{T^e\} = \{Q^e\} + \{q^e\} \tag{2.5.6}$$

The matrix $[K^e]$ is called the *coefficient matrix*, or conductivity matrix in the present context. We note that $K_{ij}^e = K_{ji}^e$ (i.e., $[K^e]$ is symmetric). The symmetry of the coefficient matrix is due to the symmetry of the bilinear form in Eq. (2.4.9a), which in turn is due to the weak form development. Equation (2.5.6) is called the *finite element model* of Eq. (2.2.1) or its weak form (2.4.7).

This completes the finite element model development. Before we discuss assembly of elements, it is informative to determine the interpolation functions ψ_i^e for certain basic elements.

2.6 Interpolation Functions

The finite element approximation $T^e(x,y)$ of $T(x,y)$ over an element Ω^e must satisfy the following conditions in order for the approximate solution to converge to the true solution:

1. $T^e(x,y)$ must be continuous as required in the weak form of the problem (i.e., all terms in the weak form are represented as nonzero values).

2. The polynomials used to represent $T^e(x,y)$ must be complete (i.e., all terms, beginning with a constant term up to the highest-order used in the polynomial should be included in the expression of $T^e(x,y)$).

3. All terms in the polynomial should be linearly independent.

The number of linearly independent terms in the representation of T^e dictates the shape and number of degrees of freedom of the element. Here we review the interpolation functions of linear triangular and rectangular elements.

An examination of the variational form (2.4.7) and the finite element matrices in Eq. (2.5.5a) shows that the ψ_i^e should be at least linear functions of x and y. The polynomial

$$T^e(x, y) = c_1^e + c_2^e x + c_3^e y \tag{2.6.1}$$

is the lowest-order polynomial that meets the requirements. It contains three linearly independent terms, and it is linear in both x and y. The polynomial is complete because the lower-order term, namely, the constant term, is included. To write the three constants (c_1^e, c_2^e, c_3^e) in terms of the nodal values of T^e, we must identify three points or nodes in the element Ω^e. The three nodes must be such that they uniquely define the geometry of the element and allow the imposition of inter-element continuity of the variable $T^e(x, y)$. Obviously, the geometric shape defined by three points in a two-dimensional domain is a triangle. Thus the polynomial in Eq. (2.6.1) is associated with a triangular element and the three nodes are identified as the vertices of the triangle.

On the other hand, the polynomial

$$T^e(x, y) = c_1^e + c_2^e x + c_3^e y + c_4^e xy \tag{2.6.2}$$

contains four linearly independent terms, and is linear in x and y, with a bilinear term in x and y. This polynomial requires an element with four nodes. It is a rectangle with nodes at the four corners of the rectangle, or a triangle with three nodes at vertices and the fourth at the centroid of the triangle. Note that the four-node triangular element does not meet the inter-element continuity of the functions approximated, and hence not used.

The interpolation functions for linear triangular and rectangular elements are given below. Higher-order two-dimensional elements (i.e., element with higher-order interpolation polynomials) will be discussed in Section 2.11.

Linear triangular element

The linear interpolation functions for the three-node triangle (see Figure 2.6.1a) are (see Reddy [6, pp. 304–307])

$$\psi_i^e(x, y) = \frac{1}{2A_e}(\alpha_i^e + \beta_i^e x + \gamma_i^e y), \quad (i = 1, 2, 3) \tag{2.6.3}$$

where A_e is the area of the triangle, and α_i^e, β_i^e, and γ_i^e are geometric constants known in terms of the nodal coordinates (x_i, y_i):

$$\alpha_i^e = x_j y_k - x_k y_j \; ; \quad \beta_i^e = y_j - y_k \; ; \quad \gamma_i^e = -(x_j - x_k) \tag{2.6.4}$$

Here the subscripts are such that $i \neq j \neq k$, and i, j, and k permute in a natural order. Note that (x, y) are the coordinates used in the governing equation (2.2.1) over the domain Ω, and are called *global coordinates*. Since the element interpolation functions are restricted to an element, it is possible to express them in a *local coordinate system*. Such local coordinate systems are useful in the numerical evaluation of the integral expressions in Eqs. (2.5.5a–c). We will return to these issues in Section 2.12.

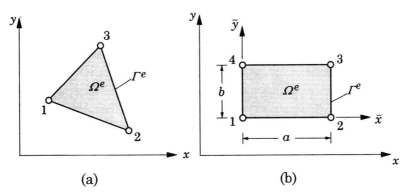

(a)   (b)

Figure 2.6.1. The linear triangular and rectangular finite elements.

The interpolation functions $\psi_i^e (i = 1, 2, \cdots, n)$ constructed such that Eq. (2.5.1) interpolates only the function at the nodes (and not its derivatives) are called the *Lagrange interpolation functions*. They satisfy the following interpolation properties:

$$\text{(i)} \quad \psi_i^e(x_j, y_j) = \delta_{ij}, \quad (i, j = 1, 2, 3); \qquad \text{(ii)} \quad \sum_{i=1}^{3} \psi_i^e(x, y) = 1 \qquad (2.6.5)$$

Note that use of the linear interpolation functions ψ_i^e of a triangle will result in the approximation of the curved surface $T(x, y)$ by a planar function $T^e = \sum_{i=1}^{3} T_i^e \psi_i^e$.

The integrals in the definition of K_{ij}^e and Q_i^e can be evaluated for given data: k_{xx}, k_{yy}, and Q. For example, for element-wise constant values of the data, i.e., $k_{xx} = k_{xx}^e$, $k_{yy} = k_{yy}^e$, and $Q = Q^e$, we have (see Reddy [6, pp. 311–313]) the following results:

$$K_{ij}^e = \frac{1}{4A_e}(k_{xx}^e \beta_i^e \beta_j^e + k_{yy}^e \gamma_i^e \gamma_j^e); \quad Q_i^e = \frac{Q_e A_e}{3} \qquad (2.6.6a)$$

where A_e is the area of the triangular element, and β_i^e and γ_i^e are known in terms of the global nodal coordinates of the element nodes, as given in Eq. (2.6.4). For a right-angled triangular element with base a and height b, and node 1 at the right angle (nodes are numbered counterclockwise), $[K^e]$ takes the form (see Reddy [6, p. 387])

$$[K^e] = \frac{k_{xx}^e}{2}\begin{bmatrix} \alpha & -\alpha & 0 \\ -\alpha & \alpha & 0 \\ 0 & 0 & 0 \end{bmatrix} + \frac{k_{yy}^e}{2}\begin{bmatrix} \beta & 0 & -\beta \\ 0 & 0 & 0 \\ -\beta & 0 & \beta \end{bmatrix} \qquad (2.6.6b)$$

where $\alpha = b/a$ and $\beta = a/b$. Of course, for cases in which the conductivities are functions of (x, y), numerical integration can be used to evaluate the coefficients (see Section 2.12).

The evaluation of boundary integrals of the type [see Eq. (2.5.5c)]

$$q_i^e = \oint_{\Gamma^e} q_n^e \psi_i^e(s) \, ds \tag{2.6.7}$$

where q_n^e is a known function of the distance s along the boundary Γ^e, involves evaluation of line integrals. It is necessary to compute such integrals only when Γ^e, or a portion of it, coincides with the boundary Γ_q of the total domain Ω on which the flux is specified. On portions of Γ^e that are in the interior of the domain Ω, q_n^e on side (i,j) of element Ω^e cancels with q_n^f on side (p,q) of element Ω^f when sides (i,j) of element Ω^e and (p,q) of element Ω^f are the same (i.e., at the interface of elements Ω^e and Ω^f). This can be viewed as the balance of the internal flux. When Γ^e falls on the boundary Γ_T of the domain Ω, q_n^e is not known there and can be determined in the post-computation. Note that the primary variable T is specified on Γ_T. For additional details, see Reddy [6, pp. 313–318].

Linear rectangular element

For a linear rectangular element, we have

$$T(\bar{x}, \bar{y}) = \sum_{i=1}^{4} T_i^e \psi_i^e(\bar{x}, \bar{y}) \tag{2.6.8}$$

where (see Reddy [6, pp. 308–311])

$$\psi_1^e = (1 - \frac{\bar{x}}{a})(1 - \frac{\bar{y}}{b}), \quad \psi_2^e = \frac{\bar{x}}{a}(1 - \frac{\bar{y}}{b})$$

$$\psi_3^e = \frac{\bar{x}}{a}\frac{\bar{y}}{b}, \qquad\qquad \psi_4^e = (1 - \frac{\bar{x}}{a})\frac{\bar{y}}{b} \tag{2.6.9}$$

and $(\bar{x}, \bar{y})$ denote the local coordinates with origin located at node 1 of the element, and (a, b) denote the horizontal and vertical dimensions of the rectangle (see Figure 2.6.1b).

The integrals in the definition of K_{ij}^e and Q_i^e can be easily evaluated over a rectangular element of sides a and b. For example, for element-wise constant values of the data, i.e., $k_{xx} = k_{xx}^e$, $k_{yy} = k_{yy}^e$, and $Q = Q^e$, we have (see Reddy [6, p. 313; p. 387]) the following results:

$$[K^e] = k_{xx}^e [S^{11}] + k_{yy}^e [S^{22}], \quad Q_i^e = \frac{Q_e ab}{4} \tag{2.6.10a}$$

where

$$[S^{11}] = \frac{1}{6}\begin{bmatrix} 2\alpha & -2\alpha & -\alpha & \alpha \\ -2\alpha & 2\alpha & \alpha & -\alpha \\ -\alpha & \alpha & 2\alpha & -2\alpha \\ \alpha & -\alpha & -2\alpha & 2\alpha \end{bmatrix} \tag{2.6.10b}$$

$$[S^{22}] = \frac{1}{6}\begin{bmatrix} 2\beta & \beta & -\beta & -2\beta \\ \beta & 2\beta & -2\beta & -\beta \\ -\beta & -2\beta & 2\beta & \beta \\ -2\beta & -\beta & \beta & 2\beta \end{bmatrix} \tag{2.6.10c}$$

and $\alpha = b/a$ and $\beta = a/b$. Again, for cases in which the conductivities are functions of (x, y), numerical integration is used to evaluate the coefficients, as discussed in Section 2.12. When the element is nonrectangular, i.e., a quadrilateral, we use coordinate transformations to represent the integrals over a square geometry (see Sections 2.12 and 3.4) and then use numerical integration to evaluate them.

2.7 Assembly of Elements

The assembly of finite elements to obtain the equations of the entire domain is based on the following two rules:

1. Continuity of the primary variable (i.e., temperature)

2. Balance of secondary variables (i.e., heat flux)

We illustrate the assembly procedure by considering a finite element mesh consisting of a triangular element and a quadrilateral element (see Figure 2.7.1).

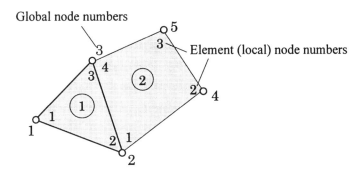

Figure 2.7.1. Global-local correspondence of nodes for assembly of elements.

Let K_{ij}^1 $(i, j = 1, 2, 3)$ denote the coefficient matrix corresponding to the triangular element, and let K_{ij}^2 $(i, j = 1, 2, 3, 4)$ denote the coefficient matrix corresponding to the quadrilateral element. The nodes of the finite element mesh are called *global nodes*. From the mesh shown in Figure 2.7.1, it is clear that the following correspondence between the global and element nodes exists: nodes 1, 2, and 3 of element 1 correspond to global nodes 1, 2, and 3, respectively. Nodes 1, 2, 3, and 4 of element 2 correspond to global nodes 2, 4, 5, and 3, respectively. Hence, the correspondence between the local and global nodal values of temperature is

$$T_1^1 = T_1 \;,\;\; T_2^1 = T_1^2 = T_2 \;,\;\; T_3^1 = T_4^2 = T_3 \;,\;\; T_2^2 = T_4 \;,\;\; T_3^2 = T_5 \qquad (2.7.1)$$

which amounts to imposing the continuity of the primary variables at the nodes common to elements 1 and 2. Note that the continuity of the primary variables at the inter-element nodes guarantees the continuity of the primary variable along the entire inter-element boundary.

Next, we consider the balance of secondary variables. At the interface between the two elements, the flux from the two elements should be equal in magnitude and opposite in sign. For the two elements in Figure 2.7.1, the interface is along the side connecting global nodes 2 and 3. Hence, the internal flux q_n^1 on side 2–3 of element 1 should balance the flux q_n^2 on side 4–1 of element 2 (recall the sign convention on q_n^e):

$$(q_n^1)_{2-3} = (q_n^2)_{4-1} \text{ or } (q_n^1)_{2-3} = (-q_n^2)_{1-4} \qquad (2.7.2)$$

In the finite element method, the above relation is imposed in a weighted-integral sense:

$$\int_{h_{23}^1} q_n^1 \psi_2^1 \, ds = -\int_{h_{14}^2} q_n^2 \psi_1^2 \, ds \qquad (2.7.3a)$$

$$\int_{h_{23}^1} q_n^1 \psi_3^1 \, ds = -\int_{h_{14}^2} q_n^2 \psi_4^2 \, ds \qquad (2.7.3b)$$

where h_{pq}^e denotes length of the side connecting node p to node q of element Ω^e.

Now we are ready to assemble the element equations to obtain the equations for the two-element mesh. The element equations of the two elements are written separately first. For the triangular element, the element equations are of the form

$$\begin{aligned}
K_{11}^1 T_1^1 + K_{12}^1 T_2^1 + K_{13}^1 T_3^1 &= Q_1^1 + q_1^1 \\
K_{21}^1 T_1^1 + K_{22}^1 T_2^1 + K_{23}^1 T_3^1 &= Q_2^1 + q_2^1 \\
K_{31}^1 T_1^1 + K_{32}^1 T_2^1 + K_{33}^1 T_3^1 &= Q_3^1 + q_3^1
\end{aligned} \qquad (2.7.4a)$$

For the rectangular element the element equations are given by

$$\begin{aligned}
K_{11}^2 T_1^2 + K_{12}^2 T_2^2 + K_{13}^2 T_3^2 + K_{14}^2 T_4^2 &= Q_1^2 + q_1^2 \\
K_{21}^2 T_1^2 + K_{22}^2 T_2^2 + K_{23}^2 T_3^2 + K_{24}^2 T_4^2 &= Q_2^2 + q_2^2 \\
K_{31}^2 T_1^2 + K_{32}^2 T_2^2 + K_{33}^2 T_3^2 + K_{34}^2 T_4^2 &= Q_3^2 + q_3^2 \\
K_{41}^2 T_1^2 + K_{42}^2 T_2^2 + K_{43}^2 T_3^2 + K_{44}^2 T_4^2 &= Q_4^2 + q_4^2
\end{aligned} \qquad (2.7.4b)$$

In order to impose the balance of secondary variables in Eq. (2.7.3), it is required that we add the second equation of element 1 to the first equation of element 2, and also add the third equation of element 1 to the fourth equation of element 2:

$$\begin{aligned}
(K_{21}^1 T_1^1 + K_{22}^1 T_2^1 + K_{23}^1 T_3^1) + (K_{11}^2 T_1^2 + K_{12}^2 T_2^2 + K_{13}^2 T_3^2 + K_{14}^2 T_4^2) \\
= (Q_2^1 + q_2^1) + (Q_1^2 + q_1^2) \qquad (2.7.5a)
\end{aligned}$$

$$\begin{aligned}
(K_{31}^1 T_1^1 + K_{32}^1 T_2^1 + K_{33}^1 T_3^1) + (K_{41}^2 T_1^2 + K_{42}^2 T_2^2 + K_{43}^2 T_3^2 + K_{44}^2 T_4^2) \\
= (Q_3^1 + q_3^1) + (Q_4^2 + q_4^2) \qquad (2.7.5b)
\end{aligned}$$

Using the local-global nodal variable correspondence in Eq. (2.7.1), we can rewrite the above equations as

$$K_{21}^1 T_1 + (K_{22}^1 + K_{11}^2)T_2 + (K_{23}^1 + K_{14}^2)T_3 + K_{12}^2 T_4 + K_{13}^2 T_5$$
$$= Q_2^1 + Q_1^2 + (q_2^1 + q_1^2) \qquad (2.7.6a)$$

$$K_{31}^1 T_1 + (K_{32}^1 + K_{41}^2)T_2 + (K_{33}^1 + K_{44}^2)T_3 + K_{42}^2 T_4 + K_{43}^2 T_5$$
$$= Q_3^1 + Q_4^2 + (q_3^1 + q_4^2) \qquad (2.7.6b)$$

Now we can impose the conditions in Eq. (2.7.3) by setting appropriate portions of the expressions in parenthesis on the right-hand side of the above equations to zero (or a specified nonzero value). In general, when several elements are connected, the assembly of the elements is carried out by putting element coefficients K_{ij}^e, Q_i^e, and q_i^e into proper locations of the global coefficient matrix and right-hand column vectors. This is done by means of the connectivity relations, i.e., correspondence of the local node number to the global node number.

The assembly procedure described above can be used to assemble elements of any shape and type. The procedure can be implemented in a computer with the help of the local-global nodal correspondence.

This completes the first four steps in the finite element modeling of the model Eq. (2.2.1). The remaining three steps of the analysis, namely, the imposition of boundary conditions, solution of equations, and postprocessing of the solution will not be discussed here, but will be reserved for subsequent chapters. It may be recalled that the derivatives of $T^e(x,y)$ will not be continuous at inter-element boundaries because continuity of the derivatives is not imposed in the model. The weak form of the equation suggests that the primary variable is T, and therefore it should be made continuous across the inter-element boundaries. If additional variables, such as the derivatives of temperature, are carried as nodal variables in the interest of making them continuous across inter-element boundaries, the degree of interpolation (or order of the element) increases. In addition, the continuity of the derivatives, which are not identified as the primary variables in the weak formulation of the problem, may violate the physical principles of the problem.

2.8 Time-Dependent Problems

2.8.1 Introduction

In this section we present the finite element model of time-dependent heat transfer problems in two dimensions. Consider the equation governing transient heat transfer in two dimensions,

$$\rho C \frac{\partial T}{\partial t} - \left[\frac{\partial}{\partial x}\left(k_{xx}\frac{\partial T}{\partial x} \right) + \frac{\partial}{\partial y}\left(k_{yy}\frac{\partial T}{\partial y} \right) \right] = Q \quad \text{in } \Omega \qquad (2.8.1)$$

We wish to analyze Eq. (2.8.1) under appropriate boundary and initial conditions. The boundary conditions for this equation are given by Eqs.

(2.2.5), and the initial condition on the temperature is given by

$$T(\mathbf{x}, 0) = T_0(\mathbf{x}) \tag{2.8.2}$$

The numerical solution of the initial-boundary value problem described by Eq. (2.8.1) involves two stages of approximation. The first stage, called *spatial discretization*, involves the development of the weak form of the equation over an element and the spatial approximation of the dependent variable (i.e., temperature) T of the problem. The three-step procedure presented in Section 2.4 for the development of the weak form is still applicable to time-dependent problems. This stage results in a set of ordinary differential equations in time among the nodal values T_j^e of the dependent variable. The second stage consists of a time approximation, called *temporal approximation*, of the ordinary differential equations (i.e., numerical integration of the equations) by finite difference schemes. This stage leads to a set of algebraic equations involving the nodal values T_j^e at time t_{n+1} $[= (n+1)\Delta t$, where n is an integer and Δt is the time increment] in terms of known values from the previous time step. The finite element model of Eq. (2.8.1) using the *space-time finite elements* can also be developed (see Reddy [6], p. 401, and [11–15]). In this procedure, time is treated as another spatial coordinate. Consequently, one should know the solution at the initial as well as final times (i.e., at the time boundaries). Here we discuss only the two-stage approximation. The semidiscretization is discussed first.

2.8.2 Semidiscretization

We assume, as usual, that the conduction heat transfer region Ω is discretized into an appropriate collection of finite elements (see Figure 2.3.1). The weak form of Eq. (2.8.1) over an element Ω^e is obtained by the standard procedure: multiply Eq. (2.8.1) with the weight function $w(x, y)$ and integrate over the element, integrate-by-parts (spatially) those terms which involve higher-order derivatives using the gradient or divergence theorem, and replace the coefficient of the weight function in the boundary integral with the secondary variable [i.e., use Eq. (2.2.5b)]. We obtain

$$0 = \int_{\Omega^e} \left[w \left(\rho C \frac{\partial T}{\partial t} - Q_s \right) + k_{xx} \frac{\partial w}{\partial x} \frac{\partial T}{\partial x} + k_{yy} \frac{\partial w}{\partial y} \frac{\partial T}{\partial y} \right] dx dy - \oint_{\Gamma^e} (q_n - q_c) w \, ds \tag{2.8.3}$$

Note that the procedure to obtain the weak form for time-dependent problems is not much different from that used for steady-state problems in Section 2.4. The difference is that all terms of the equations may be functions of time. Also, no integration by parts with respect to time is used, and the weight function w is not a function of time.

The semidiscrete finite element model is obtained from Eq. (2.8.3) by substituting a finite element approximation for the dependent variable, T. In selecting the approximation for T, we assume that the time dependence can be separated from the spatial variation,

$$T(\mathbf{x}, t) \simeq \sum_{j=1}^{n_e} T_j^e(t) \psi_j^e(\mathbf{x}) \tag{2.8.4}$$

The ith differential equation (in time) of the finite element model is obtained by substituting $w = \psi_i^e(\mathbf{x})$ and replacing T by the expression in Eq. (2.8.4):

$$0 = \sum_{j=1}^{n_e} \left(M_{ij}^e \frac{dT_j^e}{dt} + K_{ij}^e T_j^e \right) - Q_i^e - q_i^e \tag{2.8.5}$$

Equation (2.8.5) can be expressed in matrix form as

$$[M^e]\{\dot{T}^e\} + [K^e]\{T^e\} = \{Q^e\} + \{q^e\} \tag{2.8.6}$$

where a superposed dot on T denotes a derivative with time ($\dot{T} = \partial T/\partial t$), and

$$M_{ij}^e = \int_{\Omega^e} \rho C \psi_i \psi_j \; dxdy$$

$$K_{ij}^e = \int_{\Omega^e} \left(k_{xx} \frac{\partial \psi_i}{\partial x} \frac{\partial \psi_j}{\partial x} + k_{yy} \frac{\partial \psi_i}{\partial y} \frac{\partial \psi_j}{\partial y} \right) \; dxdy$$

$$Q_i^e = \int_{\Omega^e} \psi_i Q(x, y, t) \; dxdy$$

$$q_i^e = \oint_{\Gamma^e} \psi_i (q_n - q_c) \; ds \tag{2.8.7}$$

The element label 'e' on ψ is omitted for brevity. It should be noted that the capacitance (or mass) matrix $[M]$ in Eq. (2.8.7) is not a diagonal matrix. It is called a *consistent (mass) matrix* because it is defined consistent with the weak formulation in Eq. (2.8.3).

2.8.3 Temporal Approximation

The ordinary differential Eq. (2.8.6) should be integrated with respect to time to obtain the transient response. Since it is not possible, in general, to integrate these equations analytically, they are further approximated in time to obtain a set of algebraic equations in terms of the nodal temperatures (i.e., a fully discretized model is obtained). In principle, it is possible to use any of the standard methods for the solution of ordinary differential equations. However, practical considerations, such as the computational cost, dictate that only the simpler time integration (or approximation) methods be considered.

The most commonly used time integration methods for Eq. (2.8.6) are all part of a one-parameter family, called the θ-family of approximation

$$\{T\}_{n+1} = \{T\}_n + \Delta t[(1 - \theta)\{\dot{T}\}_n + \theta\{\dot{T}\}_{n+1}] \;, \; 0 \le \theta \le 1 \tag{2.8.8}$$

Using Eq. (2.8.8) in Eq. (2.8.6), we can transform the ordinary differential equations into a set of algebraic equations at time t_{n+1} :

$$[\hat{K}]_{n+1}\{T\}_{n+1} = \{\hat{F}\}_{n,n+1} \qquad (2.8.9)$$

where

$$[\hat{K}]_{n+1} = [M] + a_1[K]_{n+1} \qquad (2.8.10a)$$

$$\{\hat{F}\}_{n,n+1} = \Delta t(\theta\{F\}_{n+1} + (1-\theta)\{F\}_n) + [\bar{K}]_n\{T\}_n \qquad (2.8.10b)$$

$$[\bar{K}]_n = ([M] - a_2[K]_n), \quad a_1 = \theta\Delta t \ , \ a_2 = (1-\theta)\Delta t \qquad (2.8.10c)$$

Equation (2.8.9), after assembly and imposition of boundary conditions, is solved at each time step for the nodal values T_j at time $t_n = (n+1)\Delta t$. At time $t = 0$ (i.e., $n = 0$), the right-hand side is computed using the initial values $\{T\}_0$; the vector $\{F\}$, which is the sum of the source vector $\{Q^e\}$ and internal flux vector $\{q^e\}$, is always known, for both times t_n and t_{n+1}, at all nodes at which the solution is unknown (because $Q(x, y, t)$ is a known function of time and the sum of q_j^e at these nodes is zero).

For different values of the parameter θ, we obtain several well-known time approximation schemes:

$\theta = 0$, the forward difference scheme (conditionally stable)

$\theta = 0.5$, the Crank–Nicolson scheme (unconditionally stable)

$\theta = \dfrac{2}{3}$, the Galerkin scheme (unconditionally stable)

$\theta = 1$, the backward difference scheme (unconditionally stable) (2.8.11)

For $\theta \geq 0.5$, the scheme is stable, and for $\theta < 0.5$ the scheme is stable only if the time step meets certain restriction (i.e., *conditionally stable schemes*) (see Reddy [6]). For the forward difference scheme the stability requirement is

$$\Delta t < \Delta t_{cr} = \frac{2}{(1-2\theta)\lambda_{\max}} \ , \ \theta < \frac{1}{2} \qquad (2.8.12)$$

where $\lambda_{\max}$ is the largest eigenvalue of the eigenvalue problem associated with the matrix equation (2.8.6):

$$|\hat{\mathbf{K}} - \lambda\mathbf{I}| = 0 \qquad (2.8.13)$$

This completes the development of the finite element model of a transient heat transfer problem in two dimensions. Further details on time-dependence will be covered in Chapter 3.

2.9 Axisymmetric Problems

In studying heat transfer problems involving cylindrical geometries, it is convenient to use the cylindrical coordinate system (r, θ, z) (see Figure 2.9.1) to formulate the problem. If the geometry, data, and boundary conditions of the problem are independent of the angular coordinate θ, the problem solution will also be independent of θ. Consequently, a three-dimensional problem is reduced to a two-dimensional one in (r, z) coordinates. Here we present the weak form for an axisymmetric problem.

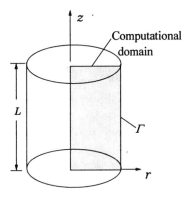

Figure 2.9.1. Domain of an axisymmetric problem.

Consider the partial differential equation governing heat transfer in an axisymmetric geometry

$$-\frac{1}{r}\frac{\partial}{\partial r}\left(rk_{rr}\frac{\partial T}{\partial r}\right) - \frac{\partial}{\partial z}\left(k_{zz}\frac{\partial T}{\partial z}\right) = Q(r, z) \qquad (2.9.1)$$

where (k_{rr}, k_{zz}) and Q are the conductivities and internal heat generation per unit volume, respectively. In developing the weak form, we integrate over the elemental volume of the axisymmetric geometry: $rdrd\theta dz$. Since the solution is independent of the θ coordinate, the integration with respect to θ yields a multiplicative constant, 2π.

Following the three-step procedure, we write the weak form of Eq. (2.9.1):

$$0 = 2\pi \int_{\Omega^e} w\left[-\frac{1}{r}\frac{\partial}{\partial r}(rk_{rr}\frac{\partial T}{\partial r}) - \frac{\partial}{\partial z}(k_{zz}\frac{\partial T}{\partial z}) - Q\right] rdrdz$$

Integrating the highest-order terms by parts, we obtain

$$0 = 2\pi \int_{\Omega^e}\left[\frac{\partial w}{\partial r}\cdot rk_{rr}\frac{\partial T}{\partial r} + \frac{\partial w}{\partial z}\cdot rk_{zz}\frac{\partial T}{\partial z} - w\cdot rQ\right] drdz$$
$$- 2\pi \oint_{\Gamma^e} w\left(rk_{rr}\frac{\partial T}{\partial r}n_r + rk_{zz}\frac{\partial T}{\partial z}n_z\right) ds$$

or

$$0 = 2\pi \int_{\Omega^e}\left[k_{rr}\frac{\partial w}{\partial r}\frac{\partial T}{\partial r} + k_{zz}\frac{\partial w}{\partial z}\frac{\partial T}{\partial z} - wQ\right] rdrdz - \oint_{\Gamma^e} wq_n\, ds \qquad (2.9.2)$$

where w is the weight function and q_n is the normal flux,

$$q_n = 2\pi r \left(k_{rr} \frac{\partial T}{\partial r} n_r + k_{zz} \frac{\partial T}{\partial z} n_z \right) \tag{2.9.3}$$

Note that the weak form (2.9.2) does not differ significantly from that developed for Eq. (2.2.1). The only difference is the presence of r in the integrand. Consequently, Eq. (2.9.2) can be obtained as a special case of Eq. (2.4.7) for $k_{xx} = 2\pi k_{rr} \cdot x$, $k_{yy} = 2\pi k_{zz} \cdot x$, and Q is replaced by $2\pi Q \cdot x$; the coordinates r and z are treated like x and y, respectively.

Let us assume that $T(r, z)$ is represented by the finite element approximation T^e over the element Ω^e :

$$T(r, z) \approx T^e(r, z) = \sum_{j=1}^{n} T_j^e \psi_j^e(r, z) \tag{2.9.4}$$

The interpolation functions $\psi^e(r, z)$ are the same as those given in Eqs. (2.6.3) and (2.6.9) for linear triangular and rectangular elements, respectively, with $x = r$ and $y = z$. Substitution of Eq. (2.9.4) for T and ψ_i^e for w into the weak form gives the ith equation of the finite element model:

$$0 = 2\pi \sum_{j=1}^{n} \int_{\Omega^e} \left[\frac{\partial \psi_i^e}{\partial r} \left(k_{rr} T_j^e \frac{\partial \psi_j^e}{\partial r} \right) + \frac{\partial \psi_i^e}{\partial z} \left(k_{zz} T_j^e \frac{\partial \psi_j^e}{\partial z} \right) - wQ \right] r \, dr \, dz$$

$$- \oint_{\Gamma^e} \psi_i^e q_n ds \tag{2.9.5}$$

or

$$0 = \sum_{j=1}^{n} K_{ij}^e T_j^e - Q_i^e - q_i^e \tag{2.9.6}$$

where

$$K_{ij}^e = 2\pi \int_{\Omega^e} \left(k_{rr} \frac{\partial \psi_i^e}{\partial r} \frac{\partial \psi_j^e}{\partial r} + k_{zz} \frac{\partial \psi_i^e}{\partial z} \frac{\partial \psi_j^e}{\partial z} \right) r \, dr \, dz \tag{2.9.7a}$$

$$Q_i^e = 2\pi \int_{\Omega^e} \psi_i^e Q \, r \, dr \, dz \tag{2.9.7b}$$

$$q_i^e = \oint_{\Gamma^e} q_n \psi_i^e \, ds \tag{2.9.7c}$$

This completes the development of the time-independent finite element model of an axisymmetric problem; the formulation for a time-dependent model follows exactly the steps as outlined for the planar case.

2.10 Convective Boundary Conditions

For heat conduction problems that involve convection heat transfer at the boundary, i.e., when heat is transferred from one medium to the surrounding medium (often, a fluid) by convection, the finite element model developed earlier requires some modification. The model to be presented allows the computation of the additional contributions to the coefficient matrix and source vector whenever the element has the convection boundary condition. For a convection boundary, the natural boundary condition is a balance of energy transfer across the boundary due to conduction and/or convection (i.e., Newton's law of cooling):

$$(k_{xx}\frac{\partial T}{\partial x}n_x + k_{yy}\frac{\partial T}{\partial y}n_y) + h_c(T - T_c) = q_n \tag{2.10.1}$$

where h_c is the convective conductance [or the convective heat transfer coefficient], T_c is the (ambient) temperature of the surrounding fluid medium, and q_n is the specified heat flux. The first term accounts for heat transfer by conduction, the second by convection, and the third accounts for the specified heat flux, if any. It is the presence of the term $h_c(T - T_c)$ that requires some modification of the weak form in Eq. (2.4.7). To include the convective boundary condition (2.10.1), the boundary integral in Eq. (2.4.5) should be modified. Instead of replacing the coefficient of w in the boundary integral with q_n, we use Eq. (2.10.1):

$$
\begin{aligned}
0 &= \int_{\Omega^e}\left(k_{xx}\frac{\partial w}{\partial x}\frac{\partial T}{\partial x} + k_{yy}\frac{\partial w}{\partial y}\frac{\partial T}{\partial y} - wQ\right)dxdy \\
&\quad - \oint_{\Gamma^e} w\left(k_{xx}\frac{\partial T}{\partial x}n_x + k_{yy}\frac{\partial T}{\partial y}n_y\right)ds \\
&= \int_{\Omega^e}\left(k_{xx}\frac{\partial w}{\partial x}\frac{\partial T}{\partial x} + k_{yy}\frac{\partial w}{\partial y}\frac{\partial T}{\partial y} - wQ\right)dxdy - \oint_{\Gamma^e} w[q_n - h_c(T - T_c)]ds
\end{aligned}
\tag{2.10.2a}
$$

or

$$B(w, T) = \ell(w) \tag{2.10.2b}$$

where w is the weight function, and $B(\cdot, \cdot)$ and $\ell(\cdot)$ are the bilinear and linear forms

$$B(w, T) = \int_{\Omega^e}\left(k_{xx}\frac{\partial w}{\partial x}\frac{\partial T}{\partial x} + k_{yy}\frac{\partial w}{\partial y}\frac{\partial T}{\partial y}\right)dxdy + \int_{\Gamma^e} h_c wT\, ds \tag{2.10.3a}$$

$$\ell(w) = \int_{\Omega^e} wQ\, dxdy + \int_{\Gamma^e} h_c wT_c\, ds + \int_{\Gamma^e} wq_n\, ds \tag{2.10.3b}$$

Note that the unknown surface temperature in the convective boundary condition has been made part of the bilinear form $B(\cdot, \cdot)$ while all the known quantities remain part of the linear form $\ell(\cdot)$. The finite element model of Eq. (2.10.2) is straightforward (see Problem 2.1; also, see Reddy [6], pp. 341–346).

The finite element model of Eq. (2.10.2a) or (2.10.2b) is valid for both conductive and convective heat transfer boundary conditions. Radiative heat transfer boundary conditions are nonlinear and will be considered in the later chapters. For problems with no convective boundary conditions, the convective contributions to the element coefficients are omitted. Indeed, these contributions have to be included only for those elements whose sides fall on the boundary with specified convection heat transfer. The contribution due to convective boundaries to the element coefficient matrix and source vector can be computed by evaluating line integrals, as discussed in Reddy [6], pp. 342–345.

2.11 Library of Finite Elements

2.11.1 Introduction

The objective of this section is to present a library of two-dimensional triangular and rectangular elements of the Lagrange family, i.e., elements over which only the function – not its derivatives – are interpolated. Once we have elements of different shapes and order at our disposal, we can choose appropriate elements and associated interpolation functions for a given problem. The interpolation functions are developed here for regularly shaped elements, called *master elements*. These elements can be used for numerical evaluation of integrals defined on irregularly shaped elements. This requires a transformation of the geometry from the actual element shape to its associated master element. We will discuss the numerical evaluation of integrals in Section 2.12.

2.11.2 Triangular Elements

The three-noded triangular element was developed in Section 2.6. Higher-order triangular elements (i.e., triangular elements with interpolation functions of higher degree) can be systematically developed with the help of the so-called *area coordinates*. For triangular elements, it is possible to construct three nondimensional coordinates $L_i(i = 1, 2, 3)$, which vary in a direction normal to the sides directly opposite each node (see Figure 2.11.1). The coordinates are defined such that

$$L_i = \frac{A_i}{A} , \quad A = \sum_{i=1}^{3} A_i \qquad (2.11.1)$$

where A_i is the area of the triangle formed by nodes j and k and an arbitrary point P in the element, and A is the total area of the element. For example, A_1 is the area of the shaded triangle which is formed by nodes 2 and 3 and point P. The point P is at a perpendicular distance of s from the side connecting nodes 2 and 3. We have $A_1 = bs/2$ and $A = bh/2$. Hence

$$L_1 = \frac{A_1}{A} = \frac{s}{h} \qquad (2.11.2)$$

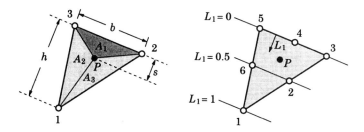

Figure 2.11.1. Definition of area coordinates L_i used for triangular elements.

Clearly, L_1 is zero on side 2–3 (hence, zero at nodes 2 and 3) and has a value of unity at node 1. Thus, L_1 is the interpolation function associated with node 1. Similarly, L_2 and L_3 are the interpolation functions associated with nodes 2 and 3, respectively. In summary, we have

$$\psi_i = L_i \tag{2.11.3}$$

The area coordinates L_i can be used to construct interpolation functions for higher-order triangular elements. For example, a higher-order element with k nodes per side (equally spaced on each side) has a total of n nodes

$$n = \sum_{i=0}^{k-1}(k - i) = k + (k - 1) + \cdots + 1 = \frac{k}{2}(k + 1) \tag{2.11.4}$$

and its degree is equal to $k-1$. The explicit forms of the interpolation functions for the linear and quadratic elements are recorded below:

$$\{\Psi^e\} = \begin{Bmatrix} L_1 \\ L_2 \\ L_3 \end{Bmatrix} \; ; \quad \{\Psi^e\} = \begin{Bmatrix} L_1(2L_1 - 1) \\ L_2(2L_2 - 1) \\ L_3(2L_3 - 1) \\ 4L_1L_2 \\ 4L_2L_3 \\ 4L_3L_1 \end{Bmatrix} \tag{2.11.5}$$

Note that the order of the interpolation functions in the above arrays corresponds to the node numbers shown in Figure 2.11.2a. Thus, the first three rows of the vectors in Eq. (2.11.5) correspond to the first three nodes of the linear and quadratic elements, which correspond to the three vertices of the triangular element. The last three rows of the second vector in Eq. (2.11.5) associated with the quadratic element correspond to the midside nodes of the triangular element. A similar node numbering scheme is used for rectangular elements, which are discussed next.

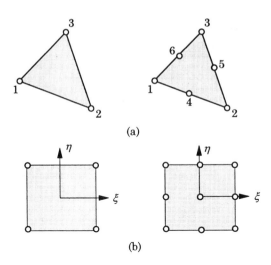

Figure 2.11.2. Linear and quadratic (a) triangular and (b) rectangular elements.

2.11.3 Rectangular Elements

The Lagrange interpolation functions associated with rectangular elements can be obtained from the tensor product of corresponding one-dimensional Lagrange interpolation functions. We take a local coordinate system (ξ, η) such that $-1 \le (\xi, \eta) \le 1$. This choice of local coordinate system is dictated by the Gauss quadrature rule used in the numerical evaluation of integrals over the element (see Section 2.12).

The linear and quadratic interpolation functions are given below (see Figure 2.11.2b for the node numbers).

$$\{\Psi^e\} = \frac{1}{4} \left\{ \begin{array}{l} (1 - \xi)(1 - \eta) \\ (1 + \xi)(1 - \eta) \\ (1 + \xi)(1 + \eta) \\ (1 - \xi)(1 + \eta) \end{array} \right\} \tag{2.11.6}$$

$$\{\Psi^e\} = \frac{1}{4} \left\{ \begin{array}{l} (1 - \xi)(1 - \eta)(-\xi - \eta - 1) + (1 - \xi^2)(1 - \eta^2) \\ (1 + \xi)(1 - \eta)(\xi - \eta - 1) + (1 - \xi^2)(1 - \eta^2) \\ (1 + \xi)(1 + \eta)(\xi + \eta - 1) + (1 - \xi^2)(1 - \eta^2) \\ (1 - \xi)(1 + \eta)(-\xi + \eta - 1) + (1 - \xi^2)(1 - \eta^2) \\ 2(1 - \xi^2)(1 - \eta) - (1 - \xi^2)(1 - \eta^2) \\ 2(1 + \xi)(1 - \eta^2) - (1 - \xi^2)(1 - \eta^2) \\ 2(1 - \xi^2)(1 + \eta) - (1 - \xi^2)(1 - \eta^2) \\ 2(1 - \xi)(1 - \eta^2) - (1 - \xi^2)(1 - \eta^2) \\ 4(1 - \xi^2)(1 - \eta^2) \end{array} \right\} \tag{2.11.7}$$

The *serendipity elements* are those rectangular elements which have no interior nodes. These elements have fewer nodes compared to the higher-order Lagrange elements. The interpolation functions of the serendipity elements are not complete, and they cannot be obtained using tensor products of one-dimensional Lagrange interpolation functions. Instead, an alternative procedure must be employed, as discussed in Reddy [6]. The interpolation functions for the 2-D quadratic serendipity element are given in Eq. (2.11.8) (see Figure 2.11.3). Although the functions are not complete, the serendipity elements have proven to be very effective in most applications.

$$
\{\Psi^e\} = \frac{1}{4}
\begin{Bmatrix}
(1 - \xi)(1 - \eta)(-\xi - \eta - 1) \\
(1 + \xi)(1 - \eta)(\xi - \eta - 1) \\
(1 + \xi)(1 + \eta)(\xi + \eta - 1) \\
(1 - \xi)(1 + \eta)(-\xi + \eta - 1) \\
2(1 - \xi^2)(1 - \eta) \\
2(1 + \xi)(1 - \eta^2) \\
2(1 - \xi^2)(1 + \eta) \\
2(1 - \xi)(1 - \eta^2)
\end{Bmatrix}
\qquad (2.11.8)
$$

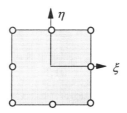

Figure 2.11.3. Quadratic rectangular serendipity element.

2.12 Numerical Integration

2.12.1 Preliminary Comments

An accurate representation of irregular domains (i.e., domains with curved boundaries) can be accomplished by the use of refined meshes and/or irregularly shaped curvilinear elements. For example, a non-rectangular region cannot be represented using rectangular elements; however, it can be represented by quadrilateral elements. Since the interpolation functions are easily derivable for a rectangular element and it is easier to evaluate integrals over rectangular geometries, we transform the finite element integral statements defined over quadrilaterals to a rectangle. The transformation results in complicated expressions for the integrands in terms of the coordinates used for the rectangular element. Therefore, numerical integration is used to evaluate such complicated integrals. The numerical integration schemes, such as the Gauss-Legendre numerical integration scheme, require the integral to be evaluated on a specific domain or with respect to a specific coordinate system. Gauss quadrature, for example, requires the integral to be expressed over a square region $\hat{\Omega}$ of dimension 2 by 2 with respect to the coordinate system, (ξ, η) to be such that $-1 \leq (\xi, \eta) \leq 1$. The transformation of the geometry and the variable coefficients of the differential equation from the problem coordinates (x, y) to the local coordinates (ξ, η) results in algebraically complex expressions, and they preclude analytical (i.e., exact) evaluation of the integrals. Thus, the transformation of a given integral expression, defined over element Ω^e, to one on the domain $\hat{\Omega}$ facilitates the numerical integration. Each element of the finite element mesh is transformed to $\hat{\Omega}$, only for the purpose of numerically evaluating the integrals (see Figure 2.12.1). The element $\hat{\Omega}$ is called a *master element*. For example, every quadrilateral element can be transformed to a square element with a side of length 2 and $-1 \leq (\xi, \eta) \leq 1$ that facilitates the use of Gauss-Legendre quadrature to evaluate integrals defined over the quadrilateral element.

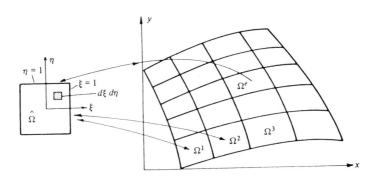

Figure 2.12.1. Transformation of arbitrarily shaped quadrilateral elements to the master rectangular element for numerical evaluation of integral expressions.

The transformation between the actual element Ω^e and the master element $\hat{\Omega}$ [or equivalently, between (x, y) and (ξ, η)] is accomplished by a coordinate transformation of the form

$$x = \sum_{j=1}^{m} x_j^e \phi_j^e(\xi, \eta) \ , \quad y = \sum_{j=1}^{m} y_j^e \phi_j^e(\xi, \eta) \qquad (2.12.1)$$

where ϕ_j denote the finite element interpolation functions of the master element $\hat{\Omega}$ (see Figure 2.12.1). The coordinates in the master element are chosen to be the natural coordinates (ξ, η) such that $-1 \leq (\xi, \eta) \leq 1$. This choice is dictated by the limits of integration in the Gauss quadrature rule, which is used to evaluate the integrals. For this case, the ϕ_j^e denote the interpolation functions of the four-node rectangular element shown in Figure

2.11.2b (i.e., $m = 4$). The transformation (2.12.1) maps a point (ξ, η) in the master element $\hat{\Omega}$ onto a point (x, y) in element Ω^e, and vice versa if the Jacobian of the transformation is positive-definite. The transformation maps the line $\xi = 1$ in $\hat{\Omega}$ to the line defined parametrically by $x = x(1, \eta)$ and $y = y(1, \eta)$ in the xy-plane. In other words, the master element $\hat{\Omega}$ is transformed, under the linear transformation, into a quadrilateral element (i.e., a four-sided element whose sides are not parallel) in the xy-plane. Conversely, every quadrilateral element of a mesh can be transformed to the same four-noded square (master) element $\hat{\Omega}$ in the (ξ, η)-plane.

In general, the dependent variable(s) of the problem are approximated by expressions of the form

$$u(x, y) = \sum_{j=1}^{n} u_j^e \psi_j^e(x, y) \tag{2.12.2}$$

The interpolation functions ψ_j^e used for the approximation of the dependent variable, in general, are different from ϕ_j^e used in the approximation of the geometry. Depending on the relative degree of approximations used for the geometry [see Eq. (2.12.1)] and the dependent variable(s) [see Eq. (2.12.2)], the finite element formulations are classified into three categories:

1. *Superparametric* $(m > n)$. The approximation used for the geometry is higher order than that used for the dependent variable.

2. *Isoparametric* $(m = n)$. Equal degree of approximation is used for both geometry and dependent variables.

3. *Subparametric* $(m < n)$. Higher-order approximation of the dependent variable is used.

In the present context, we use only the isoparametric formulations.

2.12.2 Coordinate Transformations

It should be noted that the transformation of a quadrilateral element of a mesh to the master element $\hat{\Omega}$ is solely for the purpose of numerically evaluating the integrals (see Figure 2.12.1). *No transformation of the physical domain or elements is involved in the finite element analysis.* The resulting algebraic equations of the finite element formulation are always in terms of the nodal values of the physical domain. Different elements of the finite element mesh can be generated from the same master element by assigning appropriate global coordinates to each of the elements. Master elements of a different order define different transformations and hence different collections of finite elements within the mesh. For example, a quadratic rectangular master element can be used to generate a mesh of quadratic curvilinear quadrilateral elements. The transformations of a master element should be such that no spurious gaps exist between elements, and no element overlaps occur. For example, consider the element coefficients

$$K_{ij}^e = \int_{\Omega^e} \left[k_{xx}(x,y)\frac{\partial \psi_i^e}{\partial x}\frac{\partial \psi_j^e}{\partial x} + k_{yy}(x,y)\frac{\partial \psi_i^e}{\partial y}\frac{\partial \psi_j^e}{\partial y} + c(x,y)\psi_i^e \psi_j^e \right] dx\,dy$$

$$(2.12.3)$$

The integrand (i.e., the expression in the square brackets under the integral) is a function of the global coordinates x and y. We must rewrite it in terms of ξ and η using the transformation (2.12.1). Note that the integrand contains not only functions but also derivatives with respect to the global coordinates (x, y). Therefore, we must relate $(\frac{\partial \psi_i^e}{\partial x}, \frac{\partial \psi_i^e}{\partial y})$ to $(\frac{\partial \psi_i^e}{\partial \xi}, \frac{\partial \psi_i^e}{\partial \eta})$ using the transformation (2.12.1).

The functions $\psi_i^e(x,y)$ can be expressed in terms of the local coordinates (ξ, η) by means of the transformation (2.12.1). Hence, by the chain rule of partial differentiation, we have

$$\frac{\partial \psi_i^e}{\partial \xi} = \frac{\partial \psi_i^e}{\partial x}\frac{\partial x}{\partial \xi} + \frac{\partial \psi_i^e}{\partial y}\frac{\partial y}{\partial \xi}$$

$$\frac{\partial \psi_i^e}{\partial \eta} = \frac{\partial \psi_i^e}{\partial x}\frac{\partial x}{\partial \eta} + \frac{\partial \psi_i^e}{\partial y}\frac{\partial y}{\partial \eta}$$

or, in matrix notation,

$$\left\{ \begin{matrix} \frac{\partial \psi_i^e}{\partial \xi} \\ \frac{\partial \psi_i^e}{\partial \eta} \end{matrix} \right\} = \begin{bmatrix} \frac{\partial x}{\partial \xi} & \frac{\partial y}{\partial \xi} \\ \frac{\partial x}{\partial \eta} & \frac{\partial y}{\partial \eta} \end{bmatrix} \left\{ \begin{matrix} \frac{\partial \psi_i^e}{\partial x} \\ \frac{\partial \psi_i^e}{\partial y} \end{matrix} \right\} \qquad (2.12.4)$$

which gives the relation between the derivatives of ψ_i^e with respect to the global and local coordinates. The matrix in Eq. (2.12.4) is called the *Jacobian matrix* of the transformation (2.12.1):

$$[J] = \begin{bmatrix} \frac{\partial x}{\partial \xi} & \frac{\partial y}{\partial \xi} \\ \frac{\partial x}{\partial \eta} & \frac{\partial y}{\partial \eta} \end{bmatrix} \qquad (2.12.5)$$

Note from the expression given for K_{ij}^e in Eq. (2.12.3) that we must relate $(\frac{\partial \psi_i^e}{\partial x}, \frac{\partial \psi_i^e}{\partial y})$ to $(\frac{\partial \psi_i^e}{\partial \xi}, \frac{\partial \psi_i^e}{\partial \eta})$, whereas Eq. (2.12.4) provides the inverse relations. Therefore, Eq. (2.12.4) must be inverted by inverting the Jacobian matrix:

$$\left\{ \begin{matrix} \frac{\partial \psi_i^e}{\partial x} \\ \frac{\partial \psi_i^e}{\partial y} \end{matrix} \right\} = [J]^{-1} \left\{ \begin{matrix} \frac{\partial \psi_i^e}{\partial \xi} \\ \frac{\partial \psi_i^e}{\partial \eta} \end{matrix} \right\} \qquad (2.12.6)$$

This requires that the Jacobian matrix $[J]$ be nonsingular.

Using the transformation (2.12.1), we can write

$$\frac{\partial x}{\partial \xi} = \sum_{j=1}^m x_j \frac{\partial \phi_j^e}{\partial \xi}, \quad \frac{\partial y}{\partial \xi} = \sum_{j=1}^m y_j \frac{\partial \phi_j^e}{\partial \xi} \qquad (2.12.7a)$$

$$\frac{\partial x}{\partial \eta} = \sum_{j=1}^{m} x_j \frac{\partial \phi_j^e}{\partial \eta} , \quad \frac{\partial y}{\partial \eta} = \sum_{j=1}^{m} y_j \frac{\partial \phi_j^e}{\partial \eta} \qquad (2.12.7b)$$

and by means of Eq. (2.12.5) one can compute the Jacobian matrix and then its inverse. Thus, given the global coordinates (x_j, y_j) of element nodes and the interpolation functions ϕ_j^e used for geometry, the Jacobian matrix can be evaluated using Eq. (2.12.5). A necessary and sufficient condition for $[J]^{-1}$ to exist is that the determinant J, called the Jacobian, be nonzero at every point (ξ, η) in $\hat{\Omega}$:

$$J \equiv \det[J] = \frac{\partial x}{\partial \xi} \frac{\partial y}{\partial \eta} - \frac{\partial x}{\partial \eta} \frac{\partial y}{\partial \xi} \neq 0. \qquad (2.12.8)$$

From Eq. (2.12.8) it is clear that the functions $\xi(x, y)$ and $\eta(x, y)$ must be continuous, differentiable, and invertible. Moreover, the transformation should be algebraically simple so that the Jacobian matrix can be easily evaluated. Transformations of the form in Eq. (2.12.1) satisfy these requirements and the requirement that no spurious gaps between elements or overlapping of elements occur.

Returning to numerical evaluation of integrals, we have from Eq. (2.12.6),

$$\left\{ \begin{array}{c} \frac{\partial \psi_i^e}{\partial x} \\ \frac{\partial \psi_i^e}{\partial y} \end{array} \right\} = [J]^{-1} \left\{ \begin{array}{c} \frac{\partial \psi_i^e}{\partial \xi} \\ \frac{\partial \psi_i^e}{\partial \eta} \end{array} \right\} \equiv [J^*] \left\{ \begin{array}{c} \frac{\partial \psi_i^e}{\partial \xi} \\ \frac{\partial \psi_i^e}{\partial \eta} \end{array} \right\} \qquad (2.12.9)$$

where J_{ij}^* is the element in position (i, j) of the inverse of the Jacobian matrix $[J]$. The element area $dA = dxdy$ in element Ω^e is transformed to

$$dA = |J| \, d\xi d\eta \qquad (2.12.10)$$

in the master element $\hat{\Omega}$.

Equations (2.12.7)–(2.12.10) provide the necessary relations to transform integral expressions on any element Ω^e to an associated master element $\hat{\Omega}$. For instance, consider the integral expression in Eq. (2.12.3), where k_{xx}, k_{yy}, and c are functions of x and y. Suppose that the finite element Ω^e can be generated by the master element $\hat{\Omega}^e$. Under the transformation (2.12.1) we can write

$$K_{ij}^e = \int_{\Omega^e} \left[k_{xx}(x, y) \frac{\partial \psi_i^e}{\partial x} \frac{\partial \psi_j^e}{\partial x} + k_{yy}(x, y) \frac{\partial \psi_i^e}{\partial y} \frac{\partial \psi_j^e}{\partial y} + c(x, y) \psi_i^e \psi_j^e \right] dxdy$$

$$= \int_{\hat{\Omega}^e} F_{ij}(\xi, \eta) \, d\xi d\eta \qquad (2.12.11)$$

The discussion presented above is valid for master elements of both rectangular and triangular geometry.

2.12.3 Integration Over a Master Rectangular Element

Integrals defined over a rectangular master element $\hat{\Omega}_R$ can be numerically evaluated using the Gauss–Legendre quadrature formulas

$$\int_{\hat{\Omega}_R} F(\xi,\eta)\ d\xi d\eta = \int_{-1}^{1}\int_{-1}^{1} F(\xi,\eta)\ d\xi d\eta \approx \sum_{I=1}^{M}\sum_{J=1}^{N} F(\xi_I,\eta_J)\ W_I W_J \quad (2.12.12)$$

where M and N denote the number of Gauss quadrature points, (ξ_I,η_J) denote the Gauss point coordinates, and W_I and W_J denote the corresponding Gauss weights as shown in Table 2.12.1 (from Table 7.2 in Reddy [6]).

Table 2.12.1. Gauss quadrature points and weights for rectangular elements.

$$\int_{-1}^{1} F(\xi)d\xi = \sum_{I=1}^{N} F(\xi_I)W_I$$

N	Points, ξ_I	Weights, W_I
1	0.0000000000	2.0000000000
2	$\pm$ 0.5773502692	1.0000000000
3	0.0000000000	0.8888888889
	$\pm$ 0.7745966692	0.5555555555
4	$\pm$ 0.3399810435	0.6521451548
	$\pm$ 0.8611363116	0.3478548451
5	0.0000000000	0.5688888889
	$\pm$ 0.5384693101	0.4786286705
	$\pm$ 0.9061798459	0.2369268850
6	$\pm$ 0.2386191861	0.4679139346
	$\pm$ 0.6612093865	0.3607615730
	$\pm$ 0.9324695142	0.1713244924

The selection of the number of Gauss points is based on the following formula: a polynomial of degree p is integrated exactly employing $N = int[(p+1)/2]+1$ integration points. In most cases, the interpolation functions are of the same degree in both ξ and η, and therefore one has $M = N$. When the integrand is of a different degree in ξ and η, the number of Gauss points is selected on the basis of the largest-degree polynomial. The minimum allowable quadrature rule is one that yields the area or volume of the element exactly. The maximum degree of the polynomial refers to the degree of the highest polynomial in ξ or η that is present in the integrands of the element matrices of the type in Eq. (2.12.3). Note that the polynomial degree of coefficients as well as J_{ij} should be accounted for in determining the total polynomial degree of the integrand. Of course, the coefficients k_{xx}, k_{yy}, and c and J_{ij} in general may not be polynomials. In those cases, their functional variations must be approximated by a suitable polynomial (for example, by a binomial series) in order to determine the polynomial degree of the integrand.

2.12.4 Integration Over a Master Triangular Element

In the preceding section we discussed numerical integration on quadrilateral elements which can be used to represent very general geometries as well as field variables in a variety of problems. Here we discuss numerical integration on triangular elements. Since quadrilateral elements can be geometrically distorted, it is possible to distort a quadrilateral element to obtain a required triangular element by moving the position of the corner nodes, and the fourth corner in the quadrilateral is merged with one of the neighboring nodes. In actual computation, this is achieved by assigning the same global node number to two corner nodes of the quadrilateral element. Thus, master triangular elements can be obtained in a natural way from associated master rectangular elements. Here we discuss the transformations from a master triangular element to an arbitrary triangular element.

We choose the unit right isosceles triangle (see Table 2.12.2) as the master element. An arbitrary triangular element Ω^e can be generated from the master triangular element $\hat{\Omega}_T$ by transformation of the form (2.12.1). The derivatives of ψ_i with respect to the global coordinates can be computed from Eq. (2.12.6), which take the form

$$\left\{ \begin{array}{c} \frac{\partial \psi_i^e}{\partial x} \\ \frac{\partial \psi_i^e}{\partial y} \end{array} \right\} = [J]^{-1} \left\{ \begin{array}{c} \frac{\partial \psi_i^e}{\partial L_1} \\ \frac{\partial \psi_i^e}{\partial L_2} \end{array} \right\} \tag{2.12.13a}$$

$$[J] = \begin{bmatrix} \frac{\partial x}{\partial L_1} & \frac{\partial y}{\partial L_1} \\ \frac{\partial x}{\partial L_2} & \frac{\partial y}{\partial L_2} \end{bmatrix} \tag{2.12.13b}$$

Note that only L_1 and L_2 are treated as linearly independent coordinates because $L_3 = 1 - L_1 - L_2$.

After transformation, integrals on $\hat{\Omega}_T$ have the form

$$\int_{\Omega^e} G(\xi, \eta) \, d\xi d\eta = \int_{\Omega^e} G(L_1, L_2, L_3) \, dL_1 dL_2 \tag{2.12.14}$$

which can be approximated by the quadrature formula

$$\int_{\Omega^e} G(L_1, L_2, L_3) \, dL_1 dL_2 \approx \sum_{I=1}^{N} G(\mathbf{S}_I) W_I \tag{2.12.15}$$

where W_I and $\mathbf{S}_I$ denote the weights and integration points of the quadrature rule. Table 2.12.2 (from Table 9.3 of Reddy [6]) contains the location of integration points and weights for one-point, three-point, and four-point quadrature rules over triangular elements. For evaluation of integrals whose integrands are polynomials of degree equal to or higher than 5 (in any of the area coordinates) the reader must consult books on numerical integration.

Table 2.12.2: Quadrature weights and points for triangular elements.

Number of integration points	Degree of polynomial Order of the residual	L_1	L_2	L_3	W	Geometric locations	
1	1 $O(h^2)$	1/3	1/3	1/3	1	a	
3	2 $O(h^3)$	1/2 1/2 0	0 1/2 1/2	1/2 0 1/2	1/3	a b c	
4	3 $O(h^4)$	1/3 0.6 0.2 0.2	1/3 0.2 0.6 0.2	1/3 0.2 0.2 0.6	-27/48 25/48 25/48 25/48	a b c d	

2.13 Modeling Considerations

Here we discuss some aspects of finite element model development. Guidelines concerning mesh generation, boundary flux representation, and imposition of boundary conditions are discussed in very general terms (see Burnett [10] and Reddy [6] for further details).

2.13.1 Mesh Generation

Generation of a finite element mesh for a given domain should follow the guidelines listed below:

1. The mesh should represent the geometry of the computational domain and boundary flux representation accurately.

2. The mesh should be such that large gradients in the solution (temperature and velocities) are adequately represented.

3. The mesh should not contain elements with very large aspect ratios and/or angular distortions.

Within the above guidelines, the mesh used can be *coarse* (i.e., few elements) or *refined* (i.e., many elements), and may consist of one or more orders and types of elements (e.g., linear and quadratic, triangular and quadrilateral). It should be noted that the choice of element type and mesh is problem dependent. An analyst with physical insight into the process being simulated can make a better choice of elements and mesh for the problem at hand than one who does not have the physical insight. One should evaluate the results obtained in the light of physical understanding and approximate analytical and/or experimental information.

Generation of meshes of a single element type (i.e., linear elements or quadratic elements) is easy, because elements of the same degree are compatible with each other. Mesh refinements involve several options. Refine the mesh by subdividing existing elements into two or more elements of the same type (see Figure 2.13.1a). This is called the *h-version mesh refinement*. Refine the mesh by replacing existing elements by elements of higher order (see Figure 2.13.1b). This type of refinement is called the *p-version mesh refinement*. The (h, p)-*version mesh refinement* is one in which elements are subdivided into two or more elements in some places and replaced by higher-order elements in other places. Generally, local mesh refinements should be such that very small elements are not placed adjacent to very large elements.

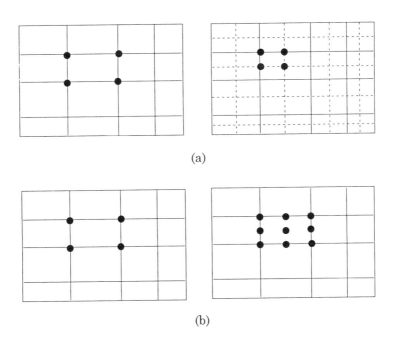

(a)

(b)

Figure 2.13.1. The two types of mesh refinements in finite element analyses: (a) h-refinement (refinement with the same order of elements); (b) p-refinement (refinement with higher-order elements).

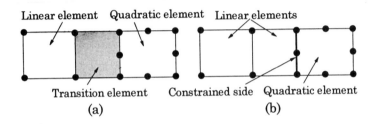

Figure 2.13.2. Connecting linear and quadratic elements with (a) transition elements, and (b) linear constraint equations.

To accomplish local mesh refinements it may be necessary to combine elements of different orders, say linear and quadratic elements. There are two ways to combine elements of different orders. One way is to use transition elements, which have different numbers of nodes on different sides of the element (see Figure 2.13.2a). Another way is to impose a condition that constrains the midside node to have the same value as that at the node of the lower-order element (see Figure 2.13.2b). Neither combination enforces inter-element continuity of the solution at every point of the entire interface.

2.13.2 Representation of Boundary Flux

When the temperature distribution in the domain is approximated by a set of finite elements, the boundary flux distribution, if known, must be replaced by a set of equivalent nodal fluxes. The accuracy of the solution depends on the element type and mesh used to represent the domain and the representation of actual flux distribution. Use of linear elements, for example, to represent a curved boundary will change the actual distribution (see Figure 2.13.3). Of course, mesh refinements by h-version or p-version will improve the representation of the specified boundary flux. Also, the vector of nodal fluxes due to a distributed heat flux should be computed using the interpolation functions of the element used to represent the solution. For example, if the temperature is approximated using the nine-node quadratic elements, the nodal contributions of a distributed heat flux should not be evaluated using four-node linear elements. This would be an inconsistent computation of nodal fluxes. Mesh refinements make the flux representation more accurate, and hence convergence of the numerical solution to the actual solution can be expected.

2.13.3 Imposition of Boundary Conditions

In most problems one encounters situations where the portion of the boundary on which flux is specified has points in common with the portion of the boundary on which the temperature is specified. In other words, at a few nodal points of the mesh, both the heat input and temperature may be specified, e.g., at an exterior corner. Such points are called *singular points*.

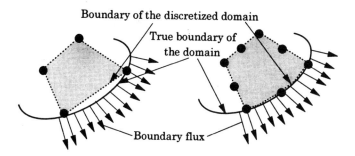

Figure 2.13.3. Representation of boundary flux in finite element analyses.

Obviously, one cannot impose both boundary conditions at the same point. As a general rule, one should impose the essential (i.e., temperature) boundary condition at the singular points and disregard the flux boundary condition. Of course, if the true situation in a problem is that the flux boundary condition is imposed and the temperature boundary condition is a result of it, then consideration must be given to the former one.

Another type of singularity one encounters in the analysis of thermal boundary value problems is the specification of two different values of a temperature or heat input at the same boundary point. In the finite element analysis, one must make a choice between the two values or take a weighted average of the two. In any case, one must note that the true boundary condition is replaced by an approximate condition. The closeness of the approximate boundary condition to the true one depends on the size of the element containing the point. It is often necessary to make a mesh refinement in the vicinity of the singular point to obtain an acceptable solution.

In most practical problems a variety of approximations are introduced before reaching the finite element simulation; additional approximations, some of which were described here, are introduced by the computational model. It is important to keep various sources of error in perspective. A feel for the relative proportions of various errors introduced into the analysis helps the analyst to make a decision on selecting a mesh and representing the boundary conditions. In summary, engineering knowledge and experience with the problem being analyzed are an essential part of any engineering analysis.

2.14 Illustrative Examples

In this section simple examples of applications of the finite element method are discussed. In the interest of simplicity, the examples are limited to heat transfer problems on rectangular domains and the use of triangular and rectangular elements. Additional examples can be found in the text book by Reddy [6]. More complicated problems, which reflect practical situations, will be addressed in the subsequent chapters.

A sample finite element program, *FEM2DHT*, that reflects the ideas presented in this chapter is discussed in Appendix A. This is a simplified version of the program *FEM2DV2* from Reddy [6]. The program is limited to linear two-dimensional problems, with linear and quadratic triangular and rectangular elements. Plane as well as axisymmetric (see Section 2.9) geometries can be modeled, both conductive and convective boundary conditions can be handled, and steady as well as transient heat transfer problems can be solved using the program. All of the results presented in this section were obtained using program *FEM2DHT* (see Appendix A for data input instructions for using the program).

2.14.1 Example 1

Problem description

Consider the conduction heat transfer in a square plate of dimension 2 cm by 2 cm, conductivity k (W/m/°C), and uniform internal heat generation of Q_0 (W/m^3). The edges of the plate are maintained at a temperature of $T_0 = 0$ (°C) (see Figure 2.14.1). We wish to determine the steady temperature distribution inside the plate using the finite element method.

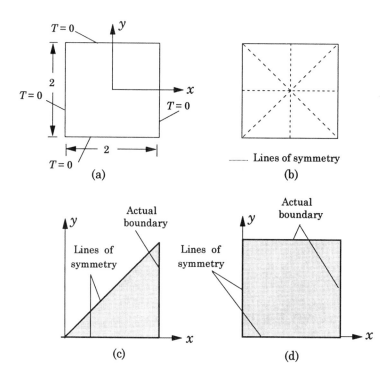

Figure 2.14.1. Heat transfer in a square domain. (a) Geometry and boundary conditions. (b) Problem symmetries. (c) Computational domain for the mesh of triangular elements. (d) Computational domain for the mesh of rectangular elements.

The problem can be expressed mathematically as one of solving the equation

$$-k\nabla^2 T = Q_0 \quad \text{or} \quad -k(\frac{\partial^2 T}{\partial x^2} + \frac{\partial^2 T}{\partial y^2}) = Q_0 \quad \text{in} \quad \Omega \qquad (2.14.1a)$$

$$T = T_0 \quad \text{on} \quad \Gamma \qquad (2.14.1b)$$

Note that the domain Ω is a unit square, and the boundary Γ consists of the four sides bounding Ω.

Equation (2.14.1) has an analytical (series) solution, which can be obtained using the method of separation of variables. We wish to compare the finite element solution of the problem with the analytical solution obtained with meshes of triangular elements and rectangular elements. Each mesh consists of only one type of element.

The problem has symmetries that can be exploited in the finite element analysis. A problem possesses symmetry of the solution about a line only when there exists a symmetry of the (1) geometry, (2) material properties, (3) source (i.e., heat generation), and (4) boundary conditions of the problem. Such a line is called the *line of symmetry*. When a line of symmetry exists in a problem, then it is sufficient to model the domain on either side of the line. Then the line of symmetry becomes a part of the boundary of the computational domain. On such boundaries, the normal derivative of the solution (i.e., derivative of the solution with respect to the coordinate normal to the line of symmetry) is zero. In the context of heat transfer problems, this implies that the temperature gradient (or heat flux) across the line of symmetry is zero:

$$q_n \equiv \frac{\partial T}{\partial n} = \hat{\mathbf{n}} \cdot \nabla T = \frac{\partial T}{\partial x} n_x + \frac{\partial T}{\partial y} n_y = 0 \qquad (2.14.2)$$

The problem at hand has the symmetry about the $x = 0$ and $y = 0$ axes; it is also symmetric about the diagonal line $x = y$ (see Figure 2.14.1b). Thus, we can use a quadrant of the domain for meshes of rectangular elements and an octant of the domain for meshes of triangular elements to analyze the problem. Of course, it is possible to mix triangular and rectangular elements to represent the domain as well as the solution, but we shall use only one type of element in each mesh and the same degree of approximation for both geometry and solution (i.e., isoparametric formulation).

Solution by linear triangular elements

Due to the symmetry along the diagonal $x = y$, we model the triangular domain shown in Figure 2.14.1c. As a first choice we use a uniform mesh of four linear triangular elements to represent the domain (see Figure 2.14.2a), and then a refined mesh (see Figure 2.14.2b) to compare the solutions. In the present case, there is no discretization error involved in the problem because the geometry is exactly represented.

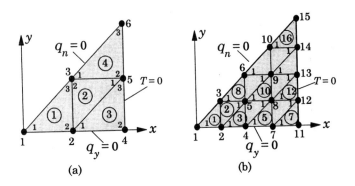

Figure 2.14.2. Discretization of the computational domain with triangular elements: (a) mesh of four elements; (b) mesh of sixteen elements.

We consider element 1 as the typical element with its local coordinate system $(\bar{x}, \bar{y})$. Suppose that the element dimensions, i.e., length and height are a and b, respectively. The coordinates of the element nodes are

$$(\bar{x}_1, \bar{y}_1) = (0,0); \ (\bar{x}_2, \bar{y}_2) = (a,0), \ (\bar{x}_3, \bar{y}_3) = (a, b) \qquad (2.14.3)$$

Hence the parameters α_i, β_i, and γ_i are given by

$$\alpha_1 = \bar{x}_2 \bar{y}_3 - \bar{x}_3 \bar{y}_2 = ab, \ \alpha_2 = \bar{x}_3 \bar{y}_1 - \bar{x}_1 \bar{y}_3 = 0, \ \alpha_3 = \bar{x}_1 \bar{y}_2 - \bar{x}_2 \bar{y}_1 = 0,$$
$$\beta_1 = \bar{y}_2 - \bar{y}_3 = -b, \ \beta_2 = \bar{y}_3 - \bar{y}_1 = b, \ \beta_3 = \bar{y}_1 - \bar{y}_2 = 0, \qquad (2.14.4)$$
$$\gamma_1 = -(\bar{x}_2 - \bar{x}_3) = 0, \ \gamma_2 = -(\bar{x}_3 - \bar{x}_1) = -a, \ \gamma_3 = -(\bar{x}_1 - \bar{x}_2) = a$$

For the mesh shown in Figure 2.14.2a, we have

$$[K^1] = [K^2] = [K^3] = [K^4]; \ \{Q^1\} = \{Q^2\} = \{Q^3\} = \{Q^4\} \qquad (2.14.5)$$

Therefore, the element coefficients K_{ij}^e and Q_i^e $(e = 1, 2, 3, 4)$ are given by (note that a denotes the length of side connecting element nodes 1 and 2) [see Eqs. (2.6.6a–c)]

$$[K^e] = \frac{k}{2} \begin{bmatrix} \alpha & -\alpha & 0 \\ -\alpha & \gamma & -\beta \\ 0 & -\beta & \beta \end{bmatrix}, \quad \{Q^e\} = \frac{Q_0 ab}{6} \begin{Bmatrix} 1 \\ 1 \\ 1 \end{Bmatrix} \qquad (2.14.6)$$

where $\alpha = \frac{b}{a}$, $\beta = \frac{a}{b}$, and $\gamma = \alpha + \beta$. The element matrix in (2.14.6) is valid for the Laplace operator $-\nabla^2$ on any right-angle triangle with sides a and b in which the right angle is at node 2, and the diagonal line of the triangle connects node 3 to node 1. Note that the off-diagonal coefficient associated with the nodes on the diagonal line is zero for a right-angle triangle. These observations can be used to write the element matrix associated with the Laplace operator on any right-angle triangle, i.e., for any element-node numbering system.

The assembled coefficient matrix for the finite element mesh is 6×6, because there are six global nodes with one unknown per node. The assembled matrix can be obtained directly by using the correspondence between the global nodes and the local nodes, as explained earlier.

The specified boundary conditions on the primary degrees of freedom of the problem are

$$T_4 = T_5 = T_6 = T_0 = 0 \tag{2.14.7}$$

The specified secondary degrees of freedom at nodes 1, 2, and 3 are all zero because the flux is zero there:

$$q_1^1 = 0, \quad q_2^1 + q_3^2 + q_1^3 = 0, \quad q_3^1 + q_2^2 + q_1^4 = 0 \tag{2.14.8}$$

Since T_4, T_5, and T_6 are known, the secondary variables at these nodes are unknown, and they can be obtained in the postcomputation.

Solution by linear rectangular elements

Note that we cannot exploit the symmetry along the diagonal $x = y$ to our advantage when we use a mesh of rectangular elements. Therefore, we use a 2×2 uniform mesh of four linear rectangular elements and a refined 4×4 mesh (see Figure 2.14.3a) to discretize a quadrant of the domain. Once again, no discretization error is introduced in the present case.

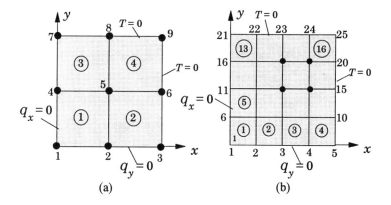

Figure 2.14.3. Discretization of the computational domain with rectangular elements: (a) mesh of four elements; (b) mesh of sixteen elements.

All elements in the mesh are identical. Therefore, we have the following element matrices [see Eqs. (2.6.10a–d)] for a typical element of the mesh ($\alpha = \frac{b}{a}$, $\beta = \frac{a}{b}$):

$$[K^e] = \frac{k}{6} \begin{bmatrix} 2(\alpha + \beta) & -2\alpha + \beta & -(\alpha + \beta) & \alpha - 2\beta \\ -2\alpha + \beta & 2(\alpha + \beta) & \alpha - 2\beta & -(\alpha + \beta) \\ -(\alpha + \beta) & \alpha - 2\beta & 2(\alpha + \beta) & -2\alpha + \beta \\ \alpha - 2\beta & -(\alpha + \beta) & -2\alpha + \beta & 2(\alpha + \beta) \end{bmatrix}, \tag{2.14.9a}$$

$$\{Q^e\} = \frac{Q_0 ab}{4} \begin{Bmatrix} 1 \\ 1 \\ 1 \end{Bmatrix} \qquad (2.14.9b)$$

The boundary conditions on the secondary variables are

$$q_1^1 = 0, \quad q_2^1 + q_1^2 = 0, \quad q_4^1 + q_1^3 = 0 \qquad (2.14.10a)$$

and the balance of secondary variables at global node 5 requires

$$q_3^1 + q_4^2 + q_2^3 + q_1^4 = 0 \qquad (2.14.10b)$$

Discussion of the results

The finite element solutions obtained by two different meshes of triangular elements and two different meshes of rectangular elements are compared in Table 2.14.1 at $x = 0$ for varying y for $k = 1, Q_0 = 1$, and $T_0 = 0$. The finite element solution obtained by 16 triangular elements (in an octant) is the most accurate when compared to the series solution. The accuracy of the triangular element mesh is due to the larger number of elements it has compared to the number of elements in the rectangular element mesh for the same size of the domain.

Table 2.14.1. Convergence of the finite element solutions to the analytical solution of Eq. (2.14.1).

y	Triangular elements		Rectangular elements		Series solution
	4 elements	16 elements	4 elements	16 elements	
0.0	0.3125	0.3013	0.3107	0.2984	0.2947
0.25	0.2709*	0.2805	0.2759*	0.2824	0.2789
0.50	0.2292	0.2292	0.2411	0.2322	0.2293
0.75	0.1146*	0.1393	0.1205*	0.1414	0.1397
1.0	0.0000	0.0000	0.0000	0.0000	0.0000

* Interpolated values.

The finite element nodal temperatures obtained with the 4 × 4 mesh of linear triangular elements in a quadrant are compared in Table 2.14.2 with the finite difference solution obtained with a 4 × 4 rectangular grid and with the exact (or series) solution. The node numbering scheme is the same as that used in Figure 2.14.3b for a rectangular element mesh. The central difference approximation is used to approximate the second derivatives in the equation. The finite element nodal temperatures are slightly more accurate, except at node 1, than those obtained with the finite difference method. One should remember that the finite element method minimizes the residual of the approximation of the differential equation over each element – not the error in the nodal values. In other words, the nodal values are adjusted such that the weighted-integral of the error in the approximation of the differential equation is minimized over each element.

Table 2.14.2: Comparison of the finite element solution (FES) with the finite difference solution (FDS) and analytical (series) solution of Eq. (2.14.1). A 4 × 4 mesh of linear triangular elements is used in the finite element method and a comparable mesh is used in the finite difference method.

Node	Analytical	FDS	Error	FES	Error
1	0.2947	0.2911	0.0036	0.3013	-0.0066
2	0.2789	0.2755	0.0034	0.2805	-0.0016
3	0.2293	0.2266	0.0027	0.2292	0.0001
4	0.1397	0.1381	0.0016	0.1392	0.0005
5	0.0000	0.0000	0.0000	0.0000	0.0000
7	0.2642	0.2609	0.0033	0.2645	-0.0003
8	0.2178	0.2151	0.0027	0.2172	0.0006
9	0.1333	0.1317	0.0016	0.1327	0.0006
10	0.0000	0.0000	0.0000	0.0000	0.0000
13	0.1811	0.1787	0.0024	0.1801	0.0010
14	0.1127	0.1110	0.0017	0.1117	0.0010
15	0.0000	0.0000	0.0000	0.0000	0.0000
19	0.0728	0.0711	0.0017	0.0715	0.0013
20	0.0000	0.0000	0.0000	0.0000	0.0000
25	0.0000	0.0000	0.0000	0.0000	0.0000

The solution T and its gradient can be computed at any interior point of the domain using the finite element approximation of $T(x,y)$ [see Eq. (2.3.1)]. We have

$$T^e(x,y) = \sum_{j=1}^{n} T_j^e \psi_j^e(x,y) \tag{2.14.11a}$$

$$\frac{\partial T^e}{\partial x} = \sum_{j=1}^{n} T_j^e \frac{\partial \psi_j^e}{\partial x} \tag{2.14.11b}$$

$$\frac{\partial T^e}{\partial y} = \sum_{j=1}^{n} T_j^e \frac{\partial \psi_j^e}{\partial y} \tag{2.14.11c}$$

where T_j^e is the value of $T^e(x,y)$ at the jth node of the element, and $\psi_j^e(x,y)$ $(j = 1,2,\cdots,n)$ are the interpolation functions. Note that for a linear triangular element the components $(\frac{\partial T^e}{\partial x}, \frac{\partial T^e}{\partial y})$ of the temperature gradient vector are constants over an entire element, whereas $\frac{\partial T^e}{\partial x}$ is linear in y and $\frac{\partial T^e}{\partial y}$ is linear in x for a linear rectangular element.

2.14.2 Example 2

Consider a cylinder of height $L = 1$ cm, radius $R = 1$ cm, thermal conductivity $k = 25$ W/(m°C), and constant internal heat generation of $Q_0 = 5 \times 10^8$ W/m^3. The top and bottom faces of the cylinder are assumed to be insulated and the boundary surface is maintained at $T_0 = 100°$C. The assumption of an insulated boundary condition at the top and bottom of the cylinder makes the heat flow one-dimensional along the radial direction. For this case, the steady temperature along the radial direction is given by

$$T(r) = T_0 + \frac{Q_0 R^2}{4k}\left(1 - \frac{r^2}{R^2}\right) \qquad (2.14.12)$$

Table 2.14.3 contains the finite element solution obtained with various finite element meshes along the r-coordinate. Only one element is used in the z-direction. The 5×1 mesh of nine-node quadratic elements gives exact values at the nodes.

Table 2.14.3. The finite element solutions of an axisymmetric heat transfer problem (1-D).

$r \times 10^{-2}$	5LR*	10LR	5QR
0.0	611.92	603.56	600.00
0.1	–	596.89	595.00
0.2	585.25	581.33	580.00
0.3	–	556.00	555.00
0.4	523.03	520.76	520.00
0.5	–	475.58	475.00
0.6	421.69	420.43	420.00
0.7	–	355.30	355.00
0.8	280.74	280.19	280.00
0.9	–	195.09	195.00
1.0	100.00	100.00	100.00

*5LR = 5 linear rectangular elements; 10LR = 10 linear rectangular elements; 5QR = 5 quadratic rectangular elements.

Next we consider a cylinder of height $L = 2$ cm, radius $R = 1$ cm, thermal conductivity $k = 25$ W/(m°C), and constant internal heat generation of $Q_0 = 5 \times 10^8$ W/m^3. The top and bottom faces and the surface of the cylinder are maintained at $T_0 = 100°$C. For this case, the steady solution is two-dimensional. Exploiting the symmetry about $z = 1$ cm, a 10×10 mesh of linear rectangular elements is used. The results are presented in Table 2.14.4.

2.14.3 Example 3

Consider a plane wall of height $b = 1$ cm, thickness $a = 1$ cm, thermal conductivity $k = 18$ W/(m°C), and constant internal heat generation of $Q_0 = 7.2 \times 10^7$ W/m^3. The boundaries at $y = 0, b$ of the domain are assumed to be insulated, the boundary at $x = 0$ is maintained at $T_0 = 50$°C, and the boundary at $x = a$ is exposed to ambient temperature $T_c = 100$°C. The film coefficient is $h_c = 200$ W/(m^2°C). The assumption of an insulated boundary condition at $y = 0, b$ of the domain makes the heat flow one-dimensional along the x-direction. For this case, the temperature is given by

$$T(x) = 50 + 5\frac{x}{a} + 200\left(1.9 - \frac{x}{a}\right)\frac{x}{a} \qquad (2.14.13)$$

Table 2.14.4. The finite element solutions of an axisymmetric heat transfer problem (2-D).

$r(z = 10^{-3})$	Temp.	$r(z = 10^{-2})$	Temp.	$z(r = 0.0)$	Temp.
0.000	195.86	0.000	504.64	0.000	100.00
0.001	195.03	0.001	499.82	0.001	195.86
0.002	193.06	0.002	488.52	0.002	274.19
0.003	189.78	0.003	469.93	0.003	337.35
0.004	185.07	0.004	443.70	0.004	387.55
0.005	178.74	0.005	409.43	0.005	426.72
0.006	170.50	0.006	366.68	0.006	456.54
0.007	159.91	0.007	314.92	0.007	478.35
0.008	146.20	0.008	253.61	0.008	493.20
0.009	127.81	0.009	182.16	0.009	501.81
0.010	100.00	0.010	100.00	0.010	504.64

For this problem, any finite element mesh would give the exact solution at the nodes. For example, the temperature values at $x/a = 0.2, 0.4, 0.6, 0.8$, and 1.0 are $119, 172, 209, 230$, and 235, respectively.

Next we consider a square domain, $a = b = 1$ cm, thermal conductivity $k = 18$ W/(m°C), and constant internal heat generation of $Q_0 = 7.2 \times 10^7$ W/m^3. The boundaries at $y = 0, b$ and $x = 0$ are maintained at $T_0 = 50$°C, and the boundary at $x = a$ is exposed to an ambient temperature of $T_c = 100$°C. The film coefficient is taken to be $h_c = 200$ W/(m^2°C). The symmetry about $y = 0.005$m allows us to use a 10×5 mesh of linear rectangular elements to solve the problem. The results are presented in Table 2.14.5.

The program *FEM2DHT* discussed in Appendix A can be used to solve a variety of linear, two-dimensional heat transfer problems with conduction and convection boundary conditions. For a Fortran source of the program, see Reddy [6].

Table 2.14.5. The finite element solutions of a two-dimensional heat transfer problem with convective boundary condition.

$r(z = 10^{-3})$	Temp.	$r(z = 10^{-2})$	Temp.	$z(r = 0.0)$	Temp.
0.000	50.000	0.000	50.000	0.000	50.000
0.001	55.768	0.001	61.396	0.001	63.002
0.002	59.303	0.002	69.702	0.002	72.765
0.003	61.717	0.003	75.714	0.003	79.991
0.004	63.416	0.004	80.055	0.004	85.279
0.005	64.623	0.005	83.178	0.005	89.109
0.006	65.481	0.006	85.405	0.006	91.848
0.007	66.083	0.007	86.964	0.007	93.763
0.008	66.500	0.008	88.015	0.008	95.038
0.009	66.796	0.009	88.664	0.009	95.789
0.010	67.135	0.010	88.960	0.010	96.067

Problems

The following problems are designed to test the understanding of the ideas presented in this chapter. Some of the exercise problems are reproduced from the textbook by Reddy [6]. For additional examples and exercise problems, the reader may consult [6].

2.1 Develop the weak form of the vector equation (2.2.3) over an element

$$-\nabla \cdot (k\nabla T) = Q \quad \text{in} \quad \Omega \qquad (i)$$

subjected to the boundary conditions of the form

$$T = \hat{T} \text{ on } \Gamma_T \text{ and } \hat{n} \cdot (k\nabla T) + q_c = q \text{ on } \Gamma_q \qquad (ii)$$

2.2 Use the weak form given in Eq. (2.10.2a) to develop the finite element model of heat transfer problems with convective boundary conditions. Show that the finite element model is of the form

$$[K^e + H^e]\{T^e\} = \{Q^e\} + \{P^e\} + \{q^e\} \qquad (i)$$

where

$$K_{ij}^e = \int_{\Omega^e} \left(k_{xx} \frac{\partial \psi_i^e}{\partial x} \frac{\partial \psi_j^e}{\partial x} + k_{yy} \frac{\partial \psi_i^e}{\partial y} \frac{\partial \psi_j^e}{\partial y} \right) dx dy \qquad (ii)$$

$$H_{ij}^e = \int_{\Gamma^e} h_c \psi_i^e \psi_j^e ds, \quad Q_i^e = \int_{\Omega^e} \psi_i^e Q \, dx dy \qquad (iii)$$

$$q_i^e = \oint_{\Gamma^e} q_n \psi_i^e ds, \quad P_i^e = \oint_{\Gamma^e} h_c T_c^e \psi_i^e ds \qquad (iv)$$

2.3 Modify the weak form (2.9.2) of an axisymmetric problem to include ~ undary condition, and develop the finite ~ e finite element model will be the same as in ~ definitions of $[H^e]$ and $\{P^e\}$.

#390 04-25-2007 11:50AM
tem(s) checked out to patron

The finite element method in heat
E: 50110014429282
TE: 05-16-07

t land State University Library
http://vikat.pdx.edu/patroninfo/

~, and P_i^e (due to the convective boundary ~ r the case of a linear triangular element.

~ ear rectangular element.

2.14.1 with the mesh shown in Figure P2.6. ~ ven in Eqs. (2.14.6) and (2.14.9).

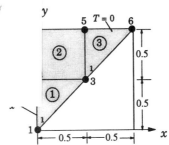

Figure P2.6

2.7 Consider steady-state heat conduction in a square region of side $2a$. Assume that the medium has a conductivity of $k = 30$ W/(m°C) and a uniform heat generation of $Q_0 = 10^7$ W/m³. For the boundary conditions and mesh shown in Figure P2.7, write the finite element algebraic equations for nodes 1, 3, and 7. Take $h_c = 60$ W/(m²°C), $T_c = 0.0$°C, $T_0 = 100$°C, $q_0 = 2 \times 10^5$ W/m², and $a = 1$ cm.

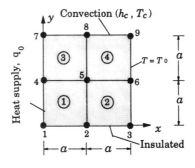

Figure P2.7

2.8 Repeat Problem 2.7 for nodes 4, 5, and 9.

2.9 For the convection heat transfer problem shown in Figure P2.9, write the four finite element equations for the unknown temperatures. Assume that the thermal conductivity of the material is $k = 5$ W/(m°C), the convection heat transfer coefficient on the left surface is $h_c = 28$ W/(m^2°C), and the internal heat generation is zero.

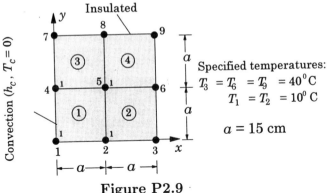

Figure P2.9

2.10 The domain shown in Figure P2.10 has its base maintained at 300°C and is exposed to convection on its remaining boundary. Write the finite element equations at nodes 7 and 10. Take $a = 2$cm, $b = 1$cm, $h_c = 40$ W/(m^2°C), $T_c = 20$°C, $k = 5$ W/(m °C), $T_c = 20$°C, and $T_0 = 300$°C.

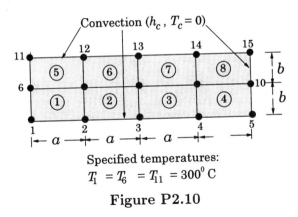

Figure P2.10

2.11 Solve the two-dimensional problem in Example 2 of Section 2.14.2 for the case in which the boundary $z = L$ is insulated. Use 5×10 mesh of linear rectangular elements in the full domain.

2.12 Repeat Problem 2.11 for the case where the surface $r = R$ is exposed to an ambient temperature of $T_c = 100$°C. Take the film coefficient to be $h_c = 200$ W/(m^2 °C).

2.13 (*Galerkin time approximation scheme* [14,15]) Consider the differential equation

$$a\frac{du}{dt} + bu = f(t), \quad t_n < t < t_{n+1} \tag{i}$$

where a and b are constants and f is a function of time t. Using linear approximation of the form

$$u(t) = u_n \psi_n(t) + u_{n+1} \psi_{n+1}(t) \tag{ii}$$

$$\psi_n(t) = \frac{(t_{n+1} - t)}{\Delta t}, \quad \psi_{n+1}(t) = \frac{(t - t_n)}{\Delta t} \qquad (iii)$$

derive the Galerkin finite element equations for a typical time element in the interval (t_n, t_{n+1}). Assuming that u_n is known, compute the solution u_{n+1} from the element equations, and then compare the result with that obtained with the θ-family of approximation in Eqs. (2.8.9) and (2.8.10).

2.14 Repeat Problem 2.13 for the case of quadratic interpolation between the stations, (t_{n-1}, t_n, t_{n+1}). Solve the third equation of the resulting finite element equations for u_{n+1} in terms of u_{n-1} and u_n, and time step Δt.

2.15 (*Collocation time approximation*) Consider the differential equation in time,

$$a\frac{du}{dt} + bu = f, \quad t_n < t < t_{n+1} \qquad (i)$$

where a and b are constants and f is a function of time. Using linear approximation of the form $u(t) = u_n\psi_n(t) + u_{n+1}\psi_{n+1}(t)$, where $\psi_n(t) = (t_{n+1} - t)/\Delta t$ and $\psi_{n+1}(t) = (t - t_n)/\Delta t$, derive the one-point collocation finite element equations (i.e., set $w = \delta(t - t_0)$, where $\delta(\cdot)$ denotes the Dirac delta function, in Eq. (i) of Problem 2.13) for a typical time element, (t_n, t_{n+1}), and compare the result with that obtained with the θ-family of approximation. Use the collocation point to be $t_0 = (1 - \theta)t_n + \theta t_{n+1}$.

References for Additional Reading

1. S. G. Mikhlin, *Variational Methods in Mathematical Physics*, (translated from the Russian by T. Boddington), Pergamon Press, Oxford (1964).

2. S. G. Mikhlin, *The Numerical Performance of Variational Methods*, (translated from the Russian by R. S. Anderssen), Wolters-Noordhoff, The Netherlands (1971).

3. B.A. Finlayson, *The Method of Weighted Residuals and Variational Principles*, Academic Press, New York (1972).

4. J. T. Oden and J. N. Reddy, *Variational Methods in Theoretical Mechanics*, Springer-Verlag, Berlin (1976); second edition (1983).

5. J. N. Reddy, *Energy and Variational Methods in Applied Mechanics*, John Wiley & Sons, New York (1984).

6. J. N. Reddy, *An Introduction to the Finite Element Method*, Second Edition, McGraw-Hill, New York (1993).

7. J. N. Reddy, *Applied Functional Analysis and Variational Methods in Engineering*, McGraw-Hill, New York (1986); reprinted by Krieger Publishing, Melbourne, Florida (1991).

8. J. N. Reddy and M. L. Rasmussen, *Advanced Engineering Analysis*, John Wiley & Sons, New York (1982); reprinted by Krieger Publishing, Melbourne, Florida (1990).

9. E. B. Becker, G. F. Carey, and J. T. Oden, *Finite Elements, an Introduction*, Vol. I, Prentice Hall, Englewood Cliffs, New Jersey (1981).

10. D. S. Burnett, *Finite Element Analysis*, Addison-Wesley, Reading, Massachusetts (1987).

11. H. S. Carslaw and J. C. Jaeger, *Conduction of Heat in Solids*, Oxford University Press, Oxford (1959).

12. J. P. Holman, *Heat Transfer*, sixth edition, McGraw-Hill, New York (1986).

13. G. G. Myers, *Analytical Methods in Conduction Heat Transfer*, McGraw-Hill, New York (1972).

14. J. H. Argyris and O. W. Scharpf, "Finite Elements in Time and Space," *Aeronautical Journal of the Royal Society*, **73**, 1041–1044 (1969).

15. W. L. Wood, "Control of Crank–Nicolson Noise in the Numerical Solution of the Heat Conduction Equation," *International Journal for Numerical Methods in Engineering*, **11**, 1059–1065 (1977).

3-D Conduction Heat Transfer

3.1 Introduction

In this chapter we present the finite element analysis of problems in 3-D conduction heat transfer and related problems in diffusion. Though conduction is the least complex mode of energy transport, it plays a major role in industrial design, manufacturing, and engineering. The prediction of temperature distributions, thermal performance, and thermal stresses in complex geometries has become a routine part of engineering analysis, due in large part to the finite element method and availability of general purpose computer programs. The conduction problem also provides a convenient framework to discuss various advanced aspects of the finite element method and numerical algorithms. The introduction to the finite element method presented in Chapter 2 aids the development presented here for the general 3-D heat conduction problem.

As described in the previous chapters, the appropriate mathematical description of thermal conduction in a 3-D solid region, Ω, is given by Eq. (1.4.4). In the interest of brevity, the subscript 's', which refers to a solid region, is omitted throughout the chapter. The Cartesian component form of the equation is [see Eq. (1.4.10)]

$$\rho C \frac{\partial T}{\partial t} = \frac{\partial}{\partial x_i} \left(k_{ij} \frac{\partial T}{\partial x_j} \right) + Q \tag{3.1.1}$$

where T is the temperature, ρ is the density, C is the specific heat, k_{ij} are the Cartesian components of the conductivity tensor (symmetric), and Q is the internal heat generation per unit volume. Summation on repeated subscripts is assumed (i.e., i and j are summed over the range of $i, j = 1, 2, 3$). All of the variables may be functions of position $\mathbf{x} = (x_1, x_2, x_3)$ and time t. We wish to solve Eq. (3.1.1) under appropriate boundary and initial conditions. The boundary conditions for this equation are given by the equations [see Eq. (1.8.10a,b)]:

$$T = f^T(s_k, t) \qquad \text{on } \Gamma_T, \text{ for } t > 0 \tag{3.1.2a}$$

$$-\left(k_{ij} \frac{\partial T}{\partial x_j} \right) n_i = q_a + q_c + q_r = f^q(s_k, t) \qquad \text{on } \Gamma_q, \text{ for } t > 0 \tag{3.1.2b}$$

Here s_k denotes the coordinates of a point on the boundary surface, q_a is an applied flux and the convective and radiative flux are

$$q_c = h_c(s_k, T, t)(T - T_c), \quad q_r = h_r(s_k, T, t)(T - T_r) \tag{3.1.2c}$$

The initial condition on the temperature is given by

$$T(x_j, 0) = T_0(x_j) \tag{3.1.3}$$

The heat transfer problem described by Eqs. (3.1.1)–(3.1.3) differs from that discussed in Chapter 2 in several respects. First, the problem considered here is for general three-dimensional geometries. Second, the boundary conditions are general enough to include conduction, convection, and radiation heat transfer between the solid regions and the surrounding environment. Third, the material coefficients, ρ, C, k_{ij}, h_c, and h_r can be functions of temperature, making the problem a nonlinear one.

As described in Section 2.8, the finite element model of the initial-boundary value problem described by Eqs. (3.1.1)–(3.1.3) is developed in two steps: (1) *spatial discretization*, in which the weak form of Eq. (3.1.1) over a typical element is developed and the spatial approximation of the dependent variable (i.e., temperature) T of the problem is assumed to obtain a set of ordinary differential equations in time among the nodal values T_j^e of the dependent variable; (2) *temporal approximation* of the ordinary differential equations obtained in the first step is carried out using some appropriate method such as a finite difference approximation. This step leads to a set of algebraic equations involving the nodal values T_j^e at time t_{n+1} [$= (n+1)\Delta t$, where n is an integer and Δt is the time increment] in terms of known values from the previous time step. Note that time-independent (steady) boundary value problems replace the second or temporal approximation step with the invocation of some type of iterative algorithm depending on the nonlinearity of the application. See Section 2.8 for an account of these two steps when applied to a two-dimensional problem. Here we develop the finite element models of three-dimensional, transient heat conduction problems using the two-step procedure described above. We begin with the development of the semidiscrete finite element model of Eqs. (3.1.1) and (3.1.2).

3.2 Semidiscrete Finite Element Model

Suppose that the conduction heat transfer region Ω is discretized into an appropriate collection of finite elements (see Figure 3.2.1). The weak form of Eqs. (3.1.1) and (3.1.2) over a typical element Ω^e is obtained by multiplying Eq. (3.1.1) with the weight function $w(\mathbf{x})$ and integrating over the element, integrating-by-parts (spatially) those terms which involve higher-order derivatives, and replacing the coefficient of the weight function in the boundary integral with the secondary variable [i.e., use Eq. (3.1.2b)]. We obtain

$$0 = \int_{\Omega^e} \left[w\left(\rho C \frac{\partial T}{\partial t} - Q \right) + k_{ij} \frac{\partial w}{\partial x_i} \frac{\partial T}{\partial x_j} \right] d\mathbf{x} + \oint_{\Gamma^e} (q_a + q_c + q_r) w \, ds \tag{3.2.1}$$

where summation on repeated subscripts, i and j, is used. Note that no integration by parts with respect to time is used, and w is not a function of time.

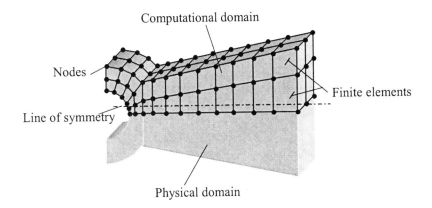

Figure 3.2.1. Finite element discretization of a region.

The semidiscrete finite element model is obtained from Eq. (3.2.1) by substituting a finite element approximation for the dependent variable, T. In selecting the approximation for T, we assume that the time dependence can be separated from the spatial variation,

$$T(\mathbf{x}, t) \simeq \sum_{i=1}^{n_e} T_i^e(t)\psi_i^e(\mathbf{x}) \tag{3.2.2a}$$

which can be written in matrix notation as

$$T(\mathbf{x}, t) = (\boldsymbol{\Psi}^e)^T \mathbf{T}^e \tag{3.2.2b}$$

where $(\boldsymbol{\Psi}^e)^T$ is a $(1 \times n_e)$ vector of interpolation functions of the element Ω^e

$$(\boldsymbol{\Psi}^e)^T = \{\psi_1^e \; \psi_2^e \; \cdots \; \psi_{n_e}^e\} \tag{3.2.2c}$$

$\mathbf{T^e}$ is a $(n_e \times 1)$ vector of nodal temperatures, superscript $(\cdot)^T$ denotes the transpose of a vector or matrix, and n_e is the number of nodal points in an element. The mth differential equation (in time) of the semidiscrete, weak-form finite element model is obtained by substituting $w(\mathbf{x}) = \psi_i^e(\mathbf{x})$ $(i = 1, 2, \cdots, n_e)$ and replacing T by the interpolation in Eq. (3.2.2):

$$0 = \sum_{j=1}^{n_e} \left(M_{ij}^e \frac{dT_j^e}{dt} + K_{ij}^e T_j^e \right) - Q_i^e + q_i^e \tag{3.2.3}$$

Equation (3.2.3) can be expressed in matrix form as

$$[M^e]\{\dot{T}^e\} + [K^e]\{T^e\} = \{Q^e\} - \{q^e\} \tag{3.2.4a}$$

or

$$\mathbf{M}^e(\mathbf{T}^e)\dot{\mathbf{T}}^e + \mathbf{K}^e(\mathbf{T}^e)\mathbf{T}^e = \mathbf{Q}^e(\mathbf{T}^e) - \mathbf{q}^e(\mathbf{T}^e) \qquad (3.2.4b)$$

where a superposed dot on T denotes a derivative with time ($\dot{T} = \partial T/\partial t$), and

$$M_{ij}^e = \int_{\Omega^e} \rho C \psi_i \psi_j \, d\mathbf{x} \;, \quad K_{ij}^e = \int_{\Omega^e} k_{mn} \frac{\partial \psi_i}{\partial x_m} \frac{\partial \psi_j}{\partial x_n} \, d\mathbf{x} \qquad (3.2.5a)$$

$$Q_i^e = \int_{\Omega^e} \psi_i Q(\mathbf{x}, t) \, d\mathbf{x} \;, \quad q_i^e = \oint_{\Gamma^e} \psi_i (q_a + q_c + q_r) \, ds \qquad (3.2.5b)$$

where summation on repeated indices ($m, n = 1, 2, 3$) is assumed, and the element label 'e' on ψ is omitted for brevity. The element coefficient matrices and vectors are written in the vector form as

$$\mathbf{M}^e = \int_{\Omega^e} \rho C \mathbf{\Psi} \mathbf{\Psi}^T \, d\mathbf{x}$$

$$\mathbf{K}^e = \int_{\Omega^e} \frac{\partial \mathbf{\Psi}}{\partial x_m} k_{mn} \frac{\partial \mathbf{\Psi}^T}{\partial x_n} \, d\mathbf{x}$$

$$\mathbf{Q}^e = \int_{\Omega^e} \mathbf{\Psi} Q \, d\mathbf{x}$$

$$\mathbf{q}^e = \oint_{\Gamma^e} \mathbf{\Psi} (q_a + q_c + q_r) \, d\mathbf{s} \qquad (3.2.6)$$

Equations (3.2.4a,b) are written in a form to show the most general nonlinear case in which material properties, boundary conditions, and volumetric sources depend on temperature. The implementation of variable material properties and coefficients will be discussed in a subsequent section.

The process of assembly for two-dimensional problems was discussed in Chapter 2; the assembly process for three-dimensional geometries is completely analogous. The assembled matrices are symbolically represented as

$$\mathbf{M} = \sum_e \mathbf{M}^e; \quad \mathbf{K} = \sum_e \mathbf{K}^e; \quad \mathbf{F} = \sum_e \mathbf{F}^e, \; \mathbf{F}^e = \mathbf{Q}^e - \mathbf{q}^e \qquad (3.2.7)$$

In Eq. (3.2.7) the sum is taken over all elements in the mesh, and the element matrices are defined by Eq. (3.2.6). Once the form of the element interpolation functions $\mathbf{\Psi}^e$ is known and the element geometry is specified, the integrals in Eq. (3.2.6) can be evaluated, and the global (assembled) equations are then constructed through use of Eq. (3.2.7):

$$\mathbf{M}(\mathbf{T})\dot{\mathbf{T}} + \mathbf{K}(\mathbf{T})\mathbf{T} = \mathbf{F}(\mathbf{T}) \qquad (3.2.8)$$

Before we consider the time or temporal approximation of Eqs. (3.2.4a,b) or Eq. (3.2.8), it is informative to discuss the geometries of elements and interpolation functions used in Eq. (3.2.2) and the numerical evaluation of the coefficients defined in Eq. (3.2.5) or (3.2.6). We will return to Eq. (3.2.8) in Section 3.7 to discuss its solution for steady-state and transient problems.

3.3 Interpolation Functions

3.3.1 Preliminary Comments

The equations for an individual element, as indicated by Eq. (3.2.6), require the specification of the element interpolation functions $\psi_i{}^e(\mathbf{x})$ for the approximation of the temperature and a description of the element geometry. In practical applications, elements based on linear and quadratic polynomials are generally used. The use of an isoparametric formulation is also standard in practice; this allows complex geometric regions to be easily and accurately described. Two-dimensional linear and quadratic elements were presented in Chapter 2. In this section, a number of commonly used three-dimensional elements are presented (see also [1–5]). Some specialty elements are described later in this chapter.

3.3.2 Hexahedral (Brick) Elements

Brick elements represent the most commonly used finite elements for three-dimensional analyses, and the straight-sided, linear eight-node brick element is the most cost-effective choice. The interpolation functions of the linear element (see Figure 3.3.1) are given in terms of the normalized coordinates (ξ, η, ζ) as follows:

$$\{\Psi^e\} = \frac{1}{8} \begin{Bmatrix} (1-\xi)(1-\eta)(1-\zeta) \\ (1+\xi)(1-\eta)(1-\zeta) \\ (1+\xi)(1+\eta)(1-\zeta) \\ (1-\xi)(1+\eta)(1-\zeta) \\ (1-\xi)(1-\eta)(1+\zeta) \\ (1+\xi)(1-\eta)(1+\zeta) \\ (1+\xi)(1+\eta)(1+\zeta) \\ (1-\xi)(1+\eta)(1+\zeta) \end{Bmatrix} \tag{3.3.1}$$

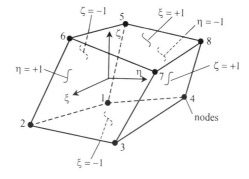

Figure 3.3.1. The linear (eight-node) brick element.

The quadratic shape functions for the twenty-node serendipity element (see Figure 3.3.2) are given by

$$
\{\Psi^e\} = \frac{1}{8}
\begin{Bmatrix}
(1-\xi)(1-\eta)(1-\zeta)(-\xi-\eta-\zeta-2) \\
(1+\xi)(1-\eta)(1-\zeta)(\xi-\eta-\zeta-2) \\
(1+\xi)(1+\eta)(1-\zeta)(\xi+\eta-\zeta-2) \\
(1-\xi)(1+\eta)(1-\zeta)(-\xi+\eta-\zeta-2) \\
(1-\xi)(1-\eta)(1+\zeta)(-\xi-\eta+\zeta-2) \\
(1+\xi)(1-\eta)(1+\zeta)(\xi-\eta+\zeta-2) \\
(1+\xi)(1+\eta)(1+\zeta)(\xi+\eta+\zeta-2) \\
(1-\xi)(1+\eta)(1+\zeta)(-\xi+\eta+\zeta-2) \\
2(1-\xi^2)(1-\eta)(1-\zeta) \\
2(1+\xi)(1-\eta^2)(1-\zeta) \\
2(1-\xi^2)(1+\eta)(1-\zeta) \\
2(1-\xi)(1-\eta^2)(1-\zeta) \\
2(1-\xi)(1-\eta)(1-\zeta^2) \\
2(1+\xi)(1-\eta)(1-\zeta^2) \\
2(1+\xi)(1+\eta)(1-\zeta^2) \\
2(1-\xi)(1+\eta)(1-\zeta^2) \\
2(1-\xi^2)(1-\eta)(1+\zeta) \\
2(1+\xi)(1-\eta^2)(1+\zeta) \\
2(1-\xi^2)(1+\eta)(1+\zeta) \\
2(1-\xi)(1-\eta^2)(1+\zeta)
\end{Bmatrix}
\tag{3.3.2}
$$

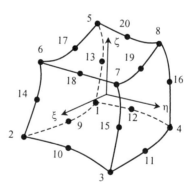

Figure 3.3.2. The (twenty-node) quadratic brick element.

As in the case of the two-dimensional quadratic (nine-node) Lagrange element, a brick element with 27 nodes may also be constructed; the shape functions are not shown for this element.

The transformation between the actual element Ω^e and the master element $\hat{\Omega}$ [or equivalently, between $(x_1, x_2, x_3) = (x, y, z)$ and (ξ, η, ζ)] is accomplished by a coordinate transformation of the form

$$
x = \sum_{i=1}^{n_g} x_i^e \phi_i^e(\xi, \eta, \zeta), \quad y = \sum_{i=1}^{n_g} y_i^e \phi_i^e(\xi, \eta, \zeta), \quad z = \sum_{i=1}^{n_g} z_i^e \phi_i^e(\xi, \eta, \zeta)
\tag{3.3.3}
$$

where n_g is the number of nodes describing the geometry of the element. If $n_g < n_e$, where n_e is the number of nodes used to interpolate the solution [see Eq. (3.2.2a)], the element is termed *subparametric*; when $n_g > n_e$ the element is called *superparametric*. The most common case, $n_g = n_e$ and $\phi_j^e = \psi_j^e$, is known as the *isoparametric formulation*. It is not uncommon to use, for example, subparametric formulation with respect to one set of dependent variables and isoparametric formulation with respect to another set (e.g., classical beam and plate bending elements; see Reddy [2]).

3.3.3 Prism Elements

A prism or wedge element is often useful in three-dimensional geometries especially for transitioning between hexahedral and tetrahedral elements. The shape functions for the six-node linear element (see Figure 3.3.3a) are given by

$$\{\Psi^e\} = \frac{1}{2} \begin{Bmatrix} L_1(1-\zeta) \\ L_2(1-\zeta) \\ L_3(1-\zeta) \\ L_1(1+\zeta) \\ L_2(1+\zeta) \\ L_3(1+\zeta) \end{Bmatrix} \tag{3.3.4}$$

The interpolation functions for the fifteen-node quadratic prism element (see Figure 3.3.3b) are

$$\{\Psi^e\} = \frac{1}{2} \begin{Bmatrix} L_1[(2L_1-1)(1-\zeta)-(1-\zeta^2)] \\ L_2[(2L_2-1)(1-\zeta)-(1-\zeta^2)] \\ L_3[(2L_3-1)(1-\zeta)-(1-\zeta^2)] \\ L_1[(2L_1-1)(1+\zeta)-(1-\zeta^2)] \\ L_2[(2L_2-1)(1+\zeta)-(1-\zeta^2)] \\ L_3[(2L_3-1)(1+\zeta)-(1-\zeta^2)] \\ 4L_1L_2(1-\zeta) \\ 4L_2L_3(1-\zeta) \\ 4L_3L_1(1-\zeta) \\ 2L_1(1-\zeta^2) \\ 2L_2(1-\zeta^2) \\ 2L_3(1-\zeta^2) \\ 4L_1L_2(1+\zeta) \\ 4L_2L_3(1+\zeta) \\ 4L_3L_1(1+\zeta) \end{Bmatrix} \tag{3.3.5}$$

The area coordinates, L_i (see Section 2.11), are used to describe the functional variation in the triangular cross section of the prism elements, while a standard normalized coordinate, ζ, describes the variation in the axial direction. Note that $L_1 + L_2 + L_3 = 1$. A coordinate transformation of the form given in (3.3.3) is easily defined to map the actual prism into a master element.

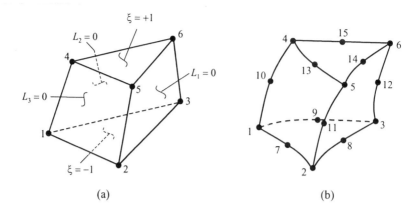

Figure 3.3.3. (a) The six-node linear prism element.   (b) The fifteen-node quadratic prism element.

3.3.4 Tetrahedral Elements

The standard tetrahedral elements are a three-dimensional version of the triangular elements. The four-node linear and ten-node quadratic tetrahedral elements are shown in Figure 3.3.4, and their interpolation functions are

$$\{\Psi^e\} = \begin{Bmatrix} L_1 \\ L_2 \\ L_3 \\ L_4 \end{Bmatrix} \; ; \quad \{\Psi^e\} = \begin{Bmatrix} L_1(2L_1 - 1) \\ L_2(2L_2 - 1) \\ L_3(2L_3 - 1) \\ L_4(2L_4 - 1) \\ 4L_1L_2 \\ 4L_2L_3 \\ 4L_3L_1 \\ 4L_1L_4 \\ 4L_2L_4 \\ 4L_3L_4 \end{Bmatrix} \qquad (3.3.6)$$

The volume coordinates, L_i, are used to describe the interpolation functions for linear and quadratic elements, where $L_1 + L_2 + L_3 + L_4 = 1$. The parametric mapping in Eq. (3.3.3) is easily defined for a master element.

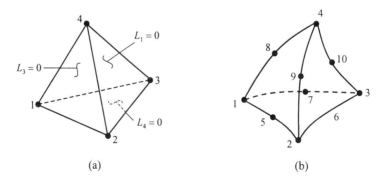

Figure 3.3.4. Linear and quadratic tetrahedral elements.

3.4 Numerical Integration

The evaluation of the various finite element coefficient matrices in (3.2.5) requires the integration of combinations of the interpolation functions and their spatial derivatives over the area or volume of the element. The numerical integration ideas described in Section 2.12 can be easily extended to three dimensions. Here we give the pertinent equations for the three-dimensional case.

For an isoparametric formulation [6], the following relations, based on the chain rule of differentiation, can be derived for brick elements:

$$\left\{\begin{array}{c}\frac{\partial \psi_i}{\partial \xi}\\ \frac{\partial \psi_i}{\partial \eta}\\ \frac{\partial \psi_i}{\partial \zeta}\end{array}\right\} = \left[\begin{array}{ccc}\frac{\partial x}{\partial \xi}&\frac{\partial y}{\partial \xi}&\frac{\partial z}{\partial \xi}\\ \frac{\partial x}{\partial \eta}&\frac{\partial y}{\partial \eta}&\frac{\partial z}{\partial \eta}\\ \frac{\partial x}{\partial \zeta}&\frac{\partial y}{\partial \zeta}&\frac{\partial z}{\partial \zeta}\end{array}\right]\left\{\begin{array}{c}\frac{\partial \psi_i}{\partial x}\\ \frac{\partial \psi_i}{\partial y}\\ \frac{\partial \psi_i}{\partial z}\end{array}\right\} = \left[\begin{array}{ccc}J_{11}&J_{12}&J_{13}\\ J_{21}&J_{22}&J_{23}\\ J_{31}&J_{32}&J_{33}\end{array}\right]\left\{\begin{array}{c}\frac{\partial \psi_i}{\partial x}\\ \frac{\partial \psi_i}{\partial y}\\ \frac{\partial \psi_i}{\partial z}\end{array}\right\} \tag{3.4.1}$$

where $[J]$ is the Jacobian of the transformation from global coordinates (x, y, z) to the local element coordinates (ξ, η, ζ). The parametric transformation defined in Eq. (3.3.3) can be used to define the components of $[J]$. Inverting Eq. (3.4.1), we obtain the global spatial derivatives of the interpolation functions in terms of the local derivatives:

$$\left\{\begin{array}{c}\frac{\partial \psi_i}{\partial x}\\ \frac{\partial \psi_i}{\partial y}\\ \frac{\partial \psi_i}{\partial z}\end{array}\right\} = \left[\begin{array}{ccc}J_{11}^*&J_{12}^*&J_{13}^*\\ J_{21}^*&J_{22}^*&J_{23}^*\\ J_{31}^*&J_{32}^*&J_{33}^*\end{array}\right]\left\{\begin{array}{c}\frac{\partial \psi_i}{\partial \xi}\\ \frac{\partial \psi_i}{\partial \eta}\\ \frac{\partial \psi_i}{\partial \zeta}\end{array}\right\} \tag{3.4.2}$$

where $[J^*]$ is the inverse of the Jacobian matrix $[J]$. The components J_{ij}^* are complicated functions of the components of $[J]$ that can in principle be obtained by analytically inverting the 3×3 Jacobian matrix. In practice, the Jacobian is usually inverted numerically at each integration point.

In performing numerical integration over the element volume, it is necessary to transform the integrand and limits of integration from the global coordinates to the local element coordinates. The differential elemental volume transforms according to

$$d\mathbf{x} = dx\, dy\, dz = ||[J]||\, d\xi\, d\eta\, d\zeta \tag{3.4.3}$$

The integration limits for the integrals transform to the limits on the local coordinates (ξ, η, ζ), i.e., -1 to $+1$. In the previous equations the (ξ, η, ζ) coordinates for a brick element were used for purposes of explanation. Similar relations for a tetrahedral element can be derived by replacing (ξ, η, ζ) with (L_1, L_2, L_3). The variable L_4 does not enter the formulae due to the relation $L_1 + L_2 + L_3 + L_4 = 1$. Hybrid coordinates, such as those utilized in the prism element, are treated in an analogous manner [5].

To illustrate further the development of a computational form for typical entries in Eq. (3.2.5) or (3.2.6), we consider the cross (xy) component in the $(1, 2)$ coefficient of the diffusion matrix $\mathbf{K^e}$ of Eq. (3.2.6):

$$(K_{12}^e)_{xy} = \int_{\Omega^e} k_{xy} \frac{\partial \psi_1}{\partial x} \frac{\partial \psi_2}{\partial y} dx dy dz \tag{3.4.4}$$

which will be evaluated for a three-dimensional, brick element. From the definitions in Eqs. (3.4.2) and (3.4.3) then

$$(K^e_{12})_{xy} = \int_{-1}^{+1} \int_{-1}^{+1} \int_{-1}^{+1} k_{xy} \left(J^*_{11} \frac{\partial \psi_1}{\partial \xi} + J^*_{12} \frac{\partial \psi_1}{\partial \eta} + J^*_{13} \frac{\partial \psi_1}{\partial \zeta} \right) \cdot$$

$$\left(J^*_{21} \frac{\partial \psi_2}{\partial \xi} + J^*_{22} \frac{\partial \psi_2}{\partial \eta} + J^*_{23} \frac{\partial \psi_2}{\partial \zeta} \right) |J| \, d\xi d\eta d\zeta \qquad (3.4.5)$$

Note that since the thermal conductivity could vary over the element, it must be treated as a function of (ξ, η, ζ) and be included in the evaluation of the integral. Variable coefficients are discussed further in Section 3.9. The above coefficient is typical of those encountered in building finite element equations. Each component of the integral in Eq. (3.4.5) is of the form

$$I = \int_{-1}^{+1} \int_{-1}^{+1} \int_{-1}^{+1} F(\xi, \eta, \zeta) \, d\xi d\eta d\zeta \qquad (3.4.6)$$

The Gauss quadrature formula (2.12.12) discussed in Section 2.12 can be readily extended to three dimensions to evaluate the integral expression I:

$$\int_{-1}^{+1} \int_{-1}^{+1} \int_{-1}^{+1} F(\xi, \eta, \zeta) \, d\xi d\eta d\zeta \approx \sum_{I=1}^{M} \sum_{J=1}^{N} \sum_{K=1}^{P} F(\xi_I, \eta_J, \zeta_K) \, W_I W_J W_K \quad (3.4.7)$$

For brick elements it is typical to employ a product rule as shown in (3.4.7) with $M = N = P = 2$ or 3. Other element types (e.g., tetrahedral elements) are also evaluated using quadrature formulas similar to (3.4.7) but slightly different in form [see Eq. (2.12.15)]. Hybrid elements, such as the prism, generally utilize combination quadrature formulas such as (2.12.15) in the triangular plane and a one-dimensional Gauss formula along the axis. For additional discussion of quadrature rules and their utilization, the reader may consult [2, 3, 5] and [7–13].

3.5 Computation of Surface Flux

The nodal flux q^e_i due to applied boundary conditions need only be computed for those element sides that coincide with the boundary of the problem domain; contributions from interior element boundaries are canceled by adjoining elements. The surface flux vector is given by

$$\mathbf{q}^e = \oint_{\Gamma^e} \mathbf{\Psi} \, q_i n_i \, ds = \oint_{\Gamma^e} \mathbf{\Psi}(q_a + q_c + q_r) \, ds \qquad (3.5.1)$$

where Γ^e denotes the boundary surface of the element and $q_i n_i$ (sum on i) is the total heat flux normal to the surface.

The computation of the indicated surface integrals is most easily carried out in the normalized or natural coordinate system for the face (edge) of an element. This requires that the elemental surface area ds, or element edge length ds, be related to the local surface coordinates. Consider the typical quadrilateral element face shown in Figure 3.5.1, where the vectors e_1 and e_2 are defined as being tangent to the curvilinear coordinates, ξ_s and η_s. The e vectors are not necessarily unit vectors; the ξ_s and η_s coordinates are assumed to be the natural coordinates for the element face. The elemental area ds in terms of the global coordinates (x, y, z) is related to an elemental area in surface coordinates by

$$ds = |\mathbf{J}_s|\, d\xi_s\, d\eta_s \tag{3.5.2}$$

where $\mathbf{J}_s$ is the Jacobian of the coordinate transformation and $|\cdot|$ indicates the determinant. The determinant of the Jacobian can be written in terms of the vectors (e_1, e_2) as

$$|\mathbf{J}_s| = |e_1 \times e_2| = \left[(e_1 \cdot e_1)(e_2 \cdot e_2) - (e_1 \cdot e_2)^2\right]^{\frac{1}{2}} \tag{3.5.3}$$

The vectors e_i can be expressed in terms of the global coordinates by

$$e_1 = \left\{ \begin{array}{c} \frac{\partial x}{\partial \xi_s} \\ \frac{\partial y}{\partial \xi_s} \\ \frac{\partial z}{\partial \xi_s} \end{array} \right\}, \quad e_2 = \left\{ \begin{array}{c} \frac{\partial x}{\partial \eta_s} \\ \frac{\partial y}{\partial \eta_s} \\ \frac{\partial z}{\partial \eta_s} \end{array} \right\} \tag{3.5.4}$$

Using the transformation (3.3.3) in Eq. (3.5.4), we can write

$$e_1 = \left\{ \begin{array}{c} \sum_{i=1}^{n_s} x_i^e \frac{\partial \hat{\psi}_i^e}{\partial \xi_s} \\ \sum_{i=1}^{n_s} y_i^e \frac{\partial \hat{\psi}_i^e}{\partial \xi_s} \\ \sum_{i=1}^{n_s} z_i^e \frac{\partial \hat{\psi}_i^e}{\partial \xi_s} \end{array} \right\}; \quad e_2 = \left\{ \begin{array}{c} \sum_{i=1}^{n_s} x_i^e \frac{\partial \hat{\psi}_i^e}{\partial \eta_s} \\ \sum_{i=1}^{n_s} y_i^e \frac{\partial \hat{\psi}_i^e}{\partial \eta_s} \\ \sum_{i=1}^{n_s} z_i^e \frac{\partial \hat{\psi}_i^e}{\partial \eta_s} \end{array} \right\} \tag{3.5.5}$$

where the $\hat{}$ indicates the restriction of the interpolation functions to an element face and n_s is the number of nodes defining the surface. The functions $\hat{\psi}_i$ may be either linear or quadratic, depending on the type of mapping used to describe the element geometry. Equation (3.5.5) provides a means for computing $|\mathbf{J}_s|$, which can be used in Eq. (3.5.2).

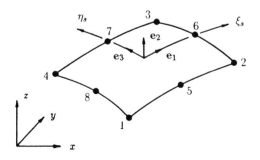

Figure 3.5.1. Nomenclature for surface flux computation.

With the above definitions, the terms in Eq. (3.5.1) can be considered individually. For the case of an applied heat flux q_a, the contribution to $\mathbf{q}^e$ is given by

$$\mathbf{q}_a = \oint_{\Gamma_e} \hat{\mathbf{\Psi}} q_a \, ds \tag{3.5.6}$$

where $\hat{\mathbf{\Psi}}$ is the vector of interpolation functions that are restricted to the element boundary. Employing Eq. (3.5.2), Eq. (3.5.6) can be written in a more explicit form

$$\mathbf{q}_a = \int_{-1}^{1} \int_{-1}^{1} \hat{\mathbf{\Psi}}(\xi, \eta) q_a(\xi, \eta) |\mathbf{J}_s| \, d\xi_s \, d\eta_s \tag{3.5.7}$$

Once the distribution of the normal flux $q_a(\xi, \eta)$ is known for the element, Eq. (3.5.7) can be evaluated using numerical quadrature as discussed in Section 2.12. The order of the quadrature used to evaluate $\mathbf{q}_a$ is typically the same as that used on the volume integrals defining the conductivity $[K^e]$ and capacitance $[M^e]$ matrices. For a three-dimensional element, the surface shape functions $\hat{\mathbf{\Psi}}$ correspond directly to the interpolation functions of the two-dimensional triangular and quadrilateral elements. In the two-dimensional case, $\hat{\mathbf{\Psi}}$ corresponds to the one-dimensional interpolation functions

$$\{\hat{\mathbf{\Psi}}^e\} = \frac{1}{2} \left\{ \begin{array}{c} (1 - \xi) \\ (1 + \xi) \end{array} \right\} \tag{3.5.8}$$

for the linear case and

$$\{\hat{\mathbf{\Psi}}^e\} = \frac{1}{2} \left\{ \begin{array}{c} (1 - \xi)\xi \\ 2(1 - \xi^2) \\ (1 + \xi)\xi \end{array} \right\} \tag{3.5.9}$$

for the quadratic case, where ξ is the normalized coordinate along the element edge. Also, for the two-dimensional case, the integral in Eq. (3.5.7) simplifies to an integral over the edge, and the definition of the incremental edge length reduces to

$$ds = |J_s| d\xi_s = \Delta d\xi_s \tag{3.5.10}$$

where

$$\Delta = \left[\left(\frac{\partial x}{\partial \xi_s} \right)^2 + \left(\frac{\partial y}{\partial \xi_s} \right)^2 \right]^{\frac{1}{2}} \tag{3.5.11}$$

The contributions to $\mathbf{q}^e$ due to convection and radiation are treated in an analogous manner. For the convection boundary condition, we have

$$\mathbf{q}_c = \oint_{\Gamma_e} \hat{\mathbf{\Psi}} q_c \, ds = \oint_{\Gamma_e} \hat{\mathbf{\Psi}} h_c (T - T_c) \, ds \tag{3.5.12}$$

where the definition in Eq. (3.1.2c) has been used. The temperature on the element surface is given by

$$T(\xi, \eta) = \left(\hat{\mathbf{\Psi}}(\xi, \eta) \right)^T \mathbf{T} \tag{3.5.13}$$

where $\hat{\mathbf{\Psi}}$ are surface or edge interpolation functions. Substituting this relation into Eq. (3.5.12) and using the definition for ds yields

$$\mathbf{q}_c = \left(\int_{-1}^{1} \int_{-1}^{1} h_c(\xi,\eta)\hat{\mathbf{\Psi}}(\xi,\eta)\hat{\mathbf{\Psi}}^T(\xi,\eta)|\mathbf{J}_s| \, d\xi_s \, d\eta_s \right) \mathbf{T}$$
$$- \int_{-1}^{1} \int_{-1}^{1} h_c(\xi,\eta)\hat{\mathbf{\Psi}}(\xi,\eta)T_c(\xi,\eta)|\mathbf{J}_s| \, d\xi_s \, d\eta_s \tag{3.5.14a}$$

$$= \mathbf{CT} - \mathbf{F}_{hc} \tag{3.5.14b}$$

where

$$\mathbf{C} = \int_{-1}^{1} \int_{-1}^{1} h_c \hat{\mathbf{\Psi}}\hat{\mathbf{\Psi}}^T |\mathbf{J}_s| \, d\xi_s \, d\eta_s \tag{3.5.15a}$$

$$\mathbf{F}_{hc} = \int_{-1}^{1} h_c T_c \hat{\mathbf{\Psi}}|\mathbf{J}_s| \, d\xi_s \, d\eta_s \tag{3.5.15b}$$

For a given variation of h_c and T_c, the matrix $\mathbf{C}$ and vector $\mathbf{F}_{hc}$ can be computed. Note that in Eq. (3.5.14b) the first term $\mathbf{CT}$ contains the vector of unknown nodal temperatures and will thus be moved to the left-hand side of the matrix Eq. (3.2.4). When h_c is a function of temperature, the problem is nonlinear. More discussion on the solution of nonlinear problems will be presented in the sequel.

A computation similar to the convective boundary condition may be carried out for the radiation boundary condition. We obtain

$$\mathbf{q}_r = \mathbf{RT} - \mathbf{F}_{hr} \tag{3.5.16}$$

where

$$\mathbf{R} = \int_{-1}^{1} \int_{-1}^{1} h_r \hat{\mathbf{\Psi}}\hat{\mathbf{\Psi}}^T |\mathbf{J}_s| \, d\xi_s d\eta_s \tag{3.5.17a}$$

$$\mathbf{F}_{hr} = \int_{-1}^{1} \int_{-1}^{1} h_r T_r \hat{\mathbf{\Psi}}|\mathbf{J}_s| \, d\xi_s d\eta_s \tag{3.5.17b}$$

Again, the term $\mathbf{RT}$ will be moved to the left-hand side of Eq. (3.3.4) while $\mathbf{F}_{hr}$ is retained in the source vector. Equations (3.5.17a,b) are highly nonlinear due to the fact that both $\mathbf{R}$ and $\mathbf{F}_{hr}$ are functions of the unknown temperature field $\mathbf{T}$, since the heat transfer coefficient, h_r, is an explicit function of temperature.

3.6 Semidiscrete Finite Element Model

Returning to the finite element model in Eq. (3.2.4b), we have

$$\mathbf{M}^e(\mathbf{T}^e)\dot{\mathbf{T}}^e + \mathbf{K}^e(\mathbf{T}^e)\mathbf{T}^e = \mathbf{Q}^e(\mathbf{T}^e) - \mathbf{q}^e(\mathbf{T}^e) \tag{3.6.1}$$

or using the definition of the boundary fluxes in Eqs. (3.5.1), (3.5.7), (3.5.14b) and (3.5.16)

$$\mathbf{M}\dot{\mathbf{T}} + \mathbf{KT} = \mathbf{Q} - \mathbf{q}_a - \mathbf{CT} + \mathbf{F}_{hc} - \mathbf{RT} + \mathbf{F}_{hr} \tag{3.6.2}$$

Rearranging the terms yields the following matrix equation of ordinary differential equations in time:

$$\mathbf{M}\dot{\mathbf{T}} + \hat{\mathbf{K}}\mathbf{T} = \hat{\mathbf{F}} \tag{3.6.3}$$

with

$$\hat{\mathbf{K}} = \mathbf{K} + \mathbf{C} + \mathbf{R}, \quad \hat{\mathbf{F}} = \mathbf{Q} - \mathbf{q}_a + \mathbf{F}_{hc} + \mathbf{F}_{hr} \tag{3.6.4}$$

In general, $\mathbf{M}, \hat{\mathbf{K}}$, and $\hat{\mathbf{F}}$ are functions of temperature and/or time.

3.7 Solution of Nonlinear Equations

3.7.1 Preliminary Comments

The discussion of solution procedures for the discretized form of the heat conduction problem naturally divides itself into two sections according to the time-independent (steady-state) or transient nature of the problem. Solution algorithms for the two types of problems are described here. For most practical conduction problems, the finite element model ultimately involves the solution of a large, sparse, symmetric system of nonlinear algebraic equations. In most cases, the nonlinear equations are linearized first and then solved, often via a direct method (e.g., Gauss elimination), which is tailored to the finite element formulation (e.g., frontal [14,15] or skyline [16,17] methods). The methods are termed direct methods because they produce a solution to the matrix system in a fixed number of steps; the solution would be exact if infinite precision arithmetic were possible. However, despite the historical popularity of direct methods, there has been a renewed and rapidly growing interest in iterative methods. Iterative methods provide an approximate solution to the matrix problem that usually improves with continued iteration; the number of iterations required to achieve a specified accuracy is not known a priori. The need to analyze extremely large three-dimensional problems and the move to parallel computer architectures have intensified the exploration of iterative methods, such as the preconditioned conjugate gradient (PCG) method. The symmetric, positive-definite matrix forms found in diffusion problems are well suited to iterative methods. These techniques have now become the standard matrix solution algorithms for many types of finite element applications. The matrix solution problem will not be discussed in depth here but may be found in books on linear algebra and matrices (e.g., see [18,19]), and in a condensed form in Appendix B.

In the following sections, a number of linearization algorithms for nonlinear algebraic equations will be described for both steady and transient problems. The iterative methods discussed here for nonlinear problems are basically linearization procedures, and these methods should not be confused with the iterative methods used for the solution of linear algebraic equations.

3.7.2 Steady-State Problems

For problems that are independent of time, the matrix equation (3.6.3) is simplified to

$$\hat{\mathbf{K}}(\mathbf{T})\mathbf{T} = \hat{\mathbf{F}}(\mathbf{T}) \tag{3.7.1}$$

which is a system of nonlinear algebraic equations. Consider first the case where $\hat{\mathbf{K}}$ and $\hat{\mathbf{F}}$ are not functions of the temperature. Then Eq. (3.7.1) reduces to a linear matrix problem, which can be solved directly, without iteration, by the standard matrix methods noted above.

When Eq. (3.7.1) is nonlinear (e.g., due to temperature-dependent properties, heat source, and/or boundary conditions), it should be linearized. For many mildly nonlinear conduction problems, a simple successive substitution iteration scheme, also known as the *Picard method*, is suitable. The Picard scheme is given by

$$\hat{\mathbf{K}}(\mathbf{T}^n)\mathbf{T}^{n+1} = \hat{\mathbf{F}}(\mathbf{T}^n) \tag{3.7.2}$$

where superscript n denotes the iteration number. By evaluating the coefficient matrix $\hat{\mathbf{K}}$ and right-hand side vector $\hat{\mathbf{F}}$ using the solution $\mathbf{T}^n$ from the previous iteration, we have linearized the problem. The linearized problem is solved for $\mathbf{T}^{n+1}$ (using any matrix solution method), which is then used to evaluate the coefficient matrix and right-hand vector for the next iteration. This procedure is continued until the root-mean-square value of the difference between the solution vectors at two consecutive iterations (normalized with respect to the current solution) is reduced to a value less than a preassigned tolerance, ϵ:

$$\frac{(\mathbf{T}^{n+1} - \mathbf{T}^n)^T(\mathbf{T}^{n+1} - \mathbf{T}^n)}{(\mathbf{T}^{n+1})^T(\mathbf{T}^{n+1})} < \epsilon^2 \tag{3.7.3}$$

The value of ϵ may be taken to be in the range of $10^{-2} - 10^{-4}$.

Typically, problems with slow variations of the material properties, heat transfer coefficients, or heat sources are effectively solved using the Picard method; convergence is normally obtained in a few iterations for problems of this type. An improvement in convergence rate can sometimes be realized by use of a relaxation formula where

$$\hat{\mathbf{K}}(\mathbf{T}^n)\mathbf{T}^\star = \hat{\mathbf{F}}(\mathbf{T}^n) \tag{3.7.4a}$$

and

$$\mathbf{T}^{n+1} = \gamma\mathbf{T}^n + (1 - \gamma)\mathbf{T}^\star \qquad 0 \le \gamma \le 1 \tag{3.7.4b}$$

Conduction problems with strong nonlinearities, such as those with radiation boundary conditions, boiling heat transfer coefficients, and Arrhenius type heat sources require iterative techniques that are more powerful than the Picard method. A fairly simple algorithm for this type of problem consists of incrementally approaching the final solution through a series of intermediate solutions. These intermediate solutions may be of physical interest or may simply be a means to obtain the required solution. The formal algorithms used to implement this procedure are termed continuation methods and can be used with either of the iterative methods cited above.

Assume that the solution of Eq. (3.7.1) depends continuously on some real parameter, λ. For heat conduction problems, λ could be the magnitude of a volumetric source or the magnitude of a boundary condition. Then Eq. (3.7.1) can be written in general as

$$\hat{\mathbf{K}}(\mathbf{T}, \lambda)\mathbf{T} = \hat{\mathbf{F}}(\mathbf{T}, \lambda) \tag{3.7.5}$$

which suggests the zeroth order continuation method

$$\hat{\mathbf{K}}(\mathbf{T}_\lambda^n, \lambda^m)\mathbf{T}_\lambda^{n+1} = \hat{\mathbf{F}}(\mathbf{T}_\lambda^n, \lambda^m) \tag{3.7.6}$$

where Eq. (3.7.6) is solved for a series of problems with increasing values of the continuation parameter $\lambda^m = \lambda^{m-1} + \Delta\lambda$. The converged solution, T_λ, at one value of λ is used as the starting solution at the next higher value of λ; the iterative method in Eq. (3.7.2) or (3.7.4a) is used at each value of λ to achieve a converged solution.

The application of second-order iteration schemes, such as Newton's method (see Section 4.6), is another approach to the highly nonlinear problem. However, for some types of nonlinearities, Newton's method produces an unsymmetric, tangent (or Jacobian) matrix. The additional computational cost and complexity of solving an unsymmetric system of equations have discouraged the use of such procedures. The most popular alternative for strongly nonlinear problems is to solve the steady-state heat transfer problem via a "false transient" procedure. Generally, implicit time integration formulas, such as those described below, are used to advance the solution through time and asymptotically approach the required steady-state solution. The cost-effectiveness of such a procedure relies heavily on the judicious choice of the integration time step; the step must be large enough to reach steady state quickly but small enough to maintain stability of the solution procedure. The combination of a simple, first-order accurate backward Euler method (see Section 4.6) with an adaptive time step control (see Gresho et al. [20]) would be an appropriate scheme for most nonlinear conduction problems of this type.

3.7.3 Transient Problems

The semidiscretization process employed in the finite element method produces a system of coupled, nonlinear ordinary differential equations of the form shown in Eq. (3.6.3) for the time-dependent conduction problem. There are a number of possible methods for solving or integrating the system in Eq. (3.6.3), including many of the classical methods used for ordinary differential equations (ODE). However, the (usually) large size of the equation system and the need to take advantage of the special (banded) form of the matrices in Eq. (3.6.3) eliminate many integration methods from consideration.

Finite element time integration methods are usually direct integration methods that employ some relatively simple discrete approximation for the time derivative. Mode superposition methods (e.g., see Bathe [21]) are also applicable to the system in Eq. (3.6.3), but unlike direct integration, this technique works with a transformed equation set. Mode superposition as an integration method will not be discussed here, though some aspects will be utilized when discussing the stability of integration methods. The discrete time approximation needed for a direct integration method may be obtained via a finite element or finite difference technique or by introducing more formal ODE ideas such as the linear multistep methods (LMS) (see Belytschko and Hughes [22]). Finite elements in time have been described by Argyris and Scharpf [23], Donea [24], Bettencourt et al. [25], and others. Because these techniques result in many of the same integration methods produced by finite difference approximations, they will not be considered here as a separate method.

The procedure for developing a finite difference-based direct integration method begins by dividing the integration interval of interest $(0 \leq t \leq t_{final})$ into a number of discrete segments denoted by $\Delta t = t^{n+1} - t^n$, where the superscript denotes the time step number. Within each interval Δt, the temperature and temperature rate are allowed to vary according to some prescription that can be expressed as a simple difference formula. For example, assume that the temperature varies linearly over the interval Δt; the time-rate of change of the temperature over the interval Δt can be approximated by the backward difference $\dot{\mathbf{T}}^{n+1} = (\mathbf{T}^{n+1} - \mathbf{T}^n)/\Delta t$. Evaluating Eq. (3.6.3) at the $n + 1$ time step and using the above definition produces

$$\frac{1}{\Delta t}\mathbf{M}\left(\mathbf{T}^{n+1} - \mathbf{T}^n\right) + \hat{\mathbf{K}}\mathbf{T}^{n+1} = \hat{\mathbf{F}}^{n+1} \tag{3.7.7}$$

Equation (3.7.7) represents a system of algebraic equations for the unknown temperatures $\mathbf{T}^{n+1}$ that can be successively solved at each new time plane.

Other difference approximations for $\dot{\mathbf{T}}$ are possible and lead to other types of recursive formulas for the solution of the temperature field as a function of time. The most popular method used for diffusion problems was introduced in Section 2.8 and consists of the following approximation:

$$\dot{\mathbf{T}}^a = \frac{1}{\Delta t}\left(\mathbf{T}^{n+1} - \mathbf{T}^n\right)$$

$$\mathbf{T}^a = (1 - \theta)\mathbf{T}^n + \theta\mathbf{T}^{n+1} \tag{3.7.8}$$

for $0 \leq \theta \leq 1$. When Eq. (3.6.3) is evaluated at $\mathbf{T}^a$ and the above definitions are used, the resulting algorithm, which goes by a variety of names including the generalized trapezoid method, the generalized Crank–Nicolson method, or the generalized midpoint rule, is given by

$$\frac{1}{\Delta t}\mathbf{M}(\mathbf{T}^a)\left(\mathbf{T}^{n+1} - \mathbf{T}^n\right) + \theta\hat{\mathbf{K}}(\mathbf{T}^a)\mathbf{T}^{n+1} = \hat{\mathbf{F}}(\mathbf{T}^a) - (1 - \theta)\hat{\mathbf{K}}(\mathbf{T}^a)\mathbf{T}^n \tag{3.7.9}$$

where superscript n indicates the time level, Δt is the time step, and θ is a parameter that determines where in the time interval, t_n to $t_n + \Delta t = t_{n+1}$, the equation will be evaluated.

The choice of $\theta = 0$ produces the Euler (forward difference) method, which is an explicit method that can be rewritten for the general nonlinear case as

$$\frac{1}{\Delta t}\mathbf{M}(\mathbf{T}^n)\mathbf{T}^{n+1} = \frac{1}{\Delta t}\mathbf{M}(\mathbf{T}^n)\mathbf{T}^n - \hat{\mathbf{K}}(\mathbf{T}^n)\mathbf{T}^n + \hat{\mathbf{F}}(\mathbf{T}^n) \tag{3.7.10}$$

and may be rearranged to clearly show its explicit nature

$$\mathbf{T}^{n+1} = \mathbf{T}^n + \Delta t \left[\mathbf{M}^{-1}(\mathbf{T}^n)\hat{\mathbf{F}}(\mathbf{T}^n) - \mathbf{M}^{-1}(\mathbf{T}^n)\hat{\mathbf{K}}(\mathbf{T}^n)\mathbf{T}^n \right] \qquad (3.7.11)$$

For effective implementation, the algorithm in Eq. (3.7.11) requires that the capacitance matrix, $\mathbf{M}$, be easily invertible, i.e., be diagonal. The reduction of the consistent capacitance matrix to a diagonal or "lumped" form can be accomplished by various procedures (e.g., row-sum technique [2,5,21,22]) for some types of finite elements (generally linear elements) but is not a universally available option. Also, the explicit nature of Eq. (3.7.11) implies a stability condition in terms of a maximum allowable time step, the thermal diffusivity of the material, and the finite element mesh spacing (see [22,26]); the stability condition can produce a prohibitively small time step for some problems. The Euler method is first-order accurate in time (i.e., $O(\Delta t)$). Despite these limitations, explicit procedures are in common use in conduction problems, especially for analyses involving short-time transient phenomena, such as thermal shocks. The main attraction of an explicit scheme is the fact that a matrix solution is not required and computer storage requirements are therefore minimal.

For the cases for which $\theta > 0$, Eq. (3.7.9) produces a family of implicit methods which require the solution of a matrix problem at each time step. For example, when $\theta = 1$, Eq. (3.7.9) reduces to the backward Euler (backward difference) method

$$\left[\frac{1}{\Delta t}\mathbf{M}(\mathbf{T}^{n+1}) + \hat{\mathbf{K}}(\mathbf{T}^{n+1}) \right] \mathbf{T}^{n+1} = \frac{1}{\Delta t}\mathbf{M}(\mathbf{T}^{n+1})\mathbf{T}^n + \hat{\mathbf{F}}(\mathbf{T}^{n+1}) \qquad (3.7.12)$$

which is a fully implicit scheme. This is an unconditionally stable method (i.e., there is no time step restriction; see a later section for a discussion of convergence and stability of numerical integration methods), and is first-order accurate in time. In the general nonlinear case, the matrix problem presented in Eq. (3.7.12) requires an iterative solution for $\mathbf{T}^{n+1}$ within each time step; the linear case can be solved without nonlinear iteration, though a matrix solution is still required. The Picard scheme would be an acceptable choice for iteratively solving Eq. (3.7.12) at time t_{n+1}. In treating the nonlinear form of (3.7.12), the use of predictor-corrector schemes, extrapolation methods, and quasilinearization can often reduce the computational effort at each step without significantly reducing the accuracy of the method. For mild nonlinearities and modest time steps, a quasilinearization of Eq. (3.7.12) yields

$$\left[\frac{1}{\Delta t}\mathbf{M}(\mathbf{T}^n) + \hat{\mathbf{K}}(\mathbf{T}^n) \right] \mathbf{T}^{n+1} = \frac{1}{\Delta t}\mathbf{M}(\mathbf{T}^n)\mathbf{T}^n + \hat{\mathbf{F}}(\mathbf{T}^n) \qquad (3.7.13)$$

which is now a single-step, non-iterative method for the solution of $\mathbf{T}^{n+1}$. An extrapolation procedure of the form

$$\mathbf{T}^* = \frac{3}{2}\mathbf{T}^n - \frac{1}{2}\mathbf{T}^{n-1} \qquad (3.7.14)$$

in conjunction with

$$\left[\frac{1}{\Delta t}\mathbf{M}(\mathbf{T}^*) + \hat{\mathbf{K}}(\mathbf{T}^*)\right]\mathbf{T}^{n+1} = \frac{1}{\Delta t}\mathbf{M}(\mathbf{T}^*)\mathbf{T}^n + \hat{\mathbf{F}}(\mathbf{T}^*) \tag{3.7.15}$$

will often produce a better solution than (3.7.13) and allow a larger time step. This form is attractive as a false transient procedure for mildly nonlinear problems. For strong nonlinearities the coupling of a forward Euler predictor (3.7.11) with (3.7.12) as a corrector is a viable option (see Gresh et al. [20]). Such methods are often used in practice with good success though the experience of the analyst is a crucial element in selecting the proper method.

We close this discussion with a brief outline of the only second-order accurate method in this family of procedures – the Crank–Nicolson method with $\theta = \frac{1}{2}$. Like the backward difference method, this is an unconditionally stable method, and it requires an iterative solution at each time step for the general nonlinear case. From Eq. (3.7.9) we have

$$\left[\frac{2}{\Delta t}\mathbf{M}(\mathbf{T}^a) + \hat{\mathbf{K}}(\mathbf{T}^a)\right]\mathbf{T}^a = \frac{2}{\Delta t}\mathbf{M}(\mathbf{T}^a)\mathbf{T}^n + \hat{\mathbf{F}}(\mathbf{T}^a) \tag{3.7.16a}$$

with the new solution recovered from the definition

$$\mathbf{T}^{n+1} = 2\mathbf{T}^a - \mathbf{T}^n \tag{3.7.16b}$$

Though Eq. (3.7.16a) has no time step restriction for stability, it does suffer from "noise" or temporal oscillations in the solution when the time step exceeds a critical value (see the later section on Convergence and Stability). Several modifications have been proposed to correct or filter this problem (see [4,27,28]). Predictor-corrector, extrapolation, and quasilinearization schemes can also be used with Eq. (3.7.16a) to reduce the computational effort at each time step. The Crank–Nicolson scheme and its variants are probably the most frequently used procedures for heat conduction problems.

Predictor-corrector methods

The time integration methods outlined above are single-step methods that work effectively when an appropriate time step Δt is selected. The time step selection, however, is crucial to accuracy though in many physical problems there is little guidance available to precisely determine a suitable value for Δt. A cost-effective solution to time step selection and a general increase in integrator performance are available through use of predictor-corrector methods. The method described here was originally developed by Gresho et al. [20].

An implicit integration method that is second-order accurate in time can be developed by combining an explicit prediction method with the Crank–Nicolson (trapezoid) method described previously. A second-order, variable step, explicit method is the Adams–Bashforth predictor given by

$$\mathbf{T}_p^{n+1} = \mathbf{T}^n + \frac{\Delta t_n}{2}\left[\left(2 + \frac{\Delta t_n}{\Delta t_{n-1}}\right)\dot{\mathbf{T}}^n - \left(\frac{\Delta t_n}{\Delta t_{n-1}}\right)\dot{\mathbf{T}}^{n-1}\right] \tag{3.7.17}$$

where $\Delta t_n = t_{n+1} - t_n$ and $\Delta t_{n-1} = t_n - t_{n-1}$. This formula can be used to predict the temperature solution vector given two "acceleration" vectors from previous times; no matrix solution is required.

A compatible corrector formula for use with Eq. (3.7.17) is available in the form of the trapezoid rule. When applied to the conduction equation, the trapezoid rule produces

$$\left[\frac{2}{\Delta t_n}\mathbf{M}(\mathbf{T}^a) + \hat{\mathbf{K}}(\mathbf{T}^a)\right]\mathbf{T}^{n+1} = \frac{2}{\Delta t_n}\mathbf{M}(\mathbf{T}^a)\mathbf{T}^n + \mathbf{M}(\mathbf{T}^a)\dot{\mathbf{T}}^n + \hat{\mathbf{F}}(\mathbf{T}^a) \quad (3.7.18)$$

which is a rearranged form of the Crank–Nicolson method given in Eq. (3.7.16a). Equation (3.7.18) is observed to be a nonlinear, algebraic system for the vector $\mathbf{T}^{n+1}$ and can again be solved using an iterative procedure such as Picard's method.

The major steps in the time integration procedure utilizing (3.7.17) and (3.7.18) are outlined here. At the beginning of each time step it is assumed that all of the required solution and "acceleration" vectors are known and the time increment for the next step has been selected. To advance the solution from time t_n to time t_{n+1} then requires the following steps:

1. A tentative solution vector, $\mathbf{T}_p^{n+1}$, is computed using the predictor Eq. (3.7.17).

2. The corrector Eq. (3.7.18) is solved for the "true" solution, $\mathbf{T}^{n+1}$. This involves the iterative solution of (3.7.18) via Picard's method. The predicted values $\mathbf{T}_p^{n+1}$ are used to estimate $\mathbf{T}^a$ and initialize the equation for the iteration procedure.

3. The "acceleration" vectors are updated using the new solution $\mathbf{T}^{n+1}$ and the "inverted" forms of the corrector formulas, that is

$$\dot{\mathbf{T}}^{n+1} = \frac{2}{\Delta t_n}\left(\mathbf{T}^{n+1} - \mathbf{T}^n\right) - \dot{\mathbf{T}}^n \quad (3.7.19)$$

4. A new integration time step is computed. The time step selection process is based on an analysis of the time truncation errors in the predictor and corrector formulas as described in the next section. If a constant time step is being used, this step is omitted.

5. Return to step 1 for next time increment.

In actual implementation the Picard iteration process in step 2 is not carried to absolute convergence. Rather, a one-step correction is employed as advocated in [20]. This procedure is quite efficient and can be very accurate provided the time step is suitably controlled.

Time step control

The implicit time integration procedures outlined above can be used with a fixed, user-specified time step or a time step that changes only at certain points during the integration interval. However, the a priori selection and modification of a reasonable integration time step can be a difficult task, especially for a complex problem. One of the benefits of using the predictor-corrector algorithms described here is that they provide a rational basis for dynamically selecting the time step.

The detailed derivation of the time step selection formula is omitted here. The reader interested in further details is referred to Gresho et al. [20]. The general ideas for the time step selection process come from the well-established procedures for solving ordinary differential equations. By comparing the time truncation errors for two time integration methods of comparable order, a formula can be developed to predict the next time step based on a user-specified error tolerance. In the present case, the time truncation errors for the explicit predictor and implicit corrector steps are analyzed.

The time step estimation formula is given by Gresho et al. [20] as

$$\Delta t_{n+1} = \Delta t_n \left(b \cdot \frac{\epsilon^t}{d_{n+1}} \right)^m \qquad (3.7.20)$$

where $m = 1/3$, $b = 3(1 + \Delta t_{n-1}/\Delta t_n)$ for this second-order scheme. The user-specified error tolerance for the integration process is ϵ^t, which typically has a default value of 0.001. The quantity d_{n+1} is an appropriate norm on the integration error, which is defined as the difference between the predicted solution and the corrected value. Generally, the following norm is used

$$d_{n+1} = \frac{1}{\sqrt{N} T_{max}} \left[\sum_{i=1}^{N} \left(T_i^{(n+1)} - T_{ip}^{(n+1)} \right)^2 \right]^{\frac{1}{2}} \qquad (3.7.21)$$

where N is the total number of nodes in the problem, T_{max} is a representative (constant) temperature scale for the problem, and T_i and T_{ip} are the corrected and predicted temperatures at the nodes.

Initialization

The predictor, Eq. (3.7.17), requires that one or more acceleration or rate vectors be available at each time in order to estimate a new solution vector. At the beginning of a transient solution these vectors are not generally available and thus a special starting procedure must be used. The approach normally taken is to use the dissipative backward Euler method for the first few steps and then switch to the standard second-order predictor-corrector methods. This procedure has the advantage that any nonphysical features of the numerical model are quickly damped by the backward Euler scheme.

For the first time step, the implicit, backward Euler scheme is used alone; the second step uses a forward Euler predictor and backward Euler corrector. Both of these steps use a fixed, user-supplied time step. At the third step, the usual predictor/corrector integration procedure begins and automatic time step selection is started. The initial time step supplied by the user to start the problem should be very conservative to prevent large time step reductions when the automatic selection procedure takes control.

Linear multi-step methods

Most of the time integration methods described above were obtained by introducing relatively simple finite difference expressions for the time derivative into the semidiscrete finite element equation.   This approach resulted in single step integration formulas in which the solution at the new time relies only on the solution data from the previous time.   Multi-step methods that rely on data from more than one previous time step are also possible. These methods may be derived from "higher-order" finite difference approximations for $\dot{\mathbf{T}}$.   However, a more formal approach with rigorous mathematical support involves the linear multi-step (LMS) methods developed for the solution of ordinary differential equations. Here we will outline a few applications of LMS to the finite element conduction problems and illustrate how some of the previous methods can be obtained as special cases of LMS. Further details on LMS are available in [29].

Consider the first-order system of equations expressed by

$$\dot{\mathbf{y}} = \mathbf{f}(\mathbf{y}, t) \tag{3.7.22}$$

A general LMS of order k for the solution of Eq. (3.7.22) is written as

$$\sum_{i=0}^{k} \left\{ \alpha_i \, \mathbf{y}^{n-i} + \Delta t \, \beta_i \, \mathbf{f}(\mathbf{y}^{n-i}, t_{n-i}) \right\} \tag{3.7.23}$$

Through the selection of appropriate values of the coefficients α_i and β_i and the order k, a number of well-known methods may be described. Note that if $\beta_0 = 0$, the resulting method is *explicit*, and if $\beta_0 \neq 0$, then the integration method is *implicit*.   For single step cases where $k = 1$, the selection of $\alpha_0 = 1$, $\alpha_1 = -1$, $\beta_0 = 0$, $\beta_1 = -1$ produces the explicit method of Eq. (3.7.9), while the choice $\alpha_0 = 1$, $\alpha_1 = -1$, $\beta_0 = -1$, $\beta_1 = 0$ corresponds to the implicit backward Euler scheme in (3.7.12).

Predictor-corrector methods are closely associated with LMS methods due to the commonality with the form in (3.7.23) for explicit and implicit integration methods.   As noted previously, the Adams–Bashforth–Moulton family of methods has been used in a finite element context by Gresho et al. [20]. An explicit two-step ($k = 2$), second-order form of (3.7.23) selects $\alpha_0 = 1$, $\alpha_1 = -1$, $\alpha_2 = 0$, $\beta_0 = 0$, $\beta_1 = -3/2$, $\beta_2 = 1/2$, which produces

$$\mathbf{y}^{n+1} = \mathbf{y}^n + \frac{3}{2}\Delta t \, \mathbf{f}(\mathbf{y}^n, t_n) - \frac{1}{2}\Delta t \, \mathbf{f}(\mathbf{y}^{n-1}, t_{n-1}) \tag{3.7.24}$$

Equation (3.7.24) is the constant time step form of (3.7.17) and would function as the predictor equation. The implicit, second-order companion to (3.7.24) is a single step method ($k = 1$) with coefficients $\alpha_0 = 1$, $\alpha_1 = -1$, $\beta_0 = -1/2$, $\beta_1 = -1/2$. This leads to

$$\mathbf{y}^{n+1} = \mathbf{y}^n + \frac{\Delta t}{2} \left\{ \mathbf{f}(\mathbf{y}^n, t_n) + \mathbf{f}(\mathbf{y}^{n+1}, t_{n+1}) \right\} \tag{3.7.25}$$

which is recognized as the trapezoidal rule. Rewriting (3.6.3) in the form of (3.7.22), we obtain

$$\dot{\mathbf{T}} = -\mathbf{M}^{-1}\mathbf{K}\mathbf{T} + \mathbf{M}^{-1}\mathbf{F} \qquad (3.7.26)$$

and substituting into the formulas in Eqs. (3.7.24) and (3.7.25) produce the predictor-corrector equations shown in (3.7.17) and (3.7.18).

Other first- and second-order methods can be derived from (3.7.23) as well as higher-order schemes. However, practical limits on function evaluations and stability considerations limit most LMS methods to those outlined here. The most significant aspect of the LMS formula is its use in the area of predictor-corrector procedures.

Convergence and stability

To conclude this section on time approximation methods, a few comments on convergence of the numerical solution to the actual solution are required. The time integration methods of the previous sections can all be written in symbolic form as

$$\mathbf{T}^{n+1} = -\mathbf{G}\,\mathbf{T}^n + \mathbf{H} \qquad (3.7.27)$$

where $\mathbf{G}$ represents the discretized form of the relevant differential operator. A time integration method is said to be *convergent* if, for fixed time t_n and time step Δt, the numerical solution $\mathbf{T}^n$ converges to its true value $\mathbf{T}(t_n)$ as $\Delta t \to 0$. To prove convergence of (3.7.27), the well-known Lax Equivalence Theorem [30] may be invoked, which allows the consistency and stability of $\mathbf{G}$ to be necessary and sufficient conditions for convergence. The numerical scheme (3.7.27) is said to be *consistent* with the original continuous problem if the errors (e.g., round-off and truncation errors) go to zero in the limit as $\Delta t \to 0$. For the diffusion problems considered here, consistency can be proved as shown in [22,26].

Since a time integration scheme is approximate, error is introduced into the solution $\mathbf{T}^{n+1}$ at each time step. The error in the solution at one time step feeds to the next time step due to the recursive use of Eq. (3.7.27). If the error is ultimately bounded, the time integration scheme is said to be *stable*. Thus, the stability of the time integration method in Eq. (3.7.27) implies that as $\mathbf{G}$ is recursively applied to each new vector $\mathbf{T}^n$, any errors that occur ultimately decay. That is, $\mathbf{G}$ is a bounded operator with a bounded spectral radius. For this discussion, the source of error is unimportant though we recognize that it may occur due to approximation of the differential equations and/or boundary conditions (truncation error) or due to finite precision computation (round-off errors).

There are a variety of techniques for assessing the stability of methods such as (3.7.27). For finite element applications, the spectral or modal methods are generally preferred. Here we will only sketch the procedure, without proof, and illustrate the stability conditions for the θ-family of integration methods.

The basic semidiscrete equation

$$\mathbf{M}\dot{\mathbf{T}} + \mathbf{K}\mathbf{T} = \mathbf{F} \qquad (3.7.28)$$

is difficult to analyze since it represents a large number of equations (degrees of freedom) that are coupled through the $\mathbf{M}$ and $\mathbf{K}$ matrices. A standard approach to this difficulty is to rewrite Eq. (3.7.28) in terms of its normal modes and thereby produce a system of uncoupled equations that are more easily studied. The generalized eigenvalue problem associated with (3.7.28) is (no sum on i)

$$\mathbf{K}\phi_i - \lambda_{(i)}\mathbf{M}\phi_i = 0 \tag{3.7.29}$$

where λ_i is the eigenvalue corresponding to the eigenvector ϕ_i. Assume that the eigenvectors are orthonormalized with respect to the capacitance matrix $\mathbf{M}$ such that

$$\phi_i^T\mathbf{M}\phi_j = \delta_{ij} \tag{3.7.30}$$

where δ_{ij} are the components of the unit tensor (Kronecker delta). Then multiplying Eq. (3.7.29) by ϕ_j^T and using Eq. (3.7.30) yield

$$\phi_j^T\mathbf{K}\phi_i - \lambda_{(i)}\phi_j^T\mathbf{M}\phi_i = 0 \tag{3.7.31}$$

or

$$\phi_j^T\mathbf{K}\phi_i - \lambda_{(i)}\delta_{ij} = 0 \tag{3.7.32}$$

which shows that the eigenvectors are also orthogonal with respect to the diffusion matrix $\mathbf{K}$.

Since the eigenvectors form a basis for the semidiscrete system in (3.7.28), the solution for $\mathbf{T}$ may be represented as a linear combination of the eigenvectors,

$$\mathbf{T}(\mathbf{x},t) = \sum_i \alpha_i(t)\phi_i(\mathbf{x}) \tag{3.7.33}$$

where α_i are the generalized coordinates. Substituting Eq. (3.7.33) into Eq. (3.7.28), premultiplying with ϕ_j^T, and using the orthogonal properties in Eqs. (3.7.30) and (3.7.32) lead to the result

$$\dot{\alpha}_i\left(\phi_j^T\mathbf{M}\phi_i\right) + \alpha_i\left(\phi_j^T\mathbf{K}\phi_i\right) = \phi_j^T\mathbf{F} \tag{3.7.34}$$

and

$$\dot{\alpha}_i\delta_{ij} + \alpha_i\lambda_{(i)}\delta_{ij} = f_j \tag{3.7.35}$$

or

$$\dot{\alpha}_j + \alpha_j\lambda_{(j)} = f_j \tag{3.7.36}$$

Note that Eq. (3.7.36) represents the decoupled modal equation for each degree of freedom (and eigenvalue) in the original system (3.7.28). If Eq. (3.7.36) can be integrated in time for each α_i, then the solution $\mathbf{T}$ can be constructed from the eigenvector expression in (3.7.33). This is the basis for mode superposition solution techniques [21,22].

The single degree of freedom equations in (3.7.36) serve another purpose and allow study of the stability properties of various time integration methods. To demonstrate this idea, the time derivative and difference approximations for the θ method, which are defined in Eq. (3.7.8), can be applied to Eq. (3.7.36) to give

$$\frac{\alpha_i^{n+1} - \alpha_i^n}{\Delta t} + \lambda_{(i)}(1-\theta)\alpha_i^n + \lambda_{(i)}\theta\alpha_i^{n+1} = f_i^a \qquad (3.7.37)$$

This can be rearranged as

$$\alpha_i^{n+1} = \frac{1 - \lambda_{(i)}\Delta t(1-\theta)}{1 + \lambda_{(i)}\Delta t\theta}\alpha_i^n + \frac{\Delta t f_i^a}{1 + \lambda_{(i)}\Delta t\theta} \qquad (3.7.38)$$

which is of the same form as given in (3.7.27) with

$$G_i = \frac{1 - \lambda_{(i)}\Delta t(1-\theta)}{1 + \lambda_{(i)}\Delta t\theta} \qquad (3.7.39)$$

As stated previously the stability of the integration method is proved by showing that the amplification matrix $\mathbf{G}$ is a bounded operator. For the single degree of freedom equations, the analogous requirement is that the amplification factor, G_i, be less than unity, i.e., $|G_i| < 1$. The restriction on G_i then is

$$-1 < \frac{1 - \lambda_i\Delta t(1-\theta)}{1 + \lambda_i\Delta t\theta} < 1 \qquad (3.7.40)$$

where θ must be between 0 and 1, and (3.7.40) must hold for all the modes, λ_i of the system. The right-hand inequality imposes no restrictions on the values $\lambda_i, \Delta t$, or θ; the left-hand inequality requires that $\lambda_i\Delta t < 2/(1-2\theta)$ when $\theta < 1/2$ but has no restrictions for $\theta \geq 1/2$. When there are no restrictions on the integration time step the method is termed *unconditionally stable* or simply *stable*. This analysis shows that all θ methods with $\theta \geq 1/2$ are unconditionally stable and these include the backward difference method ($\theta = 1$) in Eq. (3.7.12) and the Crank–Nicolson method ($\theta = 1/2$) in Eq. (3.7.16). The conditionally stable θ method is the Euler method ($\theta = 0$) in Eq. (3.7.10), where the time step is limited by $\lambda_i\Delta t_{crit} < 2$.

To ensure stability for all modes in the system, the time step must be limited by the largest eigenvalue in the system or $\Delta t_{crit} < 2/\lambda_{max}$. Recall that the λ_i are related to the system matrices $\mathbf{M}$ and $\mathbf{K}$; it can be shown that $\lambda_{max} \sim \alpha/h_{min}^2$, where $\alpha = k/\rho C$ is the thermal diffusivity (k is the thermal conductivity, ρ is the density, and C is the specific heat) and h_{min} is the characteristic dimension of the smallest finite element in the mesh. Thus for an explicit method the time step is limited by $\Delta t_{crit} < \bar{C}h_{min}^2$ (where $\bar{C}$ is a constant), which shows the strong influence of mesh size (refinement) on performance of the time integration method. Though the unconditionally stable methods ($\theta \geq 1/2$) have no stability restrictions on the time step, there are time step ranges which will produce a nonphysical oscillatory response in the solution. The oscillation limit corresponds to the point $\lambda_i\Delta t_{osc}$ where $G_i = 0$. As an example, for the Crank–Nicolson method ($\theta = 1/2$) the oscillation limit is $\lambda_i \Delta t_{osc} = 2$, and for $\Delta t > 2/\lambda_i$ the solution will exhibit an oscillatory decay. Oscillatory behavior can be filtered to produce acceptable solutions as shown by Wood and Lewis [27] and Wood [28].

3.8 Radiation Solution Algorithms

The enclosure radiation problem described in Section 1.10 is often required as part of a complete thermal analysis and must be solved in conjunction with the finite element heat conduction equations outlined above. An energy balance for each surface in the enclosure leads to the following system of equations [see Eqs. (1.10.3) and (1.10.4)]

$$\sum_{j=1}^{N} [\delta_{kj} - (1 - \epsilon_k)F_{k-j}] \, \bar{q}_j^0 = \epsilon_k \sigma \, \bar{T}_k^4 \qquad (3.8.1)$$

$$\bar{q}_k = \bar{q}_k^0 - \sum_{j=1}^{N} F_{k-j} \, \bar{q}_j^0 \qquad (3.8.2)$$

where $\bar{q}_j^0$ is the outgoing radiative flux for each surface, $\bar{q}_k$ is the net flux from each surface $\bar{q}_k = \bar{Q}_k/A_k$, $\bar{Q}_j$ is the net energy loss from each surface, δ_{kj} is the unit tensor, σ is the Stefan–Boltzmann constant, and F_{k-j} are radiation view (configuration) factors. The overbar denotes a quantity that is uniform over each surface. When the surface temperatures of all surfaces are known, Eq. (3.8.1) forms a set of linear algebraic equations for the unknown, outgoing surface fluxes, $\bar{q}_j^0$. That is, Eq. (3.8.1) can be written as the matrix problem

$$\mathbf{A}(\bar{\mathbf{T}})\bar{\mathbf{q}}^0 = \mathbf{B}\bar{\mathbf{T}}^4 = \mathbf{F}(\bar{\mathbf{T}}) \qquad (3.8.3)$$

and (3.8.2) rewritten as
$$\bar{\mathbf{q}} = (\mathbf{I} - \mathbf{F}_{vf})\bar{\mathbf{q}}^0 \qquad (3.8.4)$$

where $\mathbf{A}$ is a function of $\bar{\mathbf{T}}$ due to the possible dependence of surface emissivities on temperature. The matrix $\mathbf{A}$ is a full matrix due to the surface to surface coupling represented by the view factors $F_{k-j} = \mathbf{F}_{vf}$. This characteristic, along with the possible temperature dependencies, suggests the use of an iterative solution method for the matrix problem in (3.8.3) rather than a direct matrix factorization.

The relationship between the discrete surface temperature $\bar{T}_k$ and flux variables $\bar{q}_k$ in Eqs. (3.8.1)–(3.8.3) and the finite element temperature and flux variables remains to be determined. An obvious assumption is for the faces (edges) of the finite elements bounding the enclosure to be used as the description of the enclosure surface. In this case the (constant) surface flux obtained from Eq. (3.8.3) can be used directly as applied boundary conditions in the finite element equations [see Eq. (3.5.6)]. Likewise, the constant surface temperature required in (3.8.1)–(3.8.3) can be obtained by appropriate evaluation of the finite element temperature distribution (shape functions) at the centroid of the element face. That is, $\bar{T} = \left(\hat{\boldsymbol{\Psi}}(\xi_0, \eta_0)\right)^T \mathbf{T}$ where (ξ_0, η_0) are the local element coordinates for the centroid of the element surface. By defining common variables for the two equation sets (3.2.8) and (3.8.3), the physical processes can be coupled and a solution strategy defined. Note that the use of element faces as an enclosure description is not the only method of coupling the equations. Another possibility is to use the nodes of the finite element model as the centroid of a control area made up of surface contributions from all the elements attached to the node. Nodal point temperatures are then common between the conduction and radiation problem, though boundary surface fluxes now require interpolation. This construction is closer to the finite volume method, and it leads to a slight increase in complexity when defining data for the view factor computation. In the following, attention is restricted to the case of element faces forming the description of the enclosure.

Having defined the common variables for the two equations a solution algorithm for the coupled problem is required. Because of the strong nonlinearity in the conduction/radiation problem (i.e., $\bar{T}^4$ boundary condition), an implicit algorithm that treats $\bar{q}$ and $\mathbf{T}$ simultaneously is the most desirable from a convergence and performance point of view. The most obvious approach for the two equation sets in two unknowns is the direct substitution of the radiative flux from (3.8.3) into the heat conduction boundary condition. Premultiplying both sides of Eq. (3.8.3) with $\mathbf{A}^{-1}$, we obtain

$$\bar{q}^0 = \mathbf{A}^{-1}\mathbf{B}\bar{\mathbf{T}}^4 \tag{3.8.5}$$

and using Eq. (3.8.4)

$$\bar{q} = (\mathbf{I} - \mathbf{F}_{vf})\mathbf{A}^{-1}\mathbf{B}\bar{\mathbf{T}}^4 \tag{3.8.6}$$

Substituting (3.8.6) into (3.5.6) as an applied flux, then the conduction Eq. (3.6.3) becomes

$$\mathbf{M}\dot{\mathbf{T}} + \hat{\mathbf{K}}\mathbf{T} = \hat{\mathbf{F}} - q_e \tag{3.8.7}$$

where the new flux term is given by

$$q_e = \oint_{\Gamma_e} \hat{\boldsymbol{\Psi}}\bar{q} \, ds = \oint_{\Gamma_e} \hat{\boldsymbol{\Psi}}(\mathbf{I} - \mathbf{F}_{vf})\mathbf{A}^{-1}\mathbf{B}\bar{\mathbf{T}}^4 \tag{3.8.8}$$

Equation (3.8.7) is a nonlinear equation in terms of the nodal point temperatures only, since $\bar{\mathbf{T}}$ is related to the surface nodes in the enclosure. This system could be solved by any of the iterative or time integration methods described in the previous section. However, there are some severe drawbacks to this algorithm. The inverse of $\mathbf{A}$ in Eq. (3.8.5) is nontrivial for any enclosure with more than a moderate number of surfaces. Recall that $\mathbf{A}$ is a full matrix as is its inverse. Also, if $\mathbf{A}$ is a function of temperature then the inverse would be required multiple times, corresponding to each iteration or each time step in the conduction solution. The matrix-matrix and matrix-vector multiplications in (3.8.8) are also a significant computational burden for the generation of the flux boundary condition. Note that Eq. (3.8.7) could be made more implicit by rewriting Eq. (3.8.8) as a standard radiation boundary condition [see Eq. (3.5.16)] where the $\mathbf{R}$ matrix contains a coefficient h_r based on $\overline{T}^3$. This does not alleviate any of the disadvantages of the algorithm. The basic method remains suitable only for enclosures with a small number of surfaces. A demonstration of this type of algorithm and details of its implementation can be found in [31] and [32].

Another implicit method leaves Eq. (3.6.3) and Eq. (3.8.3) as distinct equation sets but solves them simultaneously, as a fully coupled system. Considering the time-independent case for simplicity, Eqs. (3.6.3) and (3.8.3) can be written as a combined matrix problem as

$$\begin{bmatrix} \hat{\mathbf{K}} & \mathbf{E} \\ \mathbf{B}^*\overline{\mathbf{T}}^3 & \mathbf{A} \end{bmatrix} \begin{Bmatrix} \mathbf{T} \\ \hat{\mathbf{q}} \end{Bmatrix} = \begin{Bmatrix} \hat{\mathbf{F}} \\ \mathbf{0} \end{Bmatrix} \tag{3.8.9}$$

where $\mathbf{E}$ contains the boundary contributions from the enclosure flux and $\hat{\mathbf{F}}$ only retains contributions from other boundary conditions and source terms; $\mathbf{B}^*$ indicates that $\mathbf{B}$ is modified when $\overline{\mathbf{T}}$ is rewritten in terms of the nodal point temperatures. The nonlinear algebraic system in (3.8.9) can be solved with any of the iterative methods described previously, though the best algorithm is a Newton (or Newton-Raphson) scheme that will be discussed below. However, first it is appropriate to list the advantages and disadvantages of this formulation, especially as compared to the scheme in (3.8.7). The disadvantages are that the algebraic system in (3.8.9) is larger than the previous method by the number of surfaces (M) in the enclosure; M can easily be on the order of the number of nodes in the problem. Also, the full matrix $\mathbf{A}$ is part of the overall matrix problem with its attendant memory requirements and increased operation count for solution. An advantage of the algorithm is the strong coupling between the radiation and conduction processes which leads to an improved rate of convergence. Avoiding the matrix inversion of $\mathbf{A}$ is the second and most important advantage. Not only is the solution of $\mathbf{A}$ less computational work than an inversion, but the case of a temperature dependent emissivity is easily accommodated.

A Picard iteration method, such as (3.7.2), converges linearly, at best. When applied to a large or very large system like (3.8.9), successive substitution would not be a cost effective method. A second-order method, like Newton's method, that converges quadratically is a more appropriate choice for highly nonlinear systems such as (3.8.9). Rewriting (3.8.9) in terms of residuals

$$\begin{Bmatrix} \mathbf{R}_T \\ \mathbf{R}_q \end{Bmatrix} = \begin{bmatrix} \hat{\mathbf{K}} & \mathbf{E} \\ \mathbf{B}^*\overline{\mathbf{T}}^3 & \mathbf{A} \end{bmatrix} \begin{Bmatrix} \mathbf{T} \\ \bar{\mathbf{q}} \end{Bmatrix} - \begin{Bmatrix} \hat{\mathbf{F}} \\ \mathbf{0} \end{Bmatrix} \tag{3.8.10}$$

Newton's method applied to (3.8.10) produces

$$\begin{bmatrix} \mathbf{J}_{TT} & \mathbf{J}_{Tq} \\ \mathbf{J}_{qT} & \mathbf{J}_{qq} \end{bmatrix}^n \begin{Bmatrix} \Delta\mathbf{T} \\ \Delta\bar{\mathbf{q}} \end{Bmatrix}_{n+1} = \begin{bmatrix} \frac{\partial\mathbf{R}_T}{\partial\mathbf{T}} & \frac{\partial\mathbf{R}_T}{\partial\bar{\mathbf{q}}} \\ \frac{\partial\mathbf{R}_q}{\partial\mathbf{T}} & \frac{\partial\mathbf{R}_q}{\partial\bar{\mathbf{q}}} \end{bmatrix}^n \begin{Bmatrix} \Delta\mathbf{T} \\ \Delta\bar{\mathbf{q}} \end{Bmatrix}^{n+1} = -\begin{Bmatrix} \mathbf{R}_T \\ \mathbf{R}_q \end{Bmatrix}^n \tag{3.8.11}$$

where the unknowns are changes in the temperature and uniform radiation flux between the $n+1$ and n iterations. The matrix problem in Eq. (3.8.11) is unsymmetric but can still be solved using an iterative matrix solver (conjugate gradient) with a preconditioner. Further details of this algorithm may be found in Hogan and Gartling [33].

The "fully coupled" algorithms listed above are still not generally applicable to large, geometrically complex, three-dimensional problems because of the significant memory and/or computational burden of the formulation. In order to reduce the computational requirements a decoupled procedure must be considered. The simplest method is a staggered or cyclic iteration between the two equation sets. Typically, the finite element conduction solution is advanced by a time step or iteration, at which point a new set of surface temperatures is available. After reducing the finite element surface temperature distribution to a set of constant surface temperatures, Eq. (3.8.3) can be solved for the surface flux and (3.8.4) used to produce the effective flux to the surface. The surface flux provides boundary conditions to the next finite element solution cycle. Alterations to this basic scheme might include multiple correction steps with either fixed or updated radiative flux or the fixing of the radiative flux for several steps if the temperature changes are small.

The cyclic solution procedure is a standard approach for this type of problem though it is not without its drawbacks. When the simulation is time dependent this type of algorithm is usually very successful, since the time stepping procedure usually produces only "small" changes in the temperature field within each time step. Incremental changes in surface temperature lead to incremental changes in surface flux, and rapid convergence of both the radiation and conduction solutions, i.e., the two equation sets are always "close" to equilibrium. Adaptive time step, predictor-corrector techniques can be used to advantage with this decoupled approach with the radiation solution being evaluated after the predictor step.

For steady solutions that are approached iteratively, the decoupled algorithm tends to be less successful. Large changes in the temperature field between iterations are typical and can easily be amplified by large variations in surface flux due to the strong (fourth power) dependence on surface temperature. Very slow convergence or more often, divergence, of the algorithm is observed. Severe underrelaxation of the solution vectors [see Eqs. (3.7.4a,b)] may improve the solution process as will incrementation of the boundary conditions and/or material properties. In many cases, the most cost-effective method for steady, enclosure radiation problems is the decoupled method with a time-dependent approach (false transient) to steady state.

The last part of the radiation solution that requires comment is the view factor matrix, $\mathbf{F_{vf}}$. View factors for use in enclosure radiation computations were defined in Section 1.10 as purely geometric factors. The computation of such factors is usually performed by specialized programs (e.g., see [34,35,36]) that are external to the basic finite element program. For plane, two-dimensional geometries with no obstructing (shadowing) surfaces, the view factor computation can be done with simplified analytic methods such as Hottel's cross-string method [37]. More complex geometries in two and three dimensions require the use of numerical procedures, such as double area integration, hemicube, or Monte Carlo techniques. A comparative study of a number of view factor algorithms is presented in [38].

3.9 Variable Properties

In all of the previous discussion we have mentioned the possible occurrence of variable material properties, boundary conditions, and heat sources. The previous sections on solution methods showed how such temperature dependence produced nonlinearities in the matrix problem and how various solution algorithms were set up to account for this behavior. In the present section three other aspects of variable properties will be discussed - methods for the implementation of variable coefficients into the finite element matrix formulation, the treatment of latent heat effects during phase change, and the use of anisotropic conductivities.

3.9.1 Temperature-Dependent Properties

Consider first the problem of evaluating finite element matrices that contain a variable coefficient. To focus the discussion, a typical diffusion term from the conduction equation will be examined. From Eq. (3.2.6) the xx-component of the diffusion matrix is

$$\mathbf{K}_{xx} = \int_{\Omega^e} \frac{\partial \mathbf{\Psi}}{\partial x} k \frac{\partial \mathbf{\Psi}^T}{\partial x} \, d\mathbf{x} \qquad (3.9.1)$$

where for simplicity an isotropic conductivity has been assumed. As seen in Section 3.4, it is typical to evaluate element integrals, such as the one in Eq. (3.9.1), by use of numerical quadrature. When the conductivity is constant the matrix in (3.9.1) is independent of temperature and may be constructed once and stored as a two-dimensional array for subsequent use in a solution procedure. However, when the conductivity is variable, the matrix in (3.9.1) must be reevaluated each time new values of the temperature are produced during the solution procedure. One standard method for treating this variable coefficient simply involves recomputing the matrix for each new set of conductivities. For each element the temperature is evaluated at the quadrature points via the basis functions, thus permitting the conductivity to be evaluated at the quadrature locations. The matrix coefficients are then computed by the standard quadrature procedure. The work involved in such a procedure can become large when considering anisotropic conductivities (three conductivity matrices in two dimensions, six conductivity matrices in three dimensions) and/or higher-order elements (higher-order quadrature schemes).

A second approach to this problem makes use of an interpolation function for the variable coefficient. Let the conductivity be approximated by

$$k = \mathbf{\Psi}^T \mathbf{k} \qquad (3.9.2)$$

where $\mathbf{\Psi}$ is an appropriate set of element basis functions and $\mathbf{k}$ is nodal point values of the conductivity. Inserting (3.9.2) into (3.9.1) gives

$$\mathbf{K}_{xx} = \int_{\Omega_e} \frac{\partial \mathbf{\Psi}}{\partial x} (\mathbf{\Psi}^T \mathbf{k}) \frac{\partial \mathbf{\Psi}^T}{\partial x} \, d\mathbf{x} \qquad (3.9.3)$$

The above integral may be evaluated (via quadrature) once and stored as a triple subscripted array (hypermatrix). During the solution process, the conductivity is evaluated at each node as a function of the nodal point temperature, and a simple inner product in (3.9.3) produces the needed two-dimensional diffusion matrix. Such a procedure eliminates the need for recomputing element-level matrices, albeit at the expense of some additional storage and I/O operations.

Though we are concerned in the present example with the treatment of variable material properties it is important to realize that both of the above procedures can be used with any type of variable coefficient. In the heat conduction problem variable heat capacities, convective and radiative heat transfer coefficients, and volumetric heat sources are all candidates for either of the described methods.

3.9.2 Phase Change Properties

One particularly difficult type of property variation is the latent heat effect that occurs in the simulation of phase change problems. As described in Section 1.9, latent heat effects can be adequately treated by use of an enthalpy method with a temperature-dependent specific heat. That is, from Eq. (1.9.6) we have

$$C^*(T) = C(T) + L \, \delta(T - T_t) \qquad (3.9.4)$$

where C^* is the effective specific heat, C is the normal specific heat, L is the latent heat, δ is the Dirac delta function, and T_t is the transition temperature. For application in a finite element procedure, Eq. (3.9.4) is replaced by

$$C^*(T) = C(T) + L \, \delta^*(T - T_t, \Delta T) \qquad (3.9.5)$$

where δ^* is the delta form function; δ^* has a large but finite value in the interval centered about T_t and is zero outside the interval (see Figure 3.9.1).

The interval ΔT is often referred to as the "mushy" zone and corresponds to the difference between the liquidus, T_l, and solidus, T_s, temperatures for the material. Note that Eq. (3.9.5) is thus an approximation for the behavior of pure materials that change phase at a specified temperature, T_t, but is accurate for non-pure substances that have truly distinct liquidus and solidus temperatures.

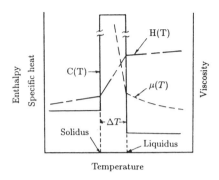

Figure 3.9.1. Definition of material properties for phase change computation.

Though the effective capacitance model described above is useful in accounting for latent heat effects, a few words of caution should be issued with regard to the time integration of this type of phase change model. Since the transition temperature interval, ΔT, is often small compared to the overall temperature variation in a conduction problem it is important to note that there are some severe practical limitations on the time integration procedure. In general, the time-stepping algorithm must be controlled such that every node that "changes phase" is forced to attain a temperature value in the interval bracketed by ΔT. If a nodal point does not "land" in the ΔT range but simply steps over this temperature interval, the latent heat effect is lost for that node and an incorrect temperature response and energy balance will result. Various methods for dealing with this difficulty have been proposed (see [39,40]) including broadening the ΔT range, placing a limit on the maximum temperature change that can occur during a time step, and using special forms of a predictor-corrector algorithm. The problem is a numerically delicate one and is still an area of active research.

A number of useful alternatives within the effective capacitance method make direct use of the enthalpy, H, versus temperature curve and differ only in how they evaluate the slope of H versus T. In one method, H is first computed at the nodes of an element (knowing T); the heat capacity at an integration point within the element is then found from

$$C_p = \left[\frac{\nabla H \cdot \nabla H}{\nabla T \cdot \nabla T} \right]^{1/2} \tag{3.9.6}$$

where ∇H and ∇T are evaluated via the element shape functions. This technique will maintain its accuracy as long as the phase boundary passes through each element and does not skip over an element. This is certainly a more robust technique than the standard effective capacitance method.

In Section 1.9 the variable heat source method for phase change problems was also outlined. The time- and temperature-dependent heat source representing the latent heat was given as

$$Q_{lh} = \rho L \delta (T - T_t) \frac{\partial T}{\partial t} \tag{3.9.7}$$

which can be rewritten using some of the previous definitions as

$$Q_{lh} = \rho L \; \delta^* (T - T_t, \Delta T) = \rho C_{lh} \frac{\partial T}{\partial t} \tag{3.9.8}$$

The effective capacitance C_{lh} is evaluated by the same methods as described above and is subject to many of the same time integration difficulties as the previous algorithm. By treating the latent heat as a source term, the overall energy balance can be more easily maintained. It is straightforward to determine which nodes pass through the ΔT interval as a result of the time step and then construct the correct source term for the subsequent time step. The integration process would typically lag the evaluation of the source term anyhow since the temperature rate is difficult to predict. Though the energy balance and individual nodal temperature response are easier to control, the temporal accuracy of the method may be limited.

3.9.3 Anisotropic Properties

In closing this section, we mention briefly the treatment of anisotropic thermal conductivities. In two dimensions, the most general physically realistic, material model is assumed to have an orthotropic conductivity, i.e., the components $k_{11} \neq k_{22}$ and $k_{12} = k_{21} = 0$ when referred to the principal material axes (x_1, x_2, x_3). Since the global finite element coordinate frame will generally not be aligned with the material axes, a coordinate transformation is required. Referring to the sketch in Figure 3.9.2 the global components of the conductivity matrix, being the components of a second-order tensor, can be obtained from (see Reddy [50], pp. 108–112)

$$\begin{bmatrix} k_{xx} & k_{xy} \\ k_{yx} & k_{yy} \end{bmatrix} = \begin{bmatrix} \cos\theta & \sin\theta \\ -\sin\theta & \cos\theta \end{bmatrix} \begin{bmatrix} k_{11} & 0 \\ 0 & k_{22} \end{bmatrix} \begin{bmatrix} \cos\theta & -\sin\theta \\ \sin\theta & \cos\theta \end{bmatrix} \tag{3.9.9a}$$

or when multiplied out $(k_{xy} = k_{yx})$

$$[k] = \begin{bmatrix} k_{11}\cos^2\theta + k_{22}\sin^2\theta & -k_{11}\sin\theta\cos\theta + k_{22}\sin\theta\cos\theta \\ -k_{11}\sin\theta\cos\theta + k_{22}\sin\theta\cos\theta & k_{11}\sin^2\theta + k_{22}\cos^2\theta \end{bmatrix} \tag{3.9.9b}$$

These last expressions give the needed conductivity components for use in the finite element diffusion matrix $\mathbf{K}$ defined in Eq. (3.2.5a). In the most general two-dimensional case, three matrices are defined by $\mathbf{K}$ for the $xx, yy,$ and xy components of the conductivity tensor as given in Eq. (3.9.9a). It is of course possible for the principal material conductivity components, k_{11} and k_{22}, to be independent functions of temperature, and for the local orientation angle, θ, to vary from one material to another and from element to element in the finite element mesh.

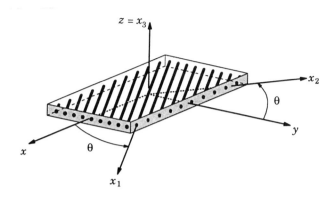

Figure 3.9.2. Notation for anisotropic conductivity.

The three-dimensional transformation for an orthotropic material has the same form but is slightly more complex. In tensor notation, the conductivity components in the global coordinate system are obtained from

$$k_{ij} = C_{in}k^p_{nm}C_{jm} \tag{3.9.10}$$

where k^p_{nm} is the conductivity tensor in the material coordinate frame and C_{in} are the direction cosines between the axes of the two coordinate systems. For an orthotropic material, k^p_{nm} is diagonal and (3.9.10) can be written as

$$[k] = \begin{bmatrix} C_{11} & C_{21} & C_{31} \\ C_{12} & C_{22} & C_{32} \\ C_{13} & C_{23} & C_{33} \end{bmatrix} \begin{bmatrix} k_{11} & 0 & 0 \\ 0 & k_{22} & 0 \\ 0 & 0 & k_{33} \end{bmatrix} \begin{bmatrix} C_{11} & C_{12} & C_{13} \\ C_{21} & C_{22} & C_{23} \\ C_{31} & C_{32} & C_{33} \end{bmatrix} = \mathbf{C}^T \mathbf{k} \mathbf{C} \tag{3.9.11}$$

which has the same form as the two-dimensional case in (3.9.9). Note that the three-dimensional form of k_{ij} now requires six matrices to define the diffusion operator $\mathbf{K}$; the principal conductivities may again be temperature dependent and their orientation may vary over the domain.

3.10 Post-Processing Operations

3.10.1 Heat Flux

An important post-processing operation in many heat conduction problems involves the computation of the heat flux from various surfaces or bodies. The diffusion flux associated with the conduction equation can be computed on an element-by-element basis. In the following only the heat flux associated with two-dimensional problems will be specifically considered since the flux in the three-dimensional case is completely analogous. Fourier's law provides the definition of the conductive heat flux as

$$q_x = -k_{xx}\frac{\partial T}{\partial x} - k_{xy}\frac{\partial T}{\partial y}, \quad q_y = -k_{yx}\frac{\partial T}{\partial x} - k_{yy}\frac{\partial T}{\partial y} \tag{3.10.1}$$

The flux components in Eq. (3.10.1) are computed by using the standard finite element approximations for T

$$T(x_i, t) = \mathbf{\Psi}^T(x_i)\mathbf{T}(t) \tag{3.10.2}$$

and the components of the conductivity tensor, $\mathbf{k}$. Also needed are the relations for the local temperature derivatives in terms of the local element coordinates and the isoparametric description of the element geometry. That is,

$$\left\{ \begin{array}{c} \frac{\partial \Psi}{\partial x} \\ \frac{\partial \Psi}{\partial y} \end{array} \right\} = \left[\begin{array}{cc} J_{11}^* & J_{12}^* \\ J_{21}^* & J_{22}^* \end{array} \right] \left\{ \begin{array}{c} \frac{\partial \Psi}{\partial \xi} \\ \frac{\partial \Psi}{\partial \eta} \end{array} \right\} \tag{3.10.3a}$$

where $J_{ij}^* = \hat{J}_{ij}^*/|J|$, and

$$\hat{J}_{11}^* = \frac{\partial y}{\partial \eta} , \quad \hat{J}_{12}^* = -\frac{\partial y}{\partial \xi} , \quad \hat{J}_{21}^* = -\frac{\partial x}{\partial \eta} , \quad \hat{J}_{22}^* = \frac{\partial x}{\partial \xi} \tag{3.10.3b}$$

Note that (x, y) are functions of the local (element) coordinates, (ξ, η) through the parametric mapping (3.3.3) (see Sections 3.4 and 3.5).

Using these definitions, the heat flux components can be expressed as

$$q_x = - k_{xx}\left(J^*{}_{11}\frac{\partial \mathbf{\Psi}^T}{\partial \xi}\mathbf{T} + J^*{}_{12}\frac{\partial \mathbf{\Psi}^T}{\partial \eta}\mathbf{T} \right)$$
$$- k_{xy}\left(J^*{}_{21}\frac{\partial \mathbf{\Psi}^T}{\partial \xi}\mathbf{T} + J^*{}_{22}\frac{\partial \mathbf{\Psi}^T}{\partial \eta}\mathbf{T} \right) \tag{3.10.4a}$$

$$q_y = - k_{yx}\left(J^*{}_{11}\frac{\partial \mathbf{\Psi}^T}{\partial \xi}\mathbf{T} + J^*{}_{12}\frac{\partial \mathbf{\Psi}^T}{\partial \eta}\mathbf{T} \right)$$
$$- k_{yy}\left(J^*{}_{21}\frac{\partial \mathbf{\Psi}^T}{\partial \xi}\mathbf{T} + J^*{}_{22}\frac{\partial \mathbf{\Psi}^T}{\partial \eta}\mathbf{T} \right) \tag{3.10.4b}$$

The definitions in (3.10.4a,b) are sufficient to define the flux components at any point (ξ_0, η_0) in the element; for optimal accuracy, the 2×2 Gauss points within the element are usually selected for the evaluation of derivatives. Extrapolation and averaging of flux values at the nodes are usually employed if a continuous flux field is required.

In addition to the local components of the flux vector, the heat flux normal to the element edge is often of importance. By definition

$$q_n = \mathbf{q} \cdot \hat{\mathbf{n}} \tag{3.10.5a}$$
$$\mathbf{q} = q_x\hat{\mathbf{e}}_x + q_y\hat{\mathbf{e}}_y \tag{3.10.5b}$$
$$\hat{\mathbf{n}} = n_x\hat{\mathbf{e}}_x + n_y\hat{\mathbf{e}}_y \tag{3.10.5c}$$

and thus

$$q_n = q_x n_x + q_y n_y \tag{3.10.6}$$

In order to employ Eq. (3.10.6), the components of the normal vector are also required; these may also be computed from the element geometry and isoparametric mapping

$$n_x = \frac{1}{\Delta}\frac{\partial y}{\partial s} \quad ; \quad n_y = -\frac{1}{\Delta}\frac{\partial x}{\partial s} \tag{3.10.7}$$

where the Jacobian of the transformation is

$$|J_s| = \Delta = \left[\left(\frac{\partial x}{\partial \xi}\right)^2 + \left(\frac{\partial y}{\partial \xi}\right)^2\right]^{\frac{1}{2}} \tag{3.10.8}$$

and ξ is the local coordinate along the edge of the element.

Relations analogous to the above may be derived for the three-dimensional case where the flux normal to an element surface is to be evaluated. The flux component definitions extend naturally with the inclusion of the z component. The flux normal to an element surface is defined by

$$\hat{\mathbf{n}} = \frac{\mathbf{e}_1 \times \mathbf{e}_2}{|J_s|} \tag{3.10.9}$$

where $|J_s| = |\mathbf{e}_1 \times \mathbf{e}_2|$ and the surface vectors $\mathbf{e}_i$ were defined in terms of the surface shape functions in Eq. (3.5.4).

3.10.2 Heat Flow Function

For two-dimensional problems, Kimura and Bejan [41] have proposed the use of a heat flow function to assist in the visualization of energy transport. The heat function is directly analogous to the stream function for incompressible fluid flow and is constructed to satisfy the steady source free form of the energy equation. In formal terms, the heat function $\mathcal{H}$ is the remaining nonzero component of a vector potential that identically satisfies a form of Eq. (3.1.1). Though the heat flow function can be used to describe both advective and diffusive processes, it will be developed here only for the conduction equation. By definition

$$q_x = -k\frac{\partial T}{\partial x} = \frac{\partial \mathcal{H}}{\partial y} \quad , \quad q_y = -k\frac{\partial T}{\partial y} = -\frac{\partial \mathcal{H}}{\partial x} \tag{3.10.10}$$

For simplicity the definitions in (3.10.10) have also assumed an isotropic conductivity though this is not a required restriction. By limiting attention to the heat conduction process, the heat function reduces to the definition for a heat flux line, i.e., a line that is everywhere tangent to the local flux vector. The change in the heat function is an exact differential such that

$$\delta\mathcal{H} = \int_A^B \mathbf{q} \cdot \mathbf{n}\, d\Gamma \tag{3.10.11}$$

where $\mathbf{n}$ is the normal to the integration path $d\Gamma$ and $\mathbf{q}$ is the flux vector along the paths which are defined in (3.10.5c) and (3.10.5b), respectively.

The computation of the change in the heat function within a finite element can be accomplished using Eq. (3.10.11) once a suitable integration path AB has been identified. An obvious choice for the integration path is along the two-dimensional element boundaries. Defining an edge interpolation for the flux components

$$q_x = \hat{\boldsymbol{\Psi}}^T \mathbf{q}_x \quad ; \quad q_y = \hat{\boldsymbol{\Psi}}^T \mathbf{q}_y \tag{3.10.12}$$

The relation for the elemental line segment is

$$d\Gamma = \Delta d\xi \tag{3.10.13}$$

where Δ is defined by the parametric mapping of the element edge given in Eq. (3.10.8). The incremental change in the heat function along an element edge can then be formulated as

$$\delta \mathcal{H} = \int_{-1}^{+1} \left[\left(\frac{\partial \hat{\boldsymbol{\Psi}}}{\partial s}^T \mathbf{y} \right) \hat{\boldsymbol{\Psi}}^T \mathbf{q}_x - \left(\frac{\partial \hat{\boldsymbol{\Psi}}}{\partial s}^T \mathbf{x} \right) \hat{\boldsymbol{\Psi}}^T \mathbf{q}_y \right] d\xi \tag{3.10.14}$$

The change in the heat function along any element boundary can be computed from Eq. (3.10.14) once the element geometry and temperature fields are specified; the flux needed in (3.10.14) is derived from the definitions in the previous section. Computation of the heat function field for an entire finite element mesh is generated by applying (3.10.14) along successive element boundaries, starting at a node for which a base value of $\mathcal{H}$ has been specified. Observe that by applying (3.10.14) to all the edges of a single element and summing the increments in $\mathcal{H}$, an estimate of the energy balance for the element can be obtained.

The calculation of the heat function for axisymmetric geometries follows a similar procedure with the appropriate definition for $\mathcal{H}$ being

$$q_r = \frac{\partial \mathcal{H}}{\partial z} \quad ; \quad q_z = -\frac{\partial \mathcal{H}}{\partial r} \tag{3.10.15}$$

and

$$\mathbf{q} = q_r \, \hat{\mathbf{e}}_r + q_z \, \hat{\mathbf{e}}_z, \quad \hat{\mathbf{n}} = n_r \, \hat{\mathbf{e}}_r + n_z \, \hat{\mathbf{e}}_z \tag{3.10.16}$$

3.11 Advanced Topics in Conduction

3.11.1 Introduction

The methods and algorithms described in the previous sections will allow the vast majority of heat conduction problems to solved accurately and efficiently. However, as finite element methods are used on problems of increasing complexity, more specialized procedures are sometimes required. In the following sections a few such techniques are described which allow the standard conduction problem to be extended to more complex geometries or more involved physical modeling.

3.11.2 Specialty Elements

The two- and three-dimensional elements described in Section 3.3 are the workhorse elements for describing and solving heat conduction problems in solid bodies or continuum regions. In many applications, special geometric features appear that are not well described by a full continuum element. Two examples of these specialty elements are bars (beams, trusses, wires, cables, etc.) and shells (membranes, panels, fins, etc.) in both two and three dimensions.

The standard three-dimensional bar element may be either a two-node or three-node, isoparametric element as shown in Figure 3.11.1. This element has a variable cross-sectional area with conduction only allowed along the axis of the element. Note that the shape of the cross-section need not be explicitly defined here, though for purposes of boundary condition application, some convention must be established. The simplest assumption is a circular cross-section with flux type boundary conditions then being applied uniformly around the circumference or uniformly over the ends of the bar. Because the surface of the bar is generally not described as a faceted surface, the computation of radiation view factors is generally not possible for this element; inclusion in the enclosure radiation problem is thus prohibited.

The shape function for the two-node element is defined by

$$\{\Psi^e\} = \frac{1}{2} \left\{ \begin{array}{c} (1-\xi) \\ (1+\xi) \end{array} \right\} \tag{3.11.1}$$

and the three-node element is described by

$$\{\Psi^e\} = \frac{1}{2} \left\{ \begin{array}{c} (1-\xi)\xi \\ 2(1-\xi^2) \\ (1+\xi)\xi \end{array} \right\} \tag{3.11.2}$$

The functions in Eqs. (3.11.1) and (3.11.2) are ordered as shown in the figure and are written in terms of the normalized coordinate ξ that varies from -1 to $+1$. The parametric mapping given in Eq. (3.3.3) still holds and relates the global coordinates (x, y, z) for the element to the local coordinate, ξ.

Bar elements for two-dimensional problems can also be defined by the shape functions in Eqs. (3.11.1) and (3.11.2). In this case, the isoparametric mapping is carried out from the (x, y) coordinates to the local coordinate ξ. The variable, cross-sectional area for the two-dimensional case reduces to a variable thickness with unit depth. The axisymmetric, two-dimensional bar can be treated in a similar manner, though it is rotated about the z axis. In both cases, these "bar" elements should be thought of as a one-dimensional conduction element in the plane of the problem. These elements are essentially two-dimensional shell elements and may be used as such. Unlike the three-dimensional bar, these two-dimensional elements can be included in a view factor computation since they have "sides" just like the continuum elements. Boundary condition application is straightforward for these element types.

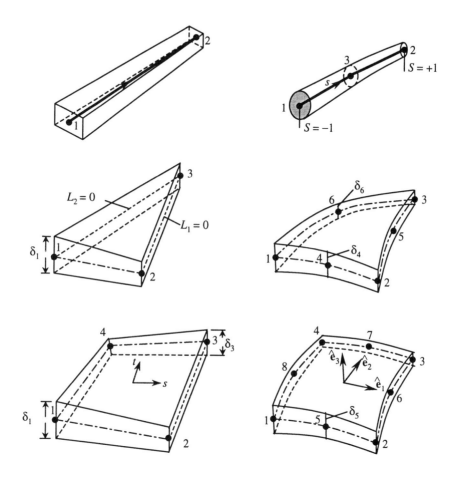

Figure 3.11.1. Three-dimensional bar and shell elements.

There are a number of different types of shell elements. The main differences between shell definitions are in the shape and the temperature approximation through the thickness of the shell. For strictly thermal applications, a standard assumption is that the shell has no temperature variation through the thickness; conduction is only allowed in the plane of the element. For thermal-stress applications it is desirable to have a good representation for the thermal gradient through the thickness of the shell and this requires a higher-order temperature approximation normal to the plane of the element. These higher-order temperature representations are generally difficult to use with continuum elements because of the need to constrain or tie the multiple temperature nodes that occur through the shell thickness to a single node in the adjacent continuum element.

Three-dimensional shell elements with both triangular and quadrilateral plane forms are in common use. Four typical element types are shown in Figure 3.11.1. Each element is assumed to have a constant temperature through its thickness. All elements usually allow the shell thickness to vary across the surface. The temperature shape function for the three-node, triangular element is defined by

$$\{\Psi^e\} = \left\{ \begin{array}{c} L_1 \\ L_2 \\ L_3 \end{array} \right\} \tag{3.11.3}$$

and the six-node, triangular shell element has the following shape functions

$$\{\Psi^e\} = \left\{ \begin{array}{c} L_1(2L_1 - 1) \\ L_2(2L_2 - 1) \\ L_3(2L_3 - 1) \\ 4L_1L_2 \\ 4L_2L_3 \\ 4L_3L_1 \end{array} \right\} \tag{3.11.4}$$

where the L_i are the standard, in-plane area coordinates that vary from 0 to $+1$. The four-node, quadrilateral shell element has shape functions of the form

$$\{\Psi^e\} = \frac{1}{4} \left\{ \begin{array}{c} (1 - \xi)(1 - \eta) \\ (1 + \xi)(1 - \eta) \\ (1 + \xi)(1 + \eta) \\ (1 - \xi)(1 + \eta) \end{array} \right\} \tag{3.11.5}$$

while the eight-node, "serendipity" shell element is defined by

$$\{\Psi^e\} = \frac{1}{4} \left\{ \begin{array}{c} (1 - \xi)(1 - \eta)(-\xi - \eta - 1) \\ (1 + \xi)(1 - \eta)(\xi - \eta - 1) \\ (1 + \xi)(1 + \eta)(\xi + \eta - 1) \\ (1 - \xi)(1 + \eta)(-\xi + \eta - 1) \\ 2(1 - \xi^2)(1 - \eta) \\ 2(1 + \xi)(1 - \eta^2) \\ 2(1 - \xi^2)(1 + \eta) \\ 2(1 - \xi)(1 - \eta^2) \end{array} \right\} \tag{3.11.6}$$

and the normalized (ξ, η) coordinates vary from -1 to $+1$. The shape functions defined in Eqs. (3.11.3)–(3.11.6) are recognized as being identical to the interpolation functions for the two-dimensional triangular and quadrilateral continuum elements. Though the interpolation of temperature within the plane of the elements is similar, the geometric representation of the continuum elements and the shell elements is quite different. The parametric mapping for any shell element is accomplished with the following definitions

$$x = \hat{\boldsymbol{\Psi}}^T \mathbf{x} + \frac{1}{2} \, r \, \hat{\boldsymbol{\Psi}}^T \delta \, \hat{\mathbf{e}}_3 \cdot \hat{\mathbf{e}}_x$$

$$y = \hat{\boldsymbol{\Psi}}^T \mathbf{y} + \frac{1}{2} \, r \, \hat{\boldsymbol{\Psi}}^T \delta \, \hat{\mathbf{e}}_3 \cdot \hat{\mathbf{e}}_y \qquad (3.11.7)$$

$$z = \hat{\boldsymbol{\Psi}}^T \mathbf{z} + \frac{1}{2} \, r \, \hat{\boldsymbol{\Psi}}^T \delta \, \hat{\mathbf{e}}_3 \cdot \hat{\mathbf{e}}_z$$

where $\hat{\boldsymbol{\Psi}}$ is the appropriate linear or quadratic interpolation within the plane. The $(\mathbf{x}, \mathbf{y}, \mathbf{z})$ are vectors of coordinates for the midplane nodes of the element, r is the normalized coordinate along the normal to the element midplane, and δ is a vector of thickness values at the nodes. The vectors $\hat{\mathbf{e}}_1$ and $\hat{\mathbf{e}}_2$ are defined as being tangent to the curvilinear coordinates (ξ, η) on the element midplane; $\hat{\mathbf{e}}_3$ is normal to the element midplane and is defined by $\hat{\mathbf{e}}_3 = \hat{\mathbf{e}}_2 \times \hat{\mathbf{e}}_1$. The unit vectors $(\hat{\mathbf{e}}_x, \hat{\mathbf{e}}_y, \hat{\mathbf{e}}_z)$ define the orientation of the global coordinate system. Note that, in general, $\hat{\mathbf{e}}_3$ varies over the surface of the element $[\hat{\mathbf{e}}_3(\xi, \eta)$ or $\hat{\mathbf{e}}_3(L_1, L_2)]$ and this variation must be accounted for in the construction of the Jacobian entries for the element mapping procedure. The element matrices for these elements are constructed using the same type of procedures as are used for continuum elements. For the constant temperature through the thickness shells, a numerical quadrature in the plane of the element is sufficient. If a higher-order temperature through the thickness is employed, the shape functions in Eqs. (3.11.3)–(3.11.6) would be altered and a full three-dimensional quadrature would be required.

Allowable boundary conditions for shell elements are basically the same as for the three-dimensional continuum elements since all "sides" of the shell are well defined. Radiation enclosures may also contain shells because each shell face is a polygon that can be easily incorporated in the view factor computation. The edge descriptions that use a midline and thickness distribution are different from the bounding surfaces that use nodes to describe the geometry; both cases could be handled by a general view factor algorithm.

3.11.3 Computational Boundary Conditions

The finite element implementation of all of the standard boundary conditions for a heat conduction problem were covered in Section 3.5. However, it is sometimes necessary to consider other implementations of the standard boundary conditions for special simulation applications. Generally these applications involve material interfaces and require more computational effort to produce a usable physical model for the heat transfer process.

Contact boundary conditions

The surface flux that was considered in Section 3.5 was derived from boundary conditions that are applied to the external boundaries of the heat conduction problem. As described in Section 1.8.3 of Chapter 1, it is also appropriate to consider "internal" flux conditions associated with a material interface and in particular, surfaces for which thermal contact resistance is important. The computational form for this internal or contact boundary condition is derived in the same manner as presented previously, though the work to obtain all of the needed data is increased.

From Eq. (1.8.12) the internal or gap heat flux between two surfaces is given by

$$q_{contact} = h_{contact}(s_k, T_{contact}, t)(T_M - T_S) \qquad (3.11.8)$$

where $h_{contact}$ is an effective heat transfer coefficient for the contacting surfaces, and $T_{contact}$ is an average temperature between the master surface temperature, T_M, and the slave surface temperature, T_S. This being of the same form as the flux conditions in (3.1.2c) the finite element form (flux vector) for the boundary condition is

$$\mathbf{q}_{cnt}(T) = \oint_{\Gamma_m} \hat{\mathbf{\Psi}} \, q_{contact} \, ds = \oint_{\Gamma_m} \hat{\mathbf{\Psi}} h_{contact}(T_M - T_S) \, ds \qquad (3.11.9)$$

or in matrix form (with interpolation for the temperature)

$$\mathbf{q}_{cnt} = \left(\int_{-1}^{1} \int_{-1}^{1} h_{contact}(\xi, \eta) \hat{\mathbf{\Psi}}(\xi, \eta) \hat{\mathbf{\Psi}}^T(\xi, \eta) |\mathbf{J}_s| \, d\xi_s d\eta_s \right) \mathbf{T}_M$$
$$- \int_{-1}^{1} \int_{-1}^{1} h_{contact}(\xi, \eta) \hat{\mathbf{\Psi}}(\xi, \eta) T_S(\xi, \eta) |\mathbf{J}_s| \, d\xi_s \, d\eta_s \qquad (3.11.10a)$$

$$= \mathbf{G} \mathbf{T}_M - \mathbf{F}_{cnt} \qquad (3.11.10b)$$

where

$$\mathbf{G} = \int_{-1}^{1} \int_{-1}^{1} h_{contact} \hat{\mathbf{\Psi}} \hat{\mathbf{\Psi}}^T |\mathbf{J}_s| \, d\xi_s \, d\eta_s \qquad (3.11.11a)$$

$$\mathbf{F}_{cnt} = \int_{-1}^{1} \int_{-1}^{1} h_{contact} T_S \hat{\mathbf{\Psi}} |\mathbf{J}_s| \, d\xi_s \, d\eta_s \qquad (3.11.11b)$$

The numerical implementation of this condition requires that master and slave sides of the contact surface be defined. Because unknown temperatures occur on both sides of the interface, each contact surface must be processed in turn as a master surface to satisfy the energy balance; the opposite or slave surface provides the reference temperature for heat transfer across the surface. For generality, the situation shown in the two-dimensional sketch of Figure 3.11.2 is considered, where the nodes and elements on each side of the contact surface are not aligned. If a node on the master surface does not have an image on the slave surface, then $h_{contact}$ is set to zero for that location and the contribution to the contact heat flux for that node is neglected.

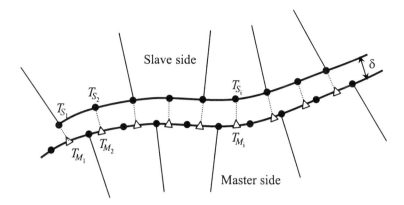

Figure 3.11.2. Sketch and nomenclature for contact boundary condition.

In developing Eqs. (13.11.10) and (3.11.11) the contact coefficient $h_{contact}$ is assumed to vary in a known manner over the contact surface. The vector $\mathbf{T}_M$ corresponds to unknown nodal point temperatures on the master surface and the coefficient matrix $\mathbf{G}$ is combined with the diffusion matrix $\mathbf{K}$ during the solution process [see Eq. (3.6.4)]. The temperatures denoted by T_S are not generally nodal point temperatures but rather interpolated temperatures from the slave surface, adjacent to the master surface nodes.

Two choices exist for the evaluation of T_S. When the geometric location of the master node on the slave surface is known, the slave temperature can be expressed in terms of the slave nodal point temperatures via the shape functions for the slave element. That is, $T_S = \hat{\mathbf{\Psi}}^T(\xi_m, \eta_m)\mathbf{T}_S$ where (ξ_m, η_m) are the local surface coordinates for the location of the master node on the slave element and $\mathbf{T}_S$ are the (unknown) nodal point temperatures on the slave surface. If this definition for T_S is used in (3.11.11b), then $\mathbf{q}_{cnt}$ in (3.11.10b) can be written as

$$\mathbf{q}_{cnt} = \mathbf{GT}_M - \mathbf{HT}_S \qquad (3.11.12)$$

where

$$\mathbf{H} = \int_{-1}^{1}\int_{-1}^{1} h_{contact}\,\hat{\mathbf{\Psi}}\hat{\mathbf{\Psi}}^T(\xi_m, \eta_m)|\mathbf{J}_s|\ d\xi_s d\eta_s \qquad (3.11.13)$$

Since the slave term is now written in terms of nodal point temperatures, the matrix $\mathbf{H}$ can also be added to the diffusion matrix during the solution process. This approach provides the most implicit method of formulating the contact condition and allows both the master and slave surface temperatures to be solved for directly. The disadvantage of this method is that an additional nodal connectivity is required to properly distribute the coefficient matrix ($\mathbf{H}$) between the master and slave nodes. For static problems this connectivity can be established once and used for each iteration or time step. In a dynamic situation, where sliding along the interface or opening and closing of a gap occurs, the additional connectivity would have to be reconstructed during the course of the solution.

The slave temperature T_S may also be evaluated by direct shape function interpolation on the slave surface from the current values of the slave nodal point temperatures. In this case, the value of T_S is inserted in the boundary condition and the vector $\mathbf{F}_{cnt}$ is constructed as shown in (3.11.11b). This method circumvents the need for additional nodal connectivity information but does introduce the need for an iterative solution for the temperature unknowns along the interface.

Central to the implementation of the contact condition is the assumed knowledge of the spatial location of the master nodes on the slave surface. Obtaining this information, especially in dynamic problems where contact surfaces may appear and disappear, can be a computationally intensive task that requires specialized algorithms. Generally, the problem specification will nominate a list of potential contact nodes and element surfaces. Various sorting methods (e.g., bin sorting, recursive bisection) can be used to group nodes and surfaces that are geometrically close and reduce the extent of the search. Ultimately each master node is tested against the possible slave surfaces to determine if the node is in contact (within some tolerance) and where on the surface (local coordinates) it is located. Testing a node for contact can be done effectively by using the isoparametric mapping for the element and writing a Newton iteration scheme to find the local coordinates of the node; the local coordinates of the master node are finally tested to determine if they coincide with the element surface.

In addition to providing a generalized surface contact boundary condition, the above formulation can provide a simple method for connecting regions with different mesh spacings. For "large" values of $h_{contact}$, Eq. (3.11.9) forces the temperature distributions on each side of the contact surface to be essentially equal. Though this method can be made to work in practice, it is not optimal as very large values of $h_{contact}$ can cause ill-conditioning of the matrix problem and difficulties in reaching convergence with an iterative matrix solution method. The constraint boundary conditions discussed next are a better alternative for connecting dissimilar mesh regions.

Multipoint constraints

For some applications it is necessary to specify the functional relationship between the temperature at one node and temperature at one or more other nodes. The enforcing of temperature continuity between coincident surfaces with dissimilar meshes and the specification of spatially periodic temperature boundary conditions are two examples of this type of constraint. Constraints between nodes at various spatial locations could also be used in some simulation situations to represent thermal controls. The general multipoint constraint condition is similar in many respects to the contact algorithm with one surface labeled the master surface and the constrained temperature surface labeled the slave surface. The locations of the slave nodes on the master surface are found and recorded using the same types of search procedures as used in the contact algorithm.

Several different methods may be used to enforce the temperature constraint condition on the system of discrete equations. Conceptually, the simplest approach involves the field equation for each slave node being replaced with a constraint equation that relates the temperature at the slave node to some function of the nodal temperatures on the master surface. In the case where temperature continuity is enforced, the slave node value is constrained to be the interpolant of the master node values. In practice this process can be awkward. Constraints must be processed after the matrix is fully assembled and may involve considerable row and column manipulations within the global matrix to process the constraint equation or condense the constrained nodal temperature. Lagrange multiplier methods, which will be discussed in some detail in a subsequent chapter, could also be used to apply the constraint condition. This scheme has the significant disadvantage of defining additional unknowns for the equation set. The preferred method is a penalty function approach that begins by writing the constraint equation as

$$\mathbf{f}_{mp} = \mathbf{C}_{mp}\mathbf{T}_{mp} - \mathbf{F}_{mp} = 0 \qquad (3.11.14)$$

where the common case has $\mathbf{F}_{mp} = 0$, which will be considered here. The matrix $\mathbf{C}_{mp}$ is a constant coefficient matrix with (typically) more columns than rows and $\mathbf{T}_{mp}$ includes both slave and master nodal point temperatures. When $\mathbf{f}_{mp} = 0$, then the constraint is satisfied and the term $\mathbf{C}_{mp}\mathbf{T}_{mp}$ could be added to the global matrix system without altering the energy equation. In general this term is not zero; the penalty method forces this term to be small by multiplying the constraint equation by a large parameter. The penalty function form of Eq. (3.11.14) is

$$\mathbf{C}_{mp}^T\lambda\mathbf{C}_{mp}\mathbf{T}_{mp} = 0 \qquad (3.11.15)$$

The matrix λ is a diagonal matrix of penalty parameters that are selected to be large enough to force the constraint to be approximately satisfied. Premultiplying Eq. (3.11.14) by $\mathbf{C}_{mp}^T$ makes the constraint system square and symmetric. Equation (3.11.15) may also be derived by formally adding a penalty function $\mathbf{f}_{mp}^T\lambda\,\mathbf{f}_{mp}$ to the variational form of the energy equation and taking a variation of the resulting augmented potential. A further discussion of penalty methods will be delayed to the next chapter.

When temperature continuity is enforced, the slave node is set to the interpolant of the master node values. That is, for each k slave node $T_S^k = \hat{\mathbf{\Psi}}^T(\xi_s, \eta_s)\mathbf{T}_M$, where (ξ_s, η_s) are the local surface coordinates for the location of the slave node on the master element and $\mathbf{T}_M$ are the (unknown) nodal point temperatures on the master surface. The coefficients of $\mathbf{C}_{mp}$ are the values of $\hat{\mathbf{\Psi}}^T(\xi_s, \eta_s)$. Once the coefficients for $\mathbf{C}_{mp}$ have been evaluated for each constraint, the matrix in (3.11.15) can be constructed and added to the global system. A connectivity between the slave node and the master nodes must be constructed; these data are usually generated as part of the search procedure.

For periodic conditions, the master/slave search procedure is complicated by the slave and master surfaces being in different spatial locations.

Typically, a coordinate transformation is specified that translates and rotates the constrained surface into the master surface, after which the search procedure may be carried out in the usual manner. Also, the slave node temperature would generally be set to the master surface temperature plus some temperature increment representing the heat flux across the periodic geometry.

Partially covered surfaces

Contact boundary conditions permit a wide variety of material interface situations to be simulated effectively, while multipoint constraints can ease the modeling of complex geometries with little loss in solution accuracy. In both cases it is possible to have a dramatic mismatch of the element faces and edges along each side of the interface. One major effect of this mismatch is to alter the ability to consistently apply flux type boundary conditions to the exposed portions of the master or slave surfaces. Shown in Figure 3.11.3 is a typical three-dimensional case of partially covered elements for which a flux type boundary condition might be required on all exposed surfaces. The question is how to numerically integrate the boundary condition specification over the arbitrary polygonal surface of the underlying element.

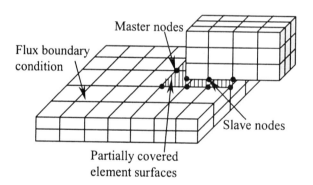

Figure 3.11.3. Sketch for partially covered element surfaces.

A method advocated by Rashid [42] begins by finding the intersection of the slave surface with the master. As a result of the intersection algorithm a series of lines is produced that describes the outline of the uncovered portion of the master surface. The surface integral that describes the flux boundary condition to the master surface can then be integrated with a special type of quadrature rule that locates evaluation points on the boundary of the exposed surface. Rashid [42] describes how the quadrature weights are derived to integrate polynomial functions to a given order of accuracy. The work involved in this algorithm is somewhat excessive for large, complex geometries and has led to a simpler and more approximate treatment. Because the flux boundary condition is evaluated or sampled at the integration points on the master surface, it might be sufficient to determine if any integration points are covered by the slave surface and then omit them from the surface quadrature. It is a relatively easy task to determine if a master element quadrature point is within the boundary of a slave surface and flag it for exclusion from the surface integration. This can produce a very rough approximation to the surface integral, especially for cases where the slave surface covers a portion of the master surface but avoids a quadrature location. However, in many cases it provides an adequate approximation to the boundary condition and has the added advantage of being computationally simple. This last point is very important for dynamic contact problems and problems with element removal or addition.

3.11.4 Bulk Nodes

The environment external to the continuum region modeled by a finite element model influences the thermal diffusion process through the flux boundary conditions given in Eqs. (3.1.2c) and (3.6.2) and in particular through the specification of the reference temperatures, T_c and T_r. In some cases, a zero-dimensional model for the external region is useful in accounting for changes in the reference temperature. Such a model is often termed a bulk node.

A bulk node is characterized by a single temperature, $T_b(t)$, and pressure, $P_b(t)$ and is defined as a general control volume, CV, bounded by a control surface, CS. Mass and energy may flow across the control surface and the control volume may be time-dependent. Processes occurring within the bulk node are assumed to be in quasi-equilibrium and be uniformly distributed within the CV. The statement of mass conservation for the CV is

$$\frac{dM}{dt} = \frac{d}{dt} \int_{CV} \rho \, dV$$

$$= \sum_{CS} \delta \dot{m}_{in} - \sum_{CS} \delta \dot{m}_{out} \qquad (3.11.16)$$

where the summations are over the segments of the control surface where an incremental mass flux, $\delta \dot{m}$, is defined. The energy conservation for the CV is

$$\frac{dE}{dt} = \frac{d}{dt} \int_{CV} \rho \mathcal{E} \, dV$$

$$= \sum_{CS} (h_0 \, \delta \dot{m} + \delta q + \delta \mathcal{P})_{in} - \sum_{CS} (h_0 \, \delta \dot{m} + \delta q + \delta \mathcal{P})_{out} \qquad (3.11.17)$$

where $h_0 \, \delta \dot{m}$ is the mass transfer energy rate, q is the thermal energy rate, and $\mathcal{P}$ is the mechanical energy rate. For the bulk node of interest here, the kinetic and potential energy changes in the mass transfer rate may be neglected. Also, shaft work and shear forces are neglected in the mechanical energy rate and conduction within the control volume is neglected in the thermal rate. The bulk node energy equation then is

$$\frac{dU}{dt} = P_b \dot{V} + \sum_{CS} (h_0 \, \delta \dot{m} + \delta q)_{in} - \sum_{CS} (h_0 \, \delta \dot{m} + \delta q)_{out} \qquad (3.11.18)$$

Here P_b is the uniform pressure for the bulk node and $\dot{V}$ is the time rate of change of the bulk node volume. Also, for an ideal substance, the internal energy is $U = MC_vT_b$ which allows the recovery of the bulk node temperature from the energy equation. For use with a finite element model, the summation over the control surface is replaced by a summation over the element surfaces bounding the bulk node. The thermal energy rates in (3.11.18) are replaced by the finite element convection and radiation boundary conditions (3.1.2c) as appropriate, where the reference temperatures are now T_b. The mass flow terms in (3.11.16) and (3.11.18) may be imposed on the bulk node to model vents or orifices associated with the geometry. These terms may also be associated with the chemical reaction of materials surrounding the bulk node. Decomposition reactions may add mass and energy to the bulk node during the reaction and/or contribute to the bulk node through the removal of elements (see Section 3.11.6). Element removal also leads to changes in the volume of the bulk node as does the Lagrangian motion of material surrounding the bulk node. The bulk node material is generally modeled as either an ideal gas or a constant pressure liquid.

The bulk node mass and energy equations, Eqs. (3.11.16) and (3.11.18), must be solved in conjunction with the finite element equations since nodal point temperatures appear in the right-hand side of Eq. (3.11.18). Steady solutions do not usually require a bulk node since a constant state can be represented by a known reference temperature in the convective or radiative boundary condition. A completely implicit (fully coupled) formulation for the bulk node requires that Eqs. (3.11.18) and (3.11.16) be discretized in time, according to the chosen finite element integration method, and added to the finite element equations (matrix) as two additional unknowns. To establish the correct relationship between the bulk node variables and the nodal point temperatures on the surface of the bulk node volume, a special connectivity must be generated which relates one bulk node to many surface nodes.

In the standard case where the geometry surrounding the bulk node is fixed, the fully coupled method is efficient and robust. Iterative matrix solvers are particularly adept at handling the large matrix bandwidth associated with the one to many connectivity of the bulk node. Direct methods for the matrix problem may need to be modified to effectively treat the increased bandwidth. Finally, when the geometry around the bulk node varies with time (e.g., element removal), the continued rebuilding of the connectivity and altering of the matrix structure make the fully coupled approach less appealing. The bulk node equations may be decoupled from the finite element equations and integrated in time with any suitable integration method. Note that the forcing functions in the ordinary differential equations in (3.11.16) and (3.11.18) may often contain rough (incremental) data that are not well-suited to sophisticated integration methods. Coupling of the bulk node variables with the finite element temperatures could occur at either the beginning or end of the common integration interval.

3.11.5 Reactive Materials

The presence of reactive materials in the heat conduction problem requires that a number of nonlinear conservation equations be solved for the chemical species in conjunction with the temperature field. The general formulation for the chemistry problem was outlined in Section 1.7. The mathematical nature (stiffness) of the kinetic equations dictates that for computational efficiency, the chemistry and thermal diffusion equations be solved independently. The solution process is formally based on an operator splitting technique as defined, for example, in [43].

Operator splitting is particularly effective in the case of condensed materials due to the form of the kinetic equations. Because diffusion of the species is neglected, the kinetic equations have no spatial gradients and reduce to ordinary differential equations that can be defined locally on each finite element. In essence, the chemical species can be viewed as state variables for each element and can be solved on an element-by-element basis. In most situations it is convenient to define all species and species equations at the integration points for each element. During a time step, the chemistry solution is advanced first using a fixed (frozen) temperature field; the temperature field is subsequently advanced over the same time interval using the recently evaluated (frozen) chemistry result. If a predictor/corrector time integration method is employed, the frozen temperature field used for the chemistry solution is the temperature produced from the predictor step. When a predictor equation is not employed for time integration, the last available temperature field is used for the chemistry solution.

The inherent stiffness of the kinetic equations requires that special integration methods be used to advance the chemistry solution in time. Stiff, ordinary differential equation (ODE) methods have been extensively studied and a number of library packages are available. The package CHEMEQ [44] developed by T. R. Young was designed specifically for chemical reaction systems and is easily incorporated into a finite element solution method. The techniques used in CHEMEQ are based on a combination of classical predictor/corrector methods and asymptotic methods for the stiff components of the system. The rate equations for each reactive finite element may be solved using their own integration time step over the global time interval of interest. The most restrictive chemistry time step for all of the reactive elements may then be used to influence the choice of the next thermal diffusion time step. A typical relation for thermal time step choice is [see Eq. (3.7.20)]

$$\Delta t_{n+1} = min\{\Delta t_{diff}, X_{chem} \times \Delta t_{chem}\} \qquad (3.11.19)$$

where Δt_{diff} is the estimated time step for the heat conduction equation and Δt_{chem} is the minimum time step estimated for the chemistry solution. The parameter X_{chem} is a user-defined scale factor that typically has a value between 10 and 100. When reactive processes are unimportant, the adaptive time integration in the chemistry integrator will produce a chemistry time step that is relatively large and Eq. (3.11.19) will allow the conduction solution to dictate the problem time scale. As the reactive process accelerates, the chemistry time step will decrease significantly and ultimately control the time step formula in (3.11.19). The transition point for control of the global time step can be dictated by the user through the X_{chem} parameter.

As noted in Section 1.9, a change of phase may be viewed as a reversible chemical reaction with material 1 going to material 2 with an appropriate change in energy. Writing the phase change reaction as two one-way reactions with reaction rates k_1 and k_2 produces

$$\mathcal{M}_1 \xrightarrow{k_1} \mathcal{M}_2$$

$$\mathcal{M}_2 \xrightarrow{k_2} \mathcal{M}_1 \tag{3.11.20}$$

where $\mathcal{M}_1$ is assumed to be the low temperature phase. Using the chemical reaction formulation from Section 1.7, the ordinary differential equations (1.7.12) describing the above reaction are

$$\frac{d\mathcal{M}_1}{dt} = R_1 = k_1\mathcal{M}_1 - k_2\mathcal{M}_2 = A_1 e^{(-E_1/RT)}\mathcal{M}_1 - A_2 e^{(-E_2/RT)}\mathcal{M}_2 \tag{3.11.21a}$$

$$\frac{d\mathcal{M}_2}{dt} = R_2 = -k_1\mathcal{M}_1 + k_2\mathcal{M}_2 = -A_1 e^{(-E_1/RT)}\mathcal{M}_1 + A_2 e^{(-E_2/RT)}\mathcal{M}_2 \tag{3.11.21b}$$

where β_1 and β_2 are assumed to be zero. Employing the substitution $\mathcal{M}_2 = 1 - \mathcal{M}_1$ in (3.11.21a), this equation may be solved analytically for $\mathcal{M}_1(T, t)$ yielding

$$\mathcal{M}_1 = \frac{k_2}{k_1 + k_2} + \frac{k_1}{k_1 + k_2} e^{-(k_1+k_2)t} \tag{3.11.22}$$

where it has been assumed that $\mathcal{M}_1$ and $\mathcal{M}_2$ range between 0 and 1 with initial conditions of $\mathcal{M}_1 = 1$ and $\mathcal{M}_2 = 0$. This is a linear first-order reaction. The constants in Eq. (3.11.22) are the pre-exponentials, A_1, A_2, and the activation energies, E_1, E_2, which may be adjusted to set the temperature and temperature interval over which the reaction or phase change occurs. The latent heat release is described by Eq. (1.7.9) and for the pair of reactions is

$$Q_r = q_1 r_1 + q_2 r_2 = q_1 k_1 \mathcal{M}_1 + q_2 k_2 \mathcal{M}_2 \tag{3.11.23}$$

The equations in (3.11.21) can be solved by the methods outlined above for the general reaction kinetics problem. When coupled with the heat release equation in (3.11.23) and the heat conduction equations for the material, an efficient method for phase change problems is produced. A comparison of this method with the standard enthalpy methods is provided in a subsequent section on example problems.

3.11.6 Material Motion

The heat conduction equation in (3.1.1) was written for a stationary region in which case the Lagrangian and Eulerian descriptions of the boundary value problem are identical (see Section 1.3.1). If motion of the conducting regions is to be considered, then a particular reference frame must be selected. The Eulerian description is usually associated with fluid-like motions that involve very large material deformations, unsteadiness, and a need to fully consider the equations of motion for the material. However, there are heat conduction problems that can be well described in an Eulerian reference frame. Generally, if the motion is continuous (steady), the kinematics are simple and essentially unidirectional and the geometry of the moving material is not complicated, then an Eulerian description may be appropriate. Problems of this type are usually related to some type of continuous processing operation on a simply shaped (cylinder, bar, slab, etc.) material region. The Eulerian form of the energy equation is altered from (3.1.1) to the advection-diffusion form of (1.3.8) or (1.4.3). That is,

$$\rho C \left(\frac{\partial T}{\partial t} + u_j \frac{\partial T}{\partial x_j} \right) = \frac{\partial}{\partial x_i} \left(k_{ij} \frac{\partial T}{\partial x_j} \right) + Q \qquad (3.11.24)$$

Since a momentum equation is not considered in the heat conduction problem, it is required that the velocity field in (3.11.24) be completely prescribed as a function of time and space; the nonconservative form of the energy equation also requires that the velocity field be divergence free. The additional advective term present in (3.11.24) adds minimal complexity to the heat conduction formulation and can be easily incorporated in a finite element formulation as an alternate energy equation. The finite element equations related to (3.11.24) will not be discussed here since they form a major element of the next two chapters on fluid mechanics and convective heat transfer. Note that Eq. (3.11.24) may be used with the standard conduction equation in different regions of the same problem since they are both referenced to the same coordinate system. Mixtures of Eulerian and Lagrangian descriptions are also permissible.

Heat conduction problems are usually described in a Lagrangian frame since the focus is on solid bodies. The incorporation of material motion in the Lagrangian description is easily accomplished as long as the material deformations are not too large; distortion of the individual finite elements may become a problem under large deformations. Again, since no equations of motion are included in the conduction problem, the kinematics for the material region must be completely specified. For rigid body motions such a prescription is straightforward; new material coordinates are directly defined by translation and rotation of the region. Material deformation is generally more complex and usually requires the solution of a solid mechanics problem for all but the simplest motions. Deformation is normally accompanied by changes in density which must be accounted for in the conduction problem. Mass conservation issues are not usually associated with heat conduction problems but must be considered when deformations are present.

Related to material motion is the situation where material is added to or removed from the original heat conduction region. This type of physical situation can be represented by the simple addition or deletion of finite elements from the original element discretization of the problem. The physical process will be well modeled if the time constant for the addition or deletion of material is short compared to the time constant for the thermal process and the details of the material alteration are unimportant. Element birth and death may be implemented by either of two methods. In the first case, elements are added to or deleted from the global element list, new connectivities are established, and entries in the global matrix are reorganized to reflect the new list of elements. This scheme has the benefit of always solving a problem of minimal size, though it suffers from the additional work needed to reorganize the element data. The second approach always builds an equation set (matrix) that corresponds to all the elements that could possible be in the problem and uses a fixed connectivity. If an element is not currently active (i.e., has not been born or has already died), then the equations for that element are zeroed out and replaced by a constant temperature constraint. The matrix problem for this case is larger than the first scheme, but there is no work to activate or deactivate an element other than the setting of an element flag.

A common problem in element birth and death is the updating of boundary conditions with changes in the element topology. If a three-dimensional element with a flux boundary condition is removed from the problem, some type of algorithm must be present to determine which newly exposed faces of the surrounding elements now have a boundary condition applied to them. One method for treating this problem of boundary condition inheritance is to tag every face of every element with information about its adjacent element face. Upon activating or deactivating an element, the face data of the adjacent elements are switched to reflect the new state of the face (either exposed or covered). For this type of method to be effective, boundary conditions must be applied to the entire region of a problem. This algorithm must also be altered, if contact or multipoint constraint conditions exist and partially covered element faces occur.

3.12 Example Problems

3.12.1 Introduction

In this section a number of heat conduction and radiation problems are presented to illustrate the various aspects of finite element analysis procedures. The first several examples are relatively simple, two-dimensional solutions that illustrate the use of anisotropic and nonlinear material properties as well as the phase change algorithm. The remaining examples are more complex and represent typical engineering problems encountered in practice. All of the problems described here were analyzed using the finite element code, COYOTE [45].

3.12.2 Temperature-Dependent Conductivity

Many materials of engineering interest have thermophysical properties, such as conductivity, that vary with temperature. This example illustrates the difference in results that can be expected when conductivity variations are included in the model. Figure 3.12.1 contains a schematic and mesh for a simple planar geometry. Each half of the region is occupied by a different isotropic material; the first material has constant conductivity, k_0, while the second material has a conductivity that varies with temperature as $k = k_0 + aT$. A constant heat flux is applied along the left edge of the domain. The upper and lower boundaries are insulated ($q_n = \mathbf{q} \cdot \mathbf{n} = 0$) as is the plane of symmetry between the two materials. All other surfaces are convectively cooled with a constant h_c and T_c. Four-node bilinear elements were used to model the problem.

Due to the variable conductivity, this problem is nonlinear and the steady-state solution was obtained using a Picard iteration. A converged solution was obtained in four iterations. Figure 3.12.2 contains a plot of the isotherms. Since the plotted isotherm bands are the same for the two materials, the difference in the temperature fields is readily observed. At the high temperature end (applied flux surface) of the region, the conductivity for the variable material is high and the temperature gradient (isotherm spacing) is therefore lower than for the constant conductivity case. At the low temperature end of the region where the conductivities are comparable, the temperature gradients are similar.

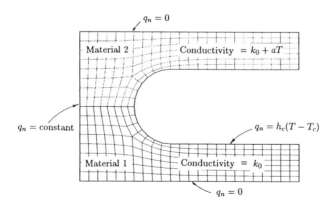

Figure 3.12.1. Schematic and mesh for the temperature-dependent conductivity problem.

3.12.3 Anisotropic Conductivity

As explained in Section 3.9, the inclusion of an anisotropic conductivity into a finite element model is relatively straightforward and may be important for the simulation of laminated, fibrous, or wire-wrapped materials. A planar square with a tilted square insert is used to demonstrate the effects of an orthotropic conductivity. Referring to Figure 3.12.3, the outer square is an isotropic material with conductivity, k. Constant temperature boundary conditions are imposed on the vertical edges of the square while the horizontal edges are insulated. The insert material is orthotropic with $k_{11} = k$ and $k_{22} = k/10$; the orientation of the material axes with respect to the global coordinate axes will be specified for each simulation.

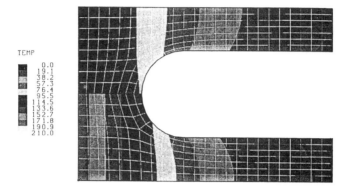

Figure 3.12.2. Isotherms for the temperature-dependent conductivity problem.

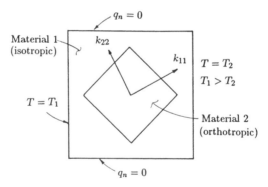

Figure 3.12.3. Geometry of the anisotropic conductivity problem.

Since the problem is linear, it may be solved directly without iteration. The first solution (see Figure 3.12.4a) corresponds to the case where the material axes for the insert material are aligned with the coordinate axes. As expected, the solution is one-dimensional since the k_{11} component for the insert is the same magnitude as the surrounding isotropic material; the k_{22} component plays no role in the solution since there is no temperature gradient imposed in the transverse direction. Figure 3.12.4b shows the temperature field for the case where the material axes of the insert are inclined upward 45° with respect to the global axes (the material axes are aligned with the mesh lines in the insert). The distortion of the temperature field due to the anisotropy is very obvious. Finally, Figure 3.12.4c illustrates the solution for the case as in (b) but with both components of the insert conductivity reduced by a factor of 10. This accentuates the difference between the two materials and leads to higher gradients and more distortion of the isotherms within the insert.

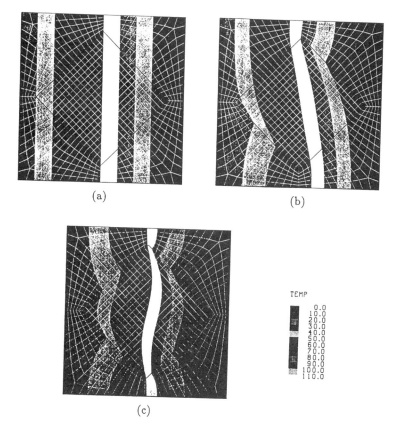

(a)

(b)

(c)

TEMP
```
0.0
10.0
20.0
30.0
40.0
50.0
60.0
70.0
80.0
90.0
100.0
110.0
```

Figure 3.12.4. Temperature fields for the anisotropic conductivity problem.

3.12.4 One-Dimensional Stefan Problem

A standard test problem for phase change algorithms is the one-dimensional Stefan problem (see Carslaw and Jaeger [46]). The problem consists of a material region, $0 \leq x \leq 4$, held initially at a uniform temperature greater than the liquidus temperature. At time zero, one face of the region $(x = 0)$ is lowered to a temperature below the solidus temperature, causing a solidification front to propagate into the material. The present example is solved using a mesh of eight-node (biquadratic) elements and the Crank–Nicolson time integration procedure. The schematic of the problem is shown in Figure 3.12.5, which also contains the finite element mesh and boundary conditions. The following material properties are used: $\rho = 1.0$, $C = 1.0$, $k = 1.08$, $L = 70.26$, $\Delta T = 2.0, 0.5$, $T_s = -0.15$, and $T_l = -2.15, -0.65$.

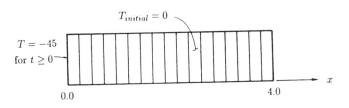

Figure 3.12.5. Schematic of the one-dimensional Stefan problem.

Figure 3.12.6 shows the computed temperature response for a point one unit from the cooled face. Computations were made for two values of the phase change temperature interval, ΔT, and are compared to the exact solution. The numerical solution is of reasonable accuracy with the smaller ΔT case being slightly closer to the exact solution. Figure 3.12.7 contains a plot of the location of the phase boundary as a function of time for the cases studied. Early in time, the numerical solution overpredicts the location of the phase boundary but comes into good agreement with the analytical solution for $t \geq 1.0$. Note that the analytical solution was developed for a single-phase transition temperature while the numerical solution requires a finite transition interval, ΔT. As ΔT gets smaller, the numerical solution becomes more difficult and more costly.

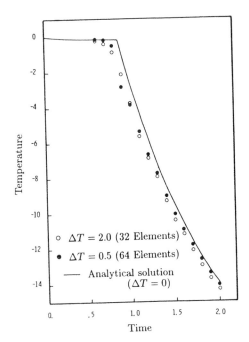

Figure 3.12.6. Temperature vs. time at $x = 1.0$ for the Stefan problem.

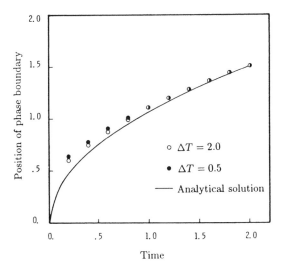

Figure 3.12.7. Location of the phase boundary vs. time for the Stefan problem.

3.12.5 Drag Bit Analysis

The next example deals with the thermal (and stress) analysis of a polycrystalline diamond compact (PDC) drag bit that is commonly used in the oil and gas drilling industry. A planar, two-dimensional model of a single stud in the drag bit body is considered (see Figure 3.12.8). The thermal boundary conditions for the model are also indicated in the sketch. Heat is input to the stud at the wearflat due to frictional heating; the heating function is given as an empirical function of cutter, rock, and operating conditions. The remaining surfaces of the stud are convectively cooled by the drilling fluid. Boundary conditions on the bit body are taken to be adiabatic since they represent planes of symmetry with adjoining cutters. Figure 3.12.9 contains two finite element meshes of eight-node quadrilateral elements used for this problem; the second mesh shows the location of a cooling nozzle and a convection boundary condition used in some of the parametric studies.

A wide variety of boundary and geometric conditions in both the steady-state and transient regimes [47] have been examined for the problem. Figures 3.12.10a and 3.12.10b contain plots of isotherms for two steady-state cases, which differ in the assumed convective heat transfer coefficient for the drilling fluid. The high temperature gradients caused by the high cooling rate are clearly seen in the figures. As is quite common in many applications, a finite element stress analysis followed the heat transfer study. Figures 3.12.11a and 3.12.11b show contour plots of the predicted plastic strains in the cutter corresponding to the temperature fields of Figures 3.12.10a and 3.12.10b, respectively, and an appropriate set of mechanical loads. Studies such as this have been used to predict the life of drill bits, improve cutter design, and provide operational limits for improved drilling performance.

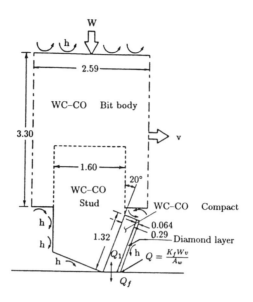

Figure 3.12.8. Schematic of the polycrystalline diamond compact (PDC) drag bit.

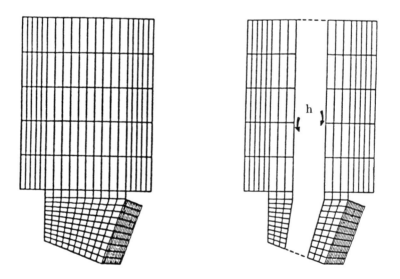

Figure 3.12.9. Finite element meshes used for the analysis of the PDC drag bit.

3.12.6 Brazing and Welding Analyses

The next two examples come from the analyses of two manufacturing processes for electrical components. Figure 3.12.12 shows a schematic and finite element mesh for one-half (upper half) of the axisymmetric geometry of interest; a three-dimensional mesh used in a subsequent analysis to verify the two-dimensional results is also shown. In manufacturing this component, two halves (upper and lower) are brought together under a modest pressure. An

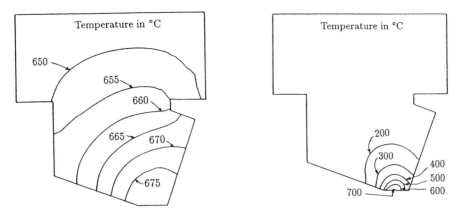

Figure 3.12.10. Steady-state isotherms for the drag bit. (a) $h_c = 5 \ \text{W}/(\text{cm}^{2\circ}\text{C})$.
(b) $h_c = 0.01 \ \text{W}/(\text{cm}^{2\circ}\text{C})$.

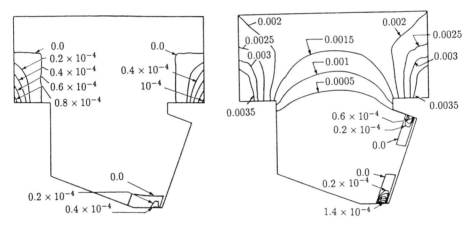

Figure 3.12.11. Plastic strain contours for the drag bit. (a) Corresponds to thermal field in Figure 3.12.10a. (b) Corresponds to thermal field in Figure 3.12.10b.

electrical current is passed through the outer flange to resistively heat the braze rings and bond the two halves of the device. In the simplified thermal model considered here the brazing process is modeled using a time-dependent (pulsed) volumetric heating of the flange and braze ring. All exterior boundaries are assumed to be adiabatic since the entire process takes place on a very short time scale. This problem was solved using four-node, bilinear elements and a modified Crank–Nicolson procedure.

Figure 3.12.13 shows a series of isotherm plots for two times during a one-pulse brazing cycle. Heating occurs over the first 0.5 seconds, during which time the temperature on the flat of the braze ring reaches $\sim 2100°F$. Later in time, the temperature in the braze ring decreases as energy is conducted toward the centerline of the component. A companion stress analysis was used to identify high stress regions and potential failure (cracking) sites in the component.

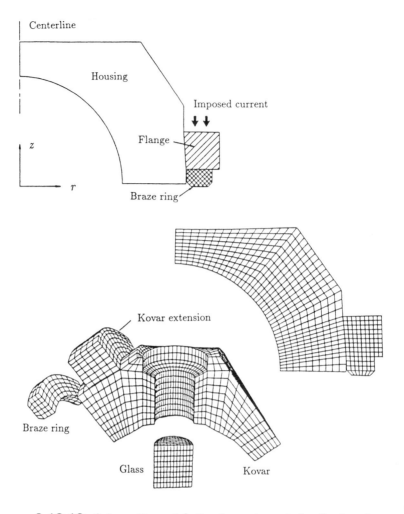

Figure 3.12.12. Schematic and finite element mesh for the brazing problem.

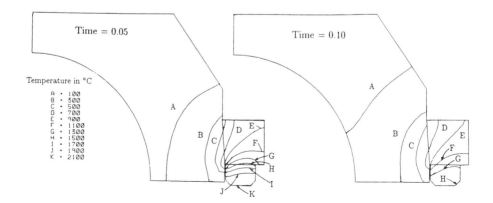

Figure 3.12.13. Isotherms at two times during the brazing analysis.

Like brazing, welding is a process that is often used in the assembly of electrical components. Figure 3.12.14 shows the finite element representation of a battery case and attached header; the header and case are welded using a gas-tungsten arc (GTA) weld. The metallic connecting pins in the battery header are fixed to the header with glass-to-metal seals and must retain an hermetic seal during the GTA process. The figure shows an enlargement of the connecting pins, seals, and header assembly. A study was performed to evaluate the benefits of providing various types of heat sinks to the header and case during the weld process. Details of the parametric study and weld parameters and modeling are available in [48]. Figure 3.12.15 contains one heat sink configuration; the mesh for this problem contained approximately 27,000 eight-node, brick elements.

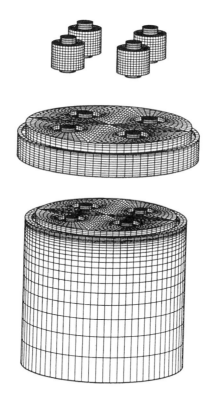

Figure 3.12.14. Finite element mesh for the battery header welding problem.

Figures 3.12.16 and 3.12.17 show temperature fields in the header at the conclusion of one revolution of the GTA welder. It is clear from the large differences in temperature at the pins that heat sinking has a strong positive benefit. The temperature gradients in the glass for the no heat sink case are sufficient to cause a tension failure (cracking) of the glass.

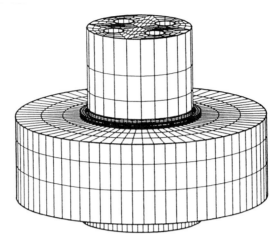

Figure 3.12.15. Finite element mesh for the heat sink used in the battery header weld.

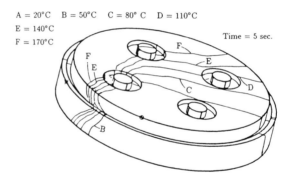

Figure 3.12.16. Temperature field in the header following weld – without heat sink (Time = 5 sec).

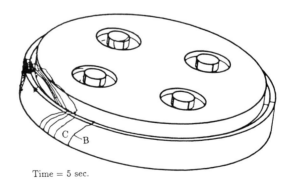

Figure 3.12.17. Temperature field in the header following weld – with heat sink (Time = 5 sec.).

3.12.7 Investment Casting

A heat transfer analysis of the cooling phase of an investment casting can provide important guidance to the design of the metal distribution system (gates, runners, etc.) for the casting and indicate potential casting problems (e.g., shrinkage). Shown in Figure 3.12.18 is a finite element mesh of a structural part to be produced in stainless steel by the investment casting process. The mesh contains ~2400 eight-node brick elements. For purposes of this preliminary analysis, the mold for the casting is not considered and the metal distribution system is also ignored. The casting model is initially assumed to be at the metal pour temperature and loses energy to the furnace walls through radiation; investment castings are poured in a vacuum. The number of element surfaces participating in radiative transfer is ~4500, which leads to approximately 20 million view factors. The view factors were computed using both the double area integration and the hemicube method. A further review of some of the numerical and modeling problems encountered in this type of simulation is available in [49].

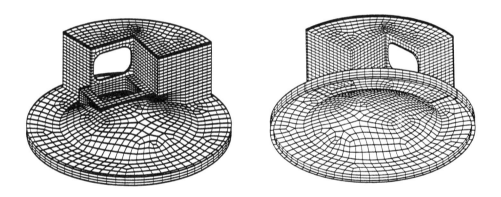

Figure 3.12.18. Finite element mesh for a cast structural part.

Shown in Figure 3.12.19 are surface temperature plots at two times during the cooling process. The temperatures in Figure 3.12.19 were computed using the full radiation model; a simplified radiation boundary condition that did not use view factors but employed a uniform σT^4 heat loss on the surface produced a significantly different temperature field.

Figure 3.12.20 contains plots showing the extent of the mushy zone (liquidus to solidus temperature interval) at two times during cooling. Such information, coupled with cooling rate data, can provide an indication of the metal microstructure and part quality.

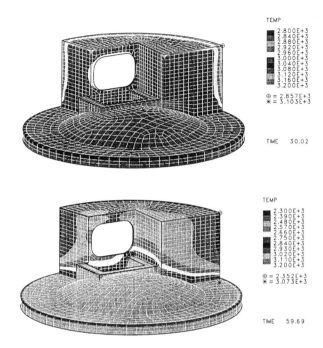

Figure 3.12.19. Temperature field during cooling of a casting using an enclosure radiation model. Temperature is in °F.

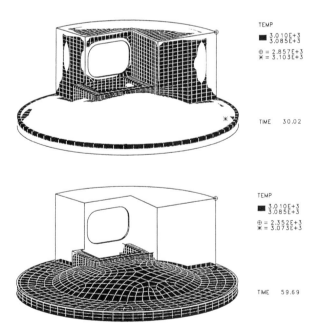

Figure 3.12.20. Liquidus to solidus temperature interval (mushy zone) during cooling of a casting. Temperature is in °F.

Problems

3.1 Show that the weak form of Eq. (3.1.1) is given by Eq. (3.2.1). Use the following component form of the Green–Gauss theorem to integrate-by-parts,

$$\int_{\Omega^e} \frac{\partial}{\partial x_i} (wF_i)\ dx = \oint_{\Gamma^e} (wF_i)n_i\ ds$$

3.2 Using the θ-family of approximation,

$$\{T\}^{n+1} = \{T\}^n + \Delta t[(1-\theta)\{\dot{T}\}^n + \theta\{\dot{T}\}^{n+1}]\ ,\ 0 \le \theta \le 1$$

in Eq. (3.2.4a), derive the fully discretized set of equations in the form

$$[\hat{K}(T^{n+1})]^{n+1}\{T\}^{n+1} = \{\hat{F}(T^n)\}$$

Define $[\hat{K}(T^{n+1})]$ and $\{\hat{F}(T^n)\}$ in terms of the original matrices, time increment Δt, and the parameter θ.

3.3 Specialize the results of Problem 3.2 to the case of $\theta = 0$ and verify Eq. (3.7.10).

3.4 Specialize the results of Problem 3.2 to the case of $\theta = 1$ and verify Eq. (3.7.12).

3.5 (*Newton's Method*) Rewrite Eq. (3.7.1) as

$$\mathbf{R}(\mathbf{T}) \equiv \hat{\mathbf{K}}(\mathbf{T})\mathbf{T} - \hat{\mathbf{F}}(\mathbf{T}) = 0 \qquad (i)$$

Newton's method is based on a truncated Taylor's series expansion of the residual $\mathbf{R}(\mathbf{T})$ about the known solution $\mathbf{T}^r$, where r denotes the iteration number:

$$0 = \mathbf{R}(\mathbf{T}^r) + \left.\frac{\partial \mathbf{R}}{\partial \mathbf{T}}\right|_{\mathbf{T}}^r \delta\mathbf{T} + O(\delta\mathbf{T})^2 \qquad (ii)$$

Here $\delta\mathbf{T} = (\mathbf{T}^{r+1} - \mathbf{T}^r)$. Omitting the terms of order two and higher, write

$$\mathbf{R}(\mathbf{T}^r) = -\left.\frac{\partial \mathbf{R}}{\partial \mathbf{T}}\right|_{\mathbf{T}}^r (\mathbf{T}^{r+1} - \mathbf{T}^r) \equiv -\mathcal{J}(\mathcal{T}^r)(\mathbf{T}^{r+1} - \mathbf{T}^r) \qquad (iii)$$

where $\mathcal{J}$ is the Jacobian matrix (also known as the tangent matrix),

$$\mathcal{J} = \left.\frac{\partial \mathbf{R}}{\partial \mathbf{T}}\right|_{\mathbf{T}}^r \qquad (iv)$$

Show that

$$\mathbf{T}^{r+1} = \mathbf{T}^r - \mathcal{J}^{-1}(\mathbf{T}^r)\mathbf{R}(\mathbf{T}^r) \qquad (v)$$

3.6 Derive the tangent matrix for the case where $\hat{\mathbf{K}}$ is defined by Eq. (3.2.5a) and k_{ij} are given by

$$k_{ij} = \delta_{ij}(k_0 + k_1 T^m)$$

where k_0 and k_1 are constants, and m is an integer.

References for Additional Reading

1. E. B. Becker, G. F. Carey, and J. T. Oden, *Finite Elements, an Introduction*, Vol. I, Prentice Hall, Englewood Cliffs, New Jersey (1981).

2. J. N. Reddy, *An Introduction to the Finite Element Method*, Second Edition, McGraw-Hill, New York (1993).

3. D. S. Burnett, *Finite Element Analysis*, Addison-Wesley, Reading, Massachusetts (1987).

4. O. C. Zienkiewicz and R. L. Taylor, *The Finite Element Method, Vol. 1: Linear Problems*, McGraw-Hill, New York (1989).

5. T. J. R. Hughes, *The Finite Element Method, Linear Static and Dynamic Finite Element Analysis*, Prentice Hall, Englewood Cliffs, New Jersey (1987).

6. I. Ergatoudis, B. M. Irons, and O. C. Zienkiewicz, "Curved, Isoparametric, 'Quadrilateral', Elements for Finite Element Analysis," *International Journal of Solids and Structures*, **4**, 31–42 (1968).

7. B. M. Irons, "Engineering Applications of Numerical Integration in Stiffness Methods," *AIAA Journal*, **14**, 2035–2037 (1966).

8. B. M. Irons, "Quadrature Rules for Brick-Based Finite Elements," *International Journal for Numerical Methods in Engineering*, **3**, 293–294 (1971).

9. G. R. Cowper, "Gaussian Quadrature Formulas for Triangles," *International Journal for Numerical Methods in Engineering*, **7**, 405–408 (1973).

10. P. C. Hammer, O. P. Marlowe, and A. H. Stroud, "Numerical Integration over Simplexes and Cones," *Mathematics Tables Aid Computation*, National Research Council, Washington, D. C., **10**, 130–137 (1956).

11. A. N. Loxan, N. Davids, and A. Levenson, "Table of the Zeros of the Legendre Polynomials of Orders 1–16 and the Weight Coefficients for Gauss' Mechanical Quadrature Formula," *Bulletin of the American Mathematical Society*, **48**, 739–743 (1942).

12. C. T. Reddy and D. J. Shippy, "Alternative Integration Formulae for Triangular Finite Elements," *International Journal for Numerical Methods in Engineering*, **17**, 133–139 (1981).

13. B. Carnahan, H. A. Luther, and J. O. Wilkes, *Applied Numerical Methods*, John Wiley & Sons, New York (1969).

14. B. M. Irons, "A Frontal Solution Program for Finite Element Analysis," *International Journal for Numerical Methods in Engineering*, **2**, 5–32 (1970).

15. P. Hood, "Frontal Solution Program for Unsymmetric Matrices," *International Journal for Numerical Methods in Engineering*, **10**, 379–399 (1976).

16. Y. Hasbani and M. Engelman, "Out-of-Core Solution of Linear Equations with Non-Symmetric Coefficient Matrix," *Computers and Fluids*, **7**, 13–31 (1979).

17. R. L. Taylor, E. L. Wilson, and S. J. Sackett, "Direct Solution of Equations by Frontal and Variable Band, Active Column Methods," in *Proceedings of the U. S.-European Workshop on Nonlinear Finite Element Analysis in Structural Mechanics*, Bochum, W. Germany (1980).

18. G. E. Forsythe and C. B. Meler, *Computer Solution of Linear Algebraic Systems*, Prentice Hall, Englewood Cliffs, New Jersey (1967).

19. G. Strang, *Linear Algebra and Its Applications*, Second Edition, Academic Press, New York (1980).

20. P. Gresho, R. Lee, R. Sani, and T. Stullich, "On the Time-Dependent FEM Solution of the Incompressible Navier–Stokes Equations in Two and Three Dimensions," in *Recent Advances in Numerical Methods in Fluids, Vol. 1*, Pineridge Press, Swansea, United Kingdom (1980).

21. K. J. Bathe, *Numerical Methods in Finite Element Analysis*, Prentice Hall, Englewood Cliffs, New Jersey (1976).

22. T. Belytschko and T. J. R. Hughes (Eds.), *Computational Methods for Transient Analysis*, North-Holland, Amsterdam (1983).

23. J. H. Argyris and O. W. Scharpf, "Finite Elements in Time and Space," *The Aeronautical Journal of the Royal Society*, **73**, 1041–1044 (1969).

24. J. Donea, "On the Accuracy of Finite Element Solutions to the Transient Heat Conduction Equation," *International Journal for Numerical Methods in Engineering*, **8**, 103–110 (1974).

25. J. M. Bettencourt, O. C. Zienkiewicz, and G. Cantin, "Consistent Use of Finite Elements in Time and the Performance of Various Recurrence Schemes for the Heat Diffusion Equation," *International Journal for Numerical Methods in Engineering*, **17**, 931–938 (1981).

26. T. J. R. Hughes, "Unconditionally Stable Algorithms for Nonlinear Heat Conduction," *Computer Methods in Applied Mechanics and Engineering*, **10**, 135–139 (1977).

27. W. L. Wood and R. W. Lewis, "A Comparison of Time Marching Schemes for the Transient Heat Conduction Equation," *International Journal for Numerical Methods in Engineering*, **9**, 679–689 (1975).

28. W. L. Wood, "Control of Crank–Nicolson Noise in the Numerical Solution of the Heat Conduction Equation," *International Journal for Numerical Methods in Engineering*, **11**, 1059–1065 (1977).

29. C. W. Gear, *Numerical Initial Value Problems in Ordinary Differential Equations*, Prentice Hall, Englewood Cliffs, New Jersey (1971).

30. P. D. Lax and R. D. Richtmyer, "Survey of the Stability of Linear Finite Difference Equations," *Communications in Pure and Applied Mathematics*, **9**, 267–293 (1956).

31. J. D. Osnes, "A Method for Efficiently Incorporating Radiative Boundaries in Finite Element Programs," in *Proceedings of the Third International Conference on Numerical Methods in Thermal Problems*, Seattle, Washington (1983).

32. A. Soria and P. Pegon, "Quasi-Newton Iterative Strategies Applied to the Heat Diffusion Equation," *International Journal for Numerical Methods in Engineering*, **30**, 661–677 (1990).

33. R. E. Hogan and D. K. Gartling, "Finite Element Solutions of Nonlinear Problems with Heat Conduction and Enclosure Radiation," in *Proceedings of the 4th ASME/JSME Thermal Engineering Joint Conference*, Lahina, Hawaii (1995).

34. A. B. Shapiro, "FACET – A Radiation View Factor Computer Code for Axisymmetric, 2D Planar and 3-D geometries with Shadowing," Lawrence Livermore National Laboratory, UCID-19887, Livermore, California (1983).

35. A. F. Emery, "VIEWC User's Manual," University of Washington, Seattle, Washington (1988).

36. M. W. Glass, "Chaparral – A Library Package for Solving Large Enclosure Radiation Heat Transfer Problems," Sandia National Laboratories Report, SAND95-2049, Albuquerque, New Mexico (1995).

37. R. Siegel and J. R. Howell, *Thermal Radiation Heat Transfer*, McGraw-Hill, New York (1972).

38. A. F. Emery, D. Johansson, M. Lobo, and A. Abrous, "A Comparative Study of Methods for Computing the Diffuse Radiation Viewfactors for Complex Structures," *Journal of Heat Transfer*, **113**, 413–422 (1991).

39. K. Morgan, R. W. Lewis, and O. C. Zienkiewicz, "An Improved Algorithm for Heat Conduction Problems with Phase Change," *International Journal for Numerical Methods in Engineering*, **12**, 1191–1195 (1978).

40. W. Randolph and K. J. Bathe, "An Efficient Algorithm for Analysis of Nonlinear Heat Transfer with Phase Changes," *International Journal for Numerical Methods in Engineering*, **18**, 119–134 (1982).

41. S. Kimura and A. Bejan, "The 'Heatline' Visualization of Convective Heat Transfer," *Journal of Heat Transfer*, **105**, 916–919 (1983).

42. M. Rashid, "The Arbitrary Local Mesh Replacement Method: An Alternative to Remeshing for Crack Propagation Analysis," *Computer Methods in Applied Mechanics and Engineering*, **154**, 133–150 (1998).

43. I. S. Wichman, "On the Use of Operator-Splitting Methods for the Equations of Combustion," *Combustion and Flame*, **83**, 240–252 (1991).

44. T. R. Young, "CHEMEQ — A Subroutine for Solving Stiff Ordinary Differential Equations," NRL Memorandum Report 4091, Naval Research Laboratory, Washington, DC (1980)

45. D. K. Gartling, R. E. Hogan, and M. W. Glass, "COYOTE — A Finite Element Computer Program for Nonlinear Heat Conduction Problems," Sandia National Laboratories Report, SAND94-1173, Albuquerque, New Mexico (1994).

46. H. S. Carslaw and J. C. Jaeger, *Conduction of Heat in Solids*, Oxford University Press, Oxford (1959).

47. D. A. Glowka and C. M. Stone, "Thermal Response of Polycrystalline Diamond Compact Cutters Under Simulated Downhole Conditions," *Society of Petroleum Engineering Journal*, 143–156 (1985).

48. G. A. Knorovsky and S. N. Burchett, "The Effect of Heat Sinks in GTA Microwelding," in *Proceedings of the Second International Conference on Trends in Welding*, American Welding Society, Gatlinburg, Tennessee (1989).

49. R. E. Hogan, M. W. Glass, P. R. Schunk, and D. K. Gartling, "Thermal Analysis of the Cooling and Solidification of Investment Castings," in *Proceedings of the First International Conference on Transport Phenomena in Processing*, Pacific Institute for Thermal Engineering, Honolulu, Hawaii (1992).

50. J. N. Reddy, *Mechanics of Laminated Composite Plates. Theory and Analysis*, CRC Press, Boca Raton, Florida (1997).

Chapter 4

Viscous Incompressible Flows

4.1 Introduction

4.1.1 Background

The field of fluid mechanics is concerned with the motion of fluids and their effect on the surroundings. A fluid state of matter is often characterized by the relative mobility of the molecules that constitute the matter. Fluids exist either in the form of a gas or a liquid; more complex materials, such as mixtures, may also have a fluid behavior. In this chapter, finite element models are developed based on the weak formulation of the equations of viscous, incompressible fluids under isothermal conditions.

The motion of a fluid is governed by the global laws of conservation of mass, momenta, and energy (see Chapter 1). These equations consist of a set of coupled, nonlinear, partial differential equations in terms of the velocity components, temperature, and pressure. When temperature effects are not important, the energy equation is uncoupled from the momentum (i.e., the Navier–Stokes) equations. Thus for isothermal flows, we need to solve only the Navier–Stokes equations and continuity equation. When the Reynolds number for the flow is very low, the nonlinear terms due to inertial effects can be neglected, resulting in a linear boundary value problem. Such a flow is termed *Stokes flow*.

4.1.2 Governing Equations

The laws governing the flow of Newtonian fluids were reviewed in Chapter 1. The equations are specialized here for viscous fluids that are subject to the assumption of incompressibility. We will use the phrases *incompressible fluid* and *incompressible flow* interchangeably, while recognizing the subtle difference in meaning. The equations are written here for a fluid region Ω both in vector as well as Cartesian component form $\mathbf{x} = (x_1, x_2, x_3)$ in an Eulerian reference frame; time is denoted by t.

Conservation of Mass:

$$\nabla \cdot \mathbf{u} = 0 \tag{4.1.1a}$$

$$\frac{\partial u_i}{\partial x_i} = 0 \tag{4.1.1b}$$

Conservation of Momentum:

$$\rho_0 \left(\frac{\partial \mathbf{u}}{\partial t} + \mathbf{u} \cdot \nabla \mathbf{u} \right) - \nabla \cdot \sigma + \rho_0 \mathbf{f} = 0 \tag{4.1.2a}$$

$$\rho_0 \left(\frac{\partial u_i}{\partial t} + u_j \frac{\partial u_i}{\partial x_j} \right) - \frac{\partial \sigma_{ij}}{\partial x_j} + \rho_0 f_i = 0 \tag{4.1.2b}$$

Constitutive Equations:

$$\sigma = \tau - P\mathbf{I} ; \quad \tau = 2\mu\mathbf{D}$$

$$\mathbf{D} = \frac{1}{2} \left[(\nabla \mathbf{u}) + (\nabla \mathbf{u})^T \right] \tag{4.1.3a}$$

$$\sigma_{ij} = \tau_{ij} - P\delta_{ij} ; \quad \tau_{ij} = \mu D_{ij}$$

$$D_{ij} = \frac{1}{2} \left(\frac{\partial u_i}{\partial x_j} + \frac{\partial u_j}{\partial x_i} \right) \tag{4.1.3b}$$

In Eqs. (4.1.1)–(4.1.3), $\mathbf{u}$ denotes the velocity vector, σ is the total stress tensor, τ is the viscous stress tensor, P is the pressure, $\mathbf{f}$ is the body force vector (per unit mass), ρ_0 is the density, and μ is the shear viscosity of the fluid. The Cartesian components of the dependent variables are obvious from the above equations. In writing the Cartesian component form of the equations, the index notation with summation on repeated indices is used. The operator D/Dt denotes the material (or convective) time derivative,

$$\frac{D}{Dt} \equiv \frac{\partial}{\partial t} + \mathbf{u} \cdot \nabla, \quad \frac{D}{Dt} \equiv \frac{\partial}{\partial t} + u_j \frac{\partial}{\partial x_j} \tag{4.1.4}$$

Equation (4.1.1a) or (4.1.1b) is also known as the *continuity equation*, incompressibility constraint, or divergence-free condition on the velocity field. Equations in (4.1.2), with the constitutive relation indicated in Eqs. (4.1.3a,b), are known as the *Navier–Stokes equations*.

The boundary conditions are given by

$$\mathbf{u} = \mathbf{f}^u \quad \text{or} \quad u_i = f_i^u(s_j, t) \quad \text{on } \Gamma_u \tag{4.1.5a}$$

$$\mathcal{T} \equiv \sigma \cdot \hat{\mathbf{n}} = \mathbf{f}^T \quad \text{or} \quad \mathcal{T}_i \equiv \sigma_{ij}(s_j, t) n_j(s_j) = f_i^T(s_j, t) \quad \text{on } \Gamma_T \tag{4.1.5b}$$

where $\hat{\mathbf{n}}$ is the unit normal to the boundary and Γ_u and Γ_T are defined in Section 1.8.

The vector form of the equations in (4.1.1)–(4.1.3) can be written in any coordinate system that is suitable for the description and solution of the problem under consideration (see Section 1.4). Using the constitutive equations and deformation rate-velocity relations (4.1.3), Eqs. (4.1.1) and (4.1.2) can be expressed solely in terms of the velocity components and pressure. These are summarized below:

$$\frac{\partial u_i}{\partial x_i} = 0 \tag{4.1.6}$$

$$\rho_0 \left(\frac{\partial u_i}{\partial t} + u_j \frac{\partial u_i}{\partial x_j} \right) - \frac{\partial}{\partial x_j} \left[-P\delta_{ij} + \mu \left(\frac{\partial u_i}{\partial x_j} + \frac{\partial u_j}{\partial x_i} \right) \right] + \rho_0 f_i = 0 \tag{4.1.7}$$

It is possible to express these equations in terms of other variables, such as the stream function and vorticity. For example, in the two-dimensional case, we could write the momentum Eq. (4.1.7) in terms of the velocity **u** and *vorticity* ω as

$$\frac{\partial \omega}{\partial t} + \mathbf{u} \cdot \nabla \omega = \nu \nabla^2 \omega \tag{4.1.8}$$

where ν is the kinematic viscosity. In two dimensions, ω is a scalar function defined by

$$\omega \equiv \frac{\partial u_1}{\partial x_2} - \frac{\partial u_2}{\partial x_1} \tag{4.1.9}$$

Notice that in deriving (4.1.8) (by taking the curl of the momentum equations and using several vector identities) the pressure has been eliminated from the momentum equation. This is a situation that is highly sought after in some computational schemes, because of the difficulties associated with the pressure variable in an incompressible flow. A three-dimensional version of the vorticity transport equation is possible, though it sees relatively little use in computation due to the complexity of the vorticity boundary conditions. Again, in two dimensions, a stream function, ψ, may be defined such that the continuity equation (4.1.6) is identically satisfied. Using both the vorticity and stream function definitions the Navier–Stokes equations can be expressed as the pair consisting of Eq. (4.1.8) and

$$\nabla^2 \psi = -\omega \tag{4.1.10}$$

with the stream function defined by

$$u_2 \equiv -\frac{\partial \psi}{\partial x_1}, \quad u_1 \equiv \frac{\partial \psi}{\partial x_2}, \quad \omega \equiv \frac{\partial u_1}{\partial x_2} - \frac{\partial u_2}{\partial x_1} \tag{4.1.11}$$

The symbol ψ should not be confused with ψ_i used for interpolation functions; it should be clear from the context of the discussion. The introduction of these variables in the two-dimensional case reduces the number of equations that have to be solved from three to two. For the three-dimensional case, the stream function and vorticity are vector functions with three components each, and thus there is no reduction in the number of variables. The pressure, if required, may be computed from a Poisson equation of the form

$$\nabla^2 P = \nabla \cdot (\mathbf{f} + \nu \nabla^2 \mathbf{u} - \mathbf{u} \cdot \nabla \mathbf{u}) \tag{4.1.12}$$

which is obtained by taking the divergence of the momentum equation. It is also possible to rewrite the momentum equation entirely in terms of the stream function and reduce the flow problem to a single equation. Unfortunately, this approach leads to a fourth-order equation that increases the complexity of any computational method. Since these alternative formulations are not as computationally convenient as the original equations, especially for the imposition of boundary conditions, we will confine our attention here to the development of finite element models for Eqs. (4.1.6) and (4.1.7).

In the present study we shall consider two different finite element models of Eqs. (4.1.6) and (4.1.7). The first one is a natural formulation in which the weak forms of Eqs. (4.1.6) and (4.1.7) are used to construct the finite element model. The resulting finite element model is termed the *velocity-pressure model* or *mixed model*. The phrase 'mixed' is used because velocity variables are mixed with the force-like variable, pressure, and both types of variables are retained in a single formulation. The second model is based on the interpretation that the continuity equation (4.1.6) is an additional relation among the velocity components (i.e., a constraint among the u_i), and the constraint is satisfied in a least-squares (i.e., approximate) sense. This particular method of including the constraint in the formulation is known as the *penalty function method*, and the model is termed the *penalty-finite element model* (see [1–4]). In this case, the pressure variable is effectively eliminated from the formulation. It is informative to note that the velocity-pressure (or mixed) formulation is the same as the Lagrange multiplier formulation, wherein the constraint is included by means of the Lagrange multiplier method. The Lagrange multiplier turns out to be the negative of the pressure.

4.2 Mixed Finite Element Model

4.2.1 Weak Form

The starting point for the development of the finite element models of Eqs. (4.1.6) and (4.1.7) is their weak statements. As discussed in Section 2.4, the weak forms are developed using a three-step procedure. These steps are briefly reviewed here. For convenience, we denote expressions on the left side of Eqs. (4.1.6) and (4.1.7) by f_1 and $\mathbf{f}_2$, respectively. The weighted-integral statements of the two equations over a typical element Ω^e are given by

$$\int_{\Omega^e} Q f_1 \, d\mathbf{x} = 0 \tag{4.2.1}$$

$$\int_{\Omega^e} \mathbf{w} \cdot \mathbf{f}_2 \, d\mathbf{x} = 0 \tag{4.2.2}$$

where $(Q, \mathbf{w})$ are weight functions, which will be equated, in the Ritz–Galerkin finite element models, to the interpolation functions used for $(P, \mathbf{u})$, respectively (see Reddy [1] for details).

To obtain the weak form of Eq. (4.1.7), we employ integration-by-parts to equally distribute integration between the dependent variables and the weight functions. However, in any problem, such trading of differentiability is subjected to the restriction that the resulting boundary expressions are physically meaningful. Otherwise, the secondary variables of the formulation may not be the quantities the physical problem admits as the boundary conditions. An examination of the boundary stress components T_i in Eq. (4.1.5) shows that the pressure term occurs as a part of the expression [see Eq. (4.1.3)]. For example, T_1 is the x_1-component of the total boundary stress, which is the sum of the viscous boundary stress and the hydrostatic boundary stress,

$$T_1 = \mu \left[2\frac{\partial u_1}{\partial x_1}n_1 + (\frac{\partial u_1}{\partial x_2} + \frac{\partial u_2}{\partial x_1})n_2 + (\frac{\partial u_1}{\partial x_3} + \frac{\partial u_3}{\partial x_1})n_3 \right] + (-Pn_1) \qquad (4.2.3)$$

Here (n_1, n_2, n_3) denote the components of the unit normal vector on the boundary. It is clear from the above expression that the entire expression in the square brackets of Eq. (4.1.7) must be integrated by parts in the momentum equations to obtain T_i in the boundary expression. No integration-by-parts is used in the continuity equation (4.1.6) because no relaxation of differentiability on u_i can be accomplished; further, the resulting boundary conditions will not be physically justifiable.

Keeping the above comments in mind, we carry out the remaining two steps of the weak formulation, and obtain the following integral statements:

$$0 = \int_{\Omega^e} Q \frac{\partial u_i}{\partial x_i} \, d\mathbf{x} \qquad (4.2.4)$$

$$0 = \int_{\Omega^e} \left[\rho_0 \left(w_i \frac{\partial u_i}{\partial t} + w_i u_j \frac{\partial u_i}{\partial x_j} \right) + \frac{\partial w_i}{\partial x_j} \left(-P\delta_{ij} + \mu(\frac{\partial u_i}{\partial x_j} + \frac{\partial u_j}{\partial x_i}) \right) \right.$$
$$\left. - \rho_0 w_i f_i \right] d\mathbf{x} - \oint_{\Gamma^e} w_i T_i \, ds \qquad (4.2.5)$$

This completes the weak form development.

4.2.2 Finite Element Model

Since we are developing the Ritz–Galerkin finite element models, the choice of the weight functions is restricted to the spaces of approximation functions used for the pressure and velocity fields. Suppose that the dependent variables (u_i, P) are approximated by expansions of the form

$$u_i(\mathbf{x}, t) = \sum_{m=1}^{M} \psi_m(\mathbf{x})u_i^m(t) = \mathbf{\Psi}^T \mathbf{u}_i \qquad (4.2.6a)$$

$$P(\mathbf{x}, t) = \sum_{l=1}^{L} \phi_l(\mathbf{x})P_l(t) = \mathbf{\Phi}^T \mathbf{P} \qquad (4.2.6b)$$

where $\boldsymbol{\Psi}$ and $\boldsymbol{\Phi}$ are (column) vectors of interpolation (or shape) functions, and $\mathbf{u}_i$ and $\mathbf{P}$ are vectors of nodal values of velocity components and pressure, respectively. The weight functions $(Q, \mathbf{w})$ have the following correspondence (see [1] for further details),

$$Q \approx \phi_l, \quad \mathbf{w} \approx \psi_m \tag{4.2.7}$$

Substitution of Eqs. (4.2.6) and (4.2.7) into Eqs. (4.2.4) and (4.2.5) results in the following finite element equations:

Continuity:

$$-\left[\int_{\Omega^e} \boldsymbol{\Phi} \frac{\partial \boldsymbol{\Psi}^T}{\partial x_i} d\mathbf{x}\right] \mathbf{u}_i = 0 \tag{4.2.8}$$

i-th Momentum:

$$\left[\int_{\Omega^e} \rho_0 \boldsymbol{\Psi} \boldsymbol{\Psi}^T d\mathbf{x}\right] \dot{\mathbf{u}}_i + \left[\int_{\Omega^e} \rho_0 \boldsymbol{\Psi} (\boldsymbol{\Psi}^T \mathbf{u}_j) \frac{\partial \boldsymbol{\Psi}^T}{\partial x_j} d\mathbf{x}\right] \mathbf{u}_i + \left[\int_{\Omega^e} \mu \frac{\partial \boldsymbol{\Psi}}{\partial x_j} \frac{\partial \boldsymbol{\Psi}^T}{\partial x_j} d\mathbf{x}\right] \mathbf{u}_i$$

$$+ \left[\int_{\Omega^e} \mu \frac{\partial \boldsymbol{\Psi}}{\partial x_j} \frac{\partial \boldsymbol{\Psi}^T}{\partial x_i} d\mathbf{x}\right] \mathbf{u}_j - \left[\int_{\Omega^e} \frac{\partial \boldsymbol{\Psi}}{\partial x_i} \boldsymbol{\Phi}^T d\mathbf{x}\right] \mathbf{P}$$

$$= \left[\int_{\Omega^e} \rho_0 \boldsymbol{\Psi} f_i d\mathbf{x}\right] + \left\{\oint_{\Gamma^e} \mathcal{T}_i \boldsymbol{\Psi} \, ds\right\} \tag{4.2.9}$$

where superscript $(\cdot)^T$ denotes a transpose of the enclosed vector or matrix. The above equations can be written symbolically in matrix form as

Continuity:

$$-\mathbf{Q}^T \mathbf{u} = 0 \tag{4.2.10}$$

Momentum:

$$\mathbf{M}\dot{\mathbf{u}} + \mathbf{C}(\mathbf{u})\mathbf{u} + \mathbf{K}\mathbf{u} - \mathbf{Q}\mathbf{P} = \mathbf{F} \tag{4.2.11}$$

where the superposed dot represents a time derivative and $\mathbf{u} = \{\mathbf{u}_1, \mathbf{u}_2, \mathbf{u}_3\}^T$. For the three-dimensional case, Eqs. (4.2.10) and (4.2.11) have the following explicit matrix form [the continuity equation (4.2.10) and momentum equation (4.2.11) are combined into one]:

$$\begin{Bmatrix} \mathbf{F}_1 \\ \mathbf{F}_2 \\ \mathbf{F}_3 \\ 0 \end{Bmatrix} = \begin{bmatrix} \mathbf{M} & 0 & 0 & 0 \\ 0 & \mathbf{M} & 0 & 0 \\ 0 & 0 & \mathbf{M} & 0 \\ 0 & 0 & 0 & 0 \end{bmatrix} \begin{Bmatrix} \dot{\mathbf{u}}_1 \\ \dot{\mathbf{u}}_2 \\ \dot{\mathbf{u}}_3 \\ \dot{\mathbf{P}} \end{Bmatrix} + \begin{bmatrix} \mathbf{C}(\mathbf{u}) & 0 & 0 & 0 \\ 0 & \mathbf{C}(\mathbf{u}) & 0 & 0 \\ 0 & 0 & \mathbf{C}(\mathbf{u}) & 0 \\ 0 & 0 & 0 & 0 \end{bmatrix} \begin{Bmatrix} \mathbf{u}_1 \\ \mathbf{u}_2 \\ \mathbf{u}_3 \\ \mathbf{P} \end{Bmatrix} +$$

$$\begin{bmatrix} 2\mathbf{K}_{11} + \mathbf{K}_{22} + \mathbf{K}_{33} & \mathbf{K}_{12} & \mathbf{K}_{31} & -\mathbf{Q}_1 \\ \mathbf{K}_{21} & \mathbf{K}_{11} + 2\mathbf{K}_{22} + \mathbf{K}_{33} & \mathbf{K}_{23} & -\mathbf{Q}_2 \\ \mathbf{K}_{31} & \mathbf{K}_{32} & \mathbf{K}_{11} + \mathbf{K}_{22} + 2\mathbf{K}_{33} & -\mathbf{Q}_3 \\ -\mathbf{Q}_1^T & -\mathbf{Q}_2^T & -\mathbf{Q}_3^T & 0 \end{bmatrix} \begin{Bmatrix} \mathbf{u}_1 \\ \mathbf{u}_2 \\ \mathbf{u}_3 \\ \mathbf{P} \end{Bmatrix} \tag{4.2.12}$$

The coefficient matrices shown in Eq. (4.2.12) are defined by

$$\mathbf{M} = \int_{\Omega^e} \rho_0 \mathbf{\Psi \Psi}^T \, d\mathbf{x} \; ; \quad \mathbf{C(u)} = \int_{\Omega^e} \rho_0 \mathbf{\Psi} (\mathbf{\Psi}^T \mathbf{u}_j) \frac{\partial \mathbf{\Psi}^T}{\partial x_j} \, d\mathbf{x}$$

$$\mathbf{K}_{ij} = \int_{\Omega^e} \mu \frac{\partial \mathbf{\Psi}}{\partial x_i} \frac{\partial \mathbf{\Psi}^T}{\partial x_j} \, d\mathbf{x} \; ; \quad \mathbf{Q}_i = \int_{\Omega^e} \frac{\partial \mathbf{\Psi}}{\partial x_i} \mathbf{\Phi}^T \, d\mathbf{x}$$

$$\mathbf{F}_i = \int_{\Omega^e} \rho_0 \mathbf{\Psi} f_i \, d\mathbf{x} + \oint_{\Gamma^e} \mathbf{\Psi} \mathcal{T}_i \, ds \tag{4.2.13}$$

where a sum on repeated indices is implied in the definition of $\mathbf{C}$.

The two sets of interpolation functions used in Eq. (4.2.6) should be of the Lagrange type, i.e., derived by interpolating only the values of the functions, and not their derivatives. There are two different finite elements associated with the two field variables (u_i, P), and hence there are two different finite element meshes corresponding to the two variables over the same domain, Ω. If one of the meshes contains the other mesh as a subset, then we choose to display the first mesh and indicate the nodal degrees of freedom associated with the nodes of a typical element of the mesh.

The interpolation used for the pressure variable should be different from that used for the velocities, because the weak forms in Eqs. (4.2.4) and (4.2.5) contain the first derivatives of the velocities u_i and no derivatives of the pressure P. In addition, the essential boundary conditions of the formulation do not include specification of the pressure; it enters the boundary conditions as a part of the natural boundary conditions. This implies that the pressure variable need not be carried as a variable that is continuous across interelement boundaries. These observations lead to the conclusion that the pressure variable should be interpolated with functions that are one order less than those used for the velocity field and that the approximation may be discontinuous (i.e., not continuous from one element to other). Thus, quadratic interpolation of u_i and discontinuous linear interpolation of P are admissible. Models that use equal interpolation of the velocities and pressure with this formulation are known to give inaccurate results (see [5–15]). This heuristic argument for unequal interpolation anticipates the formal theory, which will be cited in Section 4.5.2 with the discussion on appropriate interpolation functions.

Returning to Eqs. (4.2.10) and (4.2.11), they can be combined into one,

$$\begin{bmatrix} \mathbf{M} & \mathbf{0} \\ \mathbf{0} & \mathbf{0} \end{bmatrix} \begin{Bmatrix} \dot{\mathbf{u}} \\ \dot{\mathbf{P}} \end{Bmatrix} + \begin{bmatrix} \mathbf{C(u)} + \mathbf{K(u)} & -\mathbf{Q} \\ -\mathbf{Q^T} & \mathbf{0} \end{bmatrix} \begin{Bmatrix} \mathbf{u} \\ \mathbf{P} \end{Bmatrix} = \begin{Bmatrix} \mathbf{F} \\ \mathbf{0} \end{Bmatrix} \tag{4.2.14}$$

or in a more symbolic format as

$$\overline{\mathbf{M}} \dot{\mathbf{U}} + \overline{\mathbf{K}} \mathbf{U} = \overline{\mathbf{F}} \tag{4.2.15a}$$

where

$$\mathbf{U} = \{\mathbf{u}_1, \mathbf{u}_2, \mathbf{u}_3, \mathbf{P}\}^T \tag{4.2.15b}$$

The general form of Eq. (4.2.15a) is the same as the nonlinear diffusion equation (3.6.3). Therefore, the time-approximation schemes discussed in Section 3.7.3 are readily applicable to the ordinary differential equations in (4.2.14) and (4.2.15a).

4.3 Penalty Finite Element Models

4.3.1 Introduction

The penalty function method, like the Lagrange multiplier method, allows us to reformulate a problem with constraints as one without constraints (see [1–4, 12–21]). The problem at hand, namely, the flow of a viscous incompressible fluid, must first be restated as a variational problem subjected to a constraint. For the purpose of describing the penalty function method, we consider the steady Stokes flow problem (i.e., without time-dependent and nonlinear terms) in two dimensions:

Conservation of mass:

$$\frac{\partial u_1}{\partial x_1} + \frac{\partial u_2}{\partial x_2} = 0 \tag{4.3.1}$$

Conservation of momentum:

$$-\frac{\partial}{\partial x_1}\left(2\mu\frac{\partial u_1}{\partial x_1}\right) - \frac{\partial}{\partial x_2}\left[\mu\left(\frac{\partial u_1}{\partial x_2} + \frac{\partial u_2}{\partial x_1}\right)\right] + \frac{\partial P}{\partial x_1} - \rho_0 f_1 = 0 \tag{4.3.2a}$$

$$-\frac{\partial}{\partial x_1}\left[\mu\left(\frac{\partial u_1}{\partial x_2} + \frac{\partial u_2}{\partial x_1}\right)\right] - \frac{\partial}{\partial x_2}\left(2\mu\frac{\partial u_2}{\partial x_2}\right) + \frac{\partial P}{\partial x_2} - \rho_0 f_2 = 0 \tag{4.3.2b}$$

The weak form of these equations, also known as the *variational problem,* is given by (see Reddy [1,2]):

$$B((w_1, w_2, Q), (u_1, u_2, P)) = \ell(w_1, w_2, Q) \tag{4.3.3}$$

where (w_1, w_2, Q) are the weight functions used for the momentum and continuity equations, respectively. Also, $B(\cdot, \cdot)$ is a *bilinear form* [i.e., an expression that is linear in (w_1, w_2, Q) and (u_1, u_2, P)] and $\ell(\cdot)$ is a *linear form* that in the present case are defined by

$$B(\cdot, \cdot) = \mu \int_{\Omega^e} \left[2\left(\frac{\partial w_1}{\partial x_1}\frac{\partial u_1}{\partial x_1} + \frac{\partial w_2}{\partial x_2}\frac{\partial u_2}{\partial x_2}\right) + \left(\frac{\partial w_1}{\partial x_2} + \frac{\partial w_2}{\partial x_1}\right)\left(\frac{\partial u_1}{\partial x_2} + \frac{\partial u_2}{\partial x_1}\right)\right] d\mathbf{x}$$

$$- \int_{\Omega^e} \left[\left(\frac{\partial w_1}{\partial x_1} + \frac{\partial w_2}{\partial x_2}\right)P + \left(\frac{\partial u_1}{\partial x_1} + \frac{\partial u_2}{\partial x_2}\right)Q\right] d\mathbf{x} \tag{4.3.4a}$$

$$\ell(\cdot) = \int_{\Omega^e} \rho_0(f_1 w_1 + f_2 w_2)\, d\mathbf{x} + \oint_{\Gamma^e} (\mathcal{T}_1 w_1 + \mathcal{T}_2 w_2)\, ds \tag{4.3.4b}$$

and (T_1, T_2) are the boundary stress components,

$$T_1 = (2\mu\frac{\partial u_1}{\partial x_1} - P)n_1 + \mu(\frac{\partial u_1}{\partial x_2} + \frac{\partial u_2}{\partial x_1})n_2$$

$$T_2 = \mu(\frac{\partial u_1}{\partial x_2} + \frac{\partial u_2}{\partial x_1})n_1 + (2\mu\frac{\partial u_2}{\partial x_2} - P)n_2 \qquad (4.3.5)$$

The finite element model based on the weak form (4.3.3) is a special case of the mixed finite element model given in Eq. (4.2.14). Equation (4.2.14) is more general than the problem at hand in that Eq. (4.2.14) is valid for time-dependent Navier–Stokes equations. In the interest of completeness, the penalty function method is described first using the steady Stokes problem. The results will then be generalized to the time-dependent Navier–Stokes equations. We describe two different penalty finite element models of flows of viscous, incompressible fluids, namely, the reduced integration penalty model and the consistent penalty model.

4.3.2 Penalty Function Method

Suppose that the velocity field (u_1, u_2) is constrained to satisfy the continuity equation (4.3.1) identically. Then the weight functions (w_1, w_2) (being virtual variations of the velocity components) also satisfy the constraint condition (continuity equation)

$$\frac{\partial w_1}{\partial x_1} + \frac{\partial w_2}{\partial x_2} = 0 \qquad (4.3.6)$$

As a result, the second integral expression in the bilinear form (4.3.4a) drops out, and the pressure (and hence the weight function Q) does not appear explicitly in the variational problem (4.3.3). The resulting variational problem now can be stated as follows: among all vectors $\mathbf{u} = (u_1, u_2)$ which satisfy the continuity equation (4.3.1), find the one that satisfies the variational problem

$$B_0((w_1, w_2), (u_1, u_2)) = \ell_0(w_1, w_2) \qquad (4.3.7)$$

for all admissible weight functions $\mathbf{w} = (w_1, w_2)$ [i.e., those which satisfy the condition in Eq. (4.3.6)]. The bilinear and linear forms are defined by

$$B_0(\mathbf{w}, \mathbf{u}) = \mu \int_{\Omega^e} \left[2 \left(\frac{\partial w_1}{\partial x_1}\frac{\partial u_1}{\partial x_1} + \frac{\partial w_2}{\partial x_2}\frac{\partial u_2}{\partial x_2} \right) + \left(\frac{\partial w_1}{\partial x_2} + \frac{\partial w_2}{\partial x_1} \right)\left(\frac{\partial u_1}{\partial x_2} + \frac{\partial u_2}{\partial x_1} \right) \right] dx$$

$$(4.3.8a)$$

$$\ell_0(\mathbf{w}) = \int_{\Omega^e} \rho_0(f_1 w_1 + f_2 w_2)\ dx + \oint_{\Gamma^e} (T_1 w_1 + T_2 w_2)\ ds \qquad (4.3.8b)$$

Equation (4.3.7) is called a constrained variational problem, because the solution vector $\mathbf{u}$ is constrained to satisfy the continuity equation. We note that the bilinear form in Eq. (4.3.8a) is symmetric: $B_0(\mathbf{w}, \mathbf{u}) = B_0(\mathbf{u}, \mathbf{w})$,

$$B_0((w_1, w_2), (u_1, u_2)) = B_0((u_1, u_2), (w_1, w_2)) \qquad (4.3.9)$$

Whenever the bilinear form of a variational problem is symmetric in its arguments, it is possible to construct a quadratic functional such that the minimum of the quadratic functional is equivalent to the variational problem (see Reddy [1,2]). The quadratic functional for the problem at hand is given by the expression

$$I_0(\mathbf{u}) = \frac{1}{2}B_0(\mathbf{u}, \mathbf{u}) - \ell_0(\mathbf{u}) \tag{4.3.10}$$

where the velocity vector $\mathbf{u}$ satisfies the continuity equation (4.3.1). Now we can state that Eqs. (4.3.1) and (4.3.2) governing the steady flow of viscous incompressible fluids are equivalent to minimizing the quadratic functional $I_0(\mathbf{u})$ subjected to the constraint

$$G(\mathbf{u}) \equiv \frac{\partial u_1}{\partial x_1} + \frac{\partial u_2}{\partial x_2} = 0 \tag{4.3.11}$$

It should be remembered that the discussion presented in this section is directed toward the reformulation of the problem as one of a constrained problem so that the penalty function method can be used. As can be seen from the above development, the advantage of the formulation is that the pressure variable does not appear in the model. However, there are certain computational complications associated with the penalty finite element model. These will be discussed shortly.

In the penalty function method, the constrained problem is reformulated as an unconstrained problem as follows: minimize the modified functional

$$I_P(\mathbf{u}) \equiv I_0(\mathbf{u}) + \frac{\gamma_e}{2}\int_{\Omega^e}[G(\mathbf{u})]^2 d\mathbf{x} \tag{4.3.12}$$

where γ_e is called the *penalty parameter*. Note that the constraint is included in a least–squares sense into the functional. Seeking the minimum of the modified functional $I_P(\mathbf{u})$ is equivalent to seeking the minimum of both $I_0(\mathbf{u})$ and $G(\mathbf{u})$, the latter with respect to the weight γ_e. The larger the value of γ_e, the more exactly the constraint is satisfied. The necessary condition for the minimum of I_P is

$$\delta I_P = 0 \tag{4.3.13}$$

We have

$$0 = \int_{\Omega^e}\left[2\mu\frac{\partial\delta u_1}{\partial x_1}\frac{\partial u_1}{\partial x_1} + \mu\frac{\partial\delta u_1}{\partial x_2}\left(\frac{\partial u_1}{\partial x_2} + \frac{\partial u_2}{\partial x_1}\right)\right]d\mathbf{x} - \int_{\Omega^e}\rho_0 f_1\delta u_1\, d\mathbf{x}$$
$$- \oint_{\Gamma^e}\delta u_1\, T_1\, ds + \int_{\Omega^e}\gamma_e\frac{\partial\delta u_1}{\partial x_1}\left(\frac{\partial u_1}{\partial x_1} + \frac{\partial u_2}{\partial x_2}\right)d\mathbf{x} \tag{4.3.14a}$$

$$0 = \int_{\Omega^e}\left[2\mu\frac{\partial\delta u_2}{\partial x_2}\frac{\partial u_2}{\partial x_2} + \mu\frac{\partial\delta u_2}{\partial x_1}\left(\frac{\partial u_1}{\partial x_2} + \frac{\partial u_2}{\partial x_1}\right)\right]d\mathbf{x} - \int_{\Omega^e}\rho_0 f_2\delta u_2\, d\mathbf{x}$$
$$- \oint_{\Gamma^e}\delta u_2\, T_2\, ds + \int_{\Omega^e}\gamma_e\frac{\partial\delta u_2}{\partial x_2}\left(\frac{\partial u_1}{\partial x_1} + \frac{\partial u_2}{\partial x_2}\right)d\mathbf{x} \tag{4.3.14b}$$

These two statements provide the weak forms for the penalty finite element model with $\delta u_1 = w_1$ and $\delta u_2 = w_2$. We note that the pressure does not appear explicitly in the weak form (4.3.14), although it is a part of the boundary stresses defined in Eq. (4.3.5). An approximation for the pressure P can be post-computed from the relation (see [1,2,4,5,22,23])

$$P = -\gamma_e \left(\frac{\partial u_1}{\partial x_1} + \frac{\partial u_2}{\partial x_2} \right) \tag{4.3.15}$$

where $\mathbf{u} = \mathbf{u}(\gamma_e)$ is the solution of Eq. (4.3.14).

The time derivative terms and nonlinear terms can be added to Eq. (4.3.14) without affecting the above discussion. For the general three-dimensional case, the weak forms are given by

$$0 = \int_{\Omega^e} \rho_0\, \delta u_1 \left[\frac{\partial u_1}{\partial t} + \left(u_1 \frac{\partial u_1}{\partial x_1} + u_2 \frac{\partial u_1}{\partial x_2} + u_3 \frac{\partial u_1}{\partial x_3} \right) - f_1 \right] d\mathbf{x}$$
$$+ \int_{\Omega^e} \left[2\mu \frac{\partial \delta u_1}{\partial x_1} \frac{\partial u_1}{\partial x_1} + \mu \frac{\partial \delta u_1}{\partial x_2} \left(\frac{\partial u_1}{\partial x_2} + \frac{\partial u_2}{\partial x_1} \right) + \mu \frac{\partial \delta u_1}{\partial x_3} \left(\frac{\partial u_1}{\partial x_3} + \frac{\partial u_3}{\partial x_1} \right) \right] d\mathbf{x}$$
$$- \oint_{\Gamma^e} \delta u_1\, T_1\, ds + \int_{\Omega^e} \gamma_e \frac{\partial \delta u_1}{\partial x_1} \left(\frac{\partial u_1}{\partial x_1} + \frac{\partial u_2}{\partial x_2} + \frac{\partial u_3}{\partial x_3} \right) d\mathbf{x} \tag{4.3.16a}$$

$$0 = \int_{\Omega^e} \rho_0\, \delta u_2 \left[\frac{\partial u_2}{\partial t} + \left(u_1 \frac{\partial u_2}{\partial x_1} + u_2 \frac{\partial u_2}{\partial x_2} + u_3 \frac{\partial u_2}{\partial x_3} \right) - f_2 \right] d\mathbf{x}$$
$$+ \int_{\Omega^e} \left[2\mu \frac{\partial \delta u_2}{\partial x_2} \frac{\partial u_2}{\partial x_2} + \mu \frac{\partial \delta u_2}{\partial x_1} \left(\frac{\partial u_1}{\partial x_2} + \frac{\partial u_2}{\partial x_1} \right) + \mu \frac{\partial \delta u_2}{\partial x_3} \left(\frac{\partial u_2}{\partial x_3} + \frac{\partial u_3}{\partial x_2} \right) \right] d\mathbf{x}$$
$$- \oint_{\Gamma^e} \delta u_2\, T_2\, ds + \int_{\Omega^e} \gamma_e \frac{\partial \delta u_2}{\partial x_2} \left(\frac{\partial u_1}{\partial x_1} + \frac{\partial u_2}{\partial x_2} + \frac{\partial u_3}{\partial x_3} \right) d\mathbf{x} \tag{4.3.16b}$$

$$0 = \int_{\Omega^e} \rho_0\, \delta u_3 \left[\frac{\partial u_3}{\partial t} + \left(u_1 \frac{\partial u_3}{\partial x_1} + u_2 \frac{\partial u_3}{\partial x_2} + u_3 \frac{\partial u_3}{\partial x_3} \right) - f_3 \right] d\mathbf{x}$$
$$+ \int_{\Omega^e} \left[2\mu \frac{\partial \delta u_3}{\partial x_3} \frac{\partial u_3}{\partial x_3} + \mu \frac{\partial \delta u_3}{\partial x_1} \left(\frac{\partial u_1}{\partial x_3} + \frac{\partial u_3}{\partial x_1} \right) + \mu \frac{\partial \delta u_3}{\partial x_2} \left(\frac{\partial u_2}{\partial x_3} + \frac{\partial u_3}{\partial x_2} \right) \right] d\mathbf{x}$$
$$- \oint_{\Gamma^e} \delta u_3\, T_3\, ds + \int_{\Omega^e} \gamma_e \frac{\partial \delta u_3}{\partial x_3} \left(\frac{\partial u_1}{\partial x_1} + \frac{\partial u_2}{\partial x_2} + \frac{\partial u_3}{\partial x_3} \right) d\mathbf{x} \tag{4.3.16c}$$

This completes the description of the penalty function method as applied to viscous incompressible flows.

4.3.3 Reduced Integration Penalty Model

The basic penalty method may be implemented by either of two procedures. Historically, the first computations were carried out by directly using the previous variational statement and some numerical techniques to make the resulting computational problem tractable. The so-called reduced integration penalty (RIP) finite element model is obtained from Eqs. (4.3.16a–c) by substituting finite element interpolation (4.2.6a) for the velocity field, and setting $\delta u_i = \boldsymbol{\Psi}$:

$$\begin{bmatrix} \mathbf{M} & 0 & 0 \\ 0 & \mathbf{M} & 0 \\ 0 & 0 & \mathbf{M} \end{bmatrix} \begin{Bmatrix} \dot{\mathbf{u}}_1 \\ \dot{\mathbf{u}}_2 \\ \dot{\mathbf{u}}_3 \end{Bmatrix} + \begin{bmatrix} \mathbf{C(u)} & 0 & 0 \\ 0 & \mathbf{C(u)} & 0 \\ 0 & 0 & \mathbf{C(u)} \end{bmatrix} \begin{Bmatrix} \mathbf{u}_1 \\ \mathbf{u}_2 \\ \mathbf{u}_3 \end{Bmatrix} +$$

$$\begin{bmatrix} 2\mathbf{K}_{11} + \mathbf{K}_{22} + \mathbf{K}_{33} & \mathbf{K}_{21} & \mathbf{K}_{31} \\ \mathbf{K}_{12} & \mathbf{K}_{11} + 2\mathbf{K}_{22} + \mathbf{K}_{33} & \mathbf{K}_{32} \\ \mathbf{K}_{31} & \mathbf{K}_{32} & \mathbf{K}_{11} + \mathbf{K}_{22} + 2\mathbf{K}_{33} \end{bmatrix} \begin{Bmatrix} \mathbf{u}_1 \\ \mathbf{u}_2 \\ \mathbf{u}_3 \end{Bmatrix}$$

$$+ \begin{bmatrix} \hat{\mathbf{K}}_{11} & \hat{\mathbf{K}}_{12} & \hat{\mathbf{K}}_{13} \\ \hat{\mathbf{K}}_{21} & \hat{\mathbf{K}}_{22} & \hat{\mathbf{K}}_{23} \\ \hat{\mathbf{K}}_{31} & \hat{\mathbf{K}}_{32} & \hat{\mathbf{K}}_{33} \end{bmatrix} \begin{Bmatrix} \mathbf{u}_1 \\ \mathbf{u}_2 \\ \mathbf{u}_3 \end{Bmatrix} = \begin{Bmatrix} \mathbf{F}_1 \\ \mathbf{F}_2 \\ \mathbf{F}_3 \end{Bmatrix} \tag{4.3.17}$$

where $\mathbf{M}, \mathbf{C(u)}, \mathbf{K}_{ij}$, and $\mathbf{F}_i$ are the same as those defined in Eq. (4.2.13), and

$$\hat{\mathbf{K}}_{ij} = \int_{\Omega^e} \gamma_e \frac{\partial \boldsymbol{\Psi}}{\partial x_i} \frac{\partial \boldsymbol{\Psi}^T}{\partial x_j} \, dx \tag{4.3.18}$$

Equation (4.3.17) can be expressed symbolically as

$$\bar{\mathbf{M}}\dot{\mathbf{u}} + \left[\bar{\mathbf{C}}(\mathbf{u}) + \bar{\mathbf{K}}^1(\mu) + \bar{\mathbf{K}}^2(\gamma) \right] \mathbf{u} = \bar{\mathbf{F}} \tag{4.3.19}$$

where $\mathbf{u} = \{\mathbf{u}_1, \mathbf{u}_2, \mathbf{u}_3\}^T$. If the matrices indicated above are all constructed with the usual quadrature techniques, the penalty function equations are not solvable. The name for the method comes from the requirement that the penalty terms in $\bar{\mathbf{K}}^2$ be constructed with a reduced quadrature rule. Further details of the implementation are given in Section 4.5.3 and in [4,5].

4.3.4 Consistent Penalty Model

An alternative implementation of the penalty method is known in the literature as the consistent penalty method, and deals more directly with the finite element model than the variational statement. In this formulation, the pressure in the momentum equation (4.3.2) is replaced with [three-dimensional version of Eq. (4.3.15)]

$$P = -\gamma_e \left(\frac{\partial u_1}{\partial x_1} + \frac{\partial u_2}{\partial x_2} + \frac{\partial u_3}{\partial x_3} \right) = -\gamma_e \frac{\partial u_i}{\partial x_i} \tag{4.3.20}$$

and Eq. (4.3.1) is replaced with Eq. (4.3.20). The finite element model of Eq. (4.3.20) is given by

$$\left[\int_{\Omega^e} \boldsymbol{\Phi}\boldsymbol{\Phi}^T d\mathbf{x} \right] \mathbf{P} = - \left[\int_{\Omega^e} \gamma_e \boldsymbol{\Phi} \frac{\partial \boldsymbol{\Psi}^T}{\partial x_i} d\mathbf{x} \right] \mathbf{u}_i \tag{4.3.21}$$

or in terms of a matrix notation

$$\mathbf{M}_p\mathbf{P} = -\gamma_e \mathbf{Q}^T \mathbf{u} \tag{4.3.22}$$

or

$$\mathbf{P} = -\gamma_e \mathbf{M}_p^{-1} \mathbf{Q}^T \mathbf{u} \qquad (4.3.23)$$

where $\mathbf{u} = \{\mathbf{u}_1, \mathbf{u}_2, \mathbf{u}_3\}^T$, $\mathbf{Q}^T = (\mathbf{Q}_1, \mathbf{Q}_2, \mathbf{Q}_3)$, and $\mathbf{Q}_i$ and $\mathbf{M}_p$ are given by

$$\mathbf{Q}_i = \int_{\Omega^e} \frac{\partial \mathbf{\Psi}}{\partial x_i} \mathbf{\Phi}^T \, d\mathbf{x} \; ; \quad \mathbf{M}_p = \int_{\Omega^e} \mathbf{\Phi}\mathbf{\Phi}^T \, d\mathbf{x} \qquad (4.3.24)$$

When Eq. (4.3.23) is substituted for the pressure into the finite element model of the momentum equation (4.2.11), we obtain

$$\mathbf{M}\dot{\mathbf{u}} + (\mathbf{C}(\mathbf{u}) + \mathbf{K} + \mathbf{K}_p)\mathbf{u} = \mathbf{F} \qquad (4.3.25)$$

with

$$\mathbf{K}_p = \gamma_e \mathbf{Q}\mathbf{M}_p^{-1}\mathbf{Q}^T \qquad (4.3.26)$$

Equation (4.2.10) is not used in conjunction with Eq. (4.3.25) because the continuity equation is included in Eq. (4.3.23) through Eq. (4.3.25).

Equation (4.3.25) has the same general form as Eq. (4.3.19). The overall size of the penalty finite element model in Eq. (4.3.19) or (4.3.26) is reduced in comparison to the mixed finite element model in Eq. (4.2.15). To recover the pressure, the inverted form of Eq. (4.3.23) is used with a known velocity field. The particular form of the penalty finite element model described here is termed a *consistent penalty model* because it is derived from the consistent, mixed finite element approximation to the momentum and penalized continuity equations. This is in contrast to the reduced integration penalty (RIP) model described earlier, where the shape functions for the pressure do not enter the penalty matrix (4.3.18). Although the equations of the two models may be mathematically identical for certain choices of the pressure approximation and reduced integration, there are other differences that affect the numerical implementation. Numerical implementation of the consistent method described here relies on the ability to efficiently construct the $\mathbf{K}_p$ matrix, i.e., invert $\mathbf{M}_p$ at the element level. This requirement restricts the choices for the basis functions, $\mathbf{\Phi}$, used to represent the pressure [4–10]. Further comments on the basis functions will be given in Section 4.5.

4.4 Finite Element Models of Porous Flow

The derivation of the finite element model for the porous flow problems of Section 1.5 follows the development of previous sections in a completely analogous manner. For completeness, we present the mixed finite element model of the porous flow equations under isothermal conditions. The penalty finite element model of the equations can also be developed, following the discussion presented in Section 4.3.

The dependent variables are again approximated by the relations in Eqs. (4.2.6a,b). Substituting them into the weak statements of Eqs. (1.5.1) and (1.5.2) and omitting the temperature terms result in the following equations:

Continuity:

$$-\left[\int_{\Omega^e} \Phi \frac{\partial \Psi^T}{\partial x_i} \, d\mathbf{x}\right] \mathbf{u}_i = 0 \tag{4.4.1}$$

Momentum:

$$\left[\int_{\Omega^e} \frac{\rho_0}{\phi} \Psi \Psi^T \, d\mathbf{x}\right] \dot{\mathbf{u}}_i + \left[\int_{\Omega^e} \frac{\rho_0 \hat{c}}{\sqrt{\kappa}} \Psi \left(\Psi^T \|\mathbf{u}\|\right) \Psi^T \, d\mathbf{x}\right] \mathbf{u}_i + \left[\int_{\Omega^e} \frac{\mu_e}{\kappa} \Psi \Psi^T \, d\mathbf{x}\right] \mathbf{u}_i$$

$$+ \left[\int_{\Omega^e} \mu_e \frac{\partial \Psi}{\partial x_j} \frac{\partial \Psi^T}{\partial x_j} \, d\mathbf{x}\right] \mathbf{u}_i + \left[\int_{\Omega^e} \mu_e \frac{\partial \Psi}{\partial x_j} \frac{\partial \Psi^T}{\partial x_i} \, d\mathbf{x}\right] \mathbf{u}_j - \left[\int_{\Omega^e} \frac{\partial \Psi}{\partial x_i} \Phi^T \, d\mathbf{x}\right] \mathbf{P}$$

$$= \left[\int_{\Omega^e} \rho_0 f_i \Psi \, d\mathbf{x}\right] + \left\{\oint_{\Gamma^e} \mathcal{T}_i \Psi \, ds\right\} \tag{4.4.2}$$

where f_i denote the body force components. The acceleration tensor is assumed to be equal to $1/\phi$. Equations (4.4.1) and (4.4.2) can be written symbolically in the matrix form as

Continuity:

$$-\tilde{\mathbf{Q}}^T \mathbf{u} = 0 \tag{4.4.3}$$

Momentum:

$$\tilde{\mathbf{M}}\dot{\mathbf{u}} + \tilde{\mathbf{C}}(\mathbf{u})\mathbf{u} + \tilde{\mathbf{A}}\mathbf{u} + \tilde{\mathbf{K}}\mathbf{u} - \tilde{\mathbf{Q}}\mathbf{P} = \tilde{\mathbf{F}} \tag{4.4.4}$$

where $\mathbf{u} = \{\mathbf{u}_1, \mathbf{u}_2, \mathbf{u}_3\}^T$ and

$$\tilde{\mathbf{M}} = \int_{\Omega^e} \frac{\rho_0}{\phi} \Psi \Psi^T \, d\mathbf{x}; \quad \tilde{\mathbf{C}}(\mathbf{u}) = \int_{\Omega^e} \frac{\rho_0 \hat{c}}{\sqrt{\kappa}} \Psi \left(\Psi^T \|\mathbf{u}\|\right) \Psi^T \, d\mathbf{x} \tag{4.4.5}$$

$$\tilde{\mathbf{A}} = \int_{\Omega^e} \frac{\mu}{\kappa} \Psi \Psi^T \, d\mathbf{x}; \quad \tilde{\mathbf{K}}_{ij} = \int_{\Omega^e} \mu_e \frac{\partial \Psi}{\partial x_i} \frac{\partial \Psi^T}{\partial x_j} \, d\mathbf{x} \tag{4.4.6}$$

$$\tilde{\mathbf{Q}}_i = \int_{\Omega^e} \frac{\partial \Psi}{\partial x_i} \Phi^T \, d\mathbf{x}; \quad \tilde{\mathbf{F}}_i = \int_{\Omega^e} \rho_0 \Psi f_i d\mathbf{x} + \oint_{\Gamma^e} \Psi \mathcal{T}_i ds \tag{4.4.7}$$

where summation on repeated indices is implied. Equations (4.4.3) and (4.4.4) can be arranged into a single matrix equation of the same form as given previously for the Navier–Stokes equations [see Eq. (4.2.14)]:

$$\begin{bmatrix} \tilde{\mathbf{M}} & 0 \\ 0 & 0 \end{bmatrix} \left\{ \begin{array}{c} \dot{\mathbf{u}} \\ \dot{\mathbf{P}} \end{array} \right\} + \begin{bmatrix} \tilde{\mathbf{C}}(\mathbf{u}) + \tilde{\mathbf{A}} + \tilde{\mathbf{K}} & -\tilde{\mathbf{Q}} \\ -\tilde{\mathbf{Q}}^T & 0 \end{bmatrix} \left\{ \begin{array}{c} \mathbf{u} \\ \mathbf{P} \end{array} \right\} = \left\{ \begin{array}{c} \tilde{\mathbf{F}} \\ 0 \end{array} \right\} \tag{4.4.8}$$

The above equation may take on a somewhat different structure when the various porous flow models are invoked. When the inertia terms are unimportant, then $\tilde{\mathbf{C}} = 0$. For the familiar Darcy flow case ($\hat{c} = 0, \mu_e = 0$), the above equation simplifies to

$$\begin{bmatrix} 0 & 0 \\ 0 & 0 \end{bmatrix} \left\{ \begin{array}{c} \dot{\mathbf{u}} \\ \dot{\mathbf{P}} \end{array} \right\} + \begin{bmatrix} \tilde{\mathbf{A}} & -\tilde{\mathbf{Q}} \\ -\tilde{\mathbf{Q}}^T & 0 \end{bmatrix} \left\{ \begin{array}{c} \mathbf{u} \\ \mathbf{P} \end{array} \right\} = \left\{ \begin{array}{c} \tilde{\mathbf{F}} \\ 0 \end{array} \right\} \tag{4.4.9}$$

The full porous model in (4.4.8) is very comparable to the Navier–Stokes equations of the previous section and shares many of the computational characteristics of the viscous flow problem. In general, these mixed and penalty finite element equations are stable and convergent with the proper choice of interpolation function. The reduced system in Eq. (4.4.9) is more questionable, since the Darcy formulation is really a potential flow and does not require a mixed finite element formulation. Notice that the two equations represented in Eq. (4.4.9) can be combined to form a single scalar equation for the pressure. That is, solving the first equation for $\mathbf{u}$ (which is the Darcy equation)

$$\mathbf{u} = \tilde{\mathbf{A}}^{-1}\tilde{\mathbf{Q}}\mathbf{P} + \tilde{\mathbf{A}}^{-1}\tilde{\mathbf{F}} \qquad (4.4.10)$$

and substituting into the continuity (second) equation produces

$$\tilde{\mathbf{Q}}^T\tilde{\mathbf{A}}^{-1}\tilde{\mathbf{Q}}\mathbf{P} = \tilde{\mathbf{Q}}^T\tilde{\mathbf{A}}^{-1}\tilde{\mathbf{F}} \qquad (4.4.11)$$

which is a Poisson equation for the pressure. If the mixed method is used here, an inconsistency arises since the pressure shape function was of lower order than the velocity interpolation. A solution based on a low order pressure approxmation from Eq. (4.4.11), when substituted into Eq. (4.4.10), would not be consistent with the generation of an accurate higher order velocity solution. Darcy type flows, when approached directly (i.e., not through reduction of a Brinkman type equation) are usually formulated as a continuum Poisson equation that resembles the heat conduction problem. The pressure would then be the primary variable to be interpolated and the resulting velocity field would be of lower order and consistent. The mixed method can be made to work with Eq. (4.4.9) though its convergence is suboptimal. Further details on this problem and a demonstration of the porous flow model can be found in [24]. This completes the development of the mixed finite element model of the porous flow equations.

4.5 Computational Considerations

4.5.1 Properties of the Matrix Equations

It is important to recognize some of the basic features of the matrix equations in (4.2.12), (4.3.18), and (4.4.8) because these characteristics will heavily influence the choice of a solution procedure for the various types of problems studied here. The given matrix equations represent the discrete analogs of the basic conservation equations with each term representing a particular physical process: $\mathbf{M}$ represents the mass and $\mathbf{C}$ represents the velocity-dependent convective transport term. The viscous diffusion term is represented by $\mathbf{K}$, and $\mathbf{Q}$ represents the pressure gradient operator and $\mathbf{Q}^\mathbf{T}$ is the divergence operator. The term $\mathbf{F}$ contains body forces and surface forces.

An inspection of the structure of the individual matrices in Eq. (4.2.13) shows that $\mathbf{M}$ and $\mathbf{K}$ are symmetric, while $\mathbf{C}$ is unsymmetric. This makes $\bar{\mathbf{K}}$ in Eq. (4.2.15a) unsymmetric. Thus, for general flows, the solution procedure must deal with an unsymmetric system. Conversely, when material properties are constant and flow velocities (e.g., Reynolds number) are sufficiently small so that a Stokes approximation is valid, the equations are symmetric and linear.

An additional feature (or difficulty) of the mixed finite element model is the presence of zeroes on the matrix diagonals corresponding to the pressure variables [see Eq. (4.2.12)]. In formal terms, Eqs. (4.2.14), (4.3.20), (4.3.27), and (4.4.8) represent a set of ordinary differential equations in time. The fact that the pressure does not appear explicitly in the continuity equation imparts a number of strong characteristics to the discrete problem. For example, the lack of a time derivative for the pressure [see Eqs. (4.2.14) and (4.4.8)] makes the system time-singular in the pressure and precludes the use of purely explicit time-integration methods. The zero entries on the diagonal cause difficulties in the solution of linear algebraic equations resulting from (4.2.14) or (4.4.8). Direct equation solving methods must use some type of pivoting strategy, while the use of iterative solvers is severely handicapped by poor convergence behavior attributable mainly to the form of the constraint equation. There are a number of other technical problems associated with the incompressibility assumption that directly influence the computational problem. Questions regarding the impossibility of impulsively started flows, proper boundary conditions for derived equations (e.g., vorticity transport equation, pressure Poisson equation), and suitable initial conditions for time-dependent computations are all derived from the special form of the continuity equation. A thorough explanation of many of these topics can be found in the works of Reddy [1–3,12], Oden [4], Oden and Carey [5], Gresho [25–27], Gresho and Sani [28] and others [6–10].

4.5.2 Choice of Interpolation Functions

In the previous sections we have presented finite element formulations whereby the continuum boundary value problems of interest are reduced to discrete (i.e., finite-dimensional) systems of algebraic equations. In the present section, a discussion of the choice of interpolation functions used in various models is presented. We will limit our attention to the mixed and penalty finite element models. In a later section, some discussion of methods that are reformulated to lessen various computational difficulties will be provided.

Of central importance to the development of a finite element model is the choice of particular elements (i.e., interpolation functions) to be included in the element library for the models at hand. For the diffusion problem considered in Section 3.3, a wide variety of element types were presented. The finite element models of conductive heat transfer as well as viscous incompressible flows (see Section 4.2.4) require only the C^0-continuous functions to approximate the field variables. Thus, any of the Lagrange and serendipity family of interpolation functions are admissible for the interpolation of the velocity field in mixed and penalty finite element models.

The choice of interpolation functions used for the pressure variable in the mixed finite element model is further constrained by the special role the pressure plays in incompressible flows; recall that the pressure can be interpreted as a Lagrange multiplier that serves to enforce the incompressibility constraint on the velocity field. From Eqs. (4.2.4) and (4.2.7) it is seen that the approximation function Φ used for pressure is the weighting function for the continuity equation (or the incompressibility constraint). It has been shown, both numerically (see Reddy [3] and Taylor and Hood [29]) and theoretically (see Sani et al. [14,15]) that, in order to prevent an overconstrained system of discrete equations, the interpolation used for pressure must be at least one order lower than that used for the velocity field (i.e., unequal order interpolation). Further, pressure need not be made continuous across elements because the pressure variable does not constitute a primary variable of the weak form presented in Eqs. (4.2.4) and (4.2.5).

Convergent finite element approximations of problems with constraints are governed by the ellipticity requirement and the *Ladyzhenskaya–Babuska–Brezzi (LBB) condition* (see Reddy [2, pp. 454–461], Oden [4], Oden and Carey [5], and others [6–10]). The mixed and penalty finite elements used for viscous incompressible fluids must satisfy the LBB condition in order that they yield convergent solutions. It is by no means a simple task to rigorously prove whether every new element developed for viscous incompressible flows satisfies the LBB condition. Discussion of the LBB theory and its application to various elements is beyond the scope of the present study. An extensive description of the LBB condition and its relation to practical incompressible finite elements can be found in the text by Gresho and Sani [28]. Finite element models that circumvent the LBB restrictions have been proposed and demonstrated. These techniques invariably lead to other types of weighted residual formulations, such as augmented Galerkin or "stabilized" methods (see, for example, Hughes et al. [30], Tezduyar [31], and Franca et al. [32]) or least-squares approaches (see Jiang et al. [33] and Carey et al. [34]). The stabilized method will be discussed briefly in a later section.

In the present discussion, two basic finite elements for use in the mixed (also valid for penalty) finite element model have been selected for discussion: a two-dimensional quadrilateral and a three-dimensional hexahedron. The interpolation functions used for the velocity field as well as pressure will be identified. Comparable triangular and tetrahedral elements have been developed but are used much less often than the quadrilateral and hexahedral geometries.

Quadrilateral element (2-D)

One of the most commonly used elements for two-dimensional flows of viscous incompressible fluids is the nine-node rectangular element shown in Figure 4.5.1. The velocity components and any auxiliary variables that are present are approximated using the biquadratic Lagrange functions [also see Eq. (2.11.7)], which are given in Eq. (4.5.1). The functions are expressed in terms of the normalized coordinates for the element, (ξ, η), which vary from -1 to $+1$.

Several different pressure approximations are available when the velocities are approximated by quadratic Lagrange functions. The first is a continuous-bilinear approximation, in which the pressure is defined at the corner nodes of the element and is made continuous across element boundaries. The bilinear interpolation functions are defined in Eq. (4.5.2).

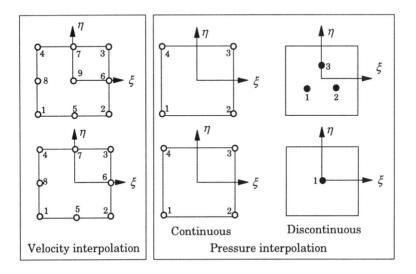

Figure 4.5.1. The quadrilateral elements used for the mixed model.

$$
\{\boldsymbol{\Psi}^e\} = \frac{1}{4}
\left\{
\begin{array}{c}
(1-\xi)(1-\eta)(-\xi-\eta-1) + (1-\xi^2)(1-\eta^2) \\
(1+\xi)(1-\eta)(\xi-\eta-1) + (1-\xi^2)(1-\eta^2) \\
(1+\xi)(1+\eta)(\xi+\eta-1) + (1-\xi^2)(1-\eta^2) \\
(1-\xi)(1+\eta)(-\xi+\eta-1) + (1-\xi^2)(1-\eta^2) \\
2(1-\xi^2)(1-\eta) - (1-\xi^2)(1-\eta^2) \\
2(1+\xi)(1-\eta^2) - (1-\xi^2)(1-\eta^2) \\
2(1-\xi^2)(1+\eta) - (1-\xi^2)(1-\eta^2) \\
2(1-\xi)(1-\eta^2) - (1-\xi^2)(1-\eta^2) \\
4(1-\xi^2)(1-\eta^2)
\end{array}
\right\}
\tag{4.5.1}
$$

$$
\{\boldsymbol{\Phi}^e\} = \frac{1}{4}
\left\{
\begin{array}{c}
(1-\xi)(1-\eta) \\
(1+\xi)(1-\eta) \\
(1+\xi)(1+\eta) \\
(1-\xi)(1+\eta)
\end{array}
\right\}
\tag{4.5.2}
$$

With this combination of velocity and pressure interpolation the element is usually designated as a $Q_2 Q_1$ element, where the first letter and subscript indicate a quadratic (second order) polynomial interpolation for the each component of velocity on the quadrilateral and the second letter and subscript refer to the linear (first order) polynomial pressure approximation on the quadrilateral.

A second pressure approximation involves a discontinuous, linear variation defined on the element by

$$\mathbf{\Phi} = \left\{ \begin{array}{c} 1 \\ x \\ y \end{array} \right\} \tag{4.5.3}$$

Here the unknowns are not nodal point values of the pressure but correspond to the coefficients in $P = a \cdot 1 + b \cdot x + c \cdot y$. In Eq. (4.5.3) the interpolation functions are written in terms of the global coordinates (x, y) for the problem; these functions could also be expressed in terms of the element coordinate system. An element with the velocity function in (4.5.1) and the pressure function in (4.5.3) is labeled a $Q_2 P_{-1}$ element where the P designation indicates a complete polynomial on the element and the negative subscript signifies the function is discontinuous at element boundaries. Notice that the P and Q designations differ in that P is a complete polynomial of degree subscript n in the spatial coordinates and Q is a polynomial (not necessarily complete) of degree subscript n in each spatial coordinate. The P designation is also used to label interpolations on simplex elements; Eq. (4.5.3) can be viewed as being defined on a triangle within the quadrilateral.

Two other pressure approximations also involve discontinuous functions on the element. If the pressure is defined at four points within the element and is not continuous at the element boundaries, then a $Q_2 Q_{-1}$ element is produced. This element is not LBB compliant but has been used with some success. Also, a $Q_2 Q_0$ element is possible where the pressure is a simple constant on each element.

An eight-node quadratic velocity element may also be defined with some of the same types of pressure interpolation. The designation $Q_2^{(8)}$ is typically used to indicate the 'serendipity' shape functions [see Eq. (2.11.8)]. Eight-node elements that have been used include the continuous pressure element $Q_2^{(8)} Q_1$, and the discontinuous pressure elements $Q_2^{(8)} Q_{-1}$ and $Q_2^{(8)} P_{-1}$.

The elements shown in Figure 4.5.1 are known to satisfy the LBB condition and thus give reliable and convergent solutions for velocity and pressure fields (see Oden [4], Le Tallec and Ruas [9], and Chapelle and Bathe [10]). Other elements that are not LBB stable may yield acceptable solutions for some problems but are not reliable for general applications. The most popular and most accurate two-dimensional quadrilateral is the $Q_2 P_{-1}$ element which has become a standard for incompressible flow problems. Though some of the eight-node elements mentioned above are fairly good, they have been supplanted by the nine-node elements.

When the element interpolation functions are written in terms of the normalized local coordinates (ξ, η), the relationship between the global coordinates (x, y) and the element coordinates (ξ, η) is defined by the transformation (see Sections 2.11 and 3.3)

$$x = \mathbf{\Upsilon}^T \mathbf{x} ; \qquad y = \mathbf{\Upsilon}^T \mathbf{y} \tag{4.5.4}$$

where $\mathbf{\Upsilon}$ is a vector of interpolation functions on the quadrilateral and the

$(\mathbf{x}, \mathbf{y})$ are vectors of coordinates, which describe the geometry of the element. This transformation is quite general, and for the nine-node quadrilateral element it allows accurate representation of straight or curved boundaries of a domain. For an isoparametric formulation, we have $\mathbf{\Upsilon} = \mathbf{\Psi}$, where $\mathbf{\Psi}$ is the interpolation functions used to approximate the velocity field. When $\mathbf{\Upsilon} = \mathbf{\Psi}$ is quadratic, a quadratic interpolation of the element boundary is possible. In a subparametric formulation, we set $\mathbf{\Upsilon} = \mathbf{\Phi}$, and only a linear interpolation of the element boundary is possible.

Hexahedral element (3-D)

A standard element used in the analysis of three-dimensional viscous flow problems is the eight-node, hexahedron (brick) shown in Figure 4.5.2. The velocity components are approximated using the following trilinear Lagrange functions [also see Eq. (3.3.1)]:

$$\{\mathbf{\Psi}^e\} = \frac{1}{8} \begin{Bmatrix} (1-\xi)(1-\eta)(1-\zeta) \\ (1+\xi)(1-\eta)(1-\zeta) \\ (1+\xi)(1+\eta)(1-\zeta) \\ (1-\xi)(1+\eta)(1-\zeta) \\ (1-\xi)(1-\eta)(1+\zeta) \\ (1+\xi)(1-\eta)(1+\zeta) \\ (1+\xi)(1+\eta)(1+\zeta) \\ (1-\xi)(1+\eta)(1+\zeta) \end{Bmatrix} \tag{4.5.5}$$

The functions are written in terms of the element natural coordinates (ξ, η, ζ).

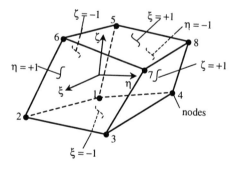

Figure 4.5.2: The eight-node brick element.

The pressure approximation for this element, being one order lower than $\mathbf{\Psi}$, is a constant (and obviously discontinuous between elements),

$$\Phi = \{c\} \tag{4.5.6}$$

This element is denoted by $Q_1 Q_0$. Higher-order elements are possible and follow the construction pattern of the quadrilateral elements. Most higher-order velocity elements have not found widespread use because of the computational expense involved for an element with a large number of unknowns.

Since the element shape functions are expressed in the normalized coordinates, the standard parametric transformation is needed

$$x = \mathbf{\Upsilon}^T \mathbf{x} \; ; \qquad y = \mathbf{\Upsilon}^T \mathbf{y} \; ; \qquad z = \mathbf{\Upsilon}^T \mathbf{z} \tag{4.5.7}$$

For an isoparametric formulation, we take $\mathbf{\Upsilon} = \mathbf{\Psi}$. Note that the eight-node brick element can only have straight edges unless a superparametric representation (i.e., $\mathbf{\Upsilon}$ is set to a triquadratic function) is employed. Superparametric descriptions are rarely used.

4.5.3 Evaluation of Element Matrices in Penalty Models

The numerical evaluation of the coefficient matrices appearing in Eq. (4.3.18) for the RIP method requires special consideration. This aspect is discussed here by considering the Stokes model for the steady-state case.

For the steady case, Eq. (4.3.20) is of the form

$$(\mu \bar{\mathbf{K}}^1 + \gamma \bar{\mathbf{K}}^2)\{\mathbf{u}\} = \{\bar{\mathbf{F}}\} \tag{4.5.8}$$

where $\bar{\mathbf{K}}^1$ is the contribution from the viscous terms and $\bar{\mathbf{K}}^2$ is from the penalty terms, which come from the incompressibility constraint. In theory, as we increase the value of γ, the conservation of mass is satisfied more exactly. However, in practice, for some large value of γ, the contribution from the viscous terms would be negligibly small compared to the penalty terms in the computer. Thus, if $\bar{\mathbf{K}}^2$ is a nonsingular (i.e., invertible) matrix, the solution of Eq. (4.3.21) for a large value of γ is trivial, $\{\mathbf{u}\} = \{\mathbf{0}\}$. While the solution satisfies the continuity equation, it does not satisfy the momentum equations. In this case the discrete problem (4.3.21) is said to be overconstrained or "locked". If $\bar{\mathbf{K}}^2$ is singular, then the sum $(\mu \bar{\mathbf{K}}^1 + \gamma \bar{\mathbf{K}}^2)$ is nonsingular (because $\bar{\mathbf{K}}^1$ is nonsingular), and a nontrivial solution to the problem can be obtained.

The numerical problem described above is eliminated by proper evaluation of the integrals defining $\bar{\mathbf{K}}^1$ and $\bar{\mathbf{K}}^2$. It is found that if the coefficients of $\bar{\mathbf{K}}^2$ (i.e., penalty terms) are evaluated using a numerical integration (see Section 2.12) rule of an order less than that required to integrate them exactly, the finite element equations (4.3.21) give acceptable solutions for the velocity field. This technique of under-integrating the penalty terms is known in the literature as *reduced* (order) *integration*. For example, if a linear rectangular element is used to approximate the velocity field in a two-dimensional problem, the matrix coefficients $\bar{\mathbf{K}}^1$, as well as $\mathbf{M}$ and $\mathbf{C}(\mathbf{u})$, are evaluated using the 2×2 Gauss quadrature, and $\bar{\mathbf{K}}^2$ are evaluated using the one-point (1×1) Gauss quadrature. The one-point quadrature yields a singular $\bar{\mathbf{K}}^2$. Therefore, Eq. (4.3.21) cannot be inverted, whereas $(\mu \bar{\mathbf{K}}^1 + \gamma \bar{\mathbf{K}}^2)$ is nonsingular and can be inverted (after assembly and imposition of boundary conditions) to obtain a good finite element solution of the original problem. When a quadratic rectangular element is used, the 3×3 Gauss quadrature is used to evaluate $\bar{\mathbf{K}}^1$, $\mathbf{M}$, and $\mathbf{C}(\mathbf{u})$, and the 2×2 Gauss quadrature is used to evaluate $\bar{\mathbf{K}}^2$. Similar comments apply to three-dimensional elements.

Unlike the RIP, the successful numerical implementation of the consistent penalty method described by Eq. (4.3.26) does not require special quadrature rules. However, the method does depend on the ability to efficiently invert $\mathbf{M}_p$ at the element level during construction of the $\mathbf{K}_p$ matrix. This requirement restricts the choices for the basis functions, $\mathbf{\Phi}$, used to represent the pressure [4–10]. If the pressure interpolation is continuous between elements, then $\mathbf{M}_p$ cannot be inverted at the element level; it is a global matrix whose inverse is a full matrix. When the pressure is approximated with a discontinuous interpolant, the $\mathbf{M}_p$ can be inverted for each element and the rest of terms in $\mathbf{K}_p$ easily constructed. Elements such as Q_2P_{-1}, Q_2Q_{-1}, and Q_1Q_0 are candidates for use with a consistent penalty method.

The choice of the penalty parameter is largely dictated by the ratio of the magnitude of penalty terms to the viscous and convective terms (or compared to the Reynolds number, R_e), the mesh, and the word length of the computer. The following range of γ is used in computations

$$\gamma = 10^4 R_e \text{ to } \gamma = 10^{12} R_e \tag{4.5.9}$$

The upper bound is often a function of the word length in the computer.

4.5.4 Pressure Calculation

The pressure is a variable of some importance in flow problems and is usually required as part of a computational solution. In the penalty function methods this requires some post-processing operations. For the consistent penalty scheme, the pressure recovery is relatively straightforward and relies on the the element level equation given in (4.3.22). That is,

$$\mathbf{M}_p\mathbf{P} = -\gamma_e\mathbf{Q}^T\mathbf{u} \tag{4.5.10}$$

which can be solved for $\mathbf{P}$ when the velocity field on the element is known. For the discontinuous interpolation functions used to construct $\mathbf{M}_p$, the solution to (4.5.10) yields a set of coefficients that allow the pressure to be evaluated at any point within the element. Typically, the pressure is evaluated at the nodes of the element and then averaged at common nodes between elements to arrive at a continuous pressure field.

The pressure recovery for the RIP method is more involved. The pressure should be computed, using Eq. (4.3.15) (or its three-dimensional equivalent), at quadrature points corresponding to the reduced integration rule. This is equivalent to using an interpolation for pressure that is one order less than the one used for the velocity field. The pressure computed using Eq. (4.3.15) at the reduced integration points, however, is not always reliable and accurate. Pressures predicted using linear elements, especially for coarse meshes, are seldom acceptable. Quadratic elements are known to yield more reliable results and in general, triangular elements perform poorly. Various techniques have been proposed in the literature to improve the accuracy of the recovered pressure fields (see [14,15,22,23]).

One alternative method for computing pressure is to use the pressure Poisson equation (obtained from the momentum equations)

$$\nabla^2 P = \nabla \cdot (\mathbf{f} + \nu \nabla^2 \mathbf{u} - \mathbf{u} \cdot \nabla \mathbf{u}) \tag{4.5.11}$$

Since Eq. (4.5.11) is similar to the heat conduction problem [with Q replaced by the right-hand side expression in Eq. (4.5.11)], it is straightforward to solve it using the finite element method. We will not discuss this method further.

Another alternative for pressure recovery was first proposed by Salonen and Aalto [35] and later used for different problems by Shiojima and Shimazaki [36]. By substituting the known velocity field and/or the shear stresses, the momentum equations can be expressed in terms of the pressure. These equations are solved using a least-squares finite element model (see [23]). Following the developments in Reference 23, a derivation of the pressure computation scheme is presented here.

Since the velocity components u_i are known from the solution of Eq. (4.3.20), their substitution into Eq. (4.1.7) results in the residual,

$$\mathbf{R} \equiv M_1 \hat{\mathbf{e}}_1 + M_2 \hat{\mathbf{e}}_2 + M_3 \hat{\mathbf{e}}_3 \tag{4.5.12}$$

where M_i is the ith component of the momentum equation and $\hat{\mathbf{e}}_i$ is the unit vector along the x_i coordinate direction:

$$M_i = \rho_0 \left(\frac{\partial u_i}{\partial t} + u_j \frac{\partial u_i}{\partial x_j} \right) - \frac{\partial}{\partial x_j} \left[-P \delta_{ij} + \mu \left(\frac{\partial u_i}{\partial x_j} + \frac{\partial u_j}{\partial x_i} \right) \right] + \rho_0 f_i \tag{4.5.13}$$

The only unknown in Eq. (4.5.12) is the pressure P. We determine P such that

$$\mathbf{R} = 0 \tag{4.5.14}$$

Equation (4.5.14) is a vector equation (i.e., there are three equations) in a single unknown, P. Hence, we require the square of the residual vector in Eq. (4.5.14) to be a minimum over an element with respect to P:

$$\frac{\partial}{\partial P} \left(\int_{\Omega^e} \mathbf{R} \cdot \mathbf{R} \, dx \right) = 0 \tag{4.5.15}$$

This method is known as the *least squares* method. Equation (4.5.15) provides the necessary weak statement for the determination of pressure:

$$\int_{\Omega^e} \mathbf{R} \cdot \frac{\partial \mathbf{R}}{\partial P} \, dx = 0 \tag{4.5.16a}$$

or

$$\int_{\Omega^e} M_i \frac{\partial M_i}{\partial P} \, d\mathbf{x} = 0 \qquad (4.5.16b)$$

Substituting $P = \sum_{i=1}^{L} P_i \phi_i$ into Eq. (4.5.16b), we obtain

$$\int_{\Omega^e} M_k \frac{\partial \phi_i}{\partial x_k} \, d\mathbf{x} = 0 \qquad (4.5.17)$$

where in Eq. (4.5.17), and in the following equations, sum on $k = 1, 2, 3$ is implied. Equation (4.5.17) gives the ith equation of the system of L equations for the nodal values $(P_1, P_2, \cdots, P_L)$. In matrix form, Eq. (4.5.17) can be expressed as

$$\mathbf{K}^* \mathbf{P} = \mathbf{F}^* \qquad (4.5.18)$$

where

$$K_{ij}^* = \int_{\Omega^e} \frac{\partial \phi_i}{\partial x_k} \frac{\partial \phi_j}{\partial x_k} \, d\mathbf{x}, \quad F_i^* = \int_{\Omega^e} \frac{\partial \phi_i}{\partial x_k} M_k \, d\mathbf{x} \qquad (4.5.19)$$

The right-hand side vector $\{\mathbf{F}^*\}$ consists of the contributions due to body forces, inertial forces, and viscous forces. The body force terms and inertial force terms are calculated at the nodes in the same manner as they were computed for velocity calculations. The viscous terms need some special treatment in their calculation. When linear elements are used, the viscous stresses are constant within the element and their derivatives vanish. The same problem arises in the finite element implementation of Eq. (4.5.11) for P because it involves second-order derivatives of the velocity field. To overcome this difficulty, the stresses are computed using the standard $2 \times 2 \times 2$ Gauss quadrature for the eight-noded (trilinear) brick elements. The stresses at the Gauss points are extrapolated to obtain the contribution of the stresses at the nodes. Such contributions from all surrounding elements are averaged to determine the nodal stresses. The main drawback of this method is that the stresses for the boundary nodes are not as accurate as those inside the domain. However, improvements are possible if the nodal stresses are calculated using an appropriate extrapolation scheme.

4.5.5 Traction Boundary Conditions

Equation (4.2.13) contains the contribution due to the boundary stresses applied to an element (sum on j)

$$\mathbf{F}_i^s = \oint_{\Gamma^e} \boldsymbol{\Psi} T_i \, ds = \oint_{\Gamma^e} \boldsymbol{\Psi} \sigma_{ij} n_j \, ds \qquad (4.5.20)$$

For example, for two-dimensional problems, Eq. (4.5.20) takes the explicit form

$$\mathbf{F}_1^s = \mathbf{F}_x^s = \oint_{\Gamma^e} \boldsymbol{\Psi} T_x \, ds = \oint_{\Gamma^e} \boldsymbol{\Psi} \sigma_{xx} n_x \, ds + \oint_{\Gamma^e} \boldsymbol{\Psi} \sigma_{xy} n_y \, ds \quad (4.5.21a)$$

$$\mathbf{F}_2^s = \mathbf{F}_y^s = \oint_{\Gamma^e} \boldsymbol{\Psi} T_y \, ds = \oint_{\Gamma^e} \boldsymbol{\Psi} \sigma_{yx} n_x \, ds + \oint_{\Gamma^e} \boldsymbol{\Psi} \sigma_{yy} n_y \, ds \quad (4.5.21b)$$

In general, the force vector for each equation contains contributions from both an applied normal stress and a shear stress. Boundary conditions could be specified directly in terms of the quantities shown in Eqs. (4.5.21a,b), i.e., the x and y components of the traction vector or the three independent components of the stress tensor. The pressure enters the calculation through the normal stress components:

$$\sigma_{xx} = -P + \tau_{xx}, \quad \sigma_{yy} = -P + \tau_{yy}, \quad \sigma_{zz} = -P + \tau_{zz}$$

As discussed in Section 3.5, the surface integrals shown in Eqs. (4.5.21a,b) can be evaluated in terms of the normalized coordinate for an element edge. Using the appropriate definition for the surface element ds

$$ds = |\mathbf{J}_s| \, d\xi_s = \left[\left(\frac{\partial \hat{\mathbf{\Upsilon}}^T}{\partial \xi_s} \mathbf{x} \right)^2 + \left(\frac{\partial \hat{\mathbf{\Upsilon}}^T}{\partial \xi_s} \mathbf{y} \right)^2 \right]^{\frac{1}{2}} d\xi_s = \Delta \, d\xi_s \qquad (4.5.22a)$$

and the components of the normal

$$n_x = \frac{1}{\Delta} \frac{\partial \hat{\mathbf{\Upsilon}}^T}{\partial \xi_s} \mathbf{y} \quad ; n_y = -\frac{1}{\Delta} \frac{\partial \hat{\mathbf{\Upsilon}}^T}{\partial \xi_s} \mathbf{x} \qquad (4.5.22b)$$

the expressions in Eqs. (4.5.21a,b) can be written as

$$\mathbf{F_x}^s = \int_{-1}^{+1} \mathbf{\Psi} \sigma_{xx} \frac{\partial \hat{\mathbf{\Upsilon}}^T}{\partial \xi_s} \mathbf{y} \, d\xi_s - \int_{-1}^{+1} \mathbf{\Psi} \sigma_{xy} \frac{\partial \hat{\mathbf{\Upsilon}}^T}{\partial \xi_s} \mathbf{x} \, d\xi_s \qquad (4.5.23a)$$

$$\mathbf{F_y}^s = \int_{-1}^{+1} \mathbf{\Psi} \sigma_{yx} \frac{\partial \hat{\mathbf{\Upsilon}}^T}{\partial \xi_s} \mathbf{y} \, d\xi_s - \int_{-1}^{+1} \mathbf{\Psi} \sigma_{yy} \frac{\partial \hat{\mathbf{\Upsilon}}^T}{\partial \xi_s} \mathbf{x} \, d\xi_s \qquad (4.5.23b)$$

Recall that $\hat{\mathbf{\Upsilon}}$ is a vector of edge functions for the element, and they may be either linear or quadratic depending on the type of mapping used to describe the geometry of the element. The stress components σ_{ij} in Eq. (4.5.23) are specified along the element boundary. The above derivation may be repeated in a completely analogous manner for the three-dimensional case using the appropriate relations for the surface element ds.

It should be noted that the traction boundary conditions are written in terms of the *total* stress components. The total normal stress components contain pressure as a part. For example, the normal stress $\sigma_{11} = \sigma_{xx}$ is given by [see Eq. (4.1.3b)]

$$\sigma_{11} = -P + 2\mu \left(\frac{\partial u_1}{\partial x_1} \right) \qquad (4.5.24)$$

In many practical cases, the viscous part of the normal stress is negligibly small and the normal stress is essentially equal to the pressure. When the viscous part is not negligible, the application of a normal stress boundary condition does not distinguish between contributions from the pressure and viscous parts but simply reflects their net effect.

The traction boundary conditions as written in Eqs. (4.5.23a,b) imply that the stress components σ_{ij} are known functions of position along the boundary. Though this is the most usual situation, other types of boundary conditions are possible. For example, a slip or "friction" boundary condition relates the shear stress to a change in velocity at the boundary [see Eq. (1.8.3)]. Along a boundary with $y = $ constant, this condition is written as

$$\sigma_{xy} = \frac{1}{\alpha}(u_x - u_b) \tag{4.5.25}$$

where u_x is the fluid velocity at the wall, u_x^s is the velocity of the boundary, and α is the slip coefficient. If this condition (or its multidimensional generalization) is substituted into Eqs. (4.5.23a,b), then a boundary matrix and vector combination is produced since the unknown fluid velocity appears in the equation. This is analogous to the convective heat transfer boundary condition in Eq. (2.10.1) or Eq. (3.5.12), and is treated in the same manner. Other types of boundary conditions, such as the surface tension condition, may involve other variables, but the basic method of implementation is similar in all cases.

In the discussion presented above, we have only considered the case where the applied tractions or stress components are specified directly in the global (x, y, z) coordinate frame. In many situations of practical interest, the stresses will only be known as components normal and tangent to the boundary surface that is not a coordinate surface. It is a fairly straightforward process to rewrite Eqs. (4.5.23a,b) so that this case is properly implemented. The situation is slightly more complicated if velocity components normal and tangent to an arbitrary noncoordinate surface are to be specified. The standard procedure in this situation is to first transform the momentum equations for the affected boundary nodes to a locally defined coordinate system oriented along the normal and tangent directions. An orthogonal transformation matrix can be constructed to operate on the appropriate element matrix equations. After transformation, the normal and tangential momentum equations can be constrained by the known values of the normal and/or tangential velocity components. After the finite element solution is obtained, the velocity components in the global coordinate frame can be obtained from the normal and tangential components computed on the boundary through a simple definition of the form (for two dimensions)

$$u_x = u_n n_x - u_t n_y , \quad u_y = u_n n_y + u_t n_x \tag{4.5.26}$$

where (u_n, u_t) denote the normal and tangential components of velocity and (n_x, n_y) are the components of the unit normal vector to the boundary. For incompressible flows, an essential part of this implementation is the definition of the local normal and tangential directions such that a divergence-free velocity field is maintained. A complete discussion of this topic can be found in the paper by Engelman et al. [37].

4.6 Solution of Nonlinear Equations

4.6.1 General Discussion

The finite element models in Eqs. (4.2.15a), (4.3.19), (4.3.25), and (4.4.8) represent a very large group of fluid mechanics problems. Unfortunately, the solution of the discrete equation set is notoriously difficult for all but the simplest flows. These numerical difficulties all have their origins in the physical description of the viscous, incompressible flow problem, i.e., the Navier–Stokes equations. The initial, boundary value problems of fluid dynamics are some of the most mathematically challenging in the physical sciences and translating them into a computational form does not simplify the problem. Viscous flows typically have a large spectrum of length and time scales, which in the computational setting implies high resolution requirements, large meshes, and very large equation sets (matrices). The description of advective transport in an Eulerian reference frame produces highly nonlinear and an unsymmetric equation set. Small fluid viscosities and even modest velocities combine to form large Reynolds numbers with the physical result being thin boundary layers. Computationally, the high Reynolds number means that the discrete equations behave very differently in different regions of the domain. Though the equations are formally elliptic over the entire domain, flow regions away from boundaries take on a very hyperbolic character. Finally, the assumption of incompressibility influences not only the formulation of the problem (mixed/penalty method, LBB stable elements) but has a strong influence on the permissible solution methods for the discrete equations. The ubiquitous nature of the incompressibility constraint is one of the reasons that make it difficult to catalog finite element solution methods for this class of problems. The form of the solution algorithm is dependent on the finite element formulation and the element choice, both of which may be altered in an attempt to satisfy incompressibility in an accurate and efficient computational algorithm.

In the following sections, the focus will be on solution methods for the equations resulting from the Galerkin finite element model (GFEM) using mixed formulation or a penalty formulation. The GFEM has proven to be very robust and well-suited to difficult flow simulations. Fixed point iteration methods, which will be described shortly, work well with this type of formulation and its associated finite elements. The fully coupled equation methods associated with the GFEM are not perfect, with the biggest drawback being their strong reliance on direct solution methods for the resulting linear equations. The matrix forms that arise from GFEM are unsymmetric, sparse, and positive, semi-definite [see Eq. (4.2.12)]. The direct solution algorithms (e.g., Gaussian elimination) for the matrix problem require some type of pivoting or equation reordering strategy to avoid problems with zero diagonal entries [see Eqs. (4.2.12) or (4.2.14)] and effect a solution. Iterative matrix solution methods (e.g., conjugate gradient) have not proven effective for these types of equations because adequate preconditioners are unavailable, the presence of the incompressibility constraint again being the major impediment.

Large, three-dimensional GFEM simulations are impractical for direct matrix solutions, and without other matrix solvers as an option, a change in solution approach is required to allow general application of the finite element method. A subsequent section describes one alternative that relies on decoupling or segregating the discrete equations prior to solution. The incompressibility constraint again appears prominently in this scheme as it forces a reformulation of the equations. However, a significant benefit of this reformulation is the ability to use either direct or iterative methods on the resulting linear equations. A final section in this catalog of solution methods outlines the "stabilized" formulations that alter the Galerkin-weighted residual form to circumvent the LBB condition and ultimately ease some computational difficulties. Notice that throughout the discussion of solution methods, the characteristics of the solution method play a large role in determining its utility. Direct and iterative methods used to solve linear systems are discussed briefly in Appendix B; this topic encompasses a large area in computational mathematics and is outside the scope of this book.

Before proceeding to the discussion of specific algorithms, it is appropriate to outline the common mathematical features of a solution technique that are basic to many solution methods. The present discussion will consider methods used for the linearization and iterative solution of the fully coupled, nonlinear equations arising from the use of the mixed or penalty finite element model for incompressible flows.

Using the symbolic format of Eq. (4.2.15), both mixed and penalty finite element models of this chapter can be written as

$$\overline{\mathbf{M}}\dot{\mathbf{U}} + \overline{\mathbf{K}}(\mathbf{U})\mathbf{U} = \overline{\mathbf{F}} \tag{4.6.1}$$

which represents a set of coupled, nonlinear, differential equations in time. For the time-dependent case, these equations are reduced to nonlinear algebraic equations by means of a time-approximation scheme, as described in Chapters 2 and 3. The resulting equations are of the form

$$\tilde{\mathbf{K}}(\mathbf{U}^*)\mathbf{U}^* = \tilde{\mathbf{F}} \tag{4.6.2}$$

which must be solved at each time step; the vector $\mathbf{U}^*$ represents the solution variables at some discrete time in the integration interval. For time-independent problems, Eq. (4.6.1) is simplified to a form such as (4.6.2) by omitting the $\dot{\mathbf{U}}$ term; the $\mathbf{U}^*$ in this case would represent the solution to the steady-state problem. Therefore, regardless of the type of problem under consideration, the solution process for the discrete system must ultimately be concerned with a set of nonlinear algebraic equations.

For convenience, Eq. (4.6.2) can be expressed simply as

$$\mathcal{R}(\mathbf{U}) = \mathbf{0} \tag{4.6.3}$$

where it is understood that $\mathcal{R}$ is in general a nonlinear operator and $\mathbf{U}$ is the vector of unknowns. Many of the solution methods of practical interest for Eq.

(4.6.3) can be regarded as variants of the method of successive approximation or functional iteration. Rewrite Eq. (4.6.3) as

$$\mathbf{U} = \mathcal{G}(\mathbf{U}) = \mathbf{U} - \mathcal{R}(\mathbf{U}) \tag{4.6.4}$$

which is usually termed a *fixed point problem* with $\mathbf{U}$ being the *fixed point* of the operator $\mathcal{G}$ (see Reddy [2]). The form of (4.6.4) suggests the following iteration scheme, where the superscript indicates the iteration number,

$$\mathbf{U}^{n+1} = \mathcal{G}(\mathbf{U}^n) = \mathbf{U}^n - \mathcal{R}(\mathbf{U}^n) \tag{4.6.5}$$

which is often termed the Picard method or the method of successive substitutions. The convergence of the Picard scheme (4.6.5) is intimately connected with the contraction mapping theorem (see Reddy [2] and Appendix C). In essence, successive substitution can be proven to converge if the operator $\mathcal{G}$ has certain properties (i.e., $\mathcal{G}$ is a *contraction mapping*) and the initial guess for $\mathbf{U}$ is sufficiently "close" to the solution. These properties are relatively difficult to prove for the discrete operators associated with the finite element equations described in this chapter. The true importance of Eq. (4.6.5) is in its suggestion of an iterative solution procedure which can be generalized to the following form

$$\mathbf{U}^{n+1} = \mathcal{G}(\mathbf{U}^n) = \mathbf{U}^n - \alpha^n \mathbf{A}(\mathbf{U}^n)\mathcal{R}(\mathbf{U}^n) \tag{4.6.6a}$$

where α^n is a constant (that may vary with the iteration number) and $\mathbf{A}(\mathbf{U}^n)$ is a nonsingular matrix which is a function of $\mathbf{U}$. The method specified by Eq. (4.6.6a) provides a recursive method for finding the fixed point of $\mathcal{G}$, i.e., the solution of $\mathcal{R}(\mathbf{U}) = \mathbf{0}$. The success of the algorithm will depend critically on the choices for the parameter α and the matrix $\mathbf{A}$. Note that the product $\mathbf{A}(\mathbf{U}^n)\mathcal{R}(\mathbf{U}^n)$ can be interpreted as a correction vector, $\mathbf{D}(\mathbf{U}^n)$, such that

$$\mathbf{U}^{n+1} = \mathbf{U}^n - \alpha^n \mathbf{A}(\mathbf{U}^n)\mathcal{R}(\mathbf{U}^n) = \mathbf{U}^n - \alpha^n \mathbf{D}(\mathbf{U}^n) \tag{4.6.6b}$$

where α is now the scalar magnitude of the correction. The basic iterative formula in (4.6.6a) defines a large class of methods, many of which are of value in the finite element context.

The correction vector $\mathbf{D}(\mathbf{x}^n)$ is often related to a gradient of $\mathcal{R}$ (or more often a gradient of $\mathcal{R}^\mathsf{T}\mathcal{R}$ (see Appendix C), and the resulting methods are termed *gradient methods*. If $\mathbf{D}$ can be constructed recursively, then (4.6.6a) leads to a number of iterative methods, such as the *method of steepest descent* and the *conjugate gradient methods*. These methods are particularly attractive for finite element analyses because the solution of a large matrix system is avoided; in some sense these are "explicit" methods since a global matrix need not be constructed. The major difficulty with these iterative schemes is the task of rationally selecting effective correction vectors. The correction vector $\mathbf{D}$ may also be constructed by defining the matrix $\mathbf{A}$; in this case the correction vector is generated from the solution of a matrix problem and the method is somewhat more "implicit" in nature. Methods in this family of algorithms include the *chord method* and *Newton's method*. These techniques are advantageous because of the many possible definitions for $\mathbf{A}$, though the requirement of a matrix solution is a significant disadvantage.

4.6.2 Fully Coupled Solution Methods

The general discussion of iterative methods for the steady-state problem will now be focused on the finite element equations for incompressible flows as generated by the mixed or penalty forms of the method of weighted residuals. These Galerkin finite element models (GFEM) are particularly important since many finite element models for flow problems are variations of this basic formulation.

For general flow problems that are independent of time, the equation set of interest is derived from (4.6.1) by setting $\dot{\mathbf{U}} = 0$. That is,

$$\overline{\mathbf{K}}(\mathbf{U})\mathbf{U} = \overline{\mathbf{F}} \qquad (4.6.7)$$

where $\mathbf{U}$ is defined as a vector containing all of the unknowns of the problem, $\overline{\mathbf{K}}$ contains the advection and diffusion terms as well as the incompressibility constraint, and $\overline{\mathbf{F}}$ contains all of the applied boundary and volumetric forcing functions. The choice of an iterative method for the nonlinear, algebraic system in (4.6.7) is governed by several considerations. The selected method should be capable of providing converged solutions for a wide range of problems with minimal sensitivity to variations in boundary conditions, material properties, and initial data, i.e., the method should have a large radius of convergence (see Appendix C). The rate of convergence should also be reasonably high for reasons of computational economy; the computational work per iterative cycle should also be "small" to increase computational efficiency. Though it is reasonable to believe that there is a "best" algorithm for each combination of characteristics in the equation set, it is impractical to have all of these solution methods available in a single computer code. Rather, a few of the most generally applicable schemes are usually made available with the hope that the characteristics of any particular problem will fall within the bounds of these few techniques. In this section the main focus is placed on methods that have proven to be of value for the most general types of flow problems; no attention is given to methods that only perform well for special types of flows. Some of these methods will be familiar from the previous chapter as they are all generally applicable to systems of nonlinear equations. Also, it must be recognized that even though the algorithms described here are used routinely in both private and commercial software, the development of more effective methods is still an active area of research.

Picard method

A particularly simple method derived directly from the fixed point problem is the successive substitution, Picard, or functional iteration method. For the system in (4.6.7) this algorithm is given by

$$\overline{\mathbf{K}}(\mathbf{U}^n)\mathbf{U}^{n+1} = \overline{\mathbf{F}}(\mathbf{U}^n) \qquad (4.6.8)$$

where the superscript indicates the iteration number. This algorithm has a reasonably large radius of convergence; the rate of convergence is generally fairly low since it is only a first-order method. An improvement in convergence rate can sometimes be obtained by use of a relaxation formula,

$$\overline{\mathbf{K}}(\mathbf{U}^n)\mathbf{U}^\star = \overline{\mathbf{F}}(\mathbf{U}^n) \tag{4.6.9a}$$

where

$$\mathbf{U}^{n+1} = \alpha\mathbf{U}^n + (1-\alpha)\mathbf{U}^\star \qquad 0 \le \alpha \le 1 \tag{4.6.9b}$$

Strongly nonlinear problems often exhibit an oscillatory convergence behavior for successive substitution, in which case the use of (4.6.9) would be of significant benefit.

Although the Picard schemes are generally slow to converge, they are often a good choice for the first several iterations in a general solution strategy. The wide radius of convergence allows the Picard method to start with a relatively poor initial solution guess, $\mathbf{U}^0$, and move the solution closer to the true solution and within the radius of convergence of higher-order, more rapidly convergent methods. Note that the starting vector $\mathbf{U}^0$ for the Picard scheme, as well as most other iterative algorithms, will by necessity be selected as a constant or zero vector; the first pass through the Picard method will produce a solution to the linear diffusion (Stokes flow) problem associated with Eq. (4.6.7). Neither of these solution vectors is particularly "close" to the solution of the nonlinear problem. The availability of a better initial estimate for $\mathbf{U}^0$ can dramatically improve the convergence behavior of the Picard and other solution methods.

Newton's method

In order to enhance the rate of convergence, a second-order method, such as Newton's method, must be considered. First rewrite Eq. (4.6.7) as

$$\mathbf{R}(\mathbf{U}) = \overline{\mathbf{K}}(\mathbf{U})\mathbf{U} - \overline{\mathbf{F}}(\mathbf{U}) = \mathbf{0} \tag{4.6.10}$$

Newton's method is based on a truncated Taylor's series expansion of $\mathbf{R}(\mathbf{U})$ about the known solution $\mathbf{U}^n$:

$$\mathbf{0} = \mathbf{R}(\mathbf{U}^n) + \frac{\partial\mathbf{R}}{\partial\mathbf{U}}\bigg|_{\mathbf{U}^n}\Delta\mathbf{U} + O(\Delta\mathbf{U})^2 \tag{4.6.11}$$

where $\Delta\mathbf{U} = (\mathbf{U}^{n+1} - \mathbf{U}^n)$. Omitting the terms of order two and higher, we obtain

$$\mathbf{R}(\mathbf{U}^n) = -\frac{\partial\mathbf{R}}{\partial\mathbf{U}}\bigg|_{\mathbf{U}^n}(\mathbf{U}^{n+1} - \mathbf{U}^n) \equiv -\mathcal{J}(\mathbf{U}^n)(\mathbf{U}^{n+1} - \mathbf{U}^n) \tag{4.6.12}$$

where $\mathcal{J}$ is the Jacobian matrix (also known as the *tangent matrix*)

$$\mathcal{J} = \frac{\partial\mathbf{R}}{\partial\mathbf{U}}\bigg|_{\mathbf{U}^n} \tag{4.6.13}$$

Solving Eq. (4.6.12) for $\mathbf{U}^{n+1}$, we obtain

$$\mathbf{U}^{n+1} = \mathbf{U}^n - \mathcal{J}^{-1}(\mathbf{U}^n)\mathbf{R}(\mathbf{U}^n) \tag{4.6.14}$$

Note the correspondence between Eq. (4.6.14) and the general procedure given in Eq. (4.6.6); the use of an α scale factor in Eq. (4.6.14) is also possible.

To illustrate some of the detail associated with Newton's method, consider the case of a two-dimensional mixed finite element model. In this example, the components of the solution vector $\mathbf{U}^T = [\mathbf{u}_1^T, \mathbf{u}_2^T, \mathbf{P}^T]$ and the components of the vector $\mathbf{R}$ are

$$\mathbf{R_1} = \mathbf{C}_1(\mathbf{u}_1)\mathbf{u}_1 + \mathbf{C}_2(\mathbf{u}_2)\mathbf{u}_1 + (2\mathbf{K}_{11} + \mathbf{K}_{22})\mathbf{u}_1 + \mathbf{K}_{12}\mathbf{u}_2 - \mathbf{Q}_1\mathbf{P} - \mathbf{F}_1$$
$$\mathbf{R_2} = \mathbf{C}_1(\mathbf{u}_1)\mathbf{u}_2 + \mathbf{C}_2(\mathbf{u}_2)\mathbf{u}_2 + \mathbf{K}_{21}\mathbf{u}_1 + (\mathbf{K}_{11} + 2\mathbf{K}_{22})\mathbf{u}_2 - \mathbf{Q}_2\mathbf{P} - \mathbf{F}_2$$
$$\mathbf{R_3} = -\mathbf{Q}_1^T\mathbf{u}_1 - \mathbf{Q}_2^T\mathbf{u}_2 \tag{4.6.15}$$

where (no sum on j)

$$\mathbf{C}_j(\mathbf{u}_j) = \int_{\Omega^e} \rho_0 \mathbf{\Psi}(\mathbf{\Psi}^T\mathbf{u}_j)\frac{\partial \mathbf{\Psi}^T}{\partial x_j}\, d\mathbf{x} \tag{4.6.16}$$

and $\mathbf{K}_{ij}$ are defined in Eq. (4.2.13). Equations in (4.6.15) correspond to the two momentum equations and the incompressibility constraint.

The key to Newton's method is the formation of the Jacobian matrix, which in this case is

$$\mathcal{J} = \left[\frac{\partial \mathbf{R}}{\partial \mathbf{U}}\right] = \begin{bmatrix} \frac{\partial \mathbf{R}_1}{\partial \mathbf{u}_1} & \frac{\partial \mathbf{R}_1}{\partial \mathbf{u}_2} & \frac{\partial \mathbf{R}_1}{\partial \mathbf{P}} \\[2mm] \frac{\partial \mathbf{R}_2}{\partial \mathbf{u}_1} & \frac{\partial \mathbf{R}_2}{\partial \mathbf{u}_2} & \frac{\partial \mathbf{R}_2}{\partial \mathbf{P}} \\[2mm] \frac{\partial \mathbf{R}_3}{\partial \mathbf{u}_1} & \frac{\partial \mathbf{R}_3}{\partial \mathbf{u}_2} & \frac{\partial \mathbf{R}_3}{\partial \mathbf{P}} \end{bmatrix} \tag{4.6.17}$$

The components of $\mathcal{J}$ can be computed directly from the definition in Eq. (4.6.13). We obtain the following nine equations:

$$\frac{\partial \mathbf{R}_1}{\partial \mathbf{u}_1} = \mathbf{C}_1(\mathbf{u}_1) + \mathbf{C}_1(1)\mathbf{u}_1 + \mathbf{C}_2(\mathbf{u}_2) + 2\mathbf{K}_{11} + \mathbf{K}_{22}$$

$$\frac{\partial \mathbf{R}_1}{\partial \mathbf{u}_2} = \mathbf{C}_2(1)\mathbf{u}_1 + \mathbf{K}_{12}, \quad \frac{\partial \mathbf{R}_1}{\partial \mathbf{P}} = -\mathbf{Q}_1, \quad \frac{\partial \mathbf{R}_2}{\partial \mathbf{u}_1} = \mathbf{C}_1(1)\mathbf{u}_2 + \mathbf{K}_{21}$$

$$\frac{\partial \mathbf{R}_2}{\partial \mathbf{u}_2} = \mathbf{C}_1(\mathbf{u}_1) + \mathbf{C}_2(\mathbf{u}_2) + \mathbf{C}_2(1)\mathbf{u}_2 + \mathbf{K}_{11} + 2\mathbf{K}_{22}$$

$$\frac{\partial \mathbf{R}_2}{\partial \mathbf{P}} = -\mathbf{Q}_2, \quad \frac{\partial \mathbf{R}_3}{\partial \mathbf{u}_1} = -\mathbf{Q}_1{}^T, \quad \frac{\partial \mathbf{R}_3}{\partial \mathbf{u}_2} = -\mathbf{Q}_2{}^T, \quad \frac{\partial \mathbf{R}_3}{\partial \mathbf{P}} = 0 \tag{4.6.18}$$

The components of $\mathcal{J}$ are combinations of the basic element matrices found in the original equations plus a few new terms that arise due to differentiation of terms that are nonlinear in the dependent variables, e.g., the advection matrices. By studying the form of the matrices $\mathbf{C}_j$ given in Eq. (4.6.18), it can be shown that the needed derivatives of these terms can be computed analytically.

All nonlinearities found in the components $\mathbf{R}_i$ of Eq. (4.6.15) were not explicitly accounted for in the derivatives defining the Jacobian matrix. For example, when the viscosity varies with velocity gradients, the $\mathbf{K}$ terms are functions of $\mathbf{u}$ and should appropriately be differentiated with respect to the velocity. The complexity of treating all possible material property and boundary condition variations in this rigorous manner is usually prohibitive and generally not warranted. These nonlinearities are usually mild enough that they do not significantly affect the convergence of the Newton algorithm when treated in a "first-order" manner using a Picard or successive substitution procedure. Therefore, a strict Jacobian formulation is normally used only for the highly nonlinear advection terms.

The Newton scheme has a superior (quadratic) rate of convergence compared to the Picard scheme in Eq. (4.6.8). However, Newton's method also has a somewhat smaller radius of convergence (i.e., is more sensitive to the initial solution vector $\mathbf{U}^0$), and therefore is often used in conjunction with a Picard scheme.

Modified and quasi-Newton methods

The major drawback to both the successive substitution and Newton algorithms is the computational expense involved in the solution of a large matrix problem at each iteration. One method for avoiding this expense in the Newton scheme is to not compute a new Jacobian matrix at each iteration but instead work with a fixed iteration operator. This procedure, termed a *modified Newton method*, can be expressed as a variant of Eq. (4.6.14),

$$\mathbf{U}^{n+1} = \mathbf{U}^n - \mathcal{J}^{-1}(\mathbf{U}^0)\mathbf{R}(\mathbf{U}^n) \tag{4.6.19}$$

After the first iteration cycle the method is very inexpensive since it only requires the reduction of the force vector and a back-substitution for each cycle. Unfortunately, the method has a poor rate of convergence and is therefore not recommended.

A method that combines the general efficiency of a modified Newton method and the approximate convergence rate of the full Newton method is the quasi-Newton procedure. Rather than working with a constant or fully updated Jacobian matrix, the quasi-Newton methods seek to approximately update the Jacobian matrix using a simple recursive procedure. The efficiency of the method comes from the ability to update the inverse of the Jacobian matrix, thus saving both the assembly and factorization time of the full method; some additional storage is required to save updates to the Jacobian matrix but this is not a serious disadvantage. A number of variants of the quasi-Newton procedure exist. The most successful scheme is based on the Broyden update [38] as implemented by Engelman et al. [39]. This scheme may also be related to the general method in (4.6.6).

Continuation methods

A common failure of all iterative methods described previously is the lack of a sufficiently large radius of convergence. Given an initial estimate of the solution vector, the iterative method may not be able to reach a converged solution, since the initial guess was in some sense "too far away" from the required result. There are two general approaches to the problem of obtaining good initial estimates for a solution vector and both of them involve some type of "tracking" procedure for the solution. The first procedure is simply the method of false transients in which the solution is followed through the use of a time parameter. The ideas associated with dynamic relaxation, which have found use in solid mechanics [40], could also be included with these techniques. Transient algorithms are covered in a following section (see also, Section 3.7.3). A second method for circumventing the problems with initial solution vectors, consists of incrementally approaching the final solution through a series of intermediate solutions. These intermediate results may be of physical interest or simply be a means to obtain the required solution. The formal algorithms used to implement such a procedure are termed *continuation (imbedding, incremental loading) methods*, and they can be used in conjunction with any of the previously described iterative procedures.

Continuation methods are similar in some respects to Newton's method and can be derived using a similar approach. Assume that the solution to Eq. (4.6.7) depends on a real parameter, λ (e.g., the Reynolds number or magnitude of a boundary condition). Then Eq. (4.6.10) can be written as

$$\mathbf{R}(\mathbf{U}, \lambda) = \overline{\mathbf{K}}(\mathbf{U}, \lambda)\mathbf{U} - \overline{\mathbf{F}}(\mathbf{U}, \lambda) = \mathbf{0} \tag{4.6.20}$$

As with Newton's method, if $\mathbf{R}$ has continuous derivatives with respect to $\mathbf{U}$ and is differentiable with respect to λ, then a truncated, two-parameter, Taylor's series of $\mathbf{R}(\mathbf{U}, \lambda)$ can be used to produce the following algorithm:

$$\left.\frac{\partial \mathbf{R}}{\partial \mathbf{U}}\right|_{\mathbf{U}, \lambda} \Delta \mathbf{U} = -\left.\frac{\partial \mathbf{R}}{\partial \lambda}\right|_{\mathbf{U}, \lambda} \Delta\lambda$$

or

$$\mathbf{U}^{n+1} = \mathbf{U}^n - \mathcal{J}^{-1}(\mathbf{U}^n, \lambda^n)\left.\frac{\partial \mathbf{R}}{\partial \lambda}\right|_{\mathbf{U}^n, \lambda^n} \Delta\lambda \tag{4.6.21}$$

where $\Delta\lambda = \lambda^{n+1} - \lambda^n$ and the Jacobian matrix is the same as that defined for Newton's method.

The solution scheme in Eq. (4.6.21) amounts to a piecewise linearization of the solution curve with respect to λ; higher order schemes could be generated. When $\Delta\lambda$ is small, it is probable that the new solution $\mathbf{U}^{n+1}$ is very close to the true solution for λ^{n+1}. However, as $\Delta\lambda$ is made larger, the predicted $\mathbf{U}^{n+1}$ will generally depart from the true solution curve. In such cases, the predicted value may be corrected via any applicable iterative method, e.g., Newton's method. Convergence for this inner iteration should be rapid since the predicted value is relatively close to the true solution.

The continuation procedure outlined here is a very powerful method and may, in fact, be used to investigate the bifurcation and stability characteristics of various flow problems. Extensive developments in this area have been made by Winters and co-workers [41–44]. For difficult problems at large Reynolds numbers, the method is also a good choice. The biggest disadvantage of the algorithm is that there is little guidance for the selection of $\Delta\lambda$ during the solution sequence. Unless specific intermediate solutions are of interest, the sequence of $\Delta\lambda$ should be made as large as possible to ensure minimum computational effort. However, the size of any given $\Delta\lambda$ is constrained by the radius of convergence requirements of the iterative method (e.g., Newton's method) used to provide the corrected solution at λ^{n+1}. In practice, adaptive continuation has been accomplished mainly through the use of heuristically based $\Delta\lambda$ selection procedures [45]. The theoretical basis for adaptive continuation methods remains an area of research [46,47], especially for applications in solid mechanics.

As a final comment on continuation methods, we note that the algorithm in (4.6.21) is a first-order continuation method. A simple and effective "zeroth-order" continuation method has also found fairly widespread use in fluid mechanics and heat transfer problems. In this case, a series of problems are solved at increasing values of the continuation parameter λ. The converged solution at one value of λ^n is used at the starting solution vector at the next higher value of $\lambda^{n+1} = \lambda^n + \Delta\lambda$ in whatever iterative scheme has been selected. The formalism in Eq. (4.6.21) is not used to track the solution curve; some of the bifurcation tracking capabilities are lost in this simplified procedure.

4.6.3 Pressure Correction/Projection Methods

All of the iteration methods discussed earlier were based on the idea of a fixed point method and were designed to work with fully coupled sets of equations, i.e., the velocity and pressure equations were all solved simultaneously. For each iteration method, the resulting linearized problem was structured such that a direct matrix solution procedure (e.g., Gauss elimination) could be effectively employed. Direct solution methods, however, are decreasingly effective as problem size increases due to excessive demands on computer resources. Most three-dimensional simulations are beyond the capabilities of a direct solution method. As noted previously, iterative matrix solutions have not proved viable with fixed point methods due to inadequate preconditioners for the positive, semi-definite equations. The requirement to consider larger and more complex flow simulations has forced reconsideration of the fully coupled equation methods and opened the search for solution algorithms that are more efficient for large-scale problems.

Following ideas developed in the finite difference community, the troublesome incompressibility constraint is removed from the equation set and replaced by a decoupled Poisson equation for the pressure. These types of methods are generally termed pressure correction or pressure projection methods; finite difference methods of this type would be labeled as SIMPLE or SIMPLER methods [48]. Note that this technique is only possible with a mixed finite element formulation; the penalty method does not allow an explicit replacement of the continuity equation.

Consider a simplified component form of the steady, two-dimensional, finite element equations given in (4.2.12)

$$K_{xx}u + K_{xy}v - Q_x P = F_x \qquad (4.6.22a)$$
$$K_{yx}u + K_{yy}v - Q_y P = F_y \qquad (4.6.22b)$$
$$-Q_x^T u - Q_y^T v = 0 \qquad (4.6.22c)$$

In writing (4.6.22) it is assumed that the nonlinear advection terms have been absorbed in the K matrices. It is now possible to manipulate the equations in (4.6.22) to form a consistent, elliptic equation for the pressure. Symbolically, Eqs. (4.6.22a,b) are solved for the velocity components, u and v, by inverting the appropriate $[K]$ matrices. The resulting expressions for u and v are then substituted into Eq. (4.6.22c) to yield

$$[Q_x^T K_{xx}^{-1} Q_x + Q_y^T K_{yy}^{-1} Q_y]P = -Q_x^T K_{xx}^{-1} f_x - Q_y^T K_{yy}^{-1} f_y \qquad (4.6.23)$$

where $f_x = F_x - K_{xy}v$ and $f_y = F_y - K_{yx}u$. Equation (4.6.23) is a discretized form of the pressure Poisson equation that can be used in place of the incompressibility constraint. The system of equations in (4.6.22a,b) and (4.6.23) forms a decoupled system that could (in theory) be solved in an iterative cycle for the unknowns (u, v) and P. Note that Eqs. (4.6.22a) and (4.6.22b) can be solved together as a single equation set or individually by component. The practical drawback to the use of Eq. (4.6.23) is the fact that the inverse $[K]$ matrices make the pressure matrix impractical to construct; the inverse of a banded matrix is a dense matrix. Fortunately, it has been found by Haroutunian et al. [49] and others that a reasonable approximation to the inverse $[K]$ matrices is sufficient to produce a usable algorithm. In particular, if the diagonal of $[K]$ is used to approximate $[K]$, then the inverse is trivial and the pressure matrix can be readily constructed.

There are a number of methods by which Eqs. (4.6.22a,b) and (4.6.23) can be employed to form a time-independent solution technique. We will only review one possible algorithm here and refer the interested reader to References 28 and 49–51. The pressure projection algorithm is defined by the following basic steps that are performed at each iteration until convergence is obtained. An approximation for the pressure is obtained from the pressure Poisson equation

$$[Q_x^T \tilde{K}_{xx}^{-1} Q_x + Q_y^T \tilde{K}_{yy}^{-1} Q_y]P^* = -Q_x^T \tilde{K}_{xx}^{-1} f_x(u^n) - Q_y^T \tilde{K}_{yy}^{-1} f_y(v^n) \qquad (4.6.24)$$

where the f vectors and approximate $\tilde{K}$ matrices are evaluated with velocities from the previous iteration n. Since P^* will generally be a poor approximation to P^{n+1}, a relaxation scheme is often applied to obtain a better estimate for pressure P

$$P^{n+1} = \alpha P^n + (1 - \alpha)P^* \qquad 0 \le \alpha \le 1 \qquad (4.6.25)$$

With a new value of the pressure (gradient), the momentum equations in (4.6.22a,b) can be solved for a velocity field. In component form then

$$K_{xx}u^* = F_x + Q_x P^{n+1} - K_{xy}v^* \tag{4.6.26}$$
$$K_{yy}v^* = F_y + Q_y P^{n+1} - K_{yx}u^* \tag{4.6.27}$$

or as a coupled system

$$\begin{bmatrix} K_{xx} & K_{xy} \\ K_{yx} & K_{yy} \end{bmatrix} \begin{Bmatrix} u^* \\ v^* \end{Bmatrix} = \begin{Bmatrix} F_x + Q_x P^{n+1} \\ F_y + Q_y P^{n+1} \end{Bmatrix} \tag{4.6.28}$$

Note that Eqs. (4.6.26)–(4.6.28) represent generally unsymmetric systems, and the left-hand side would be evaluated using u^n and v^n where appropriate. The velocities computed in Eqs. (4.6.26)–(4.6.28) are not divergence free. The last step of the algorithm is the projection step where the velocities are forced to satisfy the discretized continuity equation. By rederiving Eq. (4.6.23) via a slightly different approach, a second pressure Poisson equation provides

$$[Q_x^T \tilde{K}_{xx}^{-1} Q_x + Q_y^T \tilde{K}_{yy}^{-1} Q_y] P^\lambda = -Q_x^T u^* - Q_y^T v^* \tag{4.6.29}$$

The quantity P^λ is actually a Lagrange multiplier that is used to adjust the velocities with the following relations

$$u^{n+1} = u^* + \tilde{K}_{xx}^{-1} Q_x P^\lambda \tag{4.6.30a}$$
$$v^{n+1} = v^* + \tilde{K}_{yy}^{-1} Q_y P^\lambda \tag{4.6.30b}$$

Equations (4.6.24)–(4.6.30) are used repetitively until convergence is attained. Other methods that use a pressure Poisson equation differ mainly in the sequence of steps and how the pressure solution is utilized. The pressure projection method outlined here generally has the best rate of convergence.

The above scheme can be implemented with a direct matrix solution method (e.g., Gauss elimination) used at each step that requires such a solution. By segregating the equations, the matrix problem for each equation is significantly made smaller than the fully coupled system and, therefore, much less of a computational burden. Significant three-dimensional GFEM problems can be processed with this algorithm. However, the most important feature of the solution algorithm is that each matrix that requires solution is of a form that is readily amenable to solution via an iterative solver, such as the conjugate gradient method. Through the application of iterative solution methods, the computational problem is further reduced in size as well as computational effort and even larger problems can be analyzed. As a final point, it must be noted that the segregated iterative algorithm is also applicable to time-dependent problems after an appropriate time approximation scheme has been introduced.

4.7 Time-Approximation Schemes

4.7.1 Preliminary Comments

Here we consider time approximation of the matrix differential equations of the type in (4.6.1):

$$\overline{\mathbf{M}}\dot{\mathbf{U}} + \overline{\mathbf{K}}(\mathbf{U})\mathbf{U} = \overline{\mathbf{F}} \tag{4.7.1}$$

Equation (4.7.1) represents a, discrete in space and continuous in time, approximation to the original system of partial differential equations. A direct time integration procedure replaces the continuous time derivative with an approximation for the history of the dependent variables over a small portion of the problem time scale. The result is an incremental procedure that advances the solution by discrete steps in time. In constructing such a procedure, questions of numerical stability and accuracy must be considered.

Though explicit integration methods have been used (see [52]) in the solution of (4.7.1), preference is usually given to implicit procedures for this class of problems. The explicit methods are plagued by several difficulties, including (1) the natural implicitness of the pressure in an incompressible flow, (2) severe time step restrictions needed to maintain stability of the integration process, (4.7.3) the problems of diagonalizing and inverting $\overline{\mathbf{M}}$ in a cost-effective manner for a variety of element types, and (4.7.4) reductions in accuracy due to the diagonalization of $\overline{\mathbf{M}}$. Implicit integration methods, though computationally expensive, are desirable due to their increased stability and the consistent treatment of the pressure.

In the following sections two implicit procedures and one explicit method will be described (see [52–54]). Both of the implicit algorithms make use of a predictor/corrector cycle to improve efficiency and accuracy; both may be used with either a fixed time step or a dynamic time step selection algorithm. The solution of the resulting nonlinear, algebraic system for each time plane is normally obtained by Newton's method, though most of the iterative methods of the previous section would also be applicable. All of these integration methods are direct extensions of the algorithms (e.g., the Crank–Nicolson method) discussed for the heat conduction problem in Chapters 2 and 3.

4.7.2 Forward/Backward Euler Schemes

A first-order integration method that is useful for the equations in (4.6.1) employs a forward Euler scheme as a predictor with the backward Euler method functioning as the corrector step. Omitting the details of the derivation, the application of the explicit, forward Euler formula to Eq. (4.6.1) yields

$$\overline{\mathbf{M}}\mathbf{U}_p^{n+1} = \overline{\mathbf{M}}\mathbf{U}^n + \Delta t_n \left[\overline{\mathbf{F}}(\mathbf{U}^n) - \overline{\mathbf{K}}(\mathbf{U}^n)\mathbf{U}^n \right] \tag{4.7.2}$$

This can be written in a form that is more suitable for computation by replacing the bracketed term with a rearranged form of Eq. (4.7.1)

$$\mathbf{U}_p^{n+1} = \mathbf{U}^n + \Delta t_n \dot{\mathbf{U}}^n \tag{4.7.3}$$

In Eqs. (4.7.2) and (4.7.3) the superscript on a vector indicates the time plane, the subscript p denotes a predicted value, and $\Delta t_n = t_{n+1} - t_n$ is the time step. By using the form shown in (4.7.3) a matrix inversion of $\overline{\mathbf{M}}$ is avoided; the "acceleration" vector $\dot{\mathbf{U}}^n$ is computed from a form of the corrector formula as shown below. Note that since the forward Euler scheme is explicit, it is applied only to the velocity and any auxiliary variables contained in the $\mathbf{U}$ vector; the pressure, being implicit, is not predicted with this formula.

The corrector step of the first-order scheme is provided by the backward Euler (or fully implicit) method. When applied to Eq. (4.7.1) this implicit method yields

$$\overline{\mathbf{M}}\mathbf{U}^{n+1} = \overline{\mathbf{M}}\mathbf{U}^n + \Delta t_n \left[\mathbf{F}(\mathbf{U}^{n+1}) - \overline{\mathbf{K}}(\mathbf{U}^{n+1})\mathbf{U}^{n+1} \right] \qquad (4.7.4)$$

or in a form more suitable for computation

$$\left[\frac{1}{\Delta t_n}\overline{\mathbf{M}} + \overline{\mathbf{K}}(\mathbf{U}^{n+1}) \right] \mathbf{U}^{n+1} = \frac{1}{\Delta t_n}\overline{\mathbf{M}}\mathbf{U}^n + \mathbf{F}(\mathbf{U}^{n+1}) \qquad (4.7.5)$$

The implicit nature of this method is evident from the form of (4.7.5). Note that the pressure is included in the vector $\mathbf{U}$ in (4.7.5).

With the discretization of the time derivative, the equation in (4.7.5) now represents a set of nonlinear algebraic equations for the solution vector $\mathbf{U}$ at time t_{n+1}. This is precisely the same type of problem that was encountered in the time-independent case (see Section 4.6) and for which a number of iterative algorithms are available. In practice, Newton's method is generally used for (4.7.4) due to its rapid convergence and the obvious need to limit the number of computations that are repeated at each time plane. A segregated iteration method would be useful for large simulations though its convergence rate is very modest. As indicated previously, the rate of convergence of Newton's method is greatly increased if the initial estimate of the vector $\mathbf{U}$ is "close" to the true solution. The solution $\mathbf{U}_p$, predicted from the explicit formula (4.7.3), provides this initial guess for the iterative procedure in a cost-effective manner. It should be noted that the use of the predictor step is not mandatory and the backward Euler scheme could be used alone to integrate the equation set. In such a situation, the first estimate of the solution at $n + 1$ would be obtained from the solution at the previous time plane and a number of iterations with Newton's method would be required to converge the solution.

4.7.3 Adams–Bashforth/Trapezoid Rule

An integration method that is second-order in time can be developed along the same lines as described above. A second-order equivalent to the forward Euler method is the variable step, Adams–Bashforth predictor given by

$$\mathbf{U}_p^{n+1} = \mathbf{U}^n + \frac{\Delta t_n}{2} \left[\left(2 + \frac{\Delta t_n}{\Delta t_{n-1}} \right) \dot{\mathbf{U}}^n - \left(\frac{\Delta t_n}{\Delta t_{n-1}} \right) \dot{\mathbf{U}}^{n-1} \right] \qquad (4.7.6)$$

where $\Delta t_n = t_{n+1} - t_n$ and $\Delta t_{n-1} = t_n - t_{n-1}$. This formula can be used to predict the solution vector (excluding the pressure) given two "acceleration" vectors from previous time planes; no matrix solution is required.

A compatible corrector formula for use with (4.7.6) is available in the form of the trapezoid rule. When applied to Eq. (4.7.1) the trapezoid rule yields the scheme,

$$\left[\frac{2}{\Delta t_n} \overline{\mathbf{M}} + \overline{\mathbf{K}}(\mathbf{U}^{n+1}) \right] \mathbf{U}^{n+1} = \frac{2}{\Delta t_n} \overline{\mathbf{M}} \mathbf{U}^n + \overline{\mathbf{M}} \dot{\mathbf{U}}^n + \overline{\mathbf{F}}(\mathbf{U}^{n+1}) \qquad (4.7.7)$$

Equation (4.7.7) is again observed to be a nonlinear, algebraic system for the vector $\mathbf{U}^{n+1}$ and can be solved using an iterative procedure such as Newton's method. All of the comments regarding the forward/backward Euler method are also applicable here.

4.7.4 Implicit Integration and Time Step Control

The actual integration process using either of the above two methods can be outlined in a few simple steps that follow closely the heat conduction procedure of Chapters 2 and 3. At the beginning of each time step it is assumed that all of the required solution and "acceleration" vectors are known and the time increment for the next step has been selected. To advance the solution from time t_n to time t_{n+1} then requires the following steps (see Section 3.7):

1. A tentative solution vector, $\mathbf{U}_p^{n+1}$, is computed using the predictor Eqs. (4.7.3) or (4.7.6). The pressure variables are not included in this prediction.

2. The corrector equations (4.7.5) or (4.7.7) are solved for the "true" solution, $\mathbf{U}^{n+1}$. This involves the iterative solution of (4.7.5) or (4.7.7) via Newton's method. The predicted values $\mathbf{U}_p^{n+1}$ are used to initialize the Jacobian matrix for the Newton iteration.

3. The "acceleration" vectors are updated using the new solution $\mathbf{U}^{n+1}$ and the "inverted" forms of the corrector formulas. For the first-order method the acceleration is computed from the backward Euler definition

$$\dot{\mathbf{U}}^{n+1} = \frac{1}{\Delta t_n} \left(\mathbf{U}^{n+1} - \mathbf{U}^n \right) \qquad (4.7.8)$$

while the second-order accelerations are derived from the trapezoid rule

$$\dot{\mathbf{U}}^{n+1} \doteq \frac{2}{\Delta t_n} \left(\mathbf{U}^{n+1} - \mathbf{U}^n \right) - \dot{\mathbf{U}}^n \qquad (4.7.9)$$

4. A new integration time step is computed. The time-step selection process is based on an analysis of the time truncation errors in the predictor and corrector formulas as described below. If a constant time step is being used, this step is omitted.

5. Return to step 1 for the next time increment.

In actual implementation the Newton iteration process in step 2 is not usually carried to absolute convergence. Rather, a one-step Newton correction is employed as advocated in [54]. This procedure is quite efficient and can be very accurate provided the time step is suitably controlled.

The adaptive control of the time step is an added benefit in the use of a predictor/corrector method. The time-step estimation formula for the above methods is

$$\Delta t_{n+1} = \Delta t_n \left(b \cdot \frac{\epsilon^t}{d_{n+1}} \right)^m \tag{4.7.10}$$

where $m = 1/2$, $b = 2$ for the first-order method and $m = 1/3$, $b = 3(1 + \Delta t_{n-1}/\Delta t_n)$ for the second-order scheme. The user-specified error tolerance for the integration process is ϵ^t, which typically has a value of 0.001. The quantity d_{n+1} is an appropriate norm on the integration error, which is defined as the difference between the predicted solution and the corrected value. Typically, the following norms are used for each velocity variable,

$$d_{n+1}^U = \frac{1}{\sqrt{N} U_{max}} \left[\sum_{i=1}^{N} \left(U_i^{(n+1)} - U_{ip}^{(n+1)} \right)^2 \right]^{\frac{1}{2}} \tag{4.7.11}$$

where N is the number of velocity nodes in the problem, U_{max} is a constant velocity scale, and U_i and U_{ip} are the corrected and predicted velocities at the nodes. The norm defined in (4.7.11) is used separately for each velocity component in the problem. During the integration process, new time steps are computed using each of the velocity norms; the resulting time steps are compared and the smallest is used for the next integration step. This procedure allows the dominant momentum equation to control the integration algorithm.

4.7.5 Explicit Integration

The primary difficulty in constructing an explicit integration method was first mentioned in Section 4.5.1, and relates to the inherently implicit nature of the pressure. From Eq. (4.2.15), it is clear that the lack of a coefficient matrix for $\dot{P}$ makes the explicit advancement of this variable impossible. As in the case of the pressure correction methods in Section 4.6.3, this difficulty can be partially circumvented by replacing the standard continuity equation with a pressure Poisson equation. With this substitution, the velocity components can be advanced through time via an explicit integrator; the pressure at each time step must be computed from the implicit Poisson equation.

The explicit procedure that is outlined here was proposed by Gresho et al. [52] and follows the general philosophy of the previously described pressure projection method. Consider the simplified, segregated form of the two-dimensional, time-dependent, finite element equations given by Eq. (4.2.11)

$$\mathbf{M}\dot{\mathbf{u}} + \mathbf{K}(\mathbf{u})\mathbf{u} - \mathbf{Q}\mathbf{P} = \mathbf{F} \tag{4.7.12}$$

$$-\mathbf{Q}^T\mathbf{u} = 0 \qquad\qquad (4.7.13)$$

where the $\mathbf{K}$ Matrix contains both the diffusive and advective terms and u contains both velocity components. To construct a Poisson equation for the pressure, operate on (4.7.12) first with $\mathbf{M}^{-1}$ and then with $\mathbf{Q}^T$; noting from (4.7.13) that $\mathbf{Q}^T\dot{\mathbf{u}} = \mathbf{0}$ allows the modified equation (4.7.12) to be written as

$$(\mathbf{Q}^T\mathbf{M}^{-1}\mathbf{Q})\mathbf{P} = \mathbf{Q}^T\mathbf{M}^{-1}(\mathbf{F} - \mathbf{K}(\mathbf{u})\mathbf{u}) \qquad\qquad (4.7.14)$$

which is an elliptic equation for the pressure P. As in Section 4.6 the key to obtaining a viable algorithm is the construction of $\mathbf{M}^{-1}$. If $\mathbf{M}$ is diagonalized or lumped, then $\mathbf{M}^{-1}$ is quite simple and the Poisson equation is readily available. Note also that the left-hand side of Eq. (4.7.14) is a constant matrix and needs only be formed and triangularized once.

The general explicit scheme described by Gresho et al. [52] utilized a forward Euler integration method for (4.7.12) with a pressure solution obtained from (4.7.14). Various numerical procedures were used in conjunction with the basic method to produce a reliable, cost-effective method. Low-order (trilinear) velocity elements were used so that $\mathbf{M}$ could be diagonalized; one-point Gauss quadrature was used to speed up construction of element matrices. So-called "balancing tensor diffusivity" was used to offset the time truncation error associated with the forward Euler scheme. Also, subcycling was introduced for Eq. (4.7.14) in which a new pressure was computed only after several explicit velocity time steps. Though an explicit time integration method can be derived as shown here, its practical implementation is quite complex. The method is only conditionally stable and is still burdened with the solution of at least one matrix problem. More recent work on explicit methods with pressure projection can be found in Gresho and Chan [53].

4.8 Stabilized Methods

4.8.1 Preliminary Comments

Computational difficulties and perceived shortcomings of the Galerkin finite element (GFEM) model have lead to investigation and development of alternate finite element models. The objective in each of these alternate models has been to avoid either or both of the two main "problems" encountered with mixed and penalty finite element models. Whether or not real problems exist with the GFEM depends on analysis objectives and what can reasonably be expected from a numerical simulation of a complex physical phenomena.

The first problem area for the standard GFEM is the occurrence of "wiggles" in the velocity field. Typically, this is a node-to-node oscillation of the velocity components emanating from a boundary or some other flow feature that produces a large velocity gradient. The wiggle problem is most pronounced at high Reynolds numbers and/or on coarse computational meshes. The inability of the finite element mesh to resolve the steep gradient results in an imbalance between the advective and diffusive terms in the equation. The weighted residual formulation, in trying to do a best fit to the solution, produces a field that oscillates about the true solution.

Historically, the wiggle problem was blamed on the centered difference approximation generated by the Galerkin method and its inability to account for an "upwind" (one-sided) representation of the advective term. A long history of research and development has lead to formulations that avoid wiggle problems and produce smooth solutions on virtually any mesh. Some of the early developers of this approach include Heinrich et al. [55], Heinrich and Zienkiewicz [56], and Christie et al. [57]. The more current methods are often attributed to Hughes and co-workers [58-60] and are usually termed streamline-upwind/Petrov–Galerkin (SUPG) methods. The arguments for wiggle stabilization are that (a) upwinding mimics the physical reality of advection, (b) smooth solutions can be obtained inexpensively on coarse meshes, (c) difficult high Reynolds number problems can be stabilized and are therefore computationally tractable, and (d) stabilization is a consistent formulation (i.e., exact solutions of the Navier–Stokes equations satisfy the stabilized equations). Opponents of wiggle suppression schemes, such as SUPG, argue that (a) it is not needed because wiggles are just a mesh problem that can be alleviated by proper mesh design or mesh adaptivity, (b) nonphysical damping is always added to the equations, lowering the physical Reynolds number, and (c) it is not certain what boundary value problem is being solved when the added stabilization terms are present. A more extensive discussion of the wiggle problem can be found in Gresho and Sani [28].

The second area of concern with the GFEM has a number of facets and revolves around the ubiquitous incompressibility constraint. The saddle-point problem generated by the mixed GFEM formulation makes the problem difficult to solve, especially in three dimensions. As noted previously, direct solvers with pivoting are well-suited to the mixed formulation but too costly for large-scale applications. Iterative solvers are presently ineffective due to deficiencies in the preconditioners. The LBB stability condition associated with the GFEM limits element choice with respect to computationally convenient choices for the velocity and pressure interpolation. A number of attempts have been made to circumvent the LBB condition and produce a more efficient method. The pressure projection algorithm of a previous section is an illustration of one such method. The pressure-stabilizing/Petrov–Galerkin (PSPG) method is another approach to avoiding the LBB requirement and is in the same spirit as the SUPG. In the following section, a general stabilization method, the Galerkin/Least-squares (GLS) method, is outlined; this formulation contains forms of both the SUPG and the PSPG.

4.8.2 Galerkin/Least-Squares Formulation

One method for enhancing the stability of the mixed or GFEM for incompressible flows involves the addition of various least-squares terms to the original Galerkin variational statement. The Galerkin/Least-squares approach is sometimes termed a residual method because the added least-squares terms are weighted residuals of the momentum equation; this form of the least-squares term implies the consistency of the method since the momentum residual is employed. The GLS is also known as a perturbation method since the added terms can be viewed as perturbations to the weighting functions. The fact that the weight and test functions are different also makes GLS a Petrov–Galerkin method. Development and popularization of the GLS methods for flow problems is primarily due to Hughes, Tezduyar and co-workers [30–32,61–63] and follows as a generalization of their work on SUPG and PSPG methods.

The usual approach to the GLS begins with the discontinuous in time Galerkin method and considers a finite element approximation for a space-time slab. Continuous polynomial interpolation is used for the spatial variation while a discontinuous time function is used within the space-time slab. The assumed time representation obviates the need to independently consider time integration methods. To simplify the present discussion of the GLS, the time independent form of the incompressible flow problem is considered which avoids the complexity of the space-time finite element formulation [31]. Using vector notation and following the weak form development of Section 4.2, the GLS variational form for the momentum and continuity equations can be written as

$$\int_\Omega \mathbf{w} \cdot (\rho_0 \mathbf{u} \cdot \nabla \mathbf{u}) \, dx + \int_\Omega \nabla \mathbf{w} \cdot (-\mathbf{P} + 2\mu \mathbf{D}(\mathbf{u})) \, dx - \int_\Omega \rho_0 \mathbf{w} \mathbf{f} dx$$

$$+ \int_\Omega Q \nabla \cdot \mathbf{u} dx + \sum_{n=1}^{nel} \int_{\Omega_n} R_{GLS} dx = - \int_\Gamma \mathbf{w} \mathcal{T}_i \, ds \qquad (4.8.1)$$

where $\mathbf{w}$ and Q are the weight functions for the momentum and continuity equations and the element residual is defined by

$$R_{GLS} = (\delta + \epsilon + \beta) \cdot (\rho_0 \mathbf{u} \cdot \nabla \mathbf{u} + \nabla P - 2\mu \mathbf{D}(\mathbf{u})) \qquad (4.8.2)$$

with

$$\delta = \tau_{supg} \mathbf{u} \cdot \nabla \mathbf{w} \qquad (4.8.3)$$

$$\epsilon = \tau_{pspg} \frac{1}{\rho_0} \nabla Q \qquad (4.8.4)$$

$$\beta = -\tau_{gls} 2\mu \mathbf{D}(\mathbf{w}) \qquad (4.8.5)$$

The τ coefficients are weighting parameters that will be discussed shortly. If the definitions for $\delta, \epsilon,$ and β are substituted into (4.8.2) and the various weighting parameters are equivalenced to a single parameter, τ, the result is

$$R_{GLS} = \tau \left[\rho_0 \mathbf{u} \cdot \nabla \mathbf{w} + \nabla Q - 2\mu \mathbf{D}(\mathbf{w})\right] \cdot \left[\rho_0 \mathbf{u} \cdot \nabla \mathbf{u} + \nabla P - 2\mu \mathbf{D}(\mathbf{u})\right] \quad (4.8.6)$$

$$R_{GLS} = \mathcal{L}(\mathbf{w}, Q) \tau \mathcal{L}(\mathbf{u}, \mathcal{P}) \qquad (4.8.7)$$

which are the standard definitions for the residual contribution to the GLS formulation. In Eq. (4.8.7) $\mathcal{L}$ is the incompressible differential operator. The splitting of the first residual into three separate contributions in (4.8.6) was done to allow the original SUPG and PSPG formulations to be easily recovered. If β and ϵ are set to zero, the stabilized SUPG method is recovered; if β and δ are set to zero, the PSPG method is recovered. Setting only β to zero achieves both SUPG and PSPG stabilization. It is apparent from the form in (4.8.6) that the GLS is a more general formulation than the original wiggle and pressure stabilization methods.

Note that in Eq. (4.8.1) the first four integrals and the right-hand side define a standard Galerkin-weighted residual method that is written in terms of global shape functions; the integrals are over the problem domain, Ω. The added residual term in (4.8.6) is defined over the interior of each element and basically contains the square of all or parts of the momentum residual. The various τ parameters are positive coefficients that have the dimension of time. The forms of these parameters are usually developed from error estimates, convergence proofs, and/or dimensional analysis. Particular constants within each parameter are selected by optimizing the method on simple problems and generalizing to multidimensions. It should be noted that the development of the τ parameters is not a unique process. For the steady problems considered here a typical τ for the GLS method is

$$\tau = \left[\left(\frac{2||\mathbf{u}||}{h} \right)^2 + \left(\frac{4\nu}{h^2} \right)^2 \right]^{-1/2} \tag{4.8.8}$$

where h is an appropriate element length. For the SUPG and PSPG formulations, the τ values are a function of an element Reynolds number and the ratio of an element length to a velocity scale. Details on these parameters may be found in [31] and [64].

The error and convergence analyses for the GLS and PSPG have demonstrated that equal-order velocity and pressure interpolation for these methods is viable and the LBB condition is not a requirement [30,61]. In the finite element implementation, convenient elements, such as the bilinear Q1/Q1, linear P1/P1, biquadratic Q2/Q2, and quadratic P2/P2, are now useful approximations while being unstable in the GFEM context. The low-order elements in this group have an advantage over their higher-order counterparts because some of the stabilization terms associated with viscous diffusion are identically zero. This considerably simplifies the element equation building process.

To complete this brief discussion of stabilized methods, the finite element equations arising from the GLS formulation will be shown in matrix form. Using the notation from Section 4.2 the momentum equation is

$$(\mathbf{C}(\mathbf{u}) + \mathbf{C}_\delta(\mathbf{u}) + \mathbf{C}_\beta(\mathbf{u}))\,\mathbf{u} + (\mathbf{K} + \mathbf{K}_\delta + \mathbf{K}_\beta)\,\mathbf{u} - (\mathbf{Q} + \mathbf{Q}_\delta + \mathbf{Q}_\beta)\,\mathbf{P} = \mathbf{F} \tag{4.8.9}$$

and the continuity equation is

$$-\mathbf{Q}^T\mathbf{u} - \mathbf{Q}_\epsilon\mathbf{P} + \mathbf{C}_\epsilon(\mathbf{u})\mathbf{u} + \mathbf{K}_\epsilon\mathbf{u} = \mathbf{0} \tag{4.8.10}$$

The subscripted terms all belong to the various types of stabilization. If any τ parameter is set to zero the corresponding term is omitted and the Galerkin form can be recovered by setting all τ parameters to zero. Finally, note that since all of the τ parameters depend on the element size, in the limit as the element size gets small, the stabilized method reduces to the Galerkin form. It is only on coarse meshes that stabilization plays a significant role in the method.

4.9 Post-Processing

4.9.1 Stress Computation

The results of interest from a flow analysis generally include fluid velocities, forces, and flow patterns (e.g., plots of velocities and pressure). Many of these items are directly available from the finite element results in terms of nodal point quantities; other quantities of interest may be derived from these primary variables. The computation of the viscous stress fields for the fluid flow follow directly from the finite element approximations of these variables, i.e., using kinematic and constitutive relations. A brief discussion of the stress calculation is presented next.

For a planar two-dimensional geometry, the components of the stress tensor $(\sigma_{xx}, \sigma_{yy}, \sigma_{xy})$ are known in terms of the pressure P and velocity components $u_1 = u_x$ and $u_2 = u_y$:

$$\sigma_{xx} = -P + 2\mu \frac{\partial u_x}{\partial x} \tag{4.9.1a}$$

$$\sigma_{yy} = -P + 2\mu \frac{\partial u_y}{\partial y} \tag{4.9.1b}$$

$$\sigma_{xy} = \mu \left(\frac{\partial u_x}{\partial y} + \frac{\partial u_y}{\partial x} \right) \tag{4.9.1c}$$

where μ is the viscosity of the fluid. Substitution of the finite element approximations (4.2.6a,b) for u_i and P into Eqs. (4.9.1a–c) yields

$$\sigma_{xx} = -\mathbf{\Phi}^T \mathbf{P} + 2\mu \frac{\partial \mathbf{\Psi}^T}{\partial x} \mathbf{u}_x \tag{4.9.2a}$$

$$\sigma_{yy} = -\mathbf{\Phi}^T \mathbf{P} + 2\mu \frac{\partial \mathbf{\Psi}^T}{\partial y} \mathbf{u}_y \tag{4.9.2b}$$

$$\sigma_{xy} = \mu \left(\frac{\partial \mathbf{\Psi}^T}{\partial y} \mathbf{u}_x + \frac{\partial \mathbf{\Psi}^T}{\partial x} \mathbf{u}_y \right) \tag{4.9.2c}$$

where $(\mathbf{P}, \mathbf{u}_x, \mathbf{u}_y)$ denote the vectors of nodal values of (P, u_x, u_y). The spatial derivatives of the interpolation functions in Eqs. (4.9.2a-c) can be converted to derivatives in terms of the local (element) coordinates (ξ, η) through the use of coordinate transformations defined in Eq. (3.3.3) or (4.5.4). When this isoparametric transformation is invoked, the expressions for the stress components become

$$\sigma_{xx} = -\mathbf{\Phi}^T \mathbf{P} + \frac{\mu}{|\mathbf{J}|} \left[J^*_{11} \frac{\partial \mathbf{\Psi}}{\partial \xi}^T \mathbf{u}_x + J^*_{12} \frac{\partial \mathbf{\Psi}}{\partial \eta}^T \mathbf{u}_x \right]$$

$$\sigma_{yy} = -\mathbf{\Phi}^T \mathbf{P} + \frac{\mu}{|\mathbf{J}|} \left[J^*_{21} \frac{\partial \mathbf{\Psi}}{\partial \xi}^T \mathbf{u}_y + J^*_{22} \frac{\partial \mathbf{\Psi}}{\partial \eta}^T \mathbf{u}_y \right] \tag{4.9.3}$$

$$\sigma_{xy} = \frac{\mu}{|\mathbf{J}|} \left[J^*_{21} \frac{\partial \mathbf{\Psi}}{\partial \xi}^T \mathbf{u}_x + J^*_{22} \frac{\partial \mathbf{\Psi}}{\partial \eta}^T \mathbf{u}_x + J^*_{11} \frac{\partial \mathbf{\Psi}}{\partial \xi}^T \mathbf{u}_y + J^*_{12} \frac{\partial \mathbf{\Psi}}{\partial \eta}^T \mathbf{u}_y \right]$$

The function $\mathbf{\Upsilon}$ [see Eqs. (4.5.4) and (4.5.7)] that occurs in the Jacobian matrix and its inverse (see Section 2.12), J^*, is defined as either a quadratic or linear interpolation function, depending on whether the particular element is isoparametric or subparametric. For axisymmetric flows, the stress field is computed in the same manner as discussed above but with the following definitions for the stress components:

$$\sigma_{rr} = -P + 2\mu \frac{\partial u_r}{\partial r} \; ; \quad \sigma_{zz} = -P + 2\mu \frac{\partial u_z}{\partial z} \tag{4.9.4a}$$

$$\sigma_{\theta\theta} = -P + 2\mu \frac{u_r}{r} \; ; \quad \sigma_{rz} = \mu \left(\frac{\partial u_z}{\partial r} + \frac{\partial u_r}{\partial z} \right) \tag{4.9.4b}$$

Computation of stresses in three-dimensional problems follows directly from the procedure presented above.

The expressions in (4.9.3) allow the stress components to be evaluated at any point (ξ_0, η_0) within an element for a known element geometry and solution field. Note that since the stresses depend on velocity gradients they are discontinuous between elements. As was the case for the heat flux evaluation, the derivative quantities are normally evaluated at the $2\times2\times2$ Gauss points for the interior of a hexahedral element, the 2×2 Gauss points for a quadrilateral element, or the 2 Gauss points on the edge of an element. These points have optimal accuracy for the shape function derivatives and correspond to a least-squares approximation for the derivative [65,66]. The stresses computed at interior integration points can extrapolated to the nodes by a simple linear extrapolation procedure, and they may be appropriately averaged between adjacent elements to produce a continuous stress field.

For some applications the vorticity field is also of interest and this quantity is computed in the same manner as the fluid stresses. By definition, the vorticity for a two-dimensional flow is given by [see Eq. (4.1.9)]

$$\omega = \frac{\partial u_y}{\partial x} - \frac{\partial u_x}{\partial y} \tag{4.9.5}$$

By using the same methods as outlined for the stress components the vorticity can be evaluated at any point within an element. For three-dimensional flows, the additional components of the vorticity vector may also be computed. However, these quantities are seldom used in practical applications.

4.9.2 Stream Function Computation

A quantity that is often useful in the graphic display of computed flow fields is the stream function. For two-dimensional incompressible flows, the stream function is the remaining nonzero component of a vector potential which satisfies the conservation of mass equation identically. By definition,

$$u_1 = u_x = \frac{\partial \psi}{\partial y} \; ; \qquad u_2 = u_y = -\frac{\partial \psi}{\partial x} \qquad (4.9.6)$$

and since the change in the stream function, $\delta \psi$, is an exact differential, then

$$\delta \psi = \int_A^B \mathbf{u} \cdot \hat{\mathbf{n}} \, ds = \int_A^B (u_x n_x + u_y n_y) \, ds \qquad (4.9.7)$$

with

$$\mathbf{u} = u_x \hat{\mathbf{e}}_x + u_y \hat{\mathbf{e}}_y; \quad \hat{\mathbf{n}} = n_x \hat{\mathbf{e}}_x + n_y \hat{\mathbf{e}}_y \qquad (4.9.8)$$

where $\hat{\mathbf{n}}$ is the unit normal to the integration path ds, $\mathbf{u}$ is the velocity vector along the path, and $\hat{\mathbf{e}}_i$ are unit vectors in the coordinate directions.

The calculation of the change in the stream function within a finite element can be carried out using (4.9.7) once a suitable integration path AB is identified. In most applications, the integration path is taken along the element boundaries. Consider the typical element boundary shown in Figure 4.9.1 with the following definitions:

$$u_x = \hat{\boldsymbol{\Psi}}^T \mathbf{u}_x \; ; \quad u_y = \hat{\boldsymbol{\Psi}}^T \mathbf{u}_y \qquad (4.9.9a)$$
$$x = \hat{\boldsymbol{\Upsilon}}^T \mathbf{x} \; ; \qquad y = \hat{\boldsymbol{\Upsilon}}^T \mathbf{y} \qquad (4.9.9b)$$

where $\hat{\boldsymbol{\Psi}}$ and $\hat{\boldsymbol{\Upsilon}}$ are interpolation (edge) functions and $\mathbf{u}_x$, $\mathbf{u}_y$, $\mathbf{x}$, and $\mathbf{y}$ are vectors of nodal point velocities and coordinates. The unit normal vector is given by

$$\hat{\mathbf{n}} = \frac{1}{\Delta} \frac{\partial y}{\partial s} \hat{\mathbf{e}}_x - \frac{1}{\Delta} \frac{\partial x}{\partial s} \hat{\mathbf{e}}_y \qquad (4.9.10)$$

with ds defined by

$$ds = \left[\left(\frac{\partial x}{\partial \hat{s}} \right)^2 + \left(\frac{\partial y}{\partial \hat{s}} \right)^2 \right]^{\frac{1}{2}} d\hat{s} = \Delta \, d\hat{s} \qquad (4.9.11a)$$

Using the definitions of Eq. (4.9.9), we obtain

$$ds = \left[\left(\frac{\partial \hat{\boldsymbol{\Upsilon}}^T}{\partial \hat{s}} \mathbf{x} \right)^2 + \left(\frac{\partial \hat{\boldsymbol{\Upsilon}}^T}{\partial \hat{s}} \mathbf{y} \right)^2 \right]^{\frac{1}{2}} d\hat{s} = \Delta \, d\hat{s} \qquad (4.9.11b)$$

Combining these relations with the definition for $\delta \psi$ produces

$$\delta \psi = \int_{-1}^{+1} \left(\frac{\partial \hat{\boldsymbol{\Upsilon}}^T}{\partial \hat{s}} \mathbf{y} \hat{\boldsymbol{\Psi}}^T \mathbf{u}_x - \frac{\partial \hat{\boldsymbol{\Upsilon}}^T}{\partial \hat{s}} \mathbf{x} \hat{\boldsymbol{\Psi}}^T \mathbf{u}_y \right) d\hat{s} \qquad (4.9.12)$$

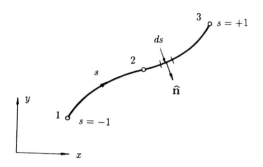

Figure 4.9.1: Definition of the element boundary for stream function computation.

The interpolation function definitions were given in Section 4.5; the vector $\hat{\Upsilon}$ can be either linear or quadratic, depending on the shape of the element boundary. The change in the stream function along any element boundary can be computed from Eq. (4.9.12) once the element geometry and velocity fields are specified. Computation of the stream function for an entire finite element mesh is generated by applying Eq. (4.9.12) along successive element boundaries, starting at a node for which a reference value of ψ has been specified.

The calculation of the stream function for axisymmetric flows follows a similar procedure with the appropriate definition for ψ, $\mathbf{u}$, and $\hat{\mathbf{n}}$

$$u_1 = u_r = \frac{1}{r}\frac{\partial \psi}{\partial z} \ ; \qquad u_2 = u_z = -\frac{1}{r}\frac{\partial \psi}{\partial r} \tag{4.9.13}$$

$$\mathbf{u} = u_r \,\hat{\mathbf{e}}_r + u_z \,\hat{\mathbf{e}}_z; \qquad \hat{\mathbf{n}} = n_r \hat{\mathbf{e}}_r + n_z \hat{\mathbf{e}}_z \tag{4.9.14}$$

A second approach to computing the stream function for a given velocity field relies on the definition of the vorticity. By substituting the stream function definition in (4.9.6) into the vorticity Eq. (4.9.5), the following elliptic partial differential equation is obtained:

$$-\nabla^2 \psi = \omega \tag{4.9.15}$$

This, being a Poisson equation, can be easily cast into a finite element model and solved for ψ, since ω is known everywhere in the domain (see Problem 4.4). The main disadvantage to this approach is that boundary conditions for ψ must be supplied for all points of the boundary and, more importantly, a separate matrix problem must be solved for each velocity field. The element-by-element procedure is computationally more efficient and it is generally the method of choice.

4.9.3 Particle Tracking

Another useful method for visualizing fluid motion involves following the movement of a massless particle that is introduced at a convenient point in the flow. The tracking of particle paths is governed by the set of ordinary differential equations

$$\frac{dx_i}{dt} = \dot{x}_i = u_i \tag{4.9.16}$$

Given the initial location of a point, $P(x_i)$, the initial value problem in (4.9.16) must be solved to describe the particle trajectory for the known flow field u_i. The integration of (4.9.16) is most easily accomplished on an element-by-element basis. Using the standard definitions for an isoparametric mapping of the element geometry and the element approximation for velocity produces

$$\mathbf{J} \begin{Bmatrix} \dot{\eta} \\ \dot{\xi} \\ \dot{\zeta} \end{Bmatrix} = \begin{Bmatrix} \mathbf{\Psi}^T(\xi,\eta,\zeta)\mathbf{u}_1 \\ \mathbf{\Psi}^T(\xi,\eta,\zeta)\mathbf{u}_2 \\ \mathbf{\Psi}^T(\xi,\eta,\zeta)\mathbf{u}_3 \end{Bmatrix} \tag{4.9.17}$$

where the Jacobian matrix $\mathbf{J}$ was defined in Section 3.4 and (ξ,η,ζ) are the local element coordinates. For a known starting location (ξ_0,η_0,ζ_0) and velocity field, Eq. (4.9.17) can be integrated by standard methods to yield the particle trajectory across the element. When the particle crosses an element boundary, the procedure is restarted with a new element Jacobian matrix and a new element velocity field.

As shown in (4.9.17) the velocity field is independent of time and the integration can be performed with the fixed values of $\mathbf{u}$. Time-dependent flows may also be computed [in which case $\mathbf{u} = \mathbf{u}(t)$]. Conceptually, the method is the same but the numerical implementation now requires interpolation to find the needed particle tracking velocities from the finite element time planes.

It is also possible to generalize this method to consider the motion of real particles in the flow. In this case, a force balance on the particle is required that accounts for drag, gravity effects, or other forces and predicts the particle velocity. The particle velocity is then used in (4.9.16) in place of the fluid velocity; the integration procedure is similar to the one shown in (4.9.17).

4.10 Advanced Topics - Free Surface Flows

4.10.1 Preliminary Comments

Many Newtonian and non-Newtonian flow problems of practical interest contain a complicating feature in the form of a free surface boundary. Boundary conditions for the free surface problem were given in Section 1.8.1 and were seen to be somewhat different for the steady-state flow problem and the case of a dynamic free surface. Likewise, the solution algorithms for steady and time-dependent free surface flows tend to be quite different since the boundary interface conditions dictate the numerical procedure. In this section we provide an overview of the types of procedures used for both static (quasi-static) free surfaces and advancing front type problems.

4.10.2 Time-Independent Free Surfaces

Steady flow problems with a free surface boundary are somewhat easier than the dynamic case since only a single location and shape of the free surface need to be computed. For some geometries, such as extrusions and jets with a simple cross section, the free surface shapes are relatively simple and the required algorithms are straightforward and easily implemented. When the free surface shape is very complex, such as found in coating flows or fluid drop problems, the numerical algorithms are necessarily more sophisticated. Steady free surface methods universally employ some type of mesh updating or mesh movement algorithm; these are surface capturing methods in which the free surface always coincides with edges (surfaces) of the finite element discretization. Such procedures allow for the accurate representation of surface phenomena, such as surface tension forces, but at the cost of mesh manipulation. The complexity of the following algorithms is directly proportional to the methods used for mesh adjustment. In the following, the derivations and discussion are limited to two-dimensional problems to simplify the notation and equations; the extension to three dimensions is conceptually the same, although the details are more complex.

The appropriate boundary conditions for a free surface involve the balance of forces in the directions normal and tangent to the surface. Neglecting drag effects from the surrounding medium, these conditions are given by Eq. (1.8.7) and are written for the two-dimensional case as

$$\sigma_{nn} = \gamma \left(\frac{1}{R} \right) - P_{amb} \tag{4.10.1}$$

$$\sigma_{ns} = 0 \tag{4.10.2}$$

where γ is the surface tension, R is the principal radius of curvature of the surface, and P_{amb} is the ambient pressure. Note that Eq. (4.10.1) is written for a planar problem; an axisymmetric problem would require consideration of a second radius of curvature. In addition to the boundary conditions (4.10.1) and (4.10.2), a kinematic condition on the surface is required; the free surface is a streamline and thus

$$u_i n_i = 0 \tag{4.10.3}$$

where n_i ($i = 1, 2, 3$) are the components of the outward normal, and u_i are the components of the velocity vector. For many applications surface tension effects are negligible. Also, without loss of generality the ambient pressure may be set to zero. Thus, the appropriate boundary conditions in many cases correspond to vanishing of the normal and tangential components of the stress vector on the free surface.

The zero stress conditions encountered for the zero surface tension case are precisely the conditions that are enforced by the "natural" boundary conditions arising in the Galerkin finite element form of the momentum equations. With these conditions automatically enforced, it remains to find a method for locating the position of the free surface. For simple free surface shapes a procedure proposed by Tanner et al. [67] and used by others (see [68,69]) can be employed.

The computation starts with an assumed free surface shape and a given finite element mesh. The appropriate boundary conditions are applied and the problem solved for a velocity and pressure field. In general, the computed velocity field will not satisfy the kinematic constraint in Eq. (4.10.3) along the assumed free surface boundary. Referring to Figure 4.10.1, the location of the free surface is updated using the following relation

$$\frac{d}{dx} f(x) = \frac{v}{u} \qquad (4.10.4)$$

where u, v are the velocity components on the surface. Equation (4.10.4) can be integrated numerically (e.g., Simpson's rule) to produce a new surface shape $f(x)$. The success of the integration is dependent on knowing at least one (fixed) point on the free surface. With a new free surface location a new finite element mesh can be constructed and the above procedure is repeated until convergence is obtained. Mesh reconstruction in this type of algorithm usually amounts to a one-dimensional rescaling of the element locations in a particular coordinate direction. More sophisticated remeshing methods could be used but are generally not cost effective since the mesh movement is relatively small. For problems with material and/or convective nonlinearities, the free surface iteration is normally intertwined with the other iterative procedures. Mapping of the current solution field to the new mesh is not normally required due to the size of the mesh movement.

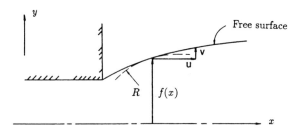

Figure 4.10.1: Nomenclature for free surface boundary computations.

When surface tension is not negligible, the procedure outlined above may still be used but with minor modifications. Instead of applying the zero normal stress boundary condition, a normal force due to the surface tension must be applied at each iteration. Again referring to Figure 4.10.1, the radius of curvature is given by

$$R = \frac{\left[1 + \left(\frac{df}{dx}\right)^2\right]^{3/2}}{\left(\frac{d^2 f}{dx^2}\right)} \tag{4.10.5}$$

which may be computed at any iteration since $f(x)$ is then known. Equations (4.10.1) and (4.10.5) allow the normal stress at the free surface to be computed and applied at each cycle of the free surface iteration. The procedure outlined above has been used to study a number of extrusion and jet flow problems with good success. However, for applications where the shape of the free surface is rather complex and the indicated procedure is not adequate, more sophisticated algorithms must be employed.

An alternative approach, developed by Saito and Scriven [70] and Scriven and Kistler [71], employs a finite element parameterization in which location parameters (degrees of freedom) for the free surface become unknowns in the general problem formulation. Though the details of implementation are fairly involved, the general idea can be obtained from the sketch in Figure 4.10.2. As indicated by the figure, nodes on the free surface are free to move along fixed lines, called spines, such that their location relative to a fixed base point is determined by a single nodal parameter, h_i. The coordinates for the free surface nodes are given by

$$x_i^{fs} = \alpha_x^i h_i + x_i^b, \quad y_i^{fs} = \alpha_y^i h_i + y_i^b \tag{4.10.6}$$

where α^i are components of the direction vector along the spine, and x_i^b and y_i^b are coordinates of the base point. Note that in an actual implementation, nodes for all the elements between the free surface and the base line would also be parameterized such that node spacing ratios would be maintained with movement of the free surface. To complete the formulation for the free surface, it is observed that the surface is composed of a series of element edges with each edge containing two or three nodes, depending on the type of finite element being employed. The isoparametric description of the element edge then has the form

$$x = \mathbf{\Upsilon}^T \mathbf{x}^{fs}, \quad y = \mathbf{\Upsilon}^T \mathbf{x}^{fs} \tag{4.10.7}$$

where $\mathbf{x}^{fs}$ and $\mathbf{y}^{fs}$ are computed using Eq. (4.10.6), and $\mathbf{\Upsilon}$ is an appropriate linear or quadratic edge interpolation function.

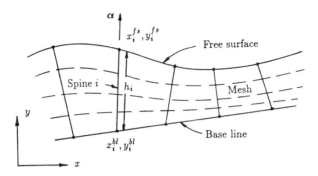

Figure 4.10.2: Nomenclature for spines used in free surface description.

With the above relations, the standard finite element discretization of the momentum and continuity equations can be implemented. The proper accounting must be made in the boundary integrals for the free surface stresses acting on the parameterized element edges. A final equation is also needed since the h parameters are now unknowns in the global equation set. The last equation needed is the kinematic condition (4.10.3) which can be written in a weighted residual form as

$$\int_{\Gamma_{fs}} \Upsilon u_n d\Gamma = \int_{\Gamma_{fs}} \Upsilon (u_x n_x + u_y n_y) d\Gamma = 0 \qquad (4.10.8)$$

where the weighting function is the interpolation function defining the free surface [see Eq. (4.10.7)]. With the proper definition for the velocities in the element and the mass consistent normals at the surface [72], the surface integrals in Eq. (4.10.8) can be evaluated and assembled to provide an equation for each boundary node. Analogous with the matrix equations (4.10.5) and (4.10.6), the finite element model for a free surface flow is then given by

$$\mathbf{C(u)u + K(u)u - QP + Bh = F} \qquad (4.10.9a)$$

$$-\mathbf{Q}^T \mathbf{u} = \mathbf{0} \qquad (4.10.9b)$$

$$\mathbf{K}_{fs}\mathbf{u} = \mathbf{0} \qquad (4.10.9c)$$

where $\mathbf{h}$ is the global vector of free surface parameters, $\mathbf{B}$ is the matrix corresponding to the surface stress conditions, and $\mathbf{K}_{fs}$ is the matrix resulting from the kinematic constraint in Eq. (4.10.8).

The solution to (4.10.9) proceeds in the usual way, with the use of Newton's method being the preferred algorithm. Other fixed point schemes such as the Picard method are not employed here since at each iteration the velocities would have to be interpolated onto the new nodal point locations defined by new $\mathbf{h}$ values. However, the Newton procedure avoids this difficulty if the full Jacobian matrix is used, i.e., the variations of all equations with respect to the free surface parameters **h** are constructed. The construction of these new entries of the Jacobian matrix is not trivial since both the integrands and the limits of integration depend on **h**. The usual methods for isoparametric mapping of the elements can be employed to simplify this task and make the Newton algorithm usable.

The method of spines is a formal method of embedding the kinematic free surface condition directly into the equations for the system. The method given in Eq. (4.10.4) uses the kinematic condition outside the basic equations as a control on the (free surface) iteration procedure. Though the spine algorithm is more general and can handle some problems of great complexity, it is not without its drawbacks. The fact that nodes are constrained to a "one-dimensional" motion along the spine limits severely the ability of the mesh to conform to a complex surface shape. Re-entrant corners are particularly difficult to describe with spines. In essence, the initial mesh design and spine placement must anticipate the final free surface shape. The method is practical in some two-dimensional applications and a limited number of three-dimensional flows.

To avoid some of these difficulties, work has progressed on changing the way the mesh is parameterized but retaining most of the other solution features of the spine approach. Christodoulou and Scriven [73] have proposed using body-fitted coordinates and elliptic mesh generation to parametrize the entire mesh. This appears to be more flexible and useful than spines and is more readily extendable to three dimensions. Elliptic mesh generation does, however, have its limitations with regard to mesh control and mesh quality. Other mesh generation methods could be used though high quality, automatic meshing of complex geometries remains an area of substantial research and development. A review of some of these free surface algorithms and mesh update methods can be found in [74].

A more recent development with regard to mesh adjustment is the use of a pseudo-structural method to define the movement of the elements [75,76]. If the meshed region is viewed as a solid body, then the element edges can be treated as springs with a given distribution of spring (stiffness) constants. When the free surface is adjusted, i.e., given a displacement, the interior nodes of the elements are adjusted to a new equilibrium position and a mesh update is completed. A more formal approach treats the meshed region as an elastic body with a distribution of elastic material parameters. When the free surface displacements are specified, an elastic boundary value problem is defined that can be solved via a finite element method. The new displacement field defines a new mesh for the subsequent resolution of the flow problem. The elasticity problem and mesh movement may be fully coupled to the flow problem as described above for the spine parameterization [77]. The elasticity problem can also be decoupled from the flow problem and solved separately in a staggered type of algorithm [78]. In this later case, a mapping or projection of the velocity field from the old mesh onto the new mesh will be required for problems with a non-zero Reynolds number.

The usual pseudo-structural approach employs the equilibrium equations for an elastic body which can be written as

$$\nabla \cdot \sigma^s + \mathbf{f}^s = \mathbf{0} \qquad (4.10.10)$$

where $\mathbf{f}^s$ is the body force vector and σ is the stress tensor for the solid. By Hooke's law, we have

$$\sigma^s = \lambda \, (tr \, \varepsilon^s)\mathbf{I} + 2\mu \, \varepsilon^s \qquad (4.10.11)$$

and the strain tensor ε^s is related to the displacement vector $\mathbf{d}$ by

$$\varepsilon^s = \frac{1}{2}\left[\nabla \mathbf{d} + (\nabla \mathbf{d})^T\right] \qquad (4.10.12)$$

In Eq. (4.10.11) λ and μ are the Lamé constants. The above equations describe a boundary value problem for the elastic response of a body and may be cast in computational form via a weak or variational formulation. A finite element approximation ultimately produces a matrix problem of the form

$$\mathbf{K}^s\mathbf{d} = \mathbf{F} \qquad (4.10.13)$$

for the nodal point displacements $\mathbf{d}$. Some of the boundary conditions for the displacements in (4.10.13) will be generated from the movement (adjustment) of the free surface to satisfy the kinematic free surface condition. The remaining boundary conditions on displacement will need to be specified so as to allow the mesh to conform to boundaries of the flow domain while minimizing mesh distortion. Once the boundary displacements are specified and a distribution of material parameters λ and μ for the solid has been selected, the problem in (4.10.13) may be solved for a new set of interior displacements. The solution to (4.10.13) provides displacements that can be used to update the locations of the nodes in the mesh. A projection of the fluid velocity field from the old mesh to the new mesh completes the iterative cycle and permits the flow problem to be solved again on a new geometry.

A simple linear elastic constitutive relation was described in (4.10.11) and this may not be suitable for many mesh moving applications. Some nonlinear relations have been suggested [77] as well as methods for selecting the distribution of elastic constants over the meshed region [78,79]. Even though the mesh movement in this approach is based on a physical principle there is no guarantee that acceptable or accurate meshes will result. Severe mesh distortions remain a problem for complex flows and if distortion cannot be controlled, remeshing of the geometry may be required. In many applications free surface analysis is still something of an art form.

4.10.3 Time-Dependent Free Surfaces

Methods for time-dependent free surface problems fall naturally into two distinct types of algorithms. The moving mesh methods, as outined in the previous section, can be extended to handle some types of dynamic free surfaces. These methods would typically be applicable to flows in which the free surface motion was limited and mesh distortion would not be a significant difficulty. When free surface motions are large, the quasi-Lagrangian methods based on mesh updating are no longer very convenient. The second type of method relies on a fixed (Eulerian) mesh and attempts to track the location of the free surface through a variable defined on the mesh. Whereas the mesh moving methods explicitly define the free surface, the fixed mesh schemes implicitly define the free surface location and motion. Physical phenomena that occur on the free surface (e.g., surface tension) are easily accommodated with the explicit moving mesh methods since the finite element variables coincide with the free surface. Surface physics are difficult to accurately include in the basic fixed mesh algorithms though complex bulk flows are well described with these methods. In the following the mesh moving algorithms are briefly updated for time dependence and then fixed mesh algorithms are described.

The mesh moving methods outlined in the previous section can, in some cases, be extended to handle time-dependent free surface problems. The pseudo-structural method is probably the most robust and most generally useful of the various update methods. To incorporate this technique into a time-dependent problem primarily requires the redefinition of the material or convected derivative in the fluid equation of motion. The convected derivative was defined in Eq. (4.1.4) and needs to be redefined as

$$\frac{D}{Dt} \equiv \left[\frac{\partial}{\partial t} + (\mathbf{u} - \mathbf{u}^m) \cdot \nabla \right] \qquad (4.10.14a)$$

$$\frac{D}{Dt} \equiv \left[\frac{\partial}{\partial t} + (u_j - u_j^m) \frac{\partial}{\partial x_j} \right] \qquad (4.10.14b)$$

where the superscript m indicates the velocity of the mesh. Use of this derivative in the fluid momentum equation (4.1.2) or in the weak form (4.2.5) allows the fluid motion to be described in a time-dependent mesh. This type of formulation is termed an Arbitrary Lagrangian Eulerian (ALE) method since the mesh motion is described in a Lagrangian frame while the fluid motion remains in an Eulerian description.

The remaining part of the ALE scheme is the algorithm for moving the mesh and generating the mesh velocity $\mathbf{u}^m$. To utilize the pseudo-structural method in the time-dependent case, the equilibrium equation (4.10.10) could be differentiated with respect to time to directly produce a stress rate and subsequently a displacement rate or mesh velocity. The equilibrium equations could also be solved in their steady form for the displacement that occurs over the time step; the displacement increment divided by the time step produces the mesh velocity. The overall ALE method may be fully coupled, in which case the fluid and pseudo-solid equations are solved simultaneously at a time step, with Newton's method providing a typical iteration procedure. The fluid and mesh motion could also be solved in a decoupled manner. Iterative convergence within a time step would be slower than the fully coupled method, though the computational work per iteration would generally be reduced.

One of the major drawbacks to ALE type methods is the problem of mesh distortion. Except in special cases, such as almost one-dimensional free surface motions or some types of periodic motions, the movement of the elements will eventually reach the point where the mesh is badly deformed and the accuracy of the finite element approximation is severely degraded. To continue the simulation the flow domain must be remeshed on the current geometry and the solution fields transferred from the old mesh to the new mesh. Automatic remeshing of a general geometry is a nontrivial task and is still the subject of current research. The transfer of field data from one mesh to another is done most simply by interpolation. Specialized programs exist that perform this function by finding new node locations in the old mesh and using the old mesh shape functions to interpolate the dependent variables [80,81]. For an incompressible flow, the interpolated velocities will not generally satisfy the incompressibility constraint; restarting the simulation with an erroneous velocity field may produce nonphysical results, especially the pressure, depending on the time integration method that is used. A mass consistent velocity field can be produced on the new mesh but at the expense of solving another matrix problem [82]. The mesh distortion and remeshing problems are sufficiently difficult and computationally expensive that ALE methods are usually limited to transient free surface flows with small motions or special geometries that avoid mesh distortion.

When large-scale free surface motions must be simulated, as in many cavity filling flows, fixed mesh methods with some type of front tracking procedure are normally the most useful. Many of the techniques that have been evaluated in a finite element context fall in the category of methods commonly known as Volume of Fluid (VOF) methods, or more correctly, Volume Tracking (VT) methods. These methods originated in finite difference applications and have evolved to a variety of algorithms that differ mainly in the details of the implementation. Rider and Kothe [83] present a summary of many of the different approaches used in VT methods. Here the method is outlined in generic form as it would be used in a finite element context.

Before describing the volume tracking algorithm, consider a simpler method that has some of the basic features of many fixed mesh methods. A fluid concentration variable, usually denoted by F, is defined as a nodal point variable on the finite element mesh. The concentration variable has a value of unity at locations within the fluid and zero at locations outside the fluid; the free surface is implicitly located between the 0 to 1 extremes. As a passive scalar concentration variable, the F function is transported by the fluid motion, without diffusion, according to the hyperbolic equation

$$\frac{\partial F}{\partial t} + u_j \frac{\partial F}{\partial x_j} = 0 \qquad (4.10.15)$$

where u_j is the fluid velocity. In a finite element framework, Eq. (4.10.15) could be discretized using a Galerkin weighted-residual formulation to produce an advection equation for the nodal values of F. Being hyperbolic, Eq. (4.10.15) requires only initial values of F. As the function F is computed at nodes throughout the mesh, the relevant fluid momentum and continuity equations are solved on an ever-changing set of grid points indicated by the condition $F = 1$. The overall method is a staggered, cyclic procedure in which the flow problem is solved on the domain defined by $F = 1$, followed by an updating of the flow domain by advancing (4.10.15) in time. Equation (4.10.15) could be solved using either an explicit or implicit time integration method, while the fluid mechanics part of the problem is solved implicitly. In either case, the concentration equation can be subcycled to improve the temporal accuracy of the free surface evolution. At the free surface (usually indicated by $F = 0.5$) the usual stress boundary conditions are applied; the standard choice is for a traction free surface as surface tension effects are somewhat difficult to incorporate in this formulation.

Unfortunately, free surface boundary conditions are not the only difficulty with the method outlined above. Theoretically, the free surface position in this algorithm is described by the propagation of a square wave profile in F on a fixed mesh; computationally, the steepest or sharpest variation in F that can be resolved is fixed by the smallest node spacing in the flow direction. In any case, it is very difficult, if not impossible, to transport a steep function across a variably spaced mesh without significant dispersion and spatial oscillation of the profile. This difficulty prohibits the accurate description of the free surface evolution and requires the modification of the above approach. The usual volume tracking algorithm eliminates the precise definition of the free surface in favor of a new variable that evaluates the volume of fluid in each element as a function of time. This new variable, the volume fraction, is defined for each element j by

$$f_j = \frac{1}{V_j} \int_{V_j} F \, dV_j \qquad (4.10.16)$$

where F is the concentration variable in (4.10.15) and V_j is the element volume; j runs from 1 to the number of elements. The volume fraction also varies between 0 and 1, with $f_j = 1$ being a filled element, $f_j = 0$ being an empty element, and $0 < f_j < 1$ being a partially filled element that contains the free surface. The VT algorithm proceeds in two stages and is a cyclic, staggered algorithm as noted above. After initialization of f_j, the fluid volumes are first advected forward in time followed by an interface reconstruction step. The fluid mechanics problem is then solved on an updated domain to complete the cycle.

Initialization consists of defining f_j for each element once the initial free surface location is specified. Given a fluid velocity field and the reconstructed location of the free surface, the volume fractions are evolved forward in time using some form of the advection equation in (4.10.15). Like the concentration variable F, the volume fraction variable must also satisfy a conservation relation

$$\frac{\partial f}{\partial t} + u_j \frac{\partial f}{\partial x_j} = 0 \qquad (4.10.17)$$

if f is viewed as a continuous function. Since the volume fraction has been defined on each element it is discontinuous and the partial differential form of

(4.10.17) is rarely used. Instead, a control volume form of (4.10.17) is defined for each element that balances the net inflow to each element with the change in volume fraction within the element. A typical element balance is

$$\frac{\partial f_j}{\partial t} = \frac{f_j^{n+1} - f_j^n}{\Delta t} = \frac{1}{V_j} \sum_{k=1}^{nfaces} q_{jk}^n \qquad (4.10.18)$$

where the flow rate for face k of the element is

$$q_{jk} = -\int_k u_i n_i \, dA \qquad (4.10.19)$$

The control volume relation in (4.10.18) has been written as an explicit time integration and is therefore subject to a time step limit that respects the hyperbolic nature of the original equation. This time step limit or Courant–Friedrich–Levy (CFL) stability limit also prevents the over-filling or emptying of an element. An implicit integration method is not practical in this cyclic type of procedure since the velocity would be required at the new time level. Note that in (4.10.19) the integral is over the area of the element face. In a partially filled or filling element, where (4.10.18) and (4.10.19) would be used, the part of the element face (area) over which a fluid velocity is defined depends on the location of the free surface within the element and within the adjoining elements. This dependence leads to the other step in the VT algorithm where the fluid interface is reconstructed from the volume fraction data. This step actually precedes the advection step but since the algorithm is cyclic, the order of explanation is not critical.

When the volume fraction variable is adopted, precise information on the location of the fluid interface within an element is lost. This information can be recovered through a reconstruction algorithm that varies depending on the particular VT method. It is important to realize that the reconstruction process is not unique and may be based on heuristic, geometric, or algebraic methods [83]. Because so many reconstruction methods are available no attempt will be made to catalog the various approaches and implementations. Rather a simple geometric method in two dimensions will be used to indicate the basic ideas.

The interface reconstruction has a limited amount of information to work with. The volume fraction of the element in question and the volume fractions of the surrounding elements form the sole basis for many algorithms. Consider the patch of elements in Figure 4.10.3, where the center element has a volume fraction of $f_e = 0.25$. If element b is filled or partially filled ($f_b > 0$) and the remaining elements are empty ($f_d = f_f = f_h = 0$), then one reconstruction method would place the fluid volume in element e along the boundary with element b in a square shape (see Figure 4.10.3a). In this particular method the partial volume fraction within an element is always square or rectangular in shape and is described by constant coordinate lines in the master or parent element. Figure 4.10.3b shows the case where an additional element is completely filled ($f_f = 1$) and the reconstruction would move the partial fluid volume to the corner of element e while maintaining the square shape. If element f had been only partially filled, element e would have remained, as shown in Figure 4.10.3a, since location is most heavily weighted toward completely filled elements. For the case where both elements b and f are partially filled the pattern would be as shown in Figure 4.10.3b since the weighting is equal between the two elements. Figure 4.10.3c illustrates the case of three surrounding elements that are partially filled or completely filled and the change in the partial volume shape in element e to a rectangle. Finally, the case where opposing elements are partially or completely filled is shown in Figure 4.10.3d and the volume fraction in element e is again rectangular. Other combinations of partially filled and filled elements are possible but have not been shown since the general pattern of the algorithm is clear. The different patterns may be cataloged according to the number of filled or partially filled neighbors. The shape and location of the reconstructed surface are then defined from a table, while the values of the master coordinate lines are computed from the value of the volume fraction. This type of pattern definition would be required for each type of element used in a VT scheme and the number of possible patterns gets quite large for three-dimensional elements. Once the fluid interface has been constructed for each element, the surface areas needed for the flow rate in (4.10.19) and the advection step can be easily established. Reference 84 has a description of many of the details associated with the implementation of the VT method.

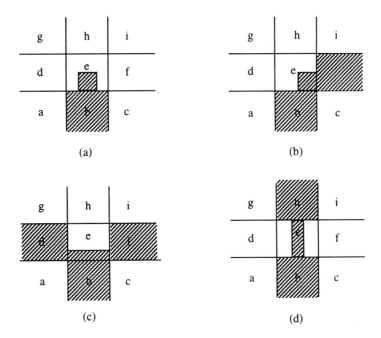

Figure 4.10.3: Examples of reconstructions for the fluid interface in a volume tracking algorithm.

Though the various volume tracking methods have been employed with good success, the overall methodology is dominated by nonunique rules governing the interface reconstruction process. Another fixed mesh method that has recently become popular is the Level Set Method (LSM). This technique has a much stronger theoretical basis than the VT methods and shows great promise in resolving very complex free surface problems. The method is well described in the text by Sethian [85] and there are many applications in the literature.

The Level Set Method is related to the simple advection method outlined previously in conjunction with the concentration equation (4.10.15). Rather than describe a concentration function, the level set method usually begins with the kinematic condition that describes the location of the free surface. In any case, the equations are the same. From Eq. (1.8.4) the free surface is defined by the function $F(x_i, t)$ and the condition

$$\frac{\partial F}{\partial t} + u_j \frac{\partial F}{\partial x_j} = 0 \qquad (4.10.20)$$

The difficulty with solving this equation as a concentration equation was the use of a function with a steep gradient to represent the free surface location. The level set method alters the F function to a smoother function that can be more easily represented on general meshes. The standard level set function is the signed distance function that is defined by

$$F(x_i, 0) = \begin{cases} +d, & x_i \in \Omega_f \\ 0, & x_i \in \Gamma \\ -d, & x_i \in \Omega_0 \end{cases} \qquad (4.10.21)$$

where d is the distance from the interface Γ, Ω_f is the fluid region, and Ω_0 is the unfilled region. A weighted residual, finite element method applied to Eq. (4.10.20) produces an equation set for the time advancement of the F function given the velocity field u_j. Using (4.10.21) as an initial condition for the function, the evolution of the free surface can be computed by tracking the zero level set of the function F. Since F is a relatively smooth function, dispersion and other computational errors should not degrade the accuracy of the front motion. Also, because F is a smooth function it is easy to take derivatives and therefore compute the unit normal vector to the free surface

$$\hat{\mathbf{n}} = \left. \frac{\nabla F}{|\nabla F|} \right|_{F=0} \qquad (4.10.22)$$

and the curvature of the free surface

$$\kappa = \nabla \cdot \hat{\mathbf{n}} = \left. \nabla \cdot \left(\frac{\nabla F}{|\nabla F|} \right) \right|_{F=0} \qquad (4.10.23)$$

Both of these quantities may be important in certain applications and they are not available from the volume tracking methods.

Two additional points that must be addressed when using a level set method are extensional velocities and reinitialization. In the advection equation for the level set it is implied that the velocity field is specified everywhere, i.e., for all the level sets, not just the zero level set. This is not usually the case in flow problems where the fluid is filling an otherwise empty cavity. Methods for defining a useful velocity away from the interface are describe by Sethian [85] and others [86]. The problem of reinitialization comes from the use of nonuniform velocity fields. When F is evolved in time using (4.10.20) the initial distance function will not generally remain a distance function. This is particularly true when fluid velocity fields are used around the free surface (zero level set). The solution to this difficulty is to reinitialize the level set to a distance function based on the current location of the zero level set. A simple iterative procedure proposed by Sussman et al. [87] produces a new signed distance function and allows the computation to proceed.

4.11 Advanced Topics - Turbulence

4.11.1 Preliminary Comments

In the previous sections, many of the essential aspects of finite element techniques for viscous incompressible flows were demonstrated. Even within this relatively narrow class of problems a number of important fluid mechanics topics were omitted. Here we will briefly consider an extension to the basic problem area and describe some aspects of turbulent flows. This type of problem is of major importance for industrial problems.

Turbulence is a highly complex physical phenomenon that is pervasive in flow problems of scientific and engineering concern. A simple, precise definition of turbulence is difficult though the phenomenon is often associated with the ideas of randomness, disorder, and chaos. Following Hinze [107], we say that turbulence is defined as an "irregular flow condition showing random variations with respect to both time and space coordinates with discernible statistical properties." Though a standard definition of turbulence may be elusive, a number of well-established characteristics of a turbulent flow can be listed. A turbulent flow is a

- highly nonlinear flow process;
- highly diffusive flow;
- three-dimensional flow;
- flow with multiple length and time scales;
- time-dependent (stochastic) phenomenon with identifiable statistical properties.

Turbulence is one of the unsolved problems in physics, especially in the sense that universally applicable mathematical models of the phenomenon are not available. Despite these theoretical difficulties, the demand for numerical computation of realistic (turbulent) flows remains high. Our interest here will, therefore, focus on the modeling and simulation of turbulence from an engineering point of view. This approach implies that the detailed resolution of a turbulent flow will be eliminated in favor of some type of averaged flow description. Turbulence effects will enter the flow description via a model that is typically based on a combination of theory and experiment.

In the following sections, the equations for the mean flow will be described followed by an outline of various types of turbulence models. Sections on finite element implementation and boundary conditions are also included.

4.11.2 Governing Equations

A majority of researchers accept the notion that, in principle, the Navier–Stokes equations are capable of fully describing a turbulent flow [108,109]. The natural question that follows this premise is if the Navier–Stokes equations are valid, why not solve them directly (via a numerical method) to obtain the needed turbulent solution? Recall from the brief introduction that turbulent flows are inherently unsteady and three-dimensional and contain a wide spectrum of length scales. To accurately simulate such a flow, the mesh resolution would have to be on the same order as the smallest length scale. Typically, the smallest eddies in a turbulent flow are $\approx O(0.001L)$ where L is the characteristic flow dimension. This translates into a three-dimensional grid with a minimum of $\approx 10^9$ mesh points. In addition, the flow must be resolved on the shortest relevant time scale, which is again associated with the smallest eddies. The computational demands of such a direct simulation of turbulence, even for simple geometries, is generally beyond the range of current computer resources. For industrially relevant problems and complex geometries, the direct simulation approach is clearly not possible with present technologies.

The standard alternative to the direct numerical simulation (DNS) approach involves the solution to some form of *averaged* Navier–Stokes equations. In most flow problems of interest it is the mean flow that is of most concern, with the turbulent fluctuations only being important in how they influence the mean flow evolution. By performing a suitable average on the instantaneous Navier–Stokes equations, a standard mean flow problem can be derived where the effects of the turbulence are relegated to a few terms that can be *modeled*. This approach forms the basis for most of the current computational work.

To outline this approach, let the instantaneous fluid velocity and pressure fields be expressed as the sum of a mean and fluctuating component. That is,

$$u_i = U_i + u_i' \qquad (4.11.1)$$

$$p = P + P' \qquad (4.11.2)$$

where the upper case represents a mean quantity and the superscript prime denotes a fluctuation. Substituting these definitions into the incompressible, viscous flow equations of Section 4.1.2 and performing an ensemble average [107,109] produces the following

$$\frac{\partial U_i}{\partial x_i} = 0 \tag{4.11.3}$$

$$\rho\left(\frac{\partial U_i}{\partial t} + U_j\frac{\partial U_i}{\partial x_j}\right) = -\frac{\partial P}{\partial x_i} + \frac{\partial}{\partial x_j}\left[\mu\left(\frac{\partial U_i}{\partial x_j} + \frac{\partial U_j}{\partial x_i}\right) - \rho\overline{u_i'u_j'}\right] + \rho g_i \tag{4.11.4}$$

The above equations are written in Cartesian component form and hence the usual summation convention is implied; the horizontal overbar indicates an averaged quantity. Equations (4.11.3) and (4.11.4) describe the behavior of the mean fluid velocity and pressure fields and are very similar in form to the laminar equations discussed extensively in a previous section. In arriving at Eqs. (4.11.3) and (4.11.4), an ensemble average was employed, although space and time averaging could also be utilized. The various averages can be related through the ergodic hypothesis. The details and subtleties of averaging are beyond this text and can be found in many standard references [109–111].

The extra term that appears in Eq. (4.11.4) is often termed the Reynolds stress and represents the effects of the turbulent velocity fluctuations on the mean flow. Note that these second-order moments form the nine components of a second-order tensor, and the symmetry considerations reduce this to a total of six independent components.

Since no additional equations were added to those appearing in Eqs. (4.11.3) and (4.11.4), the appearance of the Reynolds stress unknowns leads to the well-known closure problem for turbulence. Through appropriate manipulation of the Navier–Stokes equations, six equations for the Reynolds stress variables may be derived. Unfortunately, these derived equations contain additional unknowns in the form of their higher-order moments. Similarly, equations derived for the third-order moments contain fourth-order moments. This problem of cascading unknowns is known as the closure problem. In order to achieve closure to the boundary value problem, the equation derivation must be halted at some level and a model introduced for the remaining unknowns. A number of turbulence models will be discussed in the next section.

4.11.3 General Turbulence Models

The mean flow boundary value problem described by Eqs. (4.11.3) and (4.11.4) requires six additional equations for the Reynolds stresses. This specification constitutes a turbulence model. There are an enormous variety of turbulence models, ranging in complexity from simple algebraic statements to descriptions involving multiple, nonlinear partial differential equations. Unfortunately, there is no universal method of classification for such models which adds greatly to the confusion within the field. Here we follow the classification scheme by Ferziger [112,113], which groups turbulence models according to the following labels:

- Correlations

- Integral methods

- One-point closure

- Two-point closure

- Large eddy simulation

- Direct numerical simulation

In this section we will briefly outline some of the salient features of the models in each category. A subsequent section focuses on the one-point closure methods which are the most prominent in current numerical work.

Correlations

The general usefulness of models based on correlations is very limited as they are available for only a few standard problem geometries that have been extensively studied by experimental methods. Though this is a relatively primitive approach, it is still used in many industrial problems. Correlations are typically developed to predict friction factors and heat transfer coefficients for various types of channel flows. Correlations generally play no role in field equation simulations.

Integral methods

Integral methods are useful when dealing with problems that have a parabolic nature, e.g., boundary layer flows. Generally, this approach utilizes a standard mean field equation that has been integrated over one coordinate to reduce the mathematical complexity of the problem. Empirical information is easily incorporated with this approach and leads to a fast and simple numerical procedure. However, these models lack generality and are limited again to those types of flows that have been extensively studied. This type of model still finds use in industry for repetitive design computations, but has largely been abandoned by the research community. Reviews of this type of turbulence modeling can be found in [109,114,115].

One-point closure

One-point closure models are the most popular turbulence models, particularly for computational work. By one-point closure it is meant that the correlations between the fluctuating velocity components, u'_j and u'_i, always employ values taken at the same physical location. Despite this restriction, the one-point methods encompass a wide variety of successful models. Additional discussion of one-point closure models is provided in Section 4.11.4.

Two-point closure

Two-point closure models remove the restriction of the one-point closure approach and explicitly introduce the effect of length scale. This approach is still in the development stage and has been limited in application to homogeneous and isotropic turbulence. An excellent introduction to this type of closure model can be found in [108] and [116].

Large eddy simulation

As the name implies, Large Eddy Simulation (LES) or Subgrid Scale Modeling (SGM) attempts to simulate the large scale motions of the flow and only model the very small scale eddies. From theory and experiment it is known that the smallest scale motions tend to be universal in their behavior and thus LES would seem to be an ideal approach for a broadly applicable turbulence model. The drawback to LES is that it is still relatively expensive because grid refinement remains a primary consideration. A good introduction to LES can be found in the works of Rogallo and Moin [117] and Yoshizawa [118].

Direct numerical simulation (DNS)

This type of turbulence model was discussed previously in Section 4.11.2.

4.11.4 One-Point Closure Turbulence Models

As noted previously, the majority of computational work, especially for industrial applications, has relied on some form of the one-point closure model. Even within this category there are a wide variety of possible models and various ways to catalog them. Shown in Figure 4.11.1 is a classification scheme that will be used in the following discussion. Of the two major branches shown in the figure, we will concentrate on the so-called eddy viscosity models. The Reynolds stress models (RSM), though generally more sophisticated than the eddy viscosity approach, lead to large systems of partial differential equations and a large number of empirical parameters. The computational burden of these models, coupled with a lack of certainty regarding the parameters, makes the RSM unattractive for present engineering applications.

Returning to the eddy viscosity models, it is important to note that this approach is based on one major assumption, the Boussinesq hypothesis. By analogy with the molecular diffusion of momentum, the Boussinesq hypothesis relates the turbulent momentum transport to the gradients of the mean velocity field. The Reynolds stresses in Eq. (4.11.4) are then expressed by

$$-\rho\overline{u_i'u_j'} = \mu_T\left(\frac{\partial U_i}{\partial x_j} + \frac{\partial U_j}{\partial x_i}\right) \tag{4.11.5}$$

where μ_T is the eddy viscosity. Unlike the molecular viscosity, μ, which is a fluid property, the eddy viscosity is a local property of the flow. When the definition in Eq. (4.11.5) is substituted into the momentum equation then the equations for the mean flow become

$$\frac{\partial U_i}{\partial x_i} = 0 \tag{4.11.6}$$

$$\rho\left(\frac{\partial U_i}{\partial t} + U_j \frac{\partial U_i}{\partial x_j}\right) = -\frac{\partial P}{\partial x_i} + \frac{\partial}{\partial x_j}\left[(\mu + \mu_T)\left(\frac{\partial U_i}{\partial x_j} + \frac{\partial U_j}{\partial x_i}\right)\right] + \rho g_i \quad (4.11.7)$$

Once the form of the eddy viscosity is specified then the mean flow can be solved in the same manner as a laminar flow since the equations are the same except for an augmented viscosity.

Though the turbulent flow problem has been reduced to a familiar set of partial differential equations, there remains the nontrivial task of specifying how the eddy viscosity varies with the flow field. Scaling arguments show that the eddy viscosity is proportional to a characteristic eddy velocity, u_e, and an eddy length, ℓ_e. That is,

$$\mu_T \propto \rho u_e \ell_e \qquad (4.11.8)$$

In general, it is easier to specify the variation of u_e and ℓ_e for a turbulent flow, and most turbulence models use these parameters or some closely related variable. From Figure 4.11.1 the type of turbulence model being used is determined by the number of equations used to specify the variation of the variables in Eq. (4.11.8).

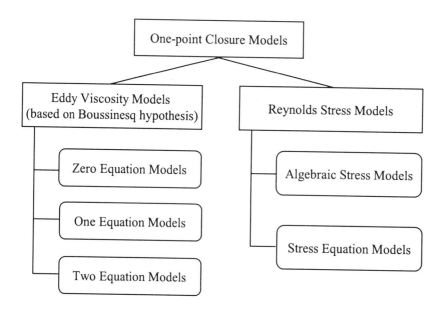

Figure 4.11.1: A classification scheme for one-point closure models.

Zero equation models

These models calculate the eddy viscosity in (4.11.8) by an algebraic prescription of u_e and ℓ_e and therefore are the simplest of all the one-point closure models. Most of these models are based on Prandtl's mixing length, which specifies ℓ_e to be the length scale across which turbulent mixing takes place. Prandtl gave the characteristic velocity u_e to be

$$u_e = \ell_e \left[\left(\frac{\partial U_i}{\partial x_j} + \frac{\partial U_j}{\partial x_i} \right) \frac{\partial U_i}{\partial x_j} \right]^{1/2} \tag{4.11.9}$$

which when substituted into (4.11.8) yields

$$\mu_T = \rho \ell_e^2 \left[\left(\frac{\partial U_i}{\partial x_j} + \frac{\partial U_j}{\partial x_i} \right) \frac{\partial U_i}{\partial x_j} \right]^{1/2} \tag{4.11.10}$$

For a number of geometrically simple flows, the variation of the mixing length, ℓ_e, is well-known and can be evaluated via simple formulas. Equation (4.11.10) then allows μ_T to be derived and the turbulence model completed. Flows that are amenable to such treatment include pipe and channel flows, jets, wakes, and boundary layers. A compilation of useful mixing length formulas can be found in [119,120]. Though the above model is quite simple, good results can be obtained if the model is employed for the types of flows for which it is strictly intended.

One equation model

The limitation of the zero equation model to simple flows, and the need to simulate more complex geometries, has led naturally to the development of more sophisticated turbulence models. As a first step in this direction the algebraic specification of u_e can be replaced with a more generally applicable transport equation. Since the characteristic velocity, u_e, is proportional to the square root of the turbulent kinetic energy, k, then

$$u_e \propto k^{1/2} \tag{4.11.11}$$

and from Eq. (4.11.8)

$$\mu_T \propto \rho k^{1/2} \ell_e \tag{4.11.12}$$

A partial differential equation for k can be derived from the Navier–Stokes equation and is given by

$$\rho \left(\frac{\partial k}{\partial t} + U_j \frac{\partial k}{\partial x_j} \right) = \frac{\partial}{\partial x_j} \left(\frac{\mu_T}{\sigma_k} \frac{\partial k}{\partial x_j} \right) + \rho G - \rho \epsilon \tag{4.11.13}$$

where G is a generation term, ϵ is the turbulent dissipation, and σ_k is a constant.

Despite the apparent sophistication involving the evaluation of k, the model still relies on an algebraic specification of the mixing length. Like the zero equation model, variations in ℓ_e are problem dependent and are well characterized for only a few types of flows. The model is marginally better than the zero equation model but is not heavily used in numerical simulations.

Two equation model

A natural evolution of the one equation model involves the replacement of the algebraic relation for mixing length with a second transport equation. Dimensional arguments lead to the proportionality

$$\ell_e \propto \frac{k^{3/2}}{\epsilon} \qquad (4.11.14)$$

where ϵ is the viscous dissipation. Substituting Eqs. (4.11.14) and (4.11.11) into Eq. (4.11.8) produces the proportionality

$$\mu_T \propto \rho \frac{k^2}{\epsilon}$$

or the Kolmogorov–Prandtl relation

$$\mu_T = C_\mu \rho \frac{k^2}{\epsilon} \qquad (4.11.15)$$

which relates the eddy viscosity directly to the turbulence variables, k and ϵ. In the two equation model, the turbulent kinetic energy is given by equation (4.11.13)

$$\rho \left(\frac{\partial k}{\partial t} + U_j \frac{\partial k}{\partial x_j} \right) = \frac{\partial}{\partial x_j} \left(\frac{\mu_T}{\sigma_k} \frac{\partial k}{\partial x_j} \right) + \rho G - \rho \epsilon \qquad (4.11.16)$$

and the turbulent dissipation is described by an equation of similar form

$$\rho \left(\frac{\partial \epsilon}{\partial t} + U_j \frac{\partial \epsilon}{\partial x_j} \right) = \frac{\partial}{\partial x_j} \left(\frac{\mu_T}{\sigma_\epsilon} \frac{\partial \epsilon}{\partial x_j} \right) + \rho \frac{\epsilon}{k} (c_1 G - c_2 \epsilon) \qquad (4.11.17)$$

where G is again a shear generation term and c_1 and c_2 are empirically derived constants as are σ_k and σ_g. Note that Eqs. (4.11.16) and (4.11.17) are obtained from manipulation of the instantaneous momentum equation. However, to provide closure, certain terms, such as G, are modeled by relating their exact form to some function of the mean flow.

The two-equation, k-ϵ model described by Eqs. (4.11.16) and (4.11.17) can be used in conjunction with the mean flow equations and the definition of μ_T given by Eq. (4.11.15), to arrive at a continuum description of turbulent flow. The equation set is highly nonlinear, with a strong coupling between the various transport equations. However, this additional mathematical and computational complexity is offset by a substantial increase in realism and reliability of the turbulence predictions. The k-ϵ model is far from universal and has a number of weaknesses, though it remains one of the most heavily used methods for flow simulation. A good discussion of this model can be found in Reference 121.

4.11.5 Finite Element Modeling of Turbulence

Once a turbulence model has been selected for a particular application, it is important to consider how a numerical technique, such as the finite element method, would be used to produce useful flow simulations. In previous chapters, various solution methods for the mean flow equations such as Eqs. (4.11.6) and (4.11.7) were described in detail and need not be reconsidered here. However, it is appropriate to outline some of the added complications that must be addressed when a turbulence model is added to the mean flow system. To limit the discussion, only the zero-equation and two-equation, k-ϵ, forms of the one-point closure models will be considered. These two approaches are the ones most commonly encountered in practical applications.

Zero equation model

The finite element implementation of a zero-equation turbulence model is straightforward since the eddy viscosity is prescribed by a simple algebraic relation, such as Eq. (4.11.10). Since the eddy viscosity varies over the flow domain, the methods introduced earlier in this chapter for modeling variable coefficients may be utilized. From Eq. (4.11.10) it is apparent that the turbulent viscosity depends on velocity gradients, which is directly analogous to the behavior of inelastic, non-Newtonian fluids (see Chapter 6). For the best accuracy, the velocity gradients should be evaluated at element integration points. The dependence of μ_T on the mean field velocity produces another source of nonlinearity in the numerical problem, though this is not of major significance for most iterative solution algorithms.

The most difficult aspect of these models comes from the specification of the mixing length over the flow domain. These empirical formulas will vary from region to region as different types of flows (e.g., boundary layers, shear layers, jets) are encountered. Also, many of the ℓ_e functions depend on distance from a boundary, a dependence that may be awkward to specify easily and uniquely in a complex geometry. Despite these inconveniences the simplicity of the model makes it a popular and useful choice for modeling.

Two equation model

The finite element implementation of a k-ϵ model is also conceptually straightforward since the turbulent transport equations (4.11.16) and (4.11.17) are of the familiar advection-diffusion type as will be discussed in Chapter 5. A weak form of the k-ϵ system can be readily constructed, and when coupled with finite element interpolations for the unknowns, a standard set of coupled ordinary differential equations can be defined. The discretized forms of Eqs. (4.11.16) and (4.11.17) will be highly nonlinear and strongly coupled to the mean flow equations. It is obvious that the size of the computational problem for a two-equation turbulent simulation has increased substantially over its laminar counterpart. Not so obvious is the fact that the numerical difficulties have also increased significantly.

The major problems that occur with the k-ϵ model come from two sources: the nonlinear behavior of the partial differential equations and boundary conditions for the turbulent transport and mean field equations. It has been observed by a number of investigators that the dissipation equation may cause instabilities in the flow simulation that lead to a poor or nonconvergence of the numerical solution algorithm. Another manifestation of these stability problems is the prediction of negative values for both k and ϵ [121–125]. This nonphysical occurrence is sometimes attributed to the inaccurate modeling of the source terms for k and ϵ. The practical solution, in many cases, involves a clipping procedure in which negative values are replaced by small positive values [121].

The sensitive nature of the k-ϵ model leads to some practical recommendations for solution methods in turbulent flows. Generally, it has been found advantageous to employ Picard iteration methods as opposed to the higher-order Newton schemes. Though the convergence rate is reduced in this case, the overall stability of the algorithm is improved. Relaxation factors greater than 0.5 have also been observed to substantially enhance the solution process. Also related to the solution stability is the grading of the computational mesh. Computations that have abrupt changes in mesh density have been reported to show poor behavior in the k and ϵ variables and in some cases can lead to divergence of the solution process. In monitoring the convergence of a turbulent flow simulation it is noteworthy that the velocity and pressure fields converge first; convergence of the k and ϵ fields is very slow and may not be monotonic.

The topic of boundary conditions for the k-ϵ model and the modeling of turbulence close to solid boundaries is quite involved and will only be summarized here. The k-ϵ model described in Eqs. (4.11.16) and (4.11.17) is known as a high Reynolds number model, since its derivation was predicated on a single length scale associated with an unbounded, fully developed flow. Close to solid boundaries, the turbulent nature of the flow changes to a low Reynolds number model, where the small eddies associated with dissipation become important. Of course, very near the wall a thin, viscous sublayer exists in which a turbulence model is no longer appropriate. From a computational viewpoint these near-wall regions are extremely challenging, since not only does the type of turbulence model vary with distance from the boundary, but all of the flow variables experience their greatest variation in this region. In most cases, attempts to solve the mean flow and k-ϵ equations all the way to the wall would be prohibitively expensive, especially for complex three-dimensional geometries.

An alternative to the above approach was developed at Imperial College [126,127] and relies on the use of universal models or wall laws to empirically estimate the flow behavior in the viscous sublayer and the buffer layer. The development of such a model allows the computational domain to be displaced from the wall and located in a fully turbulent region where the k-ϵ model is valid. Boundary conditions for the mean flow variables, k and ϵ, at the edge of the computational domain are provided by the wall laws. A wall law is typically a "universal" velocity profile that is derived from a combination of theory and experimental observation; standard profiles for k and ϵ near a boundary can also be derived. These parameterized profiles provide values of the mean field shear stress, k and ϵ, that can be directly applied as boundary conditions to the computational domain. A good description of wall laws and modeling is available in [121,127]. The major disadvantage of this approach is that the computational domain must be offset from the wall by a distance that depends on the wall law and the flow field. The location of the domain may therefore have to be altered during or after the solution process. Despite this difficulty, the wall law approach is the only practical way to model turbulent flows in the vicinity of a boundary.

An improvement to the wall law formulation for finite element applications was developed in conjunction with the FIDAP code [128]. A standard representation of a universal velocity profile (wall law) consists of a linear profile for the viscous sublayer, a log-linear profile in the buffer region, and a logarithmic profile in the fully developed turbulent region. It is possible to develop an interpolation or shape function that represents this composite profile and therefore construct a special finite element that represents the wall law region of a turbulent flow. Though the problem with proper location of the edge of the special wall element remains, the development of such an element solves many other problems involving mesh generation and the specification of boundary conditions on complex surfaces. The precursor to this work, and some of the first work on specialized elements for fluid mechanics, can be found in References 123 and 129.

Though the primary discussion has been related to boundary conditions at solid walls, it must be remembered that the k-ϵ equations require boundary values to be specified on all computational boundaries. Especially troublesome in this regard are k and ϵ values for inlet boundaries, since to a large extent these values will set the turbulence level for the entire flow. Haroutunian [121] provided a method for consistently generating these values.

4.12 Numerical Examples

4.12.1 Preliminary Comments

The true test of any numerical method is its performance in the solution of problems of engineering and scientific interest. In this section, a small sampling of flow problems solved using the finite element models developed herein are presented. No attempt is made to examine the problems in depth but rather the intent is to illustrate the variety and, to some extent, the complexity of problems that can be solved using the finite element method. The examples were solved using the computer codes FEM2DHT, NACHOS II [88], FIDAP [84] and GOMA [77]. Program FEM2DHT is a modified version of FEM2DV2 [1] and contains the reduced integration penalty finite element model for the Stokes problem, NACHOS II contains the mixed finite element model for the Navier–Stokes equations, FIDAP contains both mixed and penalty finite element models of the Navier–Stokes equations, and GOMA contains both mixed and stabilized finite element models with an emphasis on moving mesh free surface problems. The objective of the first several examples is to evaluate the accuracy of the penalty and mixed finite element models in the light of available analytical or numerical results and to illustrate the effect of the penalty parameter on the accuracy of the solutions.

4.12.2 Fluid Squeezed between Parallel Plates

Consider the (Stokes) flow of a viscous incompressible material squeezed between two long parallel plates (see Figure 4.12.1a). When the length of the plates is very large compared to both the width of and the distance between the plates, we have a case of plane flow (in the plane formed by the width of and the distance between the plates). Although this is a moving boundary problem, we wish to determine the velocity and pressure fields for a fixed distance between the plates, assuming that a state of plane flow exists. An approximate (analytical) solution to this two-dimensional problem is provided by Nadai [89](also see [1,2]).

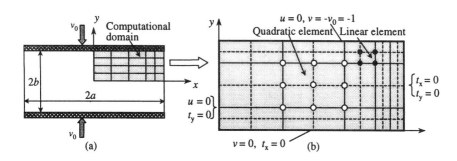

Figure 4.12.1: Domain and mesh for fluid squeezed between parallel plates.

Let V_0 be the velocity with which the two plates are moving toward each other (i.e., squeezing out the fluid), and let 2b and 2a denote, respectively, the distance between and the width of the plates (see Figure 4.12.1a). Due to the biaxial symmetry present in the problem, it suffices to model only a quadrant of the domain. A 5×3 nonuniform mesh of nine-node quadratic elements is used in the mixed model, and a 10×6 mesh of the four-node linear elements and 5×3 mesh of nine-node quadratic elements are used in the penalty model (see Figure 4.12.1b). The nonuniform mesh, with smaller elements near the free surface (i.e., at $x = a$), is used to approximate accurately the singularity in the shear stress at the point $(a, b) = (6, 2)$. The mesh used for the penalty model has exactly the same number of nodes as the mesh used for the mixed model. There are no specified nonzero secondary variables in the problem. The computer program FEM2DV2 (penalty model) from [1] is used to analyze the problem. The reduced integration rule suggested earlier for the evaluation of penalty terms is used.

The velocities $u_1(x,0) = u(x,0)$ obtained with the two models compare well with the analytical solution as shown in Table 4.12.1. The nine-node element gives very good results for both the penalty and mixed models. The influence of the penalty parameter on the accuracy of the solution is clear from the results. Whether the element is linear or quadratic, it is necessary to use a large value of the penalty parameter. For an IBM 3090 computer with double precision arithmetic, a value of $\gamma = 10^{14}$ was the upper limit beyond which the contribution from the viscous terms was not recognized in comparison to the penalty terms (and the coefficient matrix, being singular, is not invertible). With the current day PCs, it is possible to obtain the level of accuracy as shown in Table 4.12.1.

Table 4.12.1: Comparison of finite element velocity solutions with the analytical solution for fluid squeezed between plates (for $y = 0$).

x	$\gamma = 1.0$ 4-node	$\gamma = 1.0$ 9-node[†]	$\gamma = 100$ 4-node	$\gamma = 100$ 9-node	$\gamma = 10^8$ 4-node	$\gamma = 10^8$ 9-node	Mixed model 9-node	Series solution
1	0.0303	0.0310	0.6563	0.6513	0.7576	0.7605	0.7497	0.7500
2	0.0677	0.0691	1.3165	1.3062	1.5135	1.4992	1.5031	1.5000
3	0.1213	0.1233	1.9911	1.9769	2.2756	2.2557	2.2561	2.2500
4	0.2040	0.2061	2.6960	2.6730	3.0541	3.0238	3.0203	3.0000
4.5	0.2611	0.2631	3.0718	3.0463	3.4648	3.4307	3.4292	3.3750
5.0	0.3297	0.3310	3.4347	3.3956	3.8517	3.8029	3.8165	3.7500
5.25	0.3674	0.3684	3.6120	3.5732	4.0441	3.9944	3.9893	3.9375
5.5	0.4060	0.4064	3.7388	3.6874	4.1712	4.1085	4.1204	4.1250
5.75	0.4438	0.4443	3.8316	3.7924	4.2654	4.2160	4.2058	4.3125
6.0	0.4793	0.4797	3.8362	3.7862	4.2549	4.1937	4.2364	4.5000

† The three-point Gauss rule for non-penalty terms and the two-point Gauss rule for penalty terms are used for quadratic elements.

Next, a 12×8 mesh of linear elements and a 6×4 mesh of quadratic elements were used to evaluate the relative accuracies of the rectangular and triangular elements for the penalty model (meshes are not shown). The 12×8 mesh of linear triangular elements with full integration of $[K^1]$ and $[K^2]$ and selective integration (i.e., full integration of $[K^1]$ and reduced integration of $[K^2]$) both give the same results for the velocity field. However, in both cases, erroneous results for pressure and stresses are obtained. The 6×4 mesh of quadratic triangular elements with full and selective integrations give the same velocity fields, while the stresses and pressure are predicted to be the same at the same quadrature points. Both the 12×8 mesh of linear rectangular elements and the 6×4 mesh of nine-node rectangular elements give good results for velocities, pressure, and stresses. The solution accuracy can be increased by using refined meshes.

Figure 4.12.2 contains plots of the velocity $u(x,y)$ for $x = 4$ and 6, and Figure 4.12.3 contains plots of pressure $P(x,y)$, for $y = $ constant, computed using the six-node quadratic triangular (T6) and nine-node rectangular (R9) elements. The finite element solutions are compared with the analytical solutions of Nadai [89]. The pressure in the penalty model was computed using Eq. (4.3.15) with the 2×2 Gauss rule for the quadratic rectangular element and the one-point formula for the quadratic triangular element. If the pressure (see Figure 4.12.3) in the penalty model were computed using the full quadrature rule for rectangular elements, we would obtain erroneous values. The linear triangular element with full as well as reduced integrations gives unstable pressures, while the quadratic triangular element with one- or two-point rules yields good results. In general, the same quadrature rule as that used for the evaluation of the penalty terms in the coefficient matrix must be used to evaluate the pressure, and one should avoid using the linear triangular element in the penalty finite element model.

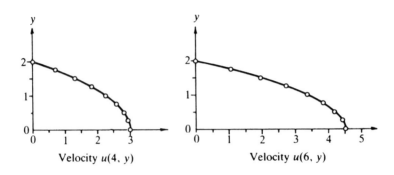

Figure 4.12.2: Velocity fields for fluid squeezed between parallel plates.

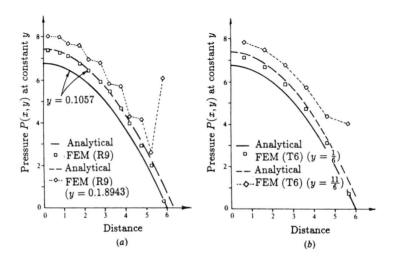

Figure 4.12.3: Pressures for fluid squeezed between parallel plates.

4.12.3 Flow of a Viscous Lubricant in a Slider Bearing

The slider (or slipper) bearing consists of a short sliding pad moving at a velocity $u = U_o$ relative to a stationary pad inclined at a small angle with respect to the stationary pad, and the small gap between the two pads is filled with a lubricant (see Figure 4.12.4a). Since the ends of the bearing are generally open, the pressure there is atmospheric, P_0. If the upper pad is parallel to the base plate, the pressure everywhere in the gap must be atmospheric (because dP/dx is a constant for flow between parallel plates), and the bearing cannot support any transverse load. If the upper pad is inclined to the base pad, a pressure distribution (in general, a function of x and y) is set up in the gap. For large values of U_o, the pressure generated can be of sufficient magnitude to support heavy loads normal to the base pad.

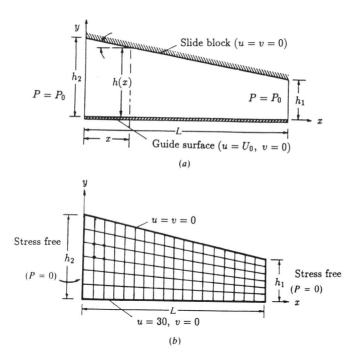

(a)

(b)

Figure 4.12.4: (a) Geometry, boundary conditions, and (b) finite element mesh (Mesh 1) for slider bearing.

Since the width of the gap and the angle of inclination are generally small, it is assumed that the pressure is not a function of y in developing an analytical solution to the problem. Assuming a two-dimensional flow and a small angle of inclination and neglecting the normal stress gradient (in comparison with the shear stress gradient), the equations governing the flow of the lubricant between the pads can be solved to give (see Schlichting [90] and Reddy [1])

$$u = \left(U_0 - \frac{1}{2\mu}h^2\frac{dP}{dx}\frac{y}{h}\right)\left(1 - \frac{y}{h}\right), \quad P = \frac{6\mu U_0 L(h_2 - h)(h - h_1)}{h^2(h_2^2 - h_1^2)} \quad (4.12.1a)$$

$$\tau_{xy} = \mu \frac{\partial u}{\partial y} = \frac{dP}{dx}\left(y - \frac{h}{2}\right) - \mu\frac{U_o}{h}, \quad h(x) = h_2 + \frac{h_1 - h_2}{L}x \qquad (4.12.1b)$$

In our computations, we choose

$$h_2 = 2h_1 = 8 \times 10^{-4}, \ L = 0.36, \ \mu = 8 \times 10^{-4}, \ U_0 = 30 \qquad (4.12.1c)$$

First, it should be pointed out that the assumption concerning the pressure not being a function of y is not necessary in the finite element analysis. We use a mesh (Mesh 1) of 18×6 linear quadrilateral elements to analyze the problem. The mesh and boundary conditions are shown in Figure 4.12.4b. Figure 4.12.5 contains plots of the pressure distributions along the length of the bearing for various values of the penalty parameter. The converged pressure is slightly higher than the approximate analytical solution.

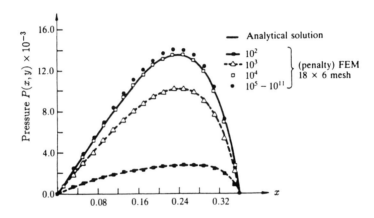

Figure 4.12.5: Pressures obtained with Mesh 1 for slider bearing.

To obtain more accurate solutions, a graded mesh (Mesh 2) of 128 linear quadrilateral elements (153 nodes), (equivalently, 32 nine-node quadrilateral elements) is used. Figures 4.12.6a and 4.12.6b contain a comparison of the finite element solutions (for $\gamma = 10^8$) with the analytical solutions for the horizontal velocity, pressure, and shear stress. The pressure was computed at the reduced Gauss points using equation (4.3.15).

4.12.4 Wall-Driven 2-D Cavity Flow

Consider the laminar flow of a viscous, incompressible fluid in a square cavity bounded by three motionless walls and a lid moving at a constant velocity in its own plane (see Figure 4.12.7). The third dimension is assumed to be long enough to have a plane flow. Singularities exist at each corner where the moving lid meets a fixed wall. This example is one which has been extensively studied by analytical, numerical, and experimental methods (see

[17,19,92–97], among others), and it is often used as a benchmark problem to test a new numerical method or formulation. In solving this problem, the mesh used should be such that the boundary layer thickness is resolved. The boundary layer thickness is of the order of $Re^{-\frac{1}{2}}$, where $Re = \rho U L/\mu$ and U is the lid velocity and L is the cavity dimension. Thus the mesh should be graded to have smaller elements near the walls.

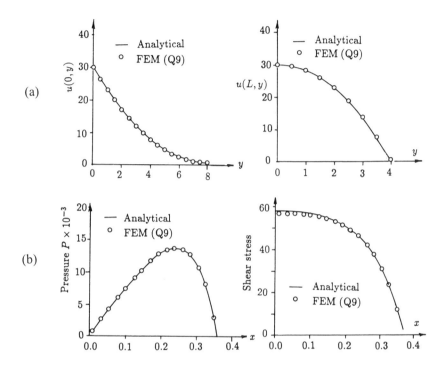

Figure 4.12.6: Velocity field, pressure, and shear stress for slider bearing.

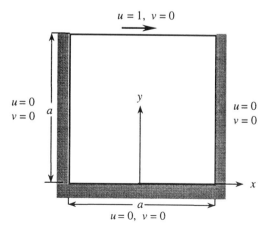

Figure 4.12.7: Geometry, boundary conditions, and finite element mesh for a wall-driven square cavity.

Figure 4.12.8 shows the horizontal velocity component along the vertical centerline of the cavity, i.e., plots of $u(0.5, y)$ vs. y, for $Re = 1, 100, 400,$ and 10^3. The results were obtained (see Reddy [97]) using a nonuniform, 14×14 mesh of linear rectangular elements. Figure 4.12.9 contains the plots of $u(0.5, y)$ vs. y for $Re = 400$ for various meshes. The effect of mesh refinement on the accuracy of the solution is clear.

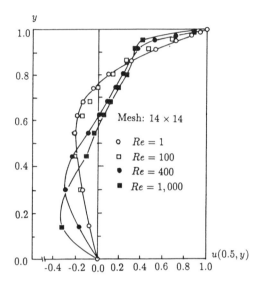

Figure 4.12.8: Horizontal velocity along the vertical centerline of the cavity for various Reynolds numbers (Mesh: 14×14).

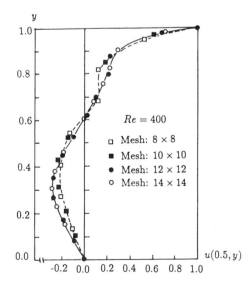

Figure 4.12.9: Horizontal velocity along the vertical centerline of the cavity for various meshes ($Re = 400$).

4.12.5 Wall-Driven 3-D Cavity Flow

The problem considered here is that of a three-dimensional cubical cavity of unit dimension [96–100]. The motion of an incompressible viscous fluid in the cavity is induced by the motion of the lid (see Figure 4.12.10). All other walls of the cavity are assumed to be stationary.

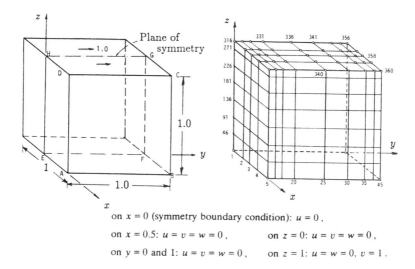

on $x = 0$ (symmetry boundary condition): $u = 0$.

on $x = 0.5$: $u = v = w = 0$,   on $z = 0$: $u = v = w = 0$,

on $y = 0$ and 1: $u = v = w = 0$,   on $z = 1$: $u = w = 0$, $v = 1$.

Figure 4.12.10: Geometry, mesh, and boundary conditions for a wall-driven cubical cavity.

Using the symmetry about the $x = 0$ plane, only half of the domain is modeled using a $4 \times 8 \times 7$ mesh of trilinear (i.e., eight-node brick) elements (see Reddy [100]). The problem is analyzed for $Re = 100$ and 400. Figures 4.12.11 and 4.12.12 contain the plots of velocity components in the midplane of the cavity. The present penalty finite element solution is compared with the finite difference solutions of Dennis et al. [98] and Agarwal [99]. Dennis et al. [98] used a second-order accurate finite difference method, whereas Agarwal [99] used a third-order accurate finite difference scheme. Both used very refined meshes. The present results agree well qualitatively with the results of Agarwal [99], although a very crude mesh was used in the finite element analysis.

4.12.6 Evaluation of the EBE Iterative Solvers

Here we use the wall-driven cavity problem to evaluate the EBE methods discussed in Appendix B. The entire discussion presented here is taken from [23]. Four nonuniform meshes, Mesh 1: $5 \times 10 \times 10$, Mesh 2: $6 \times 12 \times 12$, Mesh 3: $7 \times 14 \times 14$, and Mesh 4: $8 \times 16 \times 16$ of trilinear elements were used. The pressure at $(0.5, 0.5, 0.0)$ is taken to be equal to zero, and it is used as an essential boundary condition for the pressure calculation according to Eq. (4.5.17). The problem was analyzed for $Re = 10$ and 100. For iterative solvers the convergence criterion for velocity and pressure calculations was taken to be $\epsilon_{vel} = 10^{-5}$ and $\epsilon_{pr} = 10^{-4}$, respectively. Figure 4.12.13 shows the finite element discretization of the domain for Mesh 3.

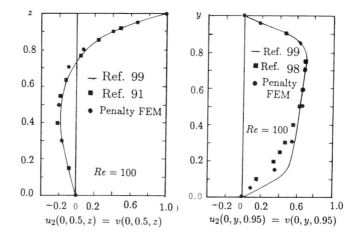

Figure 4.12.11: Velocity fields for a wall-driven cubical cavity, $Re = 100$.

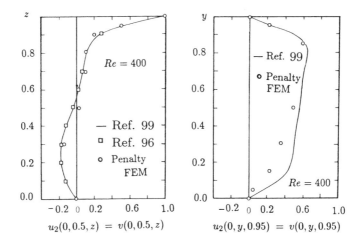

Figure 4.12.12: Velocity fields for a wall-driven cubical cavity, $Re = 400$.

It is observed that the convergence rate of the GMRES solver showed no oscillatory behavior but was slow compared to the ORTHOMIN and the ORTHORES solvers. The number of iterations required to arrive at a converged solution for $Re = 10$ for each solver is given in Table 4.12.2. The frontal solver took 3 iterations for each mesh. The number in the parenthesis denotes the number of iterations required to solve the system of equations for each nonlinear iteration. The numbers in parenthesis in the second row for each mesh represent the number of iterations required for pressure calculations. From Table 4.12.2 it is clear that the number of iterations required by the GMRES algorithm is nearly 3.5 times more than those required by the ORTHOMIN or the ORTHORES solvers for the same error tolerances. This reflects in the CPU time required to obtain a converged solution.

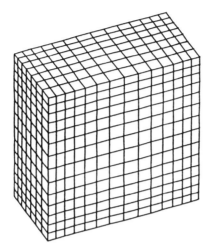

Figure 4.12.13: Finite element mesh (Mesh 3) used for the wall-driven cubical cavity.

Table 4.12.2: Number of iterations for different meshes and solvers.

Mesh	Elements	Equations	GMRES	ORTHOMIN	ORTHORES
1	500	2,178	2(1,580+85) (19+5)	3(313+135+1) (19+9+1)	3(313+137+1) (19+9+1)
2	864	3,549	2(2,472+137) (24+6)	3(462+243+1) (24+8+1)	3(463+245+1) (24+8+1)
3	1,372	5,400	2(3,383+165) (35+6)	3(600+267+1) (29+9+1)	3(600+269+1) (29+9+1)
4	2,048	7,803	2(3,811+136) (44+6)	3(789+316+1) (34+10+1)	3(790+317+1) (34+10+1)

Figure 4.12.14 shows the CPU time requirements as a function of the number of elements for different solvers. It is clear that the frontal solver requires more CPU time. This is due to the fact that it is a direct solver and performs an elimination operation which is computationally intensive. Also, the convergence of iterative solvers depends to some extent on the initial guess. Once the solution for the linear problem (Stokes) is obtained, it is used as the initial guess for the second iteration and thereon. This results in a faster convergence from the second iteration onward for iterative solvers, as shown in the Table 4.12.2. However, this is not true for the direct solvers because they require the same amount of CPU time for each iteration. Hence the CPU time requirements for direct solvers are enormous. The CPU times for ORTHOMIN were slightly less than those of ORTHORES. This is due to

the fact that the GMRES algorithm needed more iterations compared to the ORTHOMIN and the ORTHORES algorithms. A similar trend was observed by [101] for the solution of convection/diffusion problems.

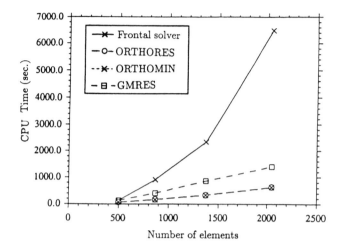

Figure 4.12.14: Comparison of CPU times for different solvers.

The storage requirements in millions of words for all four solvers for the four meshes are given in Table 4.12.3. From this table it is clear that the storage requirements for ORTHOMIN and ORTHORES are less than those for the GMRES and frontal solvers. The frontal solver requires a very large memory because the equations are stored before back substitution. The length of the equations depends on the front width. The front width for each of the meshes is given in parenthesis in Table 4.12.3.

Table 4.12.3: Memory requirements (in millions of words) for various methods.

Mesh	Elements	Equations	Frontal	GMRES	ORTHOMIN	ORTHORES
1	500	2,178	2.702 (270)	0.822	0.715	0.719
2	864	3,549	3.740 (354)	1.257	1.226	1.233
3	1,372	5,400	7.227 (450)	2.206	1.926	1.947
4	2,048	7,803	12.146 (558)	3.269	2.878	2.893

The solution for $Re = 100$ was obtained incrementally by using the flow field from the $Re = 10$ solution as the initial solution. The CPU times are given in Table 4.12.4 for Mesh 3 calculations. The difference in the CPU times for the $Re = 100$ and $Re = 10$ solutions gives the time used to obtain the $Re = 100$ solution. From Table 4.12.4 it is clear that the iterative solvers converge much faster and need less CPU time because of a good estimate of the initial guess for the next Reynolds number flow field calculation.

Table 4.12.4: Comparison of CPU times (in seconds) taken by various methods.

S. No.	Re	Frontal	GMRES	ORTHOMIN	ORTHORES
1	10	2,345	860	338	341
2	$10\&10^2$	5,489	1,046	619	626

It is clear that both the ORTHOMIN and the ORTHORES solvers with an element-by-element data structure are superior to the GMRES and frontal solvers.

4.12.7 Backward Facing Step

The laminar flow over a two-dimensional, backward facing step is frequently used as a standard test problem for numerical methods. The problem outlined here was developed as a benchmark for the testing of outflow boundary conditions [102]. The geometry, boundary conditions, and a representative mesh are shown in Figure 4.12.15. The flow problem was solved for a Reynolds number ($Re = U_{avg}H/\nu$) of 800 using zeroth-order continuation and Newton's methods. Mesh refinements from 6×120 to 40×800 elements were used to ensure mesh independence of the solution; the velocity field is approximated using the biquadratic interpolation functions, and the linear, discontinuous pressure approximation is used.

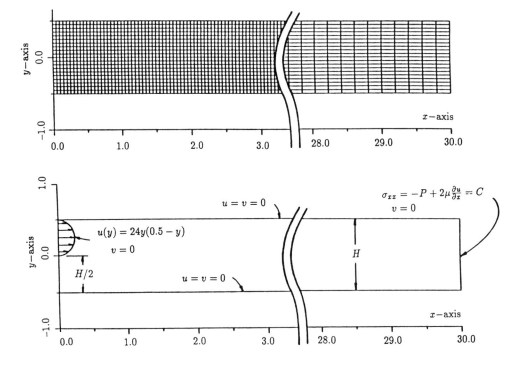

Figure 4.12.15: Schematic and mesh for backward facing step problem.

Figure 4.12.16 contains contour plots of the stream function, pressure, and vorticity obtained on a grid that extended 60 step heights downstream. Two recirculation regions are predicted with their position and intensity corresponding closely to previous numerical solutions. Other simulations were performed on computational domains that were truncated at 30 and 14 step heights downstream. In each case, the traction boundary condition shown in Figure 4.12.15 was specified at outflow.

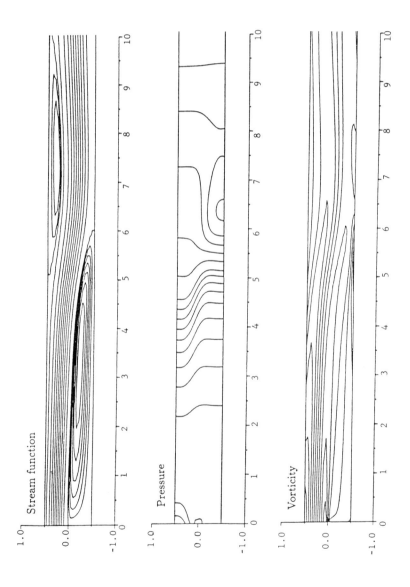

Figure 4.12.16: Contour plots for flow over a backward step; long computational domain.

Figure 4.12.17 shows a streamline plot of a result for the shortest domain; the comparison with the long domain result in Figure 4.12.16 is very good, especially considering that the upper wall recirculation zone is truncated by the outflow boundary. A comparison of axial velocities at this location between the solutions obtained with the long and short domains indicates less than 10% difference. This demonstrates the utility and accuracy of the traction boundary condition for use in short domain internal flow problems.

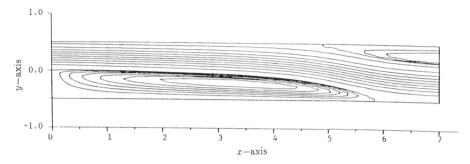

Figure 4.12.17: Contour plots for flow over a backward step; short computational domain.

4.12.8 Flow Past a Submarine

As an example of a large, three-dimensional viscous flow of engineering relevance we consider the flow past an axisymmetric submarine body with an attached fairwater or sail. This simulation was part of a benchmark exercise and consisted of predicting the steady flow past the submarine shape for $Re = 1.2 \times 10^7$. Reference 103 gives details on the entire simulation program.

Figure 4.12.18a contains several views of the surface mesh and surrounding finite element grid. Clockwise from upper left, Figure 4.12.18a illustrates the surface mesh for the complete body and enlarged views of the bow, the hull/fairwater juncture, and the stern tip, respectively. The mesh contained approximately 52,000 eight-node, hexahedral trilinear velocity elements in which the pressure was constant within an element. Figure 4.12.18b contains the three-dimensional finite element grid used for the numerical computation of steady, turbulent flow ($Re = 1.2 \times 10^7$) past the axisymmetric hull with attached fairwater. This mesh contains 57,523 nodes and 52,295 trilinear brick elements. The lower plot illustrates the relative position of the submarine within the computational domain. At the Reynolds number of interest the flow is clearly turbulent and the simulation employed a standard k-ϵ turbulence model. Wall functions were used to model the viscous sublayer along solid boundaries and provide boundary conditions for the k and ϵ variables. The equations were solved using a pressure correction method outlined in Section 4.6. The individual momentum equations, pressure Poisson equation, and k-ϵ equations were solved sequentially using an early version of the segregated solver developed for the FIDAP code [84]. Discussion of the segregated solver was presented in Section 4.6 and will be expanded in Section 5.5.3.

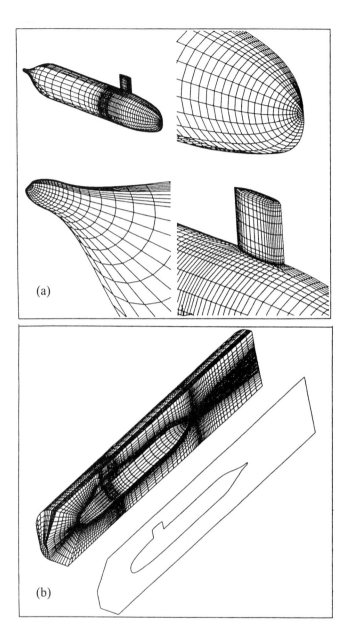

Figure 4.12.18: (a) Finite element discretization for the surface of the axisymmetric hull with attached fairwater. (b) Three-dimensional finite element grid used for the numerical computation of steady, turbulent flow ($Re = 1.2 \times 10^7$) past the axisymmetric hull with attached fairwater.

Pressure results for the submarine flow at zero angle of attack are shown as surface contours in Figure 4.12.19. A comparison of this result with an inviscid, potential flow solution showed excellent quantitative agreement. This similarity in results is to be expected since there is virtually no separated flow on the model geometry. One noticeable feature of a viscous flow is the development of secondary flows at internal corners. Shown in Figure 4.12.20 are velocity vectors in a plane normal to the longitudinal axis of the model and located at several stations downstream of the fairwater. The circulation pattern due to the horseshoe vortex that forms at the hull-fairwater junction is clearly visible. Also visible is the vortex emanating from the tip of the fairwater. Further details on this flow and the flow over other hydrodynamic bodies are available in [103].

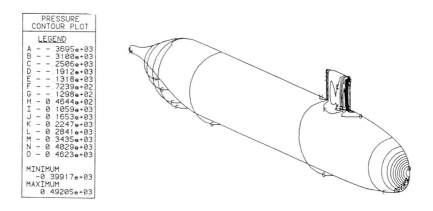

Figure 4.12.19: Pressure contour lines on the surface of the computational domain: $0°$ attack, 15 contour lines span the pressure extrema.

4.12.9 Crystal Growth from the Melt

As an illustration of the use of spines to compute a free surface shape, the problem of crystal growth from a melt using the Czochralski process is considered. This example is a brief excerpt from the extensive study performed by Sackinger [104] and Sackinger et al. [105]. Figure 4.12.21 shows a schematic of the axisymmetric crystal growth model. There are two free surfaces in this problem: the melt/gas interface denoted by H_1 and the melt/crystal interface marked by H_0. Both surfaces are parameterized with a spine technique and the location of each boundary is computed in a fully coupled manner with the momentum, continuity, and energy equations for each material region. Note that the radius of the crystal is also an unknown and was computed as part of the solution. This simulation also included the effect of Lorentz forces on the melt, rotation of the crucible and/or the crystal, and thermocapillary forces on the free surface H_1. The steady form of this problem is well suited to the spine technique, since the movement of the nodes on each free surface is essentially one dimensional. Note that the spines associated with H_0 and H_1 are permitted to slide along the r coordinate as a result of variations in the crystal radius, R.

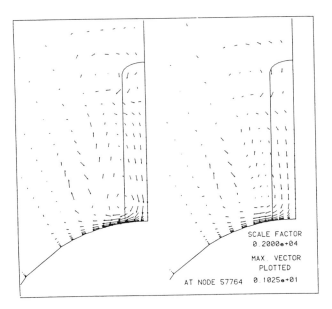

Figure 4.12.20: Velocity vectors depicting the development of the horseshoe vortex which emanates from the hull/fairwater juncture. Velocity vectors in a plane normal to the mean flow direction are illustrated for axial locations corresponding to two (left) and four (right) fairwater chord lengths downstream from the fairwater trailing edge. The mean flow velocity is graphically removed from these illustrations. The fairwater outline is superimposed upon each plot for visual perspective.

Figure 4.12.22 contains typical finite element meshes of nine-node, quadrilateral elements. Contour plots for the streamlines and isotherms for two cases with very different melt pool volumes are shown in Figures 4.12.23 and 4.12.24. The free surface shapes are quite different for these two cases, as are the temperature and flow fields.

4.12.10 Mold Filling

The filling of a cavity with a liquid presents a particularly difficult challenge for a computational algorithm. The present example considers the problem of filling a flat plate mold with a liquid metal in an experimental investment casting setup. This simulation was performed by R. C. Givler (personal communication) using the commercial finite element code, ProCAST [106]. The free surface algorithm in ProCAST is a refined version of the VOF method described in Section 4.10.3.

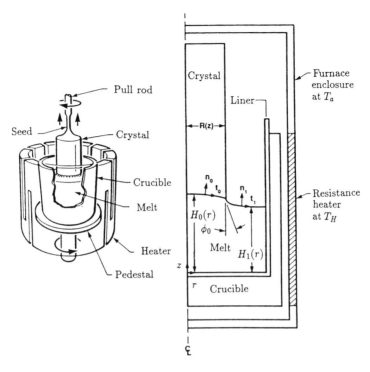

Figure 4.12.21: Schematic of the crystal growth problem.

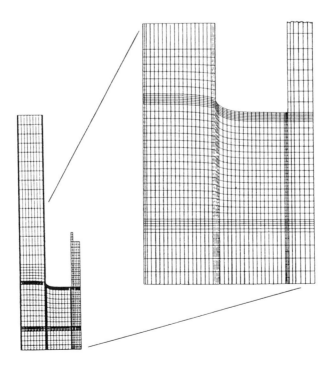

Figure 4.12.22: Typical finite element meshes for the free surface problem.

$T_H = 1835.13$ K

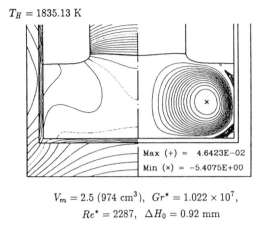

$V_m = 2.5$ (974 cm^3), $Gr^* = 1.022 \times 10^7$,
$Re^* = 2287$, $\Delta H_0 = 0.92$ mm

Figure 4.12.23: Isotherms and streamlines for the crystal growth problem (Melt volume $V_m = 2.5$ (974 cm^3), $Gr = 1.022 \times 10^7$, $Re = 2,287$, and $\Delta H_0 = 0.92$ mm).

$T_H = 1970.71$ K

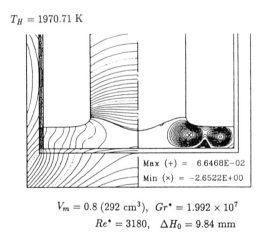

$V_m = 0.8$ (292 cm^3), $Gr^* = 1.992 \times 10^7$
$Re^* = 3180$, $\Delta H_0 = 9.84$ mm

Figure 4.12.24: Isotherms and streamlines for the crystal growth problem (Melt volume $V_m = 0.8$ (292 cm^3), $Gr = 1.992 \times 10^7$, $Re = 3,180$, and $\Delta H_0 = 9.84$ mm).

Figure 4.12.25 contains the mesh used for the filling simulation. Liquid metal (stainless steel) enters the computational domain through the circular cross section at the top of the edge gate; the fluid temperature and pressure are specified at inflow. For this particular simulation the heat transfer problem is also considered in which the metal loses energy to the mold and the mold is cooled via radiation to the surroundings.

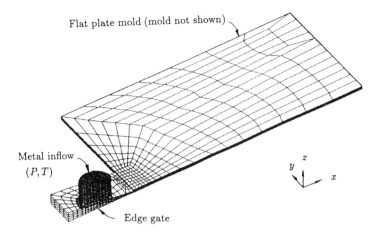

Figure 4.12.25: Schematic and mesh for mold filling (mold not shown).

Figures 4.12.26 and 4.12.27 show the pressure fields at two times during the filling process as viewed from both the top and bottom of the model. It is quite noticeable that the molten metal runs along the bottom of the mold in a film that is less than the channel thickness. Filling of the mold is completed when the fluid reaches the far end of the mold, is turned toward the inlet, and fills the upper part of the channel. Note that the gate thickness is less than the channel thickness, so this behavior is not unexpected.

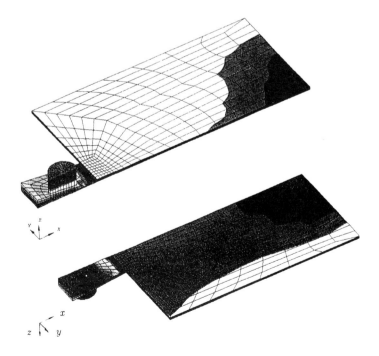

Figure 4.12.26: Top and bottom views of pressure field and free surface during filling; time = 1.45.

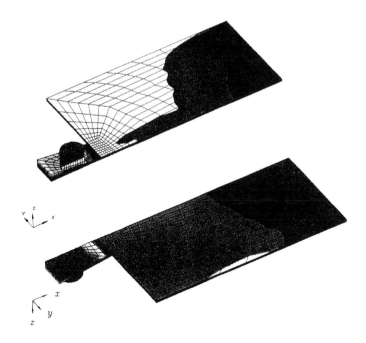

Figure 4.12.27: Top and bottom views of pressure field and free surface during filling; time $= 2.09$.

Problems

4.1 Consider the vector equations (4.1.1a) and (4.1.2a). Develop the weak statements of the equations in vector form.

4.2 Consider Eq. (4.1.2a) in cylindrical coordinates (r, θ, z). For axisymmetric viscous incompressible flows (i.e., the flow field is independent of the θ coordinate), and when the convective (nonlinear) terms are neglected, we have

$$\frac{1}{r}\frac{\partial}{\partial r}(r\sigma_r) - \frac{\sigma_\theta}{r} + \frac{\partial \sigma_{rz}}{\partial z} + f_r = \rho\frac{\partial u}{\partial t} \qquad (i)$$

$$\frac{1}{r}\frac{\partial}{\partial r}(r\sigma_{rz}) + \frac{\partial \sigma_{zz}}{\partial z} + f_z = \rho\frac{\partial w}{\partial t} \qquad (ii)$$

$$\frac{1}{r}\frac{\partial}{\partial r}(ru) + \frac{\partial w}{\partial z} = 0 \qquad (iii)$$

where

$$\sigma_r = -P + 2\mu\frac{\partial u}{\partial r}, \qquad \sigma_\theta = -P + 2\mu\frac{u}{r} \qquad (iv)$$

$$\sigma_z = -P + 2\mu\frac{\partial w}{\partial z}, \qquad \sigma_{rz} = \mu(\frac{\partial u}{\partial z} + \frac{\partial w}{\partial r}) \qquad (v)$$

Develop the semidiscrete mixed finite element model of the equations.

4.3 Develop the semidiscrete penalty finite element model of the equations in Problem 4.2.

4.4 The equations governing unsteady slow flow of viscous, incompressible fluids in the xy-plane can be written in terms of vorticity ω and stream function ψ :

$$\rho\frac{\partial \omega}{\partial t} - \mu\nabla^2\omega = 0$$

$$-\omega - \nabla^2\psi = 0$$

Develop the semidiscrete finite element model of the equations. Discuss the meaning of the secondary variables. Use the α-family of approximation to reduce the ordinary differential equations to algebraic equations.

4.5 Compute the tangent coefficient matrix for the penalty finite element model in Eq. (4.3.17).

References for Additional Reading

1. J. N. Reddy, *An Introduction to the Finite Element Method*, Second Edition, McGraw-Hill, New York (1993).

2. J. N. Reddy, *Applied Functional Analysis and Variational Methods in Engineering*, McGraw-Hill, New York (1986); reprinted by Krieger Publishing, Melbourne, Florida (1991).

3. J. N. Reddy, "On Penalty Function Methods in the Finite Element Analysis of Flow Problems," *International Journal for Numerical Methods in Fluids*, **2**, 151–171 (1982).

4. J. T. Oden, "RIP Methods for Stokesian Flows," in R. H. Gallagher, O. C. Zienkiewicz, J. T. Oden, and D. Norrie (eds.), *Finite Element Method in Flow Problems*, Vol. IV, John Wiley & Sons, London (1982).

5. J. T. Oden and G. F. Carey, *Finite Elements, Mathematical Aspects*, Vol. IV, Prentice Hall, Englewood Cliffs, New Jersey (1983).

6. F. Brezzi and M. Fortin, *Mixed and Hybrid Finite Element Methods*, Springer-Verlag, Berlin (1991).

7. F. Brezzi and K. J. Bathe, "The inf-sup Condition, Equivalent Forms and Applications," in *Reliability of Methods for Engineering Analysis*, K. J. Bathe and D. R. J. Owen (eds.), Pineridge Press, Swansea, U.K. (1986).

8. D. S. Malkus and E. T. Olsen, "Obtaining Error Estimates for Optimally Constrained Incompressible Finite Elements," *Computer Methods in Applied Mechanics and Engineering*, **45**, 331–353 (1984).

9. P. Le Tallac and V. Ruas, "On the Convergence of the Bilinear-Velocity Constant-Pressure Finite Element Method in Viscous Flow," *Computer Methods in Applied Mechanics and Engineering*, **54**, 235–243 (1986).

10. D. Chapelle and K. J. Bathe, "The inf-sup Test," *Computers & Structures,* **47**(4/5), 537–545 (1993).

11. J. N. Reddy, "On the Accuracy and Existence of Solutions to Primitive Variable Models of Viscous Incompressible Fluids," *International Journal of Engineering Science,* **16**, 921–929 (1978).

12. J. N. Reddy, "On the Finite Element Method with Penalty for Incompressible Fluid Flow Problems," in *The Mathematics of Finite Elements and Applications III,* J. R. Whiteman (ed.), Academic Press, New York (1979).

13. R. Temam, *Theory and Numerical Analysis of the Navier–Stokes Equations,* North-Holland, Amsterdam (1977).

14. R. L. Sani, P. M. Gresho, R. L. Lee, and D. F. Griffiths, "The Cause and Cure (?) of the Spurious Pressures Generated by Certain FEM Solutions of the Incompressible Navier–Stokes Equations: Part 1," *International Journal for Numerical Methods in Fluids,* **1**, 17–43 (1981).

15. R. L. Sani, P. M. Gresho, R. L. Lee, and D. F. Griffiths, "The Cause and Cure (?) of the Spurious Pressures Generated by Certain FEM Solutions of the Incompressible Navier–Stokes Equations: Part 2," *International Journal for Numerical Methods in Fluids,* **1**, 171–204 (1981).

16. T. J. R. Hughes, R. L. Taylor, and J. F. Levy, "A Finite Element Method for Incompressible Viscous Flows," in *Proceedings of the Second International Symposium on Finite Element Methods in Flow Problems,* S. Margherita, Italy (1976).

17. T. J. R. Hughes, W. K. Liu, and A. Brooks, "Review of Finite Element Analysis of Incompressible Viscous Flows by Penalty Function Formulation," *Journal of Computational Physics,* **30**, 1–60 (1979).

18. M. Bercovier and M. Engelman, "A Finite Element for the Numerical Solution of Viscous Incompressible Flows," *Journal of Computational Physics,* **30**, 181–201 (1979).

19. M. Engelman, R. Sani, P. M. Gresho, and M. Bercovier, "Consistent vs. Reduced Integration Penalty Methods for Incompressible Media Using Several Old and New Elements," *International Journal for Numerical Methods in Fluids,* **2**, 25–42 (1982).

20. R. S. Marshall, J. C. Heinrich, and O. C. Zienkiewicz, "Natural Convection in a Square Enclosure by a Finite Element, Penalty Function Method, Using Primitive Fluid Variables," *Numerical Heat Transfer,* **1**, 315–330 (1978).

21. J. C. Heinrich and M. Strada, "Penalty Finite Element Analysis of Natural Convection at High Rayleigh Numbers," in *Finite Elements in Fluids, Vol. 4,* R. H. Gallagher, D. Norrie, J. T. Oden, and O. C. Zienkiewicz (eds.), John Wiley, New York (1982).

22. M. P. Reddy and J. N. Reddy, "Finite-Element Analysis of Flows of Non-Newtonian Fluids in Three-Dimensional Enclosures," *International Journal of Non-Linear Mechanics*, **27**, 9–26 (1992).

23. M. P. Reddy, J. N. Reddy, and H. U. Akay, "Penalty Finite Element Analysis of Incompressible Flows Using Element by Element Solution Algorithms," *Computer Methods in Applied Mechanics and Engineering*, **100**, 169–205 (1992).

24. D. K. Gartling, C. E. Hickox and R. C. Givler, "Simulation of Viscous and Porous Flow Problems," *Computational Fluid Dynamics*, **7**, 23–48 (1996).

25. P. M. Gresho, "Some Current CFD Issues Relevant to the Incompressible Navier–Stokes Equations," *Computer Methods in Applied Mechanics and Engineering*, **87**, 201–252 (1991).

26. P. M. Gresho, "Incompressible Fluid Dynamics: Some Fundamental Formulation Issues," *Annual Review of Fluid Mechanics*, **23**, 413–453 (1991).

27. P. M. Gresho, "Some Interesting Issues in Incompressible Fluid Dynamics, Both in the Continuum and in Numerical Simulation," in *Advances in Applied Mechanics*, J. W. Hutchinson and T. Y. Wu (eds.), **28**, 46–140 (1992).

28. P. M. Gresho and R. L. Sani, *Incompressible Flow and the Finite Element Method*, John Wiley & Sons, London (1998).

29. C. Taylor and P. Hood, "A Numerical Solution of the Navier–Stokes Equations Using FEM Techniques," *Computers and Fluids*, **1**, 73–100 (1973).

30. T. J. R. Hughes, L. P. Franca, and M. Balestra, "A New Finite Element Formulation for Computational Fluid Dynamics, V. Circumventing the Babuska–Brezzi Condition: A Stable Petrov–Galerkin Formulation for the Stokes Problem Accommodating Equal-Order Interpolations," *Computer Methods in Applied Mechanics and Engineering*, **59**, 85–99 (1986).

31. T. E. Tezduyar, "Stabilized Finite Element Formulations for Incompressible Flow Computations," in *Advances in Applied Mechanics*, J. W. Hutchinson and T. Y. Wu (eds.), **28**, 1–44 (1992).

32. L. P. Franca, T. J. R. Hughes, and R. Stenberg, "Stabilized Finite Element Methods," in *Incompressible Computational Fluid Dynamics*, M. D. Gunzburger and R. A. Nicolaides (eds), Cambridge University Press, Cambridge, U. K., 87–107 (1993).

33. B.-N Jiang, T. L. Lin, and L. A. Povinelli, "Large-Scale Computation of Incompressible Viscous Flow by the Least-Squares Finite Element Method," *Computer Methods in Applied Mechanics and Engineering*, **114**, 213–231 (1994).

34. G. F. Carey, A. I. Pehlivanov, Y. Shen, A. Bose, and K. C. Wang, "Least-Squares Finite Elements for Fluid Flow and Transport," *International Journal for Numerical Methods in Fluids*, **27**, 97–107 (1998).

35. E. M. Salonen and J. Aalto, "A Pressure Determination Scheme," *Proceedings of the Fifth International Conference on Numerical Methods in Laminar and Turbulent Flow,* C. Taylor, M. D. Olson, P. M. Gresho, and W. G. Habashi (eds.), Pineridge Press, Swansea, U.K. (1985).

36. T. Shiojima and Y. Shimazaki, "A Pressure-Smoothing Scheme for Incompressible Flow Problems," *International Journal for Numerical Methods in Fluids,* **9**, 557–567 (1989).

37. M. Engelman, R. L. Sani, and P. M. Gresho, "The Implementation of Normal and/or Tangential Boundary Conditions in Finite Element Codes for Incompressible Fluid Flow," *International Journal for Numerical Methods in Fluids,* **2**, 225–238 (1982).

38. G. C. Broyden, "A Class of Methods for Solving Nonlinear Simultaneous Equations," *Mathematics of Computation,* **19**, 577–593 (1965).

39. M. S. Engelman, G. Strang, and K. J. Bathe, "The Application of Quasi-Newton Methods in Fluid Mechanics," *International Journal for Numerical Methods in Engineering,* **17**, 707–718 (1981).

40. P. Underwood, "Dynamic Relaxation," in *Computational Methods for Transient Analysis,* T. Belytschko and T. J. R. Hughes (eds.), North-Holland, Amsterdam, 245–265 (1983).

41. C. P. Jackson and K. H. Winters, "A Finite Element Study of the Bernard Problem Using Parameter-Stepping and Bifurcation Search,"*International Journal of Numerical Methods in Fluids,* **4**, 127–145 (1984).

42. K. H. Winters, K. A. Cliffe, and C. P. Jackson, "The Prediction of Instabilities Using Bifurcation Theory," in *Numerical Methods for Transient and Coupled Problems,* R. W. Lewis et al. (eds.), John Wiley, New York, 179–198 (1987).

43. K. H. Winters and R. O. Jack, "Anomalous Convection at Low Prandtl Number," *Communications in Applied Numerical Methods,* **5**, 401–404 (1989).

44. K. Winters, "Bifurcation and Stability: A Computational Approach," *Computational Physics Communications,* **65**, 299–309 (1991).

45. W. F. Schmidt, "Adaptive Step Size Selection for Use with the Continuation Method," *International Journal for Numerical Methods in Engineering,* **12**, 677–694 (1978).

46. W. C. Rheinboldt, "On the Solution of Some Nonlinear Equations Arising in the Application of Finite Element Methods," in *Proceedings of the Conference on The Mathematics of Finite Elements and Applications,* J. R. Whiteman (ed.), Brunel University, Academic Press, London (1975).

47. W. C. Rheinboldt, "Numerical Analysis of Continuation Methods for Nonlinear Structural Problems," *Computers and Structures,* **13**, 103–113 (1981).

48. S. V. Patankar, *Numerical Heat Transfer and Fluid Flow*, McGraw-Hill, New York (1980).

49. V. Haroutunian, M. S. Engelman, and I. Hasbani, "Segregated Finite Element Algorithms for the Numerical Solution of Large-Scale Incompressible Flow Problems," *International Journal for Numerical Methods in Fluids*, **17**, 323–348 (1993).

50. A. C. Benim and W. Zinser, "A Segregated Formulation of Navier–Stokes Equations with Finite Elements," *Computer Methods in Applied Mechanics and Engineering*, **57**, 223–237 (1986).

51. J. G. Rice and R. J. Schnipke, "An Equal-Order Velocity-Pressure Formulation that Does Not Exhibit Spurious Pressure Modes," *Computer Methods in Applied Mechanics and Engineering*, **58**, 135–149 (1986).

52. P. M. Gresho, S. T. Chan, R. L. Lee, and C. D. Upson, "A Modified Finite Element Method for Solving the Time-Dependent, Incompressible Navier–Stokes Equations. Part 1: Theory," *International Journal for Numerical Methods in Fluids*, **4**, 557–598 (1984).

53. P. M. Gresho and S.T. Chan, "On the Theory of Semi-Implicit Projection Methods for Viscous Incompressible Flow and Its Implementation Via a Finite Element Method That Introduces a Nearly Consistent Mass Matrix, Part 1–Theory," *International Journal for Numerical Methods in Fluids*, **11**, 587–620 (1990).

54. P. M. Gresho, R. L. Lee, S. T. Chan, and R. L. Sani, "Solution of the Time-Dependent Incompressible Navier–Stokes and Boussinesq Equations Using the Galerkin Finite Element Method," in *Proc. IUTAM Symposium on Approximation Methods for Navier–Stokes Problems*, Paderborn, W. Germany, Springer-Verlag, Berlin (1979).

55. J. C. Heinrich, P. S. Huyakorn, O. C. Zienkiewicz, and A. R. Mitchell, "An 'Upwind' Finite Element Scheme for Two-Dimensional Convective Transport Equation," *International Journal for Numerical Methods in Engineering*, **11**, 131–143 (1977).

56. J. C. Heinrich and O. C. Zienkiewicz, "The Finite Element Method and 'Upwinding' Techniques in the Numerical Solutions of Convection Dominated Flow Problems," in *Finite Element Methods for Convection Dominated Flows*, T. J. R. Hughes (ed.), ASME, AMD **34**, New York, 105–136 (1979).

57. I. Christie, D. F. Griffiths, A. R. Mitchell, and O. C. Zienkiewicz, "Finite Element Methods for Second Order Differential Equations with Significant First Derivatives," *International Journal for Numerical Methods in Engineering*, **10**, 1389–1396 (1976).

58. T. J. R. Hughes and A. N. Brooks, "A Multi-Dimensional Upwind Scheme with No Crosswind Diffusion," in *Finite Element Methods for Convection Dominated Flows*, T. J. R. Hughes (ed.), ASME, AMD **34**, NY, 19–35 (1979).

59. A. N. Brooks and T. J. R. Hughes, "Streamline-Upwind/Petrov–Galerkin Formulations for Convection Dominated Flows with Particular Emphasis on Incompressible Navier–Stokes Equation," *Computer Methods in Applied Mechanics and Engineering,* **32**, 199–259 (1982).

60. T. J. R. Hughes and T. E. Tezduyar, "Finite Element Methods for First-Order Hyperbolic Systems with Particular Emphasis on the Compressible Euler Equations," *Computer Methods in Applied Mechanics and Engineering,* **45**, 217–284 (1982).

61. T. J. R. Hughes and L. P. Franca, "A New Finite Element Formulation for Computational Fluid Dynamics: VII. The Stokes Problem with Various Well-posed Boundary Conditions: Symmetric Formulations that Converge for All Velocity/Pressure Spaces," *Computer Methods in Applied Mechanics and Engineering,* **65**, 85–96 (1987).

62. T. J. R. Hughes, L. P. Franca, and G. M. Hulbert, "A New Finite Element Formulation for Computational Fluid Dynamics: VIII. The Galerkin/Least-squares Method for Advective-Diffusive Equations," *Computer Methods in Applied Mechanics and Engineering,* **73**, 173–189 (1989).

63. T. E. Tezduyar, R. Shih, S. Mittal, and S. E. Ray, "Incompressible Flow Computations with Stabilized Bilinear and Linear Equal-Order-Interpolation Velocity-Pressure Elements," University of Minnesota Supercomputer Institute Research Report, UMSI 90/165, Minneapolis (1990).

64. F. Shakib, "Finite Element Analysis of the Compressible Euler and Navier–Stokes Equations," Ph.D. Thesis, Stanford University, Stanford (1988).

65. D. J. Naylor, "Stresses in Nearly Incompressible Materials by Finite Elements with Application to the Calculation of Excess Pore Pressures," *International Journal for Numerical Methods in Engineering,* **8**, 443–460 (1974).

66. E. Hinton, F. C. Scott, and R. E. Ricketts, "Local Least Squares Stress Smoothing for Parabolic Isoparametric Elements," *International Journal for Numerical Methods in Engineering,* **9**, 235–256 (1975).

67. R. I. Tanner, R. E. Nickell, and R. W. Bilger, "Finite Element Methods for the Solution of Some Incompressible Non-Newtonian Fluid Mechanics Problems with Free Surfaces," *Computer Methods in Applied Mechanics and Engineering,* **6**, 155–174 (1975).

68. K. R. Reddy and R. I. Tanner, "Finite Element Solution of Viscous Jet Flows with Surface Tension," *Computers and Fluids,* **6**, 83–91 (1978).

69. O. C. Zienkiewicz, P. C. Jain, and E. Oñate, "Flow of Solids During Forming and Extrusion: Some Aspects of Numerical Solutions," *International Journal of Solids and Structures,* **14**, 15–38 (1978).

70. H. Saito and L. E. Scriven, "Study of Coating Flow by the Finite Element Method," *Journal of Computational Physics,* **42**, 53–76 (1981).

71. L. E. Scriven and S. F. Kistler, "Coating Flow Theory by Finite Element and Asymptotic Analysis of the Navier–Stokes System," in *Finite Element Flow Analysis*, T. Kawai (ed.), University of Tokyo Press, Tokyo, 503–510 (1982).

72. M. Engelman, R. L. Sani, and P. M. Gresho, "The Implementation of Normal and/or Tangential Boundary Conditions in Finite Element Codes for Incompressible Fluid Flow," *International Journal for Numerical Methods in Fluids*, **2**, 225–238 (1982).

73. K. N. Christodoulou and L. E. Scriven, "Discretization of Free Surface Flows and Other Moving Boundary Problems," *Journal of Computational Physics*, **99**, 39–55 (1992).

74. P. M. Schweizer and S. F. Kistler (eds.), *Liquid Film Coating, Scientific Principles and Their Technical Implications*, Van Nostrand Reinhold, New York (1994).

75. D. R. Lynch and W. G. Gray, " Finite Element Simulation of Flow in Deforming Regions," *Journal of Computational Physics*, **36**, 135–153 (1980).

76. D. R. Lynch, "Unified Approach to Simulation on Deforming Elements with Application to Phase Change Problems," *Journal of Computational Physics*, **47**, 387–411 (1982).

77. P. R. Schunk, P. A. Sackinger, R. R. Rao, K. S. Chen, R. A. Cairncross, T. A. Baer, and D. A. Labreche, "GOMA 2.0 – A Full-Newton Finite Element Program for Free and Moving Boundary Problems with Coupled Fluid/Solid Momentum, Energy, Mass and Chemical Species Transport: User's Guide," Sandia National Laboratories Report, SAND97-2404, Albuquerque, New Mexico (1998).

78. A. A. Johnson and T. E. Tezduyar, "Mesh Update Strategies in Parallel Finite Element Computations of Flow Problems with Moving Boundaries and Interfaces," *Computer Methods in Applied Mechanics and Engineering*, **119**, 73–94 (1994).

79. G. Chiandussi, G. Bugeda, and E. Oñate, "A Simple Method for Automatic Updating of Finite Element Meshes," *Communications in Numerical Methods in Engineering*, **16**, 1–19 (2000).

80. D. K. Gartling, "MERLIN II – A Computer Program to Transfer Solution Data Between Finite Element Meshes," Sandia National Laboratories Report, SAND89-2989, Albuquerque, New Mexico (1991).

81. K. Jansen, F. Shakib, and T. J. R. Hughes, "Fast Projection Algorithms for Unstructured Meshes," in *Computational Nonlinear Mechanics in Aerospace Engineering*, S. Atluri (ed.), AIAA, New York, 175–204 (1992).

82. R. L. Sani, P. M. Gresho, D. R. Tuerpe, and R. L. Lee, "The Imposition of Incompressibility Constraints via Variational Adjustment of Velocity Fields," in *Proceedings of the First International Conference on Numerical Methods*

in Laminar and Turbulent Flow, C. Taylor, K. Morgan, and C. A. Brebbia (eds.), Pentech Press, London, 983–994 (1978).

83. W. J. Rider and D. B. Kothe, "Reconstructing Volume Tracking," *Journal of Computational Physics*, **141**, 112–152 (1998).

84. FIDAP Manual, Ver. 7.5, Fluid Dynamics International, Evanston, IL (1995).

85. J. A. Sethian, *Level Set Methods and Fast Marching Methods*, Second Edition, Cambridge University Press, Cambridge, United Kingdom (1999).

86. D. Adalsteinsson and J. A. Sethian, "The Fast Construction of Extensional Velocities in Level Set Methods," *Journal of Computational Physics*, **148**, 2–22 (1999).

87. M. Sussman, P. Smereka, and S. Osher, "A Level Set Approach for Computing Solutions to Incompressible Two-Phase Flow," *Journal of Computational Physics*, **114**, 146–159 (1994).

88. D. K. Gartling, "NACHOS II – A Finite Element Computer Program for Incompressible Flow Problems," Sandia National Laboratories Report, SAND86-1816 and SAND86–1817, Albuquerque, New Mexico (1987).

89. A. Nadai, *Theory of Flow and Fracture of Solids, Vol. II*, McGraw-Hill, New York (1963).

90. H. Schlichting, *Boundary-Layer Theory* (translated by J. Kestin), Seventh Edition, McGraw-Hill, New York (1979).

91. O. R. Burggraf, "Analytical and Numerical Studies of the Structure of Steady Separated Flows," *Journal of Fluid Mechanics*, **24**(1), 113–151 (1966).

92. F. Pan and A. Acrivos, "Steady Flow in Rectangular Cavities," *Journal of Fluid Mechanics*, **28**, 643 (1967).

93. R. T. Cheng, "Numerical Solution of the Navier–Stokes Equations by the Finite Element Method," *Physics of Fluids*, **15** (12), 2098–2105 (1972).

94. G. D. Mallison and G. de Vahl Davis, "The Method of False Transient for the Solution of Coupled Elliptic Equations, " *Journal of Computational Physics*, **12**, 435–461 (1967).

95. G. de Vahl Davis and G. D. Mallison, "An Evaluation of Upwind and Central Difference Approximations by a Study of Recirculating Flow," *Computers and Fluids*, **4**, 29–43 (1976).

96. H. Takami and K. Kuwahara, "Numerical Study of Three-Dimensional Flow Within a Cubic Cavity," *Journal of Physical Society of Japan*, **37**, 1695–1698 (1974).

97. J. N. Reddy, "Penalty Finite Element Methods for the Solution of Advection and Free Convection Flows," *Finite Element Methods in Engineering*, A. P. Kabaila and V. A. Pulmano (eds.), University of New South Wales, Sydney,

Australia, 583–598 (1979).

98. S. C. R. Dennis, D. B. Ingham, and R. N. Cook, "Finite Difference Methods for Calculating Steady Incompressible Flows in Three Dimensions," *Journal of Computational Physics,* **33**, 325–339 (1979).

99. R. K. Agarwal, "A Third-Order Accurate Upwind Scheme for Navier–Stokes Solutions in Three Dimensions," McDonnell Douglas Research Laboratory, Report No. MDRL 81–20, St. Louis, Missouri (1981).

100. J. N. Reddy, "Penalty-Finite-Element Analysis of 3-D Navier–Stokes Equations," *Computer Methods in Applied Mechanics and Engineering,* **35**, 87–106 (1982).

101. H. P. Langtangen and A. Tevito, "A Numerical Comparison of Conjugate Gradient-Like Methods," *Communications in Applied Numerical Methods,* **34**, 793–798 (1988).

102. D. K. Gartling, "A Test Problem for Outflow Boundary Conditions—Flow Over a Backward-Facing Step," *International Journal for Numerical Methods in Fluids,* **11**, 953–967 (1990).

103. R. C. Givler, D. K. Gartling, M. S. Engelman, and V. Haroutunian, "Navier–Stokes Simulations of Flow Past Three-Dimensional Submarine Models," *Computer Methods in Applied Mechanics and Engineering,* **87**, 175–200 (1991).

104. P. A. Sackinger, "Flows and Transitions During Solidification: Simulation of Hydrodynamics, Heat Transfer and Free Boundaries in Czochralski Growth," Ph.D. Dissertation, MIT, Cambridge, Massachusetts (1989).

105. P. A. Sackinger, R. A. Brown, and J. J. Derby, "A Finite Element Method for Analysis of Fluid Flow, Heat Transfer and Free Interfaces in Czochralski Crystal Growth," *International Journal for Numerical Methods in Fluids,* **9**, 453–492 (1989).

106. ProCAST User's Manual, Version 2.0, *UES, Inc.,* Dayton, Ohio (1992).

107. J. O. Hinze, *Turbulence,* McGraw-Hill, New York (1975).

108. D. C. Leslie, *Developments in the Theory of Turbulence,* Oxford University Press, Oxford (1984).

109. P. Bradshaw (ed.), *Turbulence,* Second Edition, Springer-Verlag, Berlin (1978).

110. G. K. Batchelor, *The Theory of Homogeneous Turbulence,* Cambridge University Press, Cambridge, U.K. (1953).

111. A. S. Monin and A. M. Yaglom, *Statistical Fluid Mechanics,* MIT Press, Cambridge, MA (1973).

112. J. H. Ferziger, "Simulation of Incompressible Turbulent Flows," *Journal of Computational Physics,* **69**, 1–48 (1987).

113. J. M. Ferziger, "Higher Level Simulations of Turbulent Flows," in *Computational Methods for Turbulent, Transonic and Viscous Flows*, J. A. Essers (ed.), 93–182, Hemisphere Publishing, Washington, D.C. (1983).

114. P. T. Harsha, "Kinetic Energy Methods," in *Handbook of Turbulence*, W. Frost and T. Moulden (ed.), 187–236, Plenum Press, New York (1977).

115. J. D. Murphy, "Turbulence Modeling," in *Encyclopedia of Fluid Mechanics*, N. Chermisinoff (ed.), **6**, 1131–1151, Gulf Publishing, Houston, Texas (1988).

116. M. Lesieur, *Turbulence in Fluids*, Martinius Nijhoff, Dordrecht, The Netherlands (1987).

117. R. S. Rogallo and P. Moin, "Numerical Simulation of Turbulent Flows," in *Annual Review of Fluid Mechanics*, **16**, 99–137 (1984).

118. A. Yoshizawa, "Large Eddy Simulation of Turbulent Flows," in *Encyclopedia of Fluid Mechanics*, N. Chermisinoff (ed.), **6**, 1277–1297, Gulf Publishing, Houston, Texas (1968).

119. W. Rodi, "Turbulence Models and Their Applications to Hydraulics," IAHR Report, Delft, The Netherlands (1984).

120. T. Cebeci and A. M. O. Smith, "Analysis of Turbulent Boundary Layers," *Applied Mathematics and Mechanics*, Academic Press, New York (1974).

121. V. Haroutunian, "Turbulent Flows with FIDAP," Fluid Dynamics International Seminar Notes, Evanston, Illinois (1988).

122. M. Nallasamy, "Turbulence Models and Their Applications to the Prediction of Internal Flows: A Review," *Computers and Fluids*, **15**, 151–194(1987).

123. R. M. Smith, "On Finite Element Calculations of Turbulent Flow Using the k-ϵ Model," *International Journal for Numerical Methods in Fluids*, **4**, 303–319 (1984).

124. R. M. Smith, "A Practical Method of Two Equation Modelling Using Finite Elements," *International Journal for Numerical Methods in Fluids*, **4**, 321–336 (1984).

125. A. G. Hutton, R. M. Smith, and S. Hickmott, "The Computation of Turbulent Flows of Industrial Complexity by Finite Element Methods – Progress and Prospects," in *Finite Elements in Fluids*, Vol. 7, R. H. Gallagher et al. (eds.), John Wiley & Sons, New York (1988).

126. B. E. Launder and D. B. Spalding, *Lectures in Mathematical Models of Turbulence*, Academic Press, London, U.K. (1972).

127. B. E. Launder and D. B. Spalding, "The Numerical Computation of Turbulent Flow," *Computer Methods in Applied Mechanics and Engineering*, **23**, 249–270 (1974).

128. FIDAP Manual, Version 6.0, Fluid Dynamics International, Evanston, Illinois (1991).

129. A. G. Hutton, "Finite Element Boundary Techniques for Improved Performance in Computing Navier–Stokes and Related Heat Transfer Problems," in *Finite Elements in Fluids*, Vol. 4, R. H. Gallagher et al. (eds.) John Wiley & Sons, New York (1982).

Chapter 5

Convective Heat Transfer

5.1 Introduction

5.1.1 Background

The general problem area of convective heat transfer is a particularly challenging one since it represents a situation involving inherently coupled problems, i.e., problems involving multiple physical phenomena. Indeed, the subject itself represents a joining of two classical areas of applied mechanics, namely, fluid mechanics and heat transfer.

Convection problems are generally divided into two major categories depending primarily on the forces that are responsible for the fluid motion. In *forced convection*, the fluid motion is due to the application of pressure or viscous forces on the fluid boundary. *Free* or *natural convection* problems are characterized by the fluid motion which is produced by temperature-induced buoyancy forces. This distinction between the types of convection is not always possible because the two types of driving forces (surface and volume) may appear in varying degrees and combinations.

Since the convective heat transfer process depends directly on the motion of a fluid medium, the Navier–Stokes equations from Chapter 4 as well as the advection-diffusion equation that describes thermal energy transport must be considered in the development of a computational scheme. These equations consist of a set of coupled partial differential equations in terms of the velocity components, temperature and pressure.

In this chapter, finite element models based on the weak formulation of the governing equations of viscous incompressible fluids in the presence of buoyancy forces and inertial effects are developed. The finite element model development is completely analogous to those developed for heat conduction in Chapter 3 and viscous incompressible flows in Chapter 4. Therefore, detailed discussions of the actual model development will not be repeated here.

5.1.2 Governing Equations

The laws describing the flow of Newtonian fluids were presented in Chapter 1 and are summarized here for ready reference. The equations are written for a fluid region Ω_f using a Cartesian coordinate system x_i in a Eulerian reference frame, with the subscript (or index) $i = 1, 2, 3$ (or 1 and 2 for two-dimensional problems), and the usual summation convention on repeated indices is used [see Eqs. (1.4.7)–(1.4.9)]; time is denoted by t. An extended form of the Boussinesq approximation is used, which allows the fluid properties to be functions of the thermodynamic state (e.g., pressure and temperature) and the density ρ to vary with temperature T according to the relation

$$\rho = \rho_0[1 - \beta(T - T_0)] \tag{5.1.1}$$

where β is the coefficient of thermal expansion and the subscript zero indicates a reference condition. The variation of density as given in (5.1.1) is permitted only in the description of the body force; the density in all other situations is assumed to be that of the reference state, ρ_0. The governing equations of convective heat transfer are summarized below.

Conservation of Mass:

$$\frac{\partial u_i}{\partial x_i} = 0 \tag{5.1.2}$$

Conservation of Momentum:

$$\rho_0 \left(\frac{\partial u_i}{\partial t} + u_j \frac{\partial u_i}{\partial x_j} \right) - \frac{\partial}{\partial x_j} \left[-P\delta_{ij} + \mu \left(\frac{\partial u_i}{\partial x_j} + \frac{\partial u_j}{\partial x_i} \right) \right] + \rho_0 g_i \beta(T - T_0) = 0 \tag{5.1.3}$$

Conservation of Energy:

$$\rho_0 C_v \left(\frac{\partial T}{\partial t} + u_j \frac{\partial T}{\partial x_j} \right) - \frac{\partial}{\partial x_i} \left(k_{ij} \frac{\partial T}{\partial x_j} \right) - Q - \Phi = 0 \tag{5.1.4a}$$

where u_i denote the velocity components, P the pressure, ρ_0 the density, g_i the gravitational force components, T the temperature, C_v the specific heat of the fluid at constant volume, Q the rate of internal heat generation, μ the shear viscosity of the fluid, k_{ij} the components of the thermal conductivity tensor, and Φ is the viscous dissipation in the fluid

$$\Phi = 2\mu D_{ij} D_{ij} ; \quad D_{ij} = \frac{1}{2} \left(\frac{\partial u_i}{\partial x_j} + \frac{\partial u_j}{\partial x_i} \right) \tag{5.1.4b}$$

Recall that Eq. (5.1.2) is also known as the continuity equation or divergence-free condition on the velocity field and Eq. (5.1.3) is known as the Navier–Stokes equations.

The boundary conditions are given by

$$u_i = f_i^u(s_k, t) \quad \text{on } \Gamma_u \tag{5.1.5a}$$

$$\mathcal{T}_i \equiv \sigma_{ij}(s_k, t) n_j(s_k) = f_i^T(s_k, t) \quad \text{on } \Gamma_T \tag{5.1.5b}$$

for the fluid mechanics part of the problem, and

$$T = f^T(s_k, t) \qquad \text{on } \Gamma_T \tag{5.1.6a}$$

$$-\left(k_{ij}\frac{\partial T}{\partial x_j}\right)n_i \equiv q_i n_i = q_c + q_r = q_a = f^q(s_k, t) \qquad \text{on } \Gamma_q \tag{5.1.6b}$$

for the heat transfer part of the problem. In Eq. (5.1.5b), σ_{ij} denote the components of the total stress tensor (i.e., viscous and hydrostatic)

$$\sigma_{ij} = \mu\left(\frac{\partial u_i}{\partial x_j} + \frac{\partial u_j}{\partial x_i}\right) - P\delta_{ij} \tag{5.1.7}$$

We will not explicitly consider the energy equation for solid regions in this chapter as it was considered in detail in Chapter 3. However, it should be recognized that the inclusion of solid body conduction regions in any convection problem (i.e., the so-called conjugate problem) is a relatively straightforward procedure. Also, as mentioned in Chapter 1, convection problems may require the addition of auxiliary variables to describe all of the physical phenomena of interest. We will not consider this case explicitly since the formulation and solution methods for transport equations follow exactly the treatment of the energy equation. Note, however, that inclusion of auxiliary transport equations may significantly increase the magnitude of the computational effort.

There are five partial differential equations expressed in terms of five unknowns (u_1, u_2, u_3, P, T) for a three-dimensional problem. Following the developments of Chapters 3 and 4, we shall consider two finite element models of Eqs. (5.1.1)–(5.1.3). The first one is the *velocity-pressure-temperature model* or *mixed model*. The second model is the *penalty-finite element model*. These models were discussed in Chapter 4 for isothermal flows. Extension of these models to convective heat transfer problems is straightforward, as will be shown in this chapter.

5.2 Mixed Finite Element Model

Let us denote expressions on the left side of the equalities in (5.1.2)–(5.1.4a) by f_1, f_2, and f_3, respectively. As discussed in Section 4.2, the weighted-integral statements of the three equations over a typical element Ω^e are given by

$$\int_{\Omega^e} w_1 f_1 \, d\mathbf{x} = 0, \qquad \int_{\Omega^e} \mathbf{w}_2 \cdot \mathbf{f}_2 \, d\mathbf{x} = 0, \qquad \int_{\Omega^e} w_3 f_3 \, d\mathbf{x} = 0 \tag{5.2.1}$$

where ($w_1, \mathbf{w}_2, w_3$) are the weight functions, which will be equated to the interpolation functions used for ($P, \mathbf{u}, T$), respectively. It should be recalled from Chapters 3 and 4 that the weighted-integral statements in Eq. (5.2.1) are reduced to weak statements by integrating the viscous and diffusion parts. All other terms are kept as they are.

Suppose that the dependent variables (T, u_i, P) are approximated by expansions of the form

$$T(\mathbf{x}, t) = \sum_{m=1}^{M} \theta_m(\mathbf{x}) T_m(t) = \boldsymbol{\Theta}^T \mathbf{T} \tag{5.2.2a}$$

$$u_i(\mathbf{x}, t) = \sum_{n=1}^{N} \psi_n(\mathbf{x}) u_i^n(t) = \boldsymbol{\Psi}^T \mathbf{u}_i \tag{5.2.2b}$$

$$P(\mathbf{x}, t) = \sum_{l=1}^{L} \phi_l(\mathbf{x}) P_l(t) = \boldsymbol{\Phi}^T \mathbf{P} \tag{5.2.2c}$$

where $\boldsymbol{\Theta}$, $\boldsymbol{\Psi}$, and $\boldsymbol{\Phi}$ are vectors of interpolation (or shape) functions, $\mathbf{T}$, $\mathbf{u}_i$, and $\mathbf{P}$ are vectors of nodal values of temperature, velocity components, and pressure, respectively, and superscript $(\cdot)^T$ denotes a transpose of the enclosed vector or matrix. The weight functions $(w_1, \mathbf{w_2}, w_3)$ have the following correspondence (see Reddy [1] for further details)

$$w_1 \approx \phi_l, \ \mathbf{w_2} \approx \psi_n, \ w_3 \approx \theta_m \tag{5.2.3}$$

Substitution of Eqs. (5.2.2) and (5.2.3) into the weak forms associated with Eq. (5.2.1) results in the following finite element equations:

Continuity:

$$-\left[\int_{\Omega^e} \boldsymbol{\Phi} \frac{\partial \boldsymbol{\Psi}^T}{\partial x_i} \, d\mathbf{x}\right] \mathbf{u}_i = 0 \tag{5.2.4}$$

Momentum:

$$\left[\int_{\Omega^e} \rho_0 \boldsymbol{\Psi}\boldsymbol{\Psi}^T \, d\mathbf{x}\right]\dot{\mathbf{u}}_i + \left[\int_{\Omega^e} \rho_0 \boldsymbol{\Psi}(\boldsymbol{\Psi}^T\mathbf{u}_j)\frac{\partial \boldsymbol{\Psi}^T}{\partial x_j} \, d\mathbf{x}\right]\mathbf{u}_i + \left[\int_{\Omega^e} \mu\frac{\partial \boldsymbol{\Psi}}{\partial x_j}\frac{\partial \boldsymbol{\Psi}^T}{\partial x_j} \, d\mathbf{x}\right]\mathbf{u}_i$$

$$+\left[\int_{\Omega^e} \mu\frac{\partial \boldsymbol{\Psi}}{\partial x_j}\frac{\partial \boldsymbol{\Psi}^T}{\partial x_i} \, d\mathbf{x}\right]\mathbf{u}_j - \left[\int_{\Omega^e} \frac{\partial \boldsymbol{\Psi}}{\partial x_i}\boldsymbol{\Phi}^T \, d\mathbf{x}\right]\mathbf{P}$$

$$= -\left[\int_{\Omega^e} \rho_0 g_i \beta \boldsymbol{\Psi}\boldsymbol{\Theta}^T \, d\mathbf{x}\right]\mathbf{T} + \left\{\int_{\Omega^e} \rho_0 g_i \beta \boldsymbol{\Psi} T_0 \, d\mathbf{x}\right\} + \left\{\oint_{\Gamma^e} \boldsymbol{\Psi} T_i \, ds\right\} \tag{5.2.5}$$

Energy:

$$\left[\int_{\Omega^e} \rho_0 C_v \boldsymbol{\Theta}\boldsymbol{\Theta}^T \, d\mathbf{x}\right]\dot{\mathbf{T}} + \left[\int_{\Omega^e} \rho_0 C_v \boldsymbol{\Theta}(\boldsymbol{\Psi}^T\mathbf{u}_j)\frac{\partial \boldsymbol{\Theta}^T}{\partial x_j} \, d\mathbf{x}\right]\mathbf{T} + \left[\int_{\Omega^e} k_{ij}\frac{\partial \boldsymbol{\Theta}}{\partial x_i}\frac{\partial \boldsymbol{\Theta}^T}{\partial x_j} \, d\mathbf{x}\right]\mathbf{T}$$

$$= \left\{\int_{\Omega^e} \boldsymbol{\Theta} Q \, d\mathbf{x}\right\} + \left\{\int_{\Omega^e} \boldsymbol{\Theta}\boldsymbol{\Phi} \, d\mathbf{x}\right\} + \left\{\oint_{\Gamma^e} \boldsymbol{\Theta} q \, ds\right\} \tag{5.2.6}$$

or

$$
\begin{bmatrix} \mathbf{M} & 0 & 0 & 0 \\ 0 & \mathbf{M} & 0 & 0 \\ 0 & 0 & \mathbf{M} & 0 \\ 0 & 0 & 0 & 0 \end{bmatrix} \begin{Bmatrix} \dot{\mathbf{u}}_1 \\ \dot{\mathbf{u}}_2 \\ \dot{\mathbf{u}}_3 \\ \dot{\mathbf{P}} \end{Bmatrix} + \begin{bmatrix} \mathbf{C(u)} & 0 & 0 & 0 \\ 0 & \mathbf{C(u)} & 0 & 0 \\ 0 & 0 & \mathbf{C(u)} & 0 \\ 0 & 0 & 0 & 0 \end{bmatrix} \begin{Bmatrix} \mathbf{u}_1 \\ \mathbf{u}_2 \\ \mathbf{u}_3 \\ \mathbf{P} \end{Bmatrix} +
$$

$$
\begin{bmatrix} \hat{\mathbf{K}}_{11} & \mathbf{K}_{21} & \mathbf{K}_{31} & -\mathbf{Q}_1 \\ \mathbf{K}_{12} & \hat{\mathbf{K}}_{22} & \mathbf{K}_{32} & -\mathbf{Q}_2 \\ \mathbf{K}_{13} & \mathbf{K}_{23} & \hat{\mathbf{K}}_{33} & -\mathbf{Q}_3 \\ -\mathbf{Q}_1^T & -\mathbf{Q}_2^T & -\mathbf{Q}_3^T & 0 \end{bmatrix} \begin{Bmatrix} \mathbf{u}_1 \\ \mathbf{u}_2 \\ \mathbf{u}_3 \\ \mathbf{P} \end{Bmatrix} = \begin{Bmatrix} \mathbf{F}_1(\mathbf{T}) \\ \mathbf{F}_2(\mathbf{T}) \\ \mathbf{F}_3(\mathbf{T}) \\ 0 \end{Bmatrix} \tag{5.2.7}
$$

$$
[\mathbf{N}]\{\dot{\mathbf{T}}\} + [\mathbf{D(u)}]\{\mathbf{T}\} + [\mathbf{L}]\{\mathbf{T}\} = \{\mathbf{G(T)}\} \tag{5.2.8}
$$

The coefficient matrices shown in Eqs. (5.2.7) and (5.2.8) are defined by

$$
\hat{\mathbf{K}}_{11} = 2\mathbf{K}_{11} + \mathbf{K}_{22} + \mathbf{K}_{33}
$$
$$
\hat{\mathbf{K}}_{22} = \mathbf{K}_{11} + 2\mathbf{K}_{22} + \mathbf{K}_{33}
$$
$$
\hat{\mathbf{K}}_{33} = \mathbf{K}_{11} + \mathbf{K}_{22} + 2\mathbf{K}_{33} \tag{5.2.9a}
$$

$$
\mathbf{M} = \int_{\Omega^e} \rho_0 \boldsymbol{\Psi} \boldsymbol{\Psi}^T \, d\mathbf{x} \; ; \quad \mathbf{C(u)} = \int_{\Omega^e} \rho_0 \boldsymbol{\Psi} (\boldsymbol{\Psi}^T \mathbf{u}_j) \frac{\partial \boldsymbol{\Psi}^T}{\partial x_j} \, d\mathbf{x}
$$

$$
\mathbf{K}_{ij} = \int_{\Omega^e} \mu \frac{\partial \boldsymbol{\Psi}}{\partial x_i} \frac{\partial \boldsymbol{\Psi}^T}{\partial x_j} \, d\mathbf{x} \; ; \quad \mathbf{Q}_i = \int_{\Omega^e} \frac{\partial \boldsymbol{\Psi}}{\partial x_i} \boldsymbol{\Phi}^T \, d\mathbf{x}
$$

$$
\mathbf{F}_i(\mathbf{T}) = -\int_{\Omega^e} \rho_0 g_i \beta \boldsymbol{\Psi} \boldsymbol{\Psi}^T \mathbf{T} \, d\mathbf{x} + \int_{\Omega^e} \rho_0 g_i \beta \boldsymbol{\Psi} T_0 \, d\mathbf{x} + \oint_{\Gamma^e} \boldsymbol{\Psi} T_i \, ds
$$

$$
\mathbf{D(u)} = \int_{\Omega^e} \rho_0 C \boldsymbol{\Psi} (\boldsymbol{\Psi}^T \mathbf{u}_j) \frac{\partial \boldsymbol{\Psi}^T}{\partial x_j} \, d\mathbf{x}
$$

$$
\mathbf{N} = \int_{\Omega^e} \rho_0 C_v \boldsymbol{\Theta} \boldsymbol{\Theta}^T \, d\mathbf{x} \; ; \quad \mathbf{L} = \int_{\Omega^e} k \frac{\partial \boldsymbol{\Theta}}{\partial x_i} \frac{\partial \boldsymbol{\Theta}^T}{\partial x_i} \, d\mathbf{x}
$$

$$
\mathbf{G} = \int_{\Omega^e} \boldsymbol{\Theta} Q \, d\mathbf{x} + \int_{\Omega^e} \boldsymbol{\Theta} \Phi \, d\mathbf{x} + \oint_{\Gamma^e} \boldsymbol{\Theta} q \, ds \tag{5.2.9b}
$$

where summation on repeated indices is implied. The above equations can be written symbolically as

Continuity:

$$
-\mathbf{Q}^T \mathbf{u} = 0 \tag{5.2.10}
$$

Momentum:

$$
\mathbf{M}\dot{\mathbf{u}} + \mathbf{C}\mathbf{u} + \mathbf{K}\mathbf{u} - \mathbf{Q}\mathbf{P} + \mathbf{B}\mathbf{T} = \mathbf{F} \tag{5.2.11}
$$

Energy:

$$
\mathbf{N}\dot{\mathbf{T}} + \mathbf{D}\mathbf{T} + \mathbf{L}\mathbf{T} = \mathbf{G} \tag{5.2.12}
$$

where the superposed dot represents a time derivative and $\mathbf{u}^T = \{\mathbf{u}_1^T, \mathbf{u}_2^T, \mathbf{u}_3^T\}$. In writing Eq. (5.2.11), the buoyancy term $\mathbf{B}$ has been separated from the general force vector $\mathbf{F}$. The expression $\mathbf{BT}$ has the meaning

$$
\mathbf{BT} = \begin{bmatrix} \mathbf{B}_1 & 0 & 0 \\ 0 & \mathbf{B}_2 & 0 \\ 0 & 0 & \mathbf{B}_3 \end{bmatrix} \begin{Bmatrix} \mathbf{T} \\ \mathbf{T} \\ \mathbf{T} \end{Bmatrix} \tag{5.2.13a}
$$

where

$$\mathbf{B}_i = \int_{\Omega^e} \rho_0 \beta g_i \mathbf{\Psi}^T \mathbf{\Theta} \, d\mathbf{x} \tag{5.2.13b}$$

Note that the solid body conduction equation of Chapter 3 is obtained from (5.2.12) by neglecting the transport term $\mathbf{D}$.

Equations (5.2.10)–(5.2.12) can be combined into a single matrix equation

$$\begin{bmatrix} \mathbf{M} & 0 & 0 \\ 0 & 0 & 0 \\ 0 & 0 & \mathbf{N} \end{bmatrix} \begin{Bmatrix} \dot{\mathbf{u}} \\ \dot{\mathbf{P}} \\ \dot{\mathbf{T}} \end{Bmatrix} + \begin{bmatrix} \mathbf{C(u)} + \mathbf{K(u,T)} & -\mathbf{Q} & \mathbf{B(T)} \\ -\mathbf{Q}^T & 0 & 0 \\ 0 & 0 & \mathbf{D(u)} + \mathbf{L(T)} \end{bmatrix} \begin{Bmatrix} \mathbf{u} \\ \mathbf{P} \\ \mathbf{T} \end{Bmatrix}$$

$$= \begin{Bmatrix} \mathbf{F(T)} \\ 0 \\ \mathbf{G(T,u)} \end{Bmatrix} \tag{5.2.14}$$

In a more symbolic format, Eq. (5.2.14) can be expressed as

$$\bar{\mathbf{M}}\dot{\mathbf{U}} + \bar{\mathbf{K}}\mathbf{U} = \bar{\mathbf{F}} \tag{5.2.15a}$$

where

$$\mathbf{U}^T = \{\mathbf{u}_1^T, \mathbf{u}_2^T, \mathbf{u}_3^T, \mathbf{P}^T, \mathbf{T}^T\} \tag{5.2.15b}$$

Note that the general form of Eq. (5.2.15a) is the same as the nonlinear diffusion Eq. (3.6.3). Therefore, the time approximations discussed in Chapter 3 are readily applicable to the ordinary differential equations in (5.2.15a).

The three sets of interpolation functions used in (5.2.2) should be of the Lagrange type, i.e., derived by interpolating only the values of the functions, and not their derivatives. There are three different finite elements associated with the three field variables (u_i, P, T). Although, in principle, one can use different interpolation functions for the velocity field and temperature, it is practical to use the same type of interpolation for the two field variables (T, u_i). The choice of interpolation functions used for the pressure variable (see Section 4.5.2) in the mixed finite element model is constrained by the special role the pressure plays in incompressible flows. In order to prevent an overconstrained system of discrete equations, the interpolation used for pressure must be at least one order lower than that used for the velocity field (i.e., unequal interpolation). Further, pressure need not be made continuous across elements because the pressure variable does not constitute a primary variable of the mixed model.

For two-dimensional flows of viscous incompressible fluids, often the Lagrange quadratic interpolation is used for the velocity components. Two different pressure approximations are available when the velocities are approximated by quadratic Lagrange functions. The first is a continuous-bilinear approximation, in which the pressure is defined at the corner nodes of the element and is made continuous across element boundaries. The second pressure approximation involves a discontinuous, linear variation defined on the element by

$$\Phi = \left\{ \begin{array}{c} 1 \\ x \\ y \end{array} \right\} \qquad (5.2.16)$$

Here the unknowns are not nodal point values of the pressure but correspond to the coefficients in $P = a \cdot 1 + b \cdot x + c \cdot y$. A standard element used in the analysis of three-dimensional viscous flow problems is the eight-node, hexahedron (brick) element, where the velocity components are approximated using the trilinear Lagrange functions and pressure is a constant (and obviously discontinuous between elements).

5.3 Penalty Finite Element Model

5.3.1 Preliminary Comments

Recall that the velocity field of viscous incompressible flows must satisfy the momentum equations (5.1.3) and in addition the continuity equation (5.1.2). Equation (5.1.2) is treated as a constraint on the velocity field. In the penalty function method, the constrained problem is reformulated as an unconstrained problem. The discussion presented in Section 4.3 is applicable here with the body force components f_i in equations (4.3.16) replaced with the buoyancy forces, $g_i \beta (T - T_0)$. The weak statements associated with the penalty formulation of convective heat transfer are as follows:

Continuity and Momentum:

$$0 = \int_{\Omega^e} \rho_0 \, \delta u_1 \left[\frac{\partial u_1}{\partial t} + \left(u_1 \frac{\partial u_1}{\partial x_1} + u_2 \frac{\partial u_1}{\partial x_2} + u_3 \frac{\partial u_1}{\partial x_3} \right) - g_1 \beta (T - T_0) \right] d\mathbf{x}$$
$$+ \int_{\Omega^e} \left[2\mu \frac{\partial \delta u_1}{\partial x_1} \frac{\partial u_1}{\partial x_1} + \mu \frac{\partial \delta u_1}{\partial x_2} \left(\frac{\partial u_1}{\partial x_2} + \frac{\partial u_2}{\partial x_1} \right) + \mu \frac{\partial \delta u_1}{\partial x_3} \left(\frac{\partial u_1}{\partial x_3} + \frac{\partial u_3}{\partial x_1} \right) \right] d\mathbf{x}$$
$$- \oint_{\Gamma^e} \delta u_1 \, T_1 \, ds + \int_{\Omega^e} \gamma_e \frac{\partial \delta u_1}{\partial x_1} \left(\frac{\partial u_1}{\partial x_1} + \frac{\partial u_2}{\partial x_2} + \frac{\partial u_3}{\partial x_3} \right) d\mathbf{x} \qquad (5.3.1a)$$

$$0 = \int_{\Omega^e} \rho_0 \, \delta u_2 \left[\frac{\partial u_2}{\partial t} + \left(u_1 \frac{\partial u_2}{\partial x_1} + u_2 \frac{\partial u_2}{\partial x_2} + u_3 \frac{\partial u_2}{\partial x_3} \right) - g_2 \beta (T - T_0) \right] d\mathbf{x}$$
$$+ \int_{\Omega^e} \left[2\mu \frac{\partial \delta u_2}{\partial x_2} \frac{\partial u_2}{\partial x_2} + \mu \frac{\partial \delta u_2}{\partial x_1} \left(\frac{\partial u_1}{\partial x_2} + \frac{\partial u_2}{\partial x_1} \right) + \mu \frac{\partial \delta u_2}{\partial x_3} \left(\frac{\partial u_2}{\partial x_3} + \frac{\partial u_3}{\partial x_2} \right) \right] d\mathbf{x}$$
$$- \oint_{\Gamma^e} \delta u_2 \, T_2 \, ds + \int_{\Omega^e} \gamma_e \frac{\partial \delta u_2}{\partial x_2} \left(\frac{\partial u_1}{\partial x_1} + \frac{\partial u_2}{\partial x_2} + \frac{\partial u_3}{\partial x_3} \right) d\mathbf{x} \qquad (5.3.1b)$$

$$0 = \int_{\Omega^e} \rho_0 \, \delta u_3 \left[\frac{\partial u_3}{\partial t} + \left(u_1 \frac{\partial u_3}{\partial x_1} + u_2 \frac{\partial u_3}{\partial x_2} + u_3 \frac{\partial u_3}{\partial x_3} \right) - g_3 \beta (T - T_0) \right] d\mathbf{x}$$
$$+ \int_{\Omega^e} \left[2\mu \frac{\partial \delta u_3}{\partial x_3} \frac{\partial u_3}{\partial x_3} + \mu \frac{\partial \delta u_3}{\partial x_1} \left(\frac{\partial u_1}{\partial x_3} + \frac{\partial u_3}{\partial x_1} \right) + \mu \frac{\partial \delta u_3}{\partial x_2} \left(\frac{\partial u_2}{\partial x_3} + \frac{\partial u_3}{\partial x_2} \right) \right] d\mathbf{x}$$
$$- \oint_{\Gamma^e} \delta u_3 \, T_3 \, ds + \int_{\Omega^e} \gamma_e \frac{\partial \delta u_3}{\partial x_3} \left(\frac{\partial u_1}{\partial x_1} + \frac{\partial u_2}{\partial x_2} + \frac{\partial u_3}{\partial x_3} \right) d\mathbf{x} \qquad (5.3.1c)$$

Energy:

$$0 = \int_{\Omega^e} \rho_0 C_v \, \delta T \left[\frac{\partial T}{\partial t} + \left(u_1 \frac{\partial T}{\partial x_1} + u_2 \frac{\partial T}{\partial x_2} + u_3 \frac{\partial T}{\partial x_3} \right) - Q \right] d\mathbf{x}$$
$$+ \int_{\Omega^e} k_{ij} \frac{\partial \delta T}{\partial x_i} \frac{\partial T}{\partial x_j} d\mathbf{x} - \oint_{\Gamma^e} \delta T \, q \, ds \qquad (5.3.2)$$

where γ_e is the penalty parameter. We note that the pressure does not appear explicitly in the weak forms (5.3.1) and (5.3.2), although it is a part of the boundary stresses, $\mathcal{T}_i$ [see Eqs. (5.1.5b) and (5.1.7)]. An approximation for the pressure can be post-computed from the relation [see Eq. (4.4.21)]

$$P = -\gamma_e \left(\frac{\partial u_1}{\partial x_1} + \frac{\partial u_2}{\partial x_2} + \frac{\partial u_3}{\partial x_3} \right) = -\gamma_e \frac{\partial u_i}{\partial x_i} \qquad (5.3.3)$$

5.3.2 Reduced Integration Penalty Model

The penalty finite element model is obtained from Eqs. (5.3.1a–c) by substituting finite element interpolation (5.2.2a) for the velocity field and $\delta u_i = \mathbf{\Psi}^T$:

$$\begin{bmatrix} \mathbf{M} & 0 & 0 \\ 0 & \mathbf{M} & 0 \\ 0 & 0 & \mathbf{M} \end{bmatrix} \begin{Bmatrix} \dot{\mathbf{u}}_1 \\ \dot{\mathbf{u}}_2 \\ \dot{\mathbf{u}}_3 \end{Bmatrix} + \begin{bmatrix} \mathbf{C(u)} & 0 & 0 \\ 0 & \mathbf{C(u)} & 0 \\ 0 & 0 & \mathbf{C(u)} \end{bmatrix} \begin{Bmatrix} \mathbf{u}_1 \\ \mathbf{u}_2 \\ \mathbf{u}_3 \end{Bmatrix} +$$

$$\begin{bmatrix} 2\mathbf{K}_{11} + \mathbf{K}_{22} + \mathbf{K}_{33} & \mathbf{K}_{21} & \mathbf{K}_{31} \\ \mathbf{K}_{12} & \mathbf{K}_{11} + 2\mathbf{K}_{22} + \mathbf{K}_{33} & \mathbf{K}_{32} \\ \mathbf{K}_{13} & \mathbf{K}_{23} & \mathbf{K}_{11} + \mathbf{K}_{22} + 2\mathbf{K}_{33} \end{bmatrix} \begin{Bmatrix} \mathbf{u}_1 \\ \mathbf{u}_2 \\ \mathbf{u}_3 \end{Bmatrix}$$

$$+ \begin{bmatrix} \hat{\mathbf{K}}_{11} & \hat{\mathbf{K}}_{12} & \hat{\mathbf{K}}_{13} \\ \hat{\mathbf{K}}_{21} & \hat{\mathbf{K}}_{22} & \hat{\mathbf{K}}_{23} \\ \hat{\mathbf{K}}_{31} & \hat{\mathbf{K}}_{32} & \hat{\mathbf{K}}_{33} \end{bmatrix} \begin{Bmatrix} \mathbf{u}_1 \\ \mathbf{u}_2 \\ \mathbf{u}_3 \end{Bmatrix} = \begin{Bmatrix} \mathbf{F}_1 \\ \mathbf{F}_2 \\ \mathbf{F}_3 \end{Bmatrix} \qquad (5.3.4)$$

where $\mathbf{M}, \mathbf{C(u)}, \mathbf{K}_{ij}$, and $\mathbf{F}_i$ are the same as those defined in Eq. (5.2.9), and

$$\hat{\mathbf{K}}_{ij} = \int_{\Omega^e} \gamma_e \frac{\partial \mathbf{\Psi}}{\partial x_i} \frac{\partial \mathbf{\Psi}^T}{\partial x_j} d\mathbf{x} \qquad (5.3.5)$$

The finite element model of the energy equation (5.2.12) remains unchanged.

The comments made in Section 4.5.3 concerning the evaluation of the integrals in $\hat{\mathbf{K}}$ of Eq. (5.3.4) should be recalled here. These coefficients (i.e., penalty terms) should be evaluated using a numerical integration rule of an order less than that required to integrate them exactly. For example, if a linear rectangular element is used to approximate the velocity field for two-dimensional problems, all matrix coefficients except the penalty terms should be evaluated using the 2×2 Gauss quadrature, and the penalty terms should be evaluated using the one-point (1×1) Gauss quadrature. When a quadratic rectangular element is used, the 3×3 Gauss quadrature is used to evaluate the non-penalty terms, and the 2×2 Gauss quadrature is used to evaluate the penalty terms. Similar comments apply to three-dimensional elements.

Concerning the post-computation of pressure in the penalty model, in general, the pressure computed from Eq. (5.3.3) at the integration points is not always reliable and accurate, and one is advised to use either the Poisson equation for pressure [see Eq. (4.5.11)] or the scheme suggested in Section 4.5.4 [see Eq. (4.5.1)].

5.3.3 Consistent Penalty Model

In this model, the finite element model of Eq. (5.3.3) is constructed first. We have

$$\left[\int_{\Omega^e} \mathbf{\Phi}\mathbf{\Phi}^T \, d\mathbf{x}\right] \mathbf{P} = -\left[\int_{\Omega^e} \gamma_e \mathbf{\Phi} \frac{\partial \mathbf{\Psi}^T}{\partial x_i} \, d\mathbf{x}\right] \mathbf{u}_i \qquad (5.3.6a)$$

or in matrix notation

$$\mathbf{M}_p \mathbf{P} = -\gamma_e \mathbf{Q}^T \mathbf{u} \qquad (5.3.6b)$$

Inverting Eq. (5.3.6b) for $\mathbf{P}$, we obtain

$$\mathbf{P} = -\gamma_e \mathbf{M}_p^{-1} \mathbf{Q}^T \mathbf{u} \qquad (5.3.7)$$

where $\mathbf{u}^T = \{\mathbf{u}_1^T, \mathbf{u}_2^T, \mathbf{u}_3^T\}$, $\mathbf{Q}^T = \{\mathbf{Q}_1^T, \mathbf{Q}_2^T, \mathbf{Q}_3^T\}$, and $\mathbf{Q}_i$ and $\mathbf{M}_p$ are given by

$$\mathbf{Q}_i = \int_{\Omega^e} \frac{\partial \mathbf{\Psi}}{\partial x_i} \mathbf{\Phi}^T \, d\mathbf{x} \; ; \quad \mathbf{M}_p = \int_{\Omega^e} \mathbf{\Phi}\mathbf{\Phi}^T \, d\mathbf{x} \qquad (5.3.8)$$

Next, Eq. (5.3.7) is substituted for the pressure into the finite element model (5.2.11) to obtain

$$\mathbf{M}\dot{\mathbf{u}} + (\mathbf{C}(\mathbf{u}) + \mathbf{K} + \mathbf{K}_p)\mathbf{u} = \mathbf{F} \qquad (5.3.9a)$$

with

$$\mathbf{K}_p = \gamma_e \mathbf{Q} \mathbf{M}_p^{-1} \mathbf{Q}^T \qquad (5.3.9b)$$

It should be noted that reduced integration is not used to evaluate $\mathbf{K}_p$. For straight-sided elements, one can show the equivalence of matrix $\hat{\mathbf{K}}$ in Eq. (5.3.4) and the matrix $\mathbf{K}_p$. The post-computation of the pressure P using Eq. (5.3.7) in the consistent penalty method yields very good results.

5.4 Finite Element Models of Porous Flow

The derivation of the finite element model for the porous flow problem follows the developments of Section 4.4 in a completely analogous manner. For completeness, we record the mixed finite element model including the buoyancy term:

Continuity:

$$-\left[\int_{\Omega^e} \mathbf{\Phi} \frac{\partial \mathbf{\Psi}^T}{\partial x_i} \, d\mathbf{x}\right] \mathbf{u}_i = 0 \qquad (5.4.1)$$

Momentum:

$$\left[\int_{\Omega^e} \frac{\rho_0}{\phi} \boldsymbol{\Psi}\boldsymbol{\Psi}^T \, dx\right]\dot{\mathbf{u}}_i + \left[\int_{\Omega^e} \frac{\rho_0 \hat{c}}{\sqrt{\kappa}} \boldsymbol{\Psi}(\boldsymbol{\Psi}^T\|\mathbf{u}\|)\boldsymbol{\Psi}^T \, dx\right]\mathbf{u}_i + \left[\int_{\Omega^e} \frac{\mu}{\kappa} \boldsymbol{\Psi}\boldsymbol{\Psi}^T \, dx\right]\dot{\mathbf{u}}_i$$

$$+ \left[\int_{\Omega^e} \mu_e \frac{\partial\boldsymbol{\Psi}}{\partial x_j}\frac{\partial\boldsymbol{\Psi}^T}{\partial x_j} \, dx\right]\mathbf{u}_i + \left[\int_{\Omega^e} \mu_e \frac{\partial\boldsymbol{\Psi}}{\partial x_j}\frac{\partial\boldsymbol{\Psi}^T}{\partial x_i} \, dx\right]\mathbf{u}_j - \left[\int_{\Omega^e} \frac{\partial\boldsymbol{\Psi}}{\partial x_i}\boldsymbol{\Phi}^T \, dx\right]\mathbf{P} =$$

$$- \left[\int_{\Omega^e} \rho_0 g_i \beta \boldsymbol{\Psi}\boldsymbol{\Theta} dx\right]\mathbf{T} + \left\{\int_{\Omega^e} \rho_0 g_i \beta \boldsymbol{\Psi} T_0 \, dx\right\} + \left\{\oint_{\Gamma^e} \boldsymbol{\Psi} T_i ds\right\} \qquad (5.4.2)$$

Energy:

$$\left[\int_{\Omega^e} (\rho C)_e \boldsymbol{\Theta}\boldsymbol{\Theta}^T \, dx\right]\dot{\mathbf{T}} + \left[\int_{\Omega^e} \rho_0 C \boldsymbol{\Theta}(\boldsymbol{\Psi}^T \mathbf{u_j})\frac{\partial\boldsymbol{\Theta}^T}{\partial x_j} \, dx\right]\mathbf{T} +$$

$$\left[\int_{\Omega^e} k_e \frac{\partial\boldsymbol{\Theta}}{\partial x_i}\frac{\partial\boldsymbol{\Theta}^T}{\partial x_i} \, dx\right]\mathbf{T} = \left\{\int_{\Omega^e} \boldsymbol{\Theta} Q \, dx\right\} + \left\{\oint_{\Gamma^e} \boldsymbol{\Theta} q_n \, ds\right\} \qquad (5.4.3)$$

Note that the dissipation term is omitted from the energy equation, and the acceleration tensor is taken to be $1/\phi$. The above equations can be written in matrix form as

Continuity:

$$-\tilde{\mathbf{Q}}^T\mathbf{u} = 0 \qquad (5.4.4)$$

Momentum:

$$\tilde{\mathbf{M}}\dot{\mathbf{u}} + \tilde{\mathbf{C}}\mathbf{u} + \tilde{\mathbf{K}}\mathbf{u} - \tilde{\mathbf{Q}}\mathbf{P} + \tilde{\mathbf{B}}\mathbf{T} = \tilde{\mathbf{F}} \qquad (5.4.5)$$

Energy:

$$\tilde{\mathbf{N}}\dot{\mathbf{T}} + \tilde{\mathbf{D}}\mathbf{T} + \tilde{\mathbf{L}}\mathbf{T} = \tilde{\mathbf{G}} \qquad (5.4.6)$$

where $\mathbf{u}^T = \{\mathbf{u}_1^T, \mathbf{u}_2^T, \mathbf{u}_3^T\}$ and

$$\tilde{\mathbf{M}} = \int_{\Omega^e} \frac{\rho_0}{\phi} \boldsymbol{\Psi}\boldsymbol{\Psi}^T \, dx; \quad \tilde{\mathbf{C}}(\mathbf{u}) = \int_{\Omega^e} \frac{\rho_0 \hat{c}}{\sqrt{\kappa}} \boldsymbol{\Psi}\boldsymbol{\Psi}^T\|\mathbf{u}\|\boldsymbol{\Psi}^T \, dx$$

$$\tilde{\mathbf{A}} = \int_{\Omega^e} \frac{\mu}{\kappa} \boldsymbol{\Psi}\boldsymbol{\Psi}^T \, dx; \quad \tilde{\mathbf{K}}_{ij} = \int_{\Omega^e} \mu_e \frac{\partial\boldsymbol{\Psi}}{\partial x_i}\frac{\partial\boldsymbol{\Psi}^T}{\partial x_j} \, dx; \quad \tilde{\mathbf{Q}}_i = \int_{\Omega^e} \frac{\partial\boldsymbol{\Psi}}{\partial x_i}\boldsymbol{\Phi}^T \, dx$$

$$\tilde{\mathbf{F}}_i(\mathbf{T}) = -\int_{\Omega^e} \rho_0 g_i \beta \boldsymbol{\Psi}\boldsymbol{\Theta} dx + \int_{\Omega^e} \rho_0 g_i \beta \boldsymbol{\Psi} T_0 \, dx + \oint_{\Gamma^e} \boldsymbol{\Psi} T_i ds$$

$$\tilde{\mathbf{D}}(\mathbf{u}) = \int_{\Omega^e} \rho_0 C \boldsymbol{\Theta}(\boldsymbol{\Psi}^T \mathbf{u_j})\frac{\partial\boldsymbol{\Theta}^T}{\partial x_j} \, dx$$

$$\tilde{\mathbf{N}} = \int_{\Omega^e} (\rho C)_e \boldsymbol{\Theta}\boldsymbol{\Theta}^T \, dx; \quad \tilde{\mathbf{L}} = \int_{\Omega^e} k_e \frac{\partial\boldsymbol{\Theta}}{\partial x_i}\frac{\partial\boldsymbol{\Theta}^T}{\partial x_i} \, dx$$

$$\tilde{\mathbf{G}} = \int_{\Omega^e} \boldsymbol{\Theta} Q \, dx + \oint_{\Gamma^e} \boldsymbol{\Theta} q_n \, ds \qquad (5.4.7)$$

where summation on repeated indices is implied. The porous flow equations (5.4.5)–(5.4.7) can be arranged into a single matrix equation of the same form as given previously for the flow equations:

$$
\begin{bmatrix} \tilde{M} & 0 & 0 \\ 0 & 0 & 0 \\ 0 & 0 & \tilde{N} \end{bmatrix} \begin{Bmatrix} \dot{u} \\ \dot{P} \\ \dot{T} \end{Bmatrix} + \begin{bmatrix} \tilde{C}(u) + \tilde{A} + \tilde{K} & -\tilde{Q} & \tilde{B}(T) \\ -\tilde{Q}^T & 0 & 0 \\ 0 & 0 & \tilde{D}(u) + \tilde{L}(T) \end{bmatrix} \begin{Bmatrix} u \\ P \\ T \end{Bmatrix}
$$

$$
= \begin{Bmatrix} \tilde{F}(T) \\ 0 \\ \tilde{G}(T, u) \end{Bmatrix} \tag{5.4.8}
$$

5.5 Solution Methods

5.5.1 General Discussion

The finite element models developed in Sections 5.2–5.4 describe a very large group of heat transfer and fluid mechanics problems ranging in complexity from simple solid body conduction through isothermal flows, forced and free convection, to mixed mode convection/conduction. The solution methods used depend on the model, computational resources, the nonlinearity of the system, and the strength of the coupling between equations (see Section 4.6).

The solution procedures described in Section 4.6 were all variants of the basic fixed point method (see Appendix C) and were generally designed to operate concurrently on all of the equations present in a finite element model. The strong coupling between equations (e.g., momentum and energy) that is characteristic of convective heat transfer problems makes these combined equation methods optimal from the standpoint of convergence rate. The disadvantage, of course, is that a very large and computationally expensive matrix problem must be treated at each iteration. The requirement to perform larger (more elements and higher dimensionality) and more complex (physical phenomena) simulations has reached the point where the usual direct matrix methods for combined equations are prohibitively expensive. A natural choice to make the matrix problem more affordable (while retaining the standard fixed point schemes) is to switch from the Gauss elimination type methods to the iterative methods, such as the preconditioned conjugate gradient (PCG) method. Unfortunately, the development of iterative methods for combined equation sets has been severely handicapped by the lack of good preconditioner techniques that can adequately treat the dominating effect of the incompressibility constraint. To make progress with current iterative methods, the combined equation approach must be sacrificed for alternative formulations of the discrete equations.

In the following sections we first review Newton's method for convective heat transfer problems, a combined equation technique that is heavily used for two-dimensional problems and some smaller three-dimensional applications. Various segregated solution methods are then reviewed. Throughout this section we will only consider the time-independent equations. From the discussion in Section 4.6 it should be recognized that any of the standard time integration procedures can be applied to the convective heat transfer equations. The resulting nonlinear algebraic equations for each time step are then analogous to the steady state problem, and they can be treated with the algorithms described in the following sections.

5.5.2 Newton's Method

The application of Newton's method to the combined equations for convective heat transfer (5.2.14) or (5.4.8) follows directly from the isothermal case described in Section 4.6.3. The iterative procedure is given by

$$\mathbf{U}^{n+1} = \mathbf{U}^n - \mathcal{J}^{-1}(\mathbf{U}^n)\mathbf{R}(\mathbf{U}^n) \tag{5.5.1}$$

where the vector of unknowns is $\mathbf{U}^T = \{\mathbf{u}_1^T, \mathbf{u}_2^T, \mathbf{u}_3^T, \mathbf{P}^T, \mathbf{T}^T\}$ for the mixed finite element model and $\mathbf{U}^T = \{\mathbf{u}_1^T, \mathbf{u}_2^T, \mathbf{u}_3^T, \mathbf{T}^T\}$ for the penalty finite element model. The definition of the Jacobian matrix $\mathcal{J}$ is expanded to include the variations in the equations with respect to the vector of nodal temperatures, $\mathbf{T}$. Thus for the two-dimensional mixed finite element model we have

$$\mathcal{J} = \frac{\partial \mathbf{R}}{\partial \mathbf{U}} = \begin{bmatrix} \frac{\partial \mathbf{R}_1}{\partial \mathbf{u}_1} & \frac{\partial \mathbf{R}_1}{\partial \mathbf{u}_2} & \frac{\partial \mathbf{R}_1}{\partial \mathbf{P}} & \frac{\partial \mathbf{R}_1}{\partial \mathbf{T}} \\[2mm] \frac{\partial \mathbf{R}_2}{\partial \mathbf{u}_1} & \frac{\partial \mathbf{R}_2}{\partial \mathbf{u}_2} & \frac{\partial \mathbf{R}_2}{\partial \mathbf{P}} & \frac{\partial \mathbf{R}_2}{\partial \mathbf{T}} \\[2mm] \frac{\partial \mathbf{R}_3}{\partial \mathbf{u}_1} & \frac{\partial \mathbf{R}_3}{\partial \mathbf{u}_2} & \frac{\partial \mathbf{R}_3}{\partial \mathbf{P}} & \frac{\partial \mathbf{R}_3}{\partial \mathbf{T}} \\[2mm] \frac{\partial \mathbf{R}_4}{\partial \mathbf{u}_1} & \frac{\partial \mathbf{R}_4}{\partial \mathbf{u}_2} & \frac{\partial \mathbf{R}_4}{\partial \mathbf{P}} & \frac{\partial \mathbf{R}_4}{\partial \mathbf{T}} \end{bmatrix} \tag{5.5.2}$$

where

$$\mathbf{R}_1 = \mathbf{C}_1(\mathbf{u}_1)\mathbf{u}_1 + \mathbf{C}_2(\mathbf{u}_2)\mathbf{u}_1 + (2\mathbf{K}_{11} + \mathbf{K}_{22})\,\mathbf{u}_1 + \mathbf{K}_{12}\mathbf{u}_2 - \mathbf{Q}_1\mathbf{P} + \mathbf{B}_1\mathbf{T} - \mathbf{F}_1$$
$$\mathbf{R}_2 = \mathbf{C}_1(\mathbf{u}_1)\mathbf{u}_2 + \mathbf{C}_2(\mathbf{u}_2)\mathbf{u}_2 + \mathbf{K}_{21}\mathbf{u}_1 + (\mathbf{K}_{11} + 2\mathbf{K}_{22})\mathbf{u}_2 - \mathbf{Q}_2\mathbf{P} + \mathbf{B}_2\mathbf{T} - \mathbf{F}_2$$
$$\mathbf{R}_3 = -\mathbf{Q}_1^T\mathbf{u}_1 - \mathbf{Q}_2^T\mathbf{u}_2$$
$$\mathbf{R}_4 = \mathbf{D}_1(\mathbf{u}_1)T + \mathbf{D}_2(\mathbf{u}_2)\mathbf{T} + (\mathbf{L}_{11} + \mathbf{L}_{22})\,\mathbf{T} - \mathbf{G} \tag{5.5.3}$$

which correspond to the two momentum equations, the incompressibility constraint, and the energy equation. The matrices appearing in Eq. (5.5.3) are defined in Eqs. (5.2.9a,b). The components of $\mathcal{J}$ are computed directly from the definitions in (5.5.3) and yield the following values:

$$\frac{\partial \mathbf{R}_1}{\partial \mathbf{u}_1} = \mathbf{C}_1(\mathbf{u}_1) + \mathbf{C}_2(\mathbf{u}_2) + \mathbf{C}_1(1)\mathbf{u}_1 + 2\mathbf{K}_{11} + \mathbf{K}_{22}$$

$$\frac{\partial \mathbf{R}_1}{\partial \mathbf{u}_2} = \mathbf{C}_2(1)\mathbf{u}_1 + \mathbf{K}_{21} \;; \quad \frac{\partial \mathbf{R}_1}{\partial \mathbf{P}} = -\mathbf{Q}_1 \;; \quad \frac{\partial \mathbf{R}_1}{\partial \mathbf{T}} = \mathbf{B}_1$$

$$\frac{\partial \mathbf{R}_2}{\partial \mathbf{u}_1} = \mathbf{C}_1(1)\mathbf{u}_2 + \mathbf{K}_{12} \;; \quad \frac{\partial \mathbf{R}_2}{\partial \mathbf{P}} = -\mathbf{Q}_2$$

$$\frac{\partial \mathbf{R}_2}{\partial \mathbf{u}_2} = \mathbf{C}_1(\mathbf{u}_1) + \mathbf{C}_2(\mathbf{u}_2) + \mathbf{C}_2(1)\mathbf{u}_2 + \mathbf{K}_{11} + 2\mathbf{K}_{22}$$

$$\frac{\partial \mathbf{R}_2}{\partial \mathbf{T}} = \mathbf{B}_2 \ ; \quad \frac{\partial \mathbf{R}_3}{\partial \mathbf{u}_1} = -\mathbf{Q}_1^T \ ; \quad \frac{\partial \mathbf{R}_3}{\partial \mathbf{u}_2} = -\mathbf{Q}_2^T \ ; \quad \frac{\partial \mathbf{R}_3}{\partial \mathbf{P}} = 0$$

$$\frac{\partial \mathbf{R}_3}{\partial \mathbf{T}} = 0 \ ; \quad \frac{\partial \mathbf{R}_4}{\partial \mathbf{u}_1} = \mathbf{D}_1(1)\mathbf{T} \ ; \quad \frac{\partial \mathbf{R}_4}{\partial \mathbf{u}_2} = \mathbf{D}_2(1)\mathbf{T} \ ; \quad \frac{\partial \mathbf{R}_4}{\partial \mathbf{P}} = 0$$

$$\frac{\partial \mathbf{R}_4}{\partial \mathbf{T}} = \mathbf{D}_1(\mathbf{u}_1) + \mathbf{D}_2(\mathbf{u}_2) + \mathbf{L}_{11} + \mathbf{L}_{22} \qquad (5.5.4)$$

The three-dimensional formulation follows directly from the above definitions.

The advantages and disadvantages of Newton's method as outlined in Section 4.6.3 for the isothermal case hold equally for the convective heat transfer system. Also, the other variations of the fixed point schemes, such as Picard iteration, modified and quasi-Newton methods, and the various continuation techniques can be directly extended to the nonisothermal situation with comparable effectiveness.

5.5.3 Segregated Equation Methods

Segregated equation methods are not specifically solution algorithms but rather an approach to decomposing (5.2.14) into a series of smaller problems that can be solved efficiently using direct matrix methods or are more amenable to PCG type methods. For convective heat transfer problems (or other transport problems) a particularly simple segregated or cyclic procedure is given by the following for the $(n + 1)$ iteration step:

Step 1 (Energy Equation):

$$\mathbf{D}(\mathbf{u}^n, \mathbf{T}^n)\mathbf{T}^{n+1} + \mathbf{L}(\mathbf{T}^n)\mathbf{T}^{n+1} = \mathbf{G}(\mathbf{u}^n, \mathbf{T}^n) \qquad (5.5.5)$$

Step 2 (Continuity and Momentum Equations):

$$-\mathbf{Q}^T\mathbf{u}^{n+1} = 0 \qquad (5.5.6)$$

$$\mathbf{C}(\mathbf{u}^n)\mathbf{u}^{n+1} + \mathbf{K}(\mathbf{T}^{n+1})\mathbf{u}^{n+1} - \mathbf{Q}\mathbf{P}^{n+1} = \mathbf{F}(\mathbf{T}^{n+1}) \qquad (5.5.7)$$

The energy equation is typically solved first in each cycle, especially when considering free convection-dominated flows since the buoyancy term drives the fluid motion; the buoyancy term is included in $\mathbf{F}$ in Eq. (5.5.7).

The procedure given in Eqs. (5.5.5)–(5.5.7), known as the cyclic Picard method, has some advantages over a combined equation method in terms of computer storage and execution time per iteration. Unfortunately, as with most variants of successive substitution, the procedure has a relatively slow rate of convergence; the convergence rate is further reduced due to the lagging of the coefficients induced by the splitting of the equations. This method has been used successfully for convection problems of moderate nonlinearity by several investigators [2–5].

Though the cyclic procedure described above offers some reduction in computational requirements, the combined momentum and continuity equations, especially in three dimensions, can still represent a very large matrix problem. Following the idea of segregating equations to its logical extreme [6], the system in (5.2.14) can be decomposed into the following series of equations:

$$(C_1(u) + K_1) u_1 - Q_1 P = F_1 \tag{5.5.8a}$$
$$(C_2(u) + K_2) u_2 - Q_2 P = F_2 \tag{5.5.8b}$$
$$Q_1^T u_1 + Q_2^T u_2 = 0 \tag{5.5.8c}$$
$$(D(u) + L) T = G \tag{5.5.8d}$$

for the case of two-dimensional, nonisothermal flow; the three-dimensional case follows the same procedure as does the inclusion of additional transport equations. The (C, D) and (K, L) matrices represent the advection and diffusion operators, respectively. A segregated solution approach to Eq. (5.5.8) would solve each individual momentum equation, the continuity equation, and the energy equation for the appropriate variable in a cyclic procedure similar to the one shown in Eqs. (5.5.5)–(5.5.7). As each variable is updated from a new solution, it is used where appropriate in the coefficients of the remaining equations, resulting in another type of cyclic Picard scheme. By reducing (5.2.14) to the series of equations in (5.5.8), a significant reduction in size and cost of the matrix problem is realized. The penalty to be paid is the slower convergence rate of the cyclic procedure. However, for very large problems there is a cross-over point for these competing effects and the segregated approach is clearly superior to standard combined equation methods.

A significant complication in the use of Eq. (5.5.8) arises due to the fact that there is no equation in the set from which the pressure may be directly obtained. The solution to this dilemma involves the construction of a pressure Poisson equation. Solving Eqs. (5.5.8a,b) symbolically for the velocities u_1 and u_2 and substituting this result into the incompressibility condition (5.5.8c) produces a consistent Poisson equation for the pressure

$$\left[Q_1^T (C_1(u) + K_1)^{-1} Q_1 + Q_2^T (C_2(u) + K_2)^{-1} Q_2 \right] P = F \tag{5.5.9}$$

where the explicit form of F is not important in this discussion. Equation (5.8.9), when used in place of the continuity equation, provides a complete set of equations for all of the unknowns. A number of possible variations exist for the sequential solution of Eqs. (5.8.8) and (5.8.9), each of which produces a slightly different algorithm. In general, the Poisson equation (5.8.9) provides an initial estimate of the pressure field which is used in the pressure gradients of the momentum equations; velocity estimates are then obtained from each momentum equation in turn. Since the velocity field is not divergence free at this point, a second application of the Poisson equation is required; this solution supplies a pressure correction that allows the velocity fields to be adjusted to satisfy incompressibility. This general algorithm is analogous to the family of SIMPLE methods developed by Patankar for finite difference methods [7]. Details regarding several variants of the method can be found in the work of Haroutunian et al. [6], who were the original developers of this technique for finite element applications. Note that another significant benefit in the usage of the segregated approach with the pressure Poisson equation is the applicability of iterative matrix solution methods for the individual equations in the sequence. The removal of the incompressibility constraint allows useful preconditioners to be found for each equation in the system. The combination of preconditioned, conjugate gradient (PCG) matrix methods with a segregated equation approach provides the most effective solution method presently available for large scale applications.

5.6 Convection with Change of Phase

There are two major difficulties in simulating melting and solidification change of phase for convection heat transfer problems: (1) accounting for the latent heat effect, and (2) the application of fluid boundary conditions on the phase boundary. The latent heat problem has been detailed in the heat conduction section and its solution via the enthalpy method or any other method carries over directly to the convection problem. The boundary condition difficulty, however, is unique to the convection problem though it has some similarities with the free surface problems of Chapter 4.

As part of the formulation of the phase change problem in Section 1.9, it was noted that on the phase boundary the fluid velocity was assumed to be zero [see Eq. (1.9.1)]. This is a particularly difficult condition to apply in standard finite element formulations for two reasons. The location of the phase boundary is not known a priori, and even if it were known, the boundary does not generally coincide with a set of element edges (nodes) to which an essential boundary condition can be applied. One method to circumvent these difficulties is to explicitly track the motion of the phase boundary and allow the finite element mesh to adjust to this motion. Such mesh movement or remeshing schemes allow the phase boundary to always remain aligned with element edges and thus simplify the application of the no-slip boundary condition. However, these procedures are not without their shortcomings. They are expensive, complex to implement for general applications, and are generally limited in application to simple phase boundary motions and shapes. The moving mesh methods described in Section 4.10.3 are candidates for application in this type of phase boundary simulation.

A second, more approximate, method for imposing the no-slip boundary condition relies on the use of a temperature-dependent fluid viscosity and the allowed variation of material properties within each finite element. Depending on the type of variable coefficient implementation that is used in the construction of element matrices, a variable viscosity is evaluated at the numerical quadrature points or nodal points within an element (see Section 3.9 for details). During each cycle of the solution process the temperature at the viscosity evaluation point is compared with the solidus (liquidus) temperature of the material. If the temperature is above the liquidus, the appropriate fluid viscosity is assigned at the evaluation point; for temperatures below the solidus temperature, the evaluation point viscosity is assigned a "large" value to "solidify" the fluid. Typically, the "solid viscosity" is set at four to five orders of magnitude larger than the normal fluid viscosity; the viscosity function is usually designed to take on a realistic, or at least convenient, variation between the liquidus and solidus temperatures (i.e., through the "mushy" zone). Thermal properties, such as the conductivity, can also be allowed to vary in some physically realistic manner across the phase change zone.

The viscosity evaluation method proposed here allows the phase boundary computation to be maintained on a subelement basis since a given element may contain both fluid and solid phases. However, this procedure is also similar to a smearing technique since a distinct phase boundary is never identified. This is in keeping with the spirit and implementation of the latent heat procedures described previously. This approach is also somewhat like the front tracking, free surface methods of Section 4.10.3. In the present case the front tracking variable is a physical variable, the temperature, rather than an artificial concentration variable [see Eq. (4.10.15)] or a volume fraction [see Eq. (4.10.17)]. The utility of the proposed approach is demonstrated in a subsequent section and in [8].

For materials that change phase over a temperature range, the simple viscosity variation method provides a first-order evaluation of the fluid mechanics and heat transfer processes through the phase change region. In some cases, a more accurate evaluation of the fluid and thermal fields can be achieved. For metal alloys, in which an interdendritic region forms between the liquidus and solidus temperatures, the so-called "mushy" zone is often characterized as a porous media. Careful study of the Darcy-Brinkman porous flow model in (1.5.2) shows that the Navier-Stokes equations can be made a subset of this model if an advective term is added to the porous flow equations. This has been theoretically justified by a number of investigators [9–11] and leads to a momentum equation of the form

$$\frac{\rho_0}{\phi}\frac{\partial u_j}{\partial t} + \frac{\rho_0}{\phi^2}u_j\frac{\partial u_i}{\partial x_j} + \left(\frac{\mu}{\kappa}\right)u_i = \frac{\partial}{\partial x_j}\left[-P\delta_{ij} + \mu_e\left(\frac{\partial u_i}{\partial x_j} + \frac{\partial u_j}{\partial x_i}\right)\right]$$
$$+ \rho_0 f_i - \rho_0 g_i \beta(T - T_0) \qquad (5.6.1)$$

where the Forchheimer term has been replaced by the advective term and the Brinkman viscosity has been equivalenced to the fluid viscosity. Recall that ϕ is the porosity or liquid volume fraction. This equation can be used throughout the fluid domain and within the mushy zone by proper adjustment of the material properties and coefficients as a function of temperature. In the liquid region $(T > T_{liq})$, the porosity is unity and the permeability κ is infinite; Eq. (5.6.1) is the standard Navier-Stokes equation. When the temperature drops below the solidus temperature $(T < T_{sol})$, the porosity and permeability are zero and the fluid viscosity is again very large, resulting in a "solidified" material. Between the liquidus and solidus temperatures, the porosity goes from unity to zero as a function of temperature according to the particular model being employed [11]. A model for the permeability as a function of porosity transitions the permeability between a very large value and zero. The consequence of all these transitions is that the flow and heat transfer problem are smoothly moved from a viscous flow description to a porous media flow description to a solidified material description as a function of temperature. The model is substantially more detailed than the simple viscosity variation and is grounded in theory. Further details of the this approach and a demonstration of its capabilities are available in [9,11-13].

5.7 Convection with Enclosure Radiation

The inclusion of enclosure radiation effects in convective heat transfer problems involves the direct application of the methods from Section 3.8. By assumption, the fluid in the enclosure is nonparticipating and the radiative exchange therefore only influences the flux and temperature on the fluid boundaries. For purposes of computation the radiation problem is decoupled from the convection/conduction part of the problem and simply provides flux boundary conditions at each iteration or time step of the simulation.

5.8 Post-Computation of Heat Flux

The results of interest from a convective heat transfer analysis generally include fluid velocities, flow patterns, temperature distributions, and heat fluxes. Most of these items are directly available from the finite element results in terms of nodal point quantities. The heat flux (or flux of an auxiliary variable), however, being a derivative of the temperature, requires an auxiliary computation. Details of a method for performing this operation were given in Section 3.10. The simplest procedure for computing the components of the heat flux vector is based on the element level shape function for the temperature. That is, since

$$T(x_i, t) = \mathbf{\Theta}^T(x_i)\mathbf{T}(t) \tag{5.8.1}$$

then the flux components are

$$q_i(x_i, t) = -k\frac{\partial \mathbf{\Theta}^T}{\partial x_i}\mathbf{T}(t) \tag{5.8.2}$$

The relation in (5.8.2) is valid for any element and allows the flux to be evaluated at any time and at any point within the element. Note that in evaluating the derivative of $\mathbf{\Theta}$, the element geometry must be considered for mapped or parametric elements [see Eq. (3.3.3)]. Maximum accuracy in the shape function derivative is obtained by evaluating the derivative at the element numerical integration points (typically, the 2×2 Gauss points in a quadrilateral element) [14]. Since fluxes are normally required on element or domain boundaries, the integration point values are extrapolated to the nodes [15] and the discontinuous element level fluxes averaged between adjacent elements. Other evaluation points within an element have been employed, such as the points midway between nodes on the element boundary, or the Gauss points on the element boundary. All of these methods provide results of acceptable accuracy provided the computational mesh is sufficiently refined. Note that a common feature of these procedures is the need to compute a unit normal on an element boundary if a total or integrated flux is to be derived. Methods for determining unique and consistent normals to element boundaries are discussed by Engelman et al. [16].

A second, less familiar, approach for computing flux quantities relies on the consistent application of the weighted residual method. This idea was proposed by Larock and Herrmann [17] and Marshall et al. [18], and refined by Gresho et al. [19]. In essence, a weighted residual equation is derived that relates the boundary flux to local heat sources and computed temperature gradients. The consistent flux equation requires a matrix solution, which for many cases can be simplified by the use of a "lumping" procedure. Though somewhat more accurate than the simpler basis function derivative method, this procedure requires more computation, especially for the advection-diffusion equation. For details of this procedure, see [19].

It is important to recall that the above methods were described for the case of a conductive heat flux as would be computed for a solid boundary or a no flow-through surface. The heat flux computation for an open flow boundary would have to include terms describing the advective transport of energy. The appropriate flux description for this case would be

$$q_i(x_i, t) = \rho C_v u_i T - k \frac{\partial T}{\partial x_i} \tag{5.8.3}$$

and using the usual interpolations for velocity and temperature produces

$$q_i(x_i, t) = \rho C_v \mathbf{\Psi}^T \mathbf{u}_i(t) - k \frac{\partial \mathbf{\Theta}^T}{\partial x_i} \mathbf{T}(t) \tag{5.8.4}$$

the computational form for the total flux. Equation (5.8.4) would be evaluated in the same manner as the conductive flux in (5.8.2) with the same extrapolation and averaging techniques.

In Section 3.10.1, the heat flow function was defined for a conduction problem in two-dimensional geometries. The definition can be extended to convective transport though the restriction to two dimensions remains. Using the fluxes defined in (5.8.3), the heat function is defined by

$$q_x = \rho C_v u_x T - k \frac{\partial T}{\partial x} = \frac{\partial \mathcal{H}}{\partial y} \tag{5.8.5}$$

$$q_y = \rho C_v u_y T - k \frac{\partial T}{\partial y} = -\frac{\partial \mathcal{H}}{\partial x} \tag{5.8.6}$$

As before, the change in the heat function is an exact differential and

$$\delta \mathcal{H} = \int_A^B \mathbf{q} \cdot \mathbf{n} \, d\Gamma \tag{5.8.7}$$

where **n** is the normal to the integration path. The computation of $\mathcal{H}$ follows from the flux definition in terms of the interpolation functions and the definition of the normal for the element. The details were given in Section 3.10.1 for both the planar and axisymmetric cases and will not be repeated here. It is worth noting again that the integration of (5.8.7) around the boundary of an element should be zero; the departure from zero indicates the error in the element energy balance.

5.9 Advanced Topics - Turbulent Heat Transfer

The previous sections have been limited to consideration of laminar flows though the extension to turbulent flows is of considerable engineering importance. This section continues and extends the discussion of Section 4.11 on isothermal turbulence modeling to the nonisothermal case. Most of the ideas and turbulence models for isothermal flows have a direct counterpart when the energy equation is included in the formulation. Here we will outline some of the most important concepts for the case of forced convection heat transfer; the theory for strongly coupled, buoyancy driven flows is somewhat less developed.

Extending the assumption given in Eqs. (4.11.1) and (4.11.2) to the instantaneous temperature, we can write

$$\theta = \Theta + \theta' \tag{5.9.1}$$

where Θ is the average or mean temperature and the primed quantity is the fluctuation. When this definition and Eq. (4.11.1) are substituted into the instantaneous energy equation and ensemble averaged, the result is an equation of the form

$$\rho C \left(\frac{\partial \Theta}{\partial t} + U_j \frac{\partial \Theta}{\partial x_j} \right) = \frac{\partial}{\partial x_j} \left(k \frac{\partial \Theta}{\partial x_j} \right) - \rho C \, \overline{u_j' \theta'} \tag{5.9.2}$$

The averaging process has introduced the three new variables $\overline{u_j' \theta'}$ which are the components of the Reynolds heat flux vector. These quantities can be modeled in a variety of ways, all of which are analogous to the methods used for the Reynolds stresses. The current status of turbulent heat transfer modeling is reviewed by Launder [29]. Considering only one-point closure models (see Section 4.12), the Reynolds heat flux can be related to the mean temperature field through a Boussinesq type hypothesis. That is, let

$$-\rho C \, \overline{u_j' \theta'} = \frac{k_T}{C} \frac{\partial \Theta}{\partial x_j} = \frac{\mu_T}{\text{Pr}_T} \frac{\partial \Theta}{\partial x_j} \tag{5.9.3}$$

which can be used in (5.9.2) to produce

$$\rho C \left(\frac{\partial \Theta}{\partial t} + U_j \frac{\partial \Theta}{\partial x_j} \right) = \frac{\partial}{\partial x_j} \left[\left(k + \frac{\mu_T}{\text{Pr}_T} \right) \frac{\partial \Theta}{\partial x_j} \right] \tag{5.9.4}$$

where k_T is the eddy conductivity and Pr_T is the turbulent Prandtl number. Once the form of the eddy conductivity or the turbulent Prandtl number is specified, then the energy equation in (5.9.4) can be solved in conjunction with the flow equations.

Methods for specifying the eddy conductivity or turbulent Prandtl number parallel the ideas for the eddy viscosity specification. Of all the possible approaches, one of the most popular is the specification of Pr_T as an empirical relation. Reynolds [30] describes approximately thirty different formulas for Pr_T, all of which have a limited applicability. Clearly, the modeling of turbulent heat transfer processes remains an active area of research, though its progress is directly related to progress made in turbulent flow modeling.

5.10 Advanced Topics - Chemically Reacting Systems

5.10.1 Preliminary Comments

The equations describing a reactive material were presented in Section 1.7. The incorporation of chemical kinetics into solid body heat conduction problems was outlined in Section 3.11.5. The complexity of the problem, where general chemical reactions are included with a nonisothermal flow problem, places this topic outside the primary scope of this text. In the following section a brief discussion of reactive flow modeling is provided to simply indicate the scope of the modeling problem.

5.10.2 Finite Element Modeling of Chemical Reactions

The development of a finite element model for chemically reactive flows follows the usual prescription. The continuity, momentum, and energy equations for the mixture are standard and were considered in Chapter 4 and previous sections of the present chapter. The conservation equations for the species are of the advection-diffusion type and can therefore be treated by the same methods as used for the energy equation. Recall that the auxiliary equations defined in Chapter 1 could be used directly as species equations. The discretized form of the species equations provides an additional set of I matrix equations that can, in principle, be added to the matrix system for the basic nonisothermal flow problem.

Though the development of the reactive flow model appears to be straightforward, the solution of the resulting equations is extremely difficult. First, from the standpoint of problem size, the addition of multistep chemistry leads to a very rapid growth in the number of degrees of freedom considered in any given problem. However, problem size alone is not the most significant difficulty. The equation set is highly coupled and very nonlinear, making the convergence by standard solution methods unpredictable. More importantly, the length and time scales for the chemical processes tend to be significantly different (smaller) than the scales for the fluid and thermal response. This disparity in scales makes the use of fully coupled solution procedures impractical. In mathematical terms, the equation system becomes very stiff. The most effective solution algorithms allow the flow problem to run on its natural time scale, while the chemistry solution is decoupled and run as a subcycled process on a shorter time scale. Obviously, some of the rigorous coupling between variables is destroyed by this technique and variable coefficients must be lagged or predicted and corrected. Despite the drawbacks, this type of approach is the most widely used for problems with "difficult" chemistry.

Chemical reactions also tend to produce large gradients in the dependent variables over small distances. Since the location, extent, and movement of such reaction zones is not known a priori, excessively refined meshes are often employed to ensure spatial accuracy. Again, this leads to a large problem size and excessive computer resources. A more efficient solution to the length scale problem is the use of adaptive meshes. An example of this technique for a reactive problem can be found in [31].

5.11 Numerical Examples

5.11.1 Preliminary Comments

The numerical examples selected for this section are intended to illustrate the performance of the finite element models presented for convective heat transfer problems. Again, the problems are not examined in full engineering detail, but are sufficiently complete to demonstrate the flexibility and capability of the recommended algorithms. All of the examples were solved using the codes NACHOS II [20] and FIDAP [21] with the mixed finite element model, unless otherwise stated. For numerical examples of natural convection problems solved by the penalty finite element models, the reader may consult the review paper by Pelletier, Reddy, and Schetz [22].

5.11.2 Concentric Tube Flow

The first example we consider is the forced convection/conduction problem of flow in a concentric tube. The problem schematic is shown in Figure 5.11.1 along with the finite element mesh of eight-noded elements and boundary conditions. The fluid in the inner tube is a medium-weight oil and the counterflowing cooling fluid is water. The central tube is made of copper and the outside concentric tube is made of steel. All material properties are assumed to be constant.

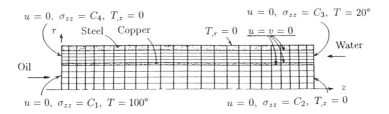

Figure 5.11.1: Schematic for concentric tube flow.

As is typical for internal flow problems, flux type boundary conditions are used at the outflow boundary for both fluid regions to allow a smooth solution field without an excessive tube length. Also, in this case a specified traction boundary condition was specified at the inflow boundary; the difference in the normal tractions across the tube length produces the driving force for the fluid. Note that the nonlinear terms are identically zero in the fluid and thermal equations in this case, and therefore the problem can be solved without iteration.

The solution to the hydrodynamic part of the problem consists of a fully developed Poiseuille flow for the oil and flow in a concentric annulus for the water. Both computed profiles are essentially exact for this model. Three cases were computed in which the oil Peclet number ($Pe = \mu C_p/k$) was varied by changing the pressure drop imposed on the tube. A typical isotherm plot is shown in Figure 5.11.2. The computed temperature profiles along the inside of the copper tube are shown in Figure 5.11.3. Further details on this analysis and other conjugate tube flow problems are available in [23].

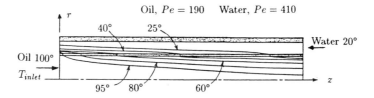

Figure 5.11.2: Isotherms for the concentric tube flow problem.

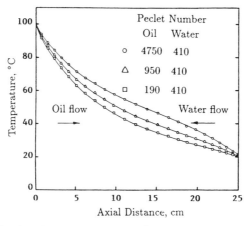

Figure 5.11.3: Axial temperature profiles for the concentric tube flow problem.

5.11.3 Tube Flow with Change of Phase

Consider the problem of a heated fluid flowing through a cold tube, the temperature of which is below the fluid solidification temperature. The problem geometry is shown in Figure 5.11.4. The fluid within the tube is initially quiescent with a uniform temperature equal to the inlet temperature. Fluid motion is initiated by imposing a specified pressure drop on the tube. The methods described in Sections 3.3 and 4.7 were used to model the latent heat release and the immobilization of the fluid at the phase boundary [8].

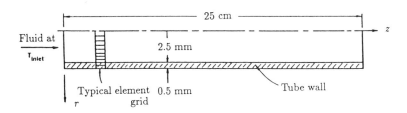

5.11.4: Schematic for tube flow with change of phase.

The initial temperature response of the fluid is due solely to conduction at the tube wall. The large initial heat loss to the tube cools the fluid below the solidus temperature, producing a solid crust on the tube wall. As the fluid velocity increases, convection along the length of the tube becomes important.

Figure 5.11.5 shows the result of a computation in which the tube eventually becomes obstructed due to the crust formation and growth. Each frame shows the phase change isotherm which corresponds to the phase boundary.

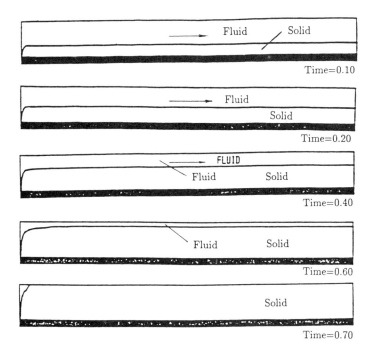

Figure 5.11.5: Motion of phase boundary for impulsively started tube flow.

Figure 5.11.6 shows velocity and temperature profiles at a station four tube diameters from the inlet. The velocity profiles show the effects of initial fluid acceleration, reduction in tube diameter due to crust growth, and finally the reduction in flow rate due to increasing viscosity.

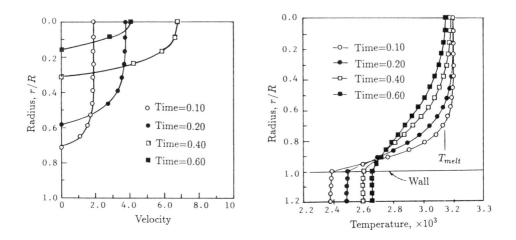

Figure 5.11.6: Velocity and temperature profiles for tube flow, $z/R = 8$.

5.11.4 Heated Cavity

A standard benchmark problem for natural convection algorithms consists of a square, closed cavity with insulated top and bottom faces and differentially heated vertical boundaries. For the present case, the contained fluid is a gas with constant material properties. A schematic of the problem with the boundary conditions and nondimensional parameters is shown in Figure 5.11.7. Since the velocity field is specified on all boundaries of the domain, the pressure field is determined only to within a constant value; the pressure level within the cavity is set by specifying a single pressure value at one node on the cavity boundary. A typical mesh consisting of eight-node elements is also shown in Figure 5.11.7.

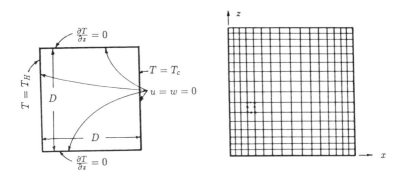

Figure 5.11.7: Schematic for natural convection in a cavity.

The cavity problem was initially solved using the cyclic Picard scheme employed in the earliest versions of the NACHOS code. More recent solutions have been obtained using Newton's method for the combined equation set and zeroth-order continuation through the Rayleigh number [$Ra = \beta g(T_H - T_C)D^3/\nu\kappa$] range of interest. Results for this problem in the form of streamline and isotherm plots are shown in Figures 5.11.8 and 5.11.9 for Rayleigh numbers of $Ra = 10^4$ and 10^6, respectively ($Pr = \nu/\kappa = 0.71$). For the lower Rayleigh number, the flow is relatively weak and the thermal field is only slightly perturbed from a conduction solution. At the higher Rayleigh number, the flow field develops a considerable structure while the thermal field becomes vertically stratified in the core of the cavity with high heat flux regions along the vertical boundaries. As required by the form of the boundary conditions and boundary value problem, the computed solutions are exactly centrosymmetric as seen in the contour plots. A detailed comparison of the present solutions with a large number of other numerical simulations is contained in [24].

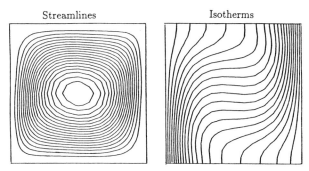

Figure 5.11.8: Streamlines and isotherms for natural convection in a square cavity ($Ra = 10^4$).

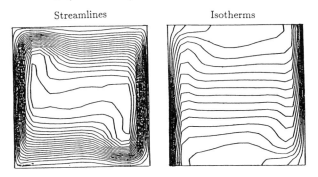

Figure 5.11.9 Streamlines and isotherms for natural convection in a square cavity ($Ra = 10^6$).

5.11.5 Solar Receiver

The schematic shown in Figure 5.11.10 represents a cross section of an annular solar receiver tube surrounded by an eccentrically located glass envelope. The inner tube carries a heat transfer fluid that is heated by a flux that varies with position around the tube; the incident flux is due to solar energy being concentrated on the tube by a parabolic trough collector. The glass envelope provides a shield to reduce the forced convection (wind) heat loss from the collector tube. Of interest in this problem is the prediction of heat loss from the inner tube due to natural convection in the annular space. Parameters such as inner tube temperature distribution, working fluid in the annulus, annular gap, and eccentricity of the envelope were varied to assess their importance on overall heat loss.

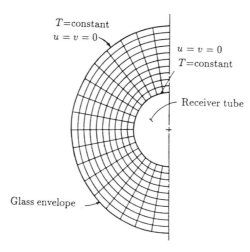

Figure 5.11.10: Mesh and boundary conditions for the annular solar receiver.

Figures 5.11.11 through 5.11.14 contain streamline and isotherm plots for an air-filled annulus for various temperature and geometric configurations. The flow pattern and heat flux distribution are quite sensitive to variations in these parameters even though the Rayleigh number is the same for all cases. The computed heat transfer results for these cases are compared with experimental data in Figure 5.11.15.

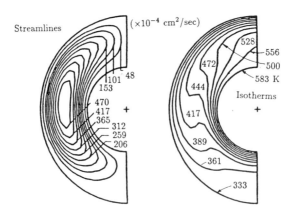

Figure 5.11.11: Streamline and isotherm plots for the solar receiver; *uniform wall temperature*, $Ra = 1.2 \times 10^4$.

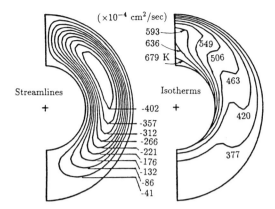

Figure 5.11.12: Streamline and isotherm plots for the solar receiver; asymmetric wall temperature, *hot on top* ($Ra = 1.2 \times 10^4$).

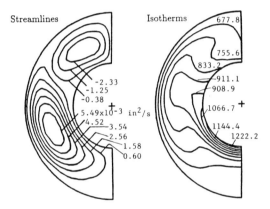

Figure 5.11.13: Streamline and isotherm plots for the solar receiver; uniform wall temperature, *hot on bottom* ($Ra = 1.2 \times 10^4$).

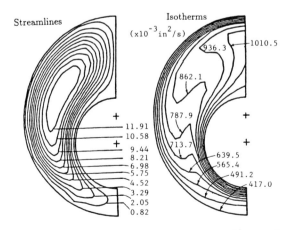

Figure 5.11.14: Streamline and isotherm plots for the solar receiver; uniform wall temperature, *eccentric geometry* ($Ra = 1.2 \times 10^4$).

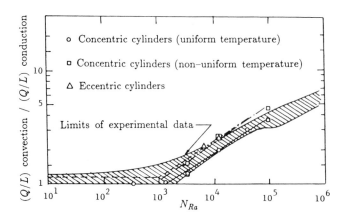

Figure 5.11.15: Heat loss by natural convection in annular spaces of a solar receiver. Computed points are shown as symbols.

5.11.6 Tube Bundle

The natural convection heat transfer from three tubes contained within a larger diameter conduit is modeled with the finite element grids shown in Figure 5.11.16. The geometry is symmetric, allowing the simulation to be performed on one-half of the conduit domain. The interior tubes have constant, but different wall temperatures; the exterior of the conduit is held at a temperature that is lower than the interior tubes. The two meshes shown in Figure 5.11.16 represent two very different mesh generation techniques. The more structured mesh was developed in three sections using a standard mapping technique. The completely unstructured quadrilateral mesh was produced using a method called "paving" [25] which works with the entire region at one time. The only input to the paving algorithm, other than a boundary description, is an element size along each boundary.

Isotherm and streamline plots for two cases are shown in Figures 5.11.17 and 5.11.18. Figure 5.11.17 illustrates the flow produced when a small temperature difference ($\Delta T_{max} = 10$) is maintained between the hottest interior tube and the exterior boundary. The second case shows the result when the temperature difference from the first case is doubled (i.e., $\Delta T_{max} = 20$). Solutions for both cases were obtained using a solution strategy with several Picard iterations followed by a Newton procedure. The computed flow fields show a characteristic plume rising from the lower cylinder with an overall circulation within the larger conduit.

Figure 5.11.16: Finite element meshes for tube bundle geometry.

Isotherms   Streamlines

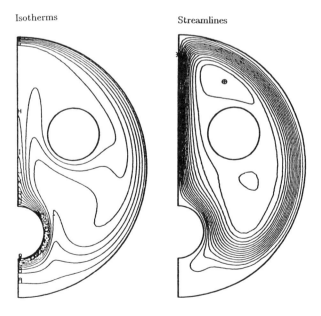

Figure 5.11.17: Contour plots for natural convection in a tube bundle, $\Delta T_{max} = 10$.

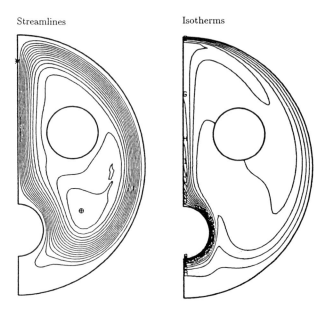

Figure 5.11.18: Contour plots for natural convection in a tube bundle, $\Delta T_{max} = 20$.

5.11.7 Volumetrically Heated Fluid

This example consists of a volumetrically heated fluid contained in an isothermal circular cylindrical container. A quadrant of the domain, boundary conditions, and finite element mesh are shown in Figure 5.11.19. At time zero, the initially quiescent, isothermal fluid is heated volumetrically at a uniform rate. The transient solution is obtained by an implicit integration scheme using the trapezoid rule with quasilinearization.

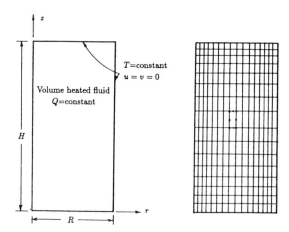

Figure 5.11.19: Computational domain, boundary conditions, and mesh used for volumetrically heated fluid in a circular cylinder.

The series of streamline and isotherm plots in Figure 5.11.20 shows the development of the flow at a moderate modified Grashof number, $Gr = g\beta l^5 Q/\nu^2 k = 2 \times 10^5$. A time history of the maximum stream function shown in Figure 5.11.21 shows the overshoot that occurs during the initial transient. The stream function vs. time plot of Figure 5.11.22 shows interaction between the two cells at a slightly higher Grashof number, $Gr = 4 \times 10^5$. A further series of plots in Figures 5.11.23 and 5.11.24 illustrates the additional dynamics present in the flow at Grashof number, $Gr = 4 \times 10^5$. The development of a secondary cell that undergoes a damped oscillation is quite evident. The development of secondary cells is common in volumetrically heated fluids. Further details of this analysis are available in [26].

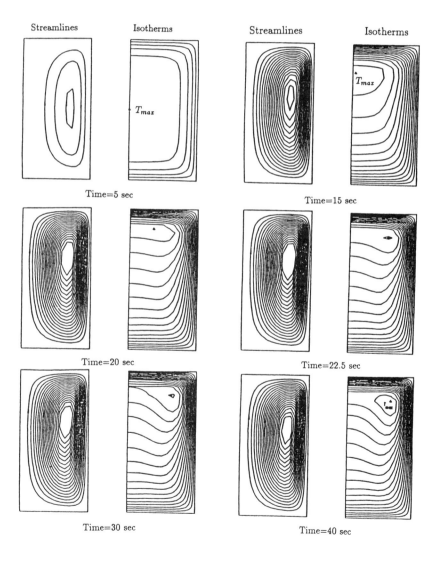

Figure 5.11.20: Streamlines and isotherms for volumetrically heated fluid $(Gr = 2 \times 10^5)$.

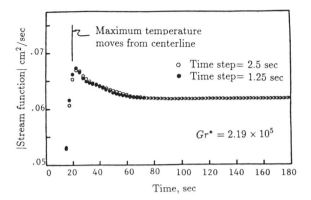

Figure 5.11.21: Stream function versus time, volumetrically heated fluid, $Gr = 2 \times 10^5$.

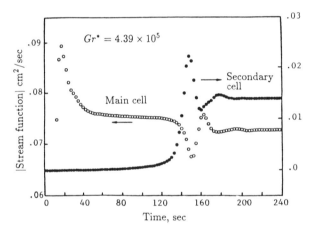

Figure 5.11.22: Stream function versus time, volumetrically heated fluid, $Gr = 4 \times 10^5$.

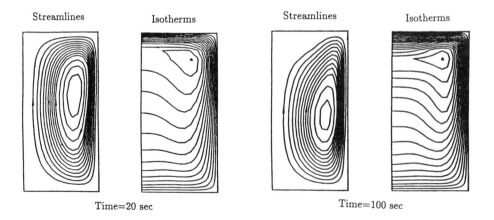

Figure 5.11.23: Streamlines and isotherms for volumetrically heated fluid $(Gr = 4 \times 10^5)$.

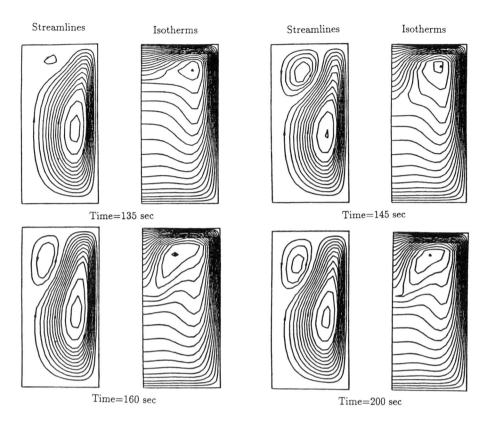

Figure 5.11.24: Streamlines and isotherms for volumetrically heated fluid $(Gr = 4 \times 10^5)$.

5.11.8 Porous/Fluid Layer

A simple problem that illustrates the use of the saturated porous media formulation in conjunction with an incompressible, viscous flow is shown in Figure 5.11.25. A rectangular geometry is divided vertically with a porous material occupying the right half of the cavity. The top and bottom boundaries of the cavity are insulated while the vertical boundaries are held at uniform but different temperatures. The steady natural convection problem in the cavity was modeled using a Brinkman model for flow in the porous medium. Newton's method was used to solve the flow problem.

For small values of the Darcy number (or permeability), $Da = \sqrt{k}/L$, the dominant fluid motion is confined to the fluid layer with little circulation occurring in the porous layer. At higher Darcy numbers, the fluid motion spans the cavity though the velocities in the porous layer are still much smaller than in the bulk fluid. Representative stream function and isotherm plots are shown in Figures 5.8.26 and 5.8.27 for extremes in the Darcy number. Further discussion on the modeling of conjugate problems and the use of various porous/layer/bulk fluid interface conditions can be found in [11] and [27].

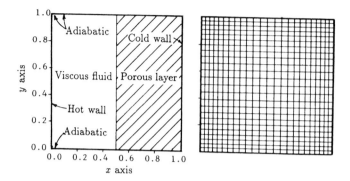

Figure 5.11.25: Schematic and mesh for a conjugate, porous/fluid layer problem.

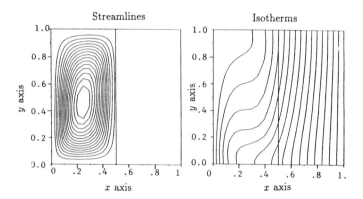

Figure 5.11.26: Contour plots for a conjugate, porous/fluid layer, $Ra = 10^5$, $Da = 10^{-5}$.

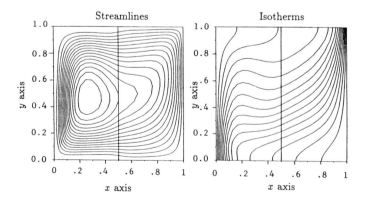

Figure 5.11.27: Contour plots for a conjugate, porous/fluid layer, $Ra = 10^5$, $Da = 10^{-3}$.

5.11.9 Curing of an Epoxy

This example demonstrates the use of both the phase change procedures and the auxiliary equation formulation to predict the gelation behavior of an epoxy. The geometry for the simulation is a simple rectangular crucible that initially contains a well-stirred mixture of epoxy resin and curing agent (see Figure 5.11.28). The crucible is held in a constant-temperature oven during the course of the curing process. Since the curing reaction is exothermic, the volumetric heat addition to the fluid is given by

$$Q = \rho \Delta H \frac{D\alpha}{Dt} = \rho \left(C_p^p - C_p^r \right) T \frac{D\alpha}{Dt} \tag{5.11.1}$$

where ΔH is the change in enthalpy of the reacting material due to its change in state, and the superscripts p and r denote the products and reactants, respectively. The variable α is an internal state variable that describes the extent of the gelation reaction; α can be interpreted as the ratio of the mass of reacted material (per unit volume) to the total mass of material (per unit volume). Therefore, α varies between 0 (unreacted) to 1 (fully reacted) with $\alpha = 0.6$ indicating the point of gelation (solidification). The time evolution of the extent of reaction variable, α, is provided by an advection-diffusion equation

$$\frac{\partial \alpha}{\partial t} + u_j \frac{\partial \alpha}{\partial x_j} = \frac{\partial}{\partial x_j} \left(D \frac{\partial \alpha}{\partial x_j} \right) + R \tag{5.11.2}$$

where R is the reaction rate. For this particular material the reaction rate is often assumed to be of second-order with Arrhenius rate constants

$$R = [A_1 exp(-E_1/\mathcal{R}T) + \alpha A_2 exp(-E_2/\mathcal{R}T)] \left(1 - \alpha \right) \tag{5.11.3}$$

where A_1, A_2 are pre-exponential factors, E_1, E_2 are activation energies, and $\mathcal{R}$ is the gas constant.

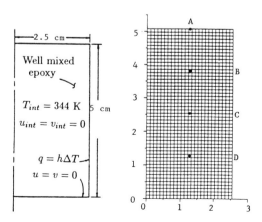

Figure 5.11.28: Schematic and mesh for epoxy curing simulation.

Equations (5.11.1) through (5.11.3) must be coupled to the standard nonisothermal, Navier–Stokes equations to describe the entire curing process. In addition, material properties, which vary with the extent of the reaction and the temperature, must be specified. The present problem was solved using the adaptive time-step version of the trapezoid rule. A finite element model with quadratic interpolation of the velocity, temperature, and extent of the reaction, and a discontinuous linear approximation of pressure is used [28]. A time history plot of the adaptive time step employed for the analysis is shown in Figure 5.11.29. It is quite evident that the size of the time step reflects the changing complexity of the flow problem. Figures 5.11.30 through 5.11.33 show contour plots of the field variables at four different times during the curing process. As with most volumetrically heated fluids, multiple convection cells are predicted during the low heat release part of the process. When the exothermic reaction begins to accelerate, gelation occurs first at the top of the crucible and proceeds downward as a planar front. A comparison of the velocity of the gel front with the limited experimental data available showed good agreement, considering the uncertainty in material properties.

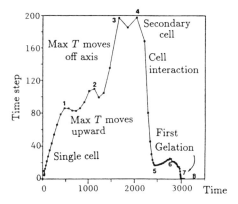

Figure 5.11.29: Time-step history for epoxy curing simulation.

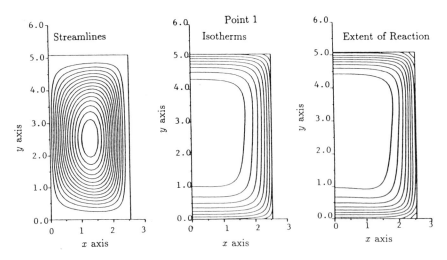

Figure 5.11.30: Contour plots for epoxy curing simulation, time = 480 sec.

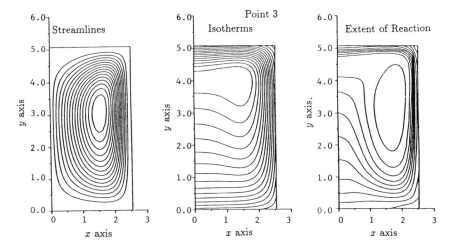

Figure 5.11.31: Contour plots for epoxy curing simulation, time = 1660 sec.

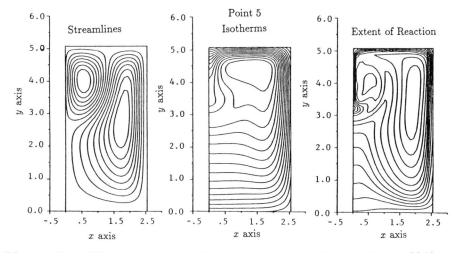

Figure 5.11.32: Contour plots for epoxy curing simulation, time = 2340 sec.

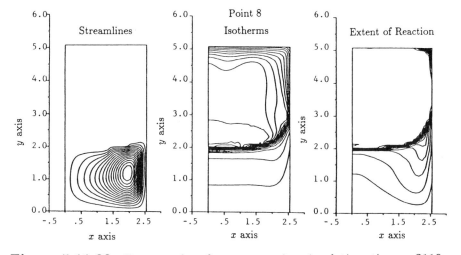

Figure 5.11.33: Contour plots for epoxy curing simulation, time = 3110 sec.

References for Additional Reading

1. J. N. Reddy, *An Introduction to the Finite Element Method*, Second Edition, McGraw-Hill, New York (1993).

2. D. K. Gartling, "Convective Heat Transfer Analysis by the Finite Element Method," *Computer Methods in Applied Mechanics and Engineering*, **12**, 365–382 (1977).

3. B. Tabarrok and R. C. Lin, "Finite Element Analysis of Free Convection Flows," *International Journal of Heat and Mass Transfer*, **20**, 953–960 (1977).

4. J. N. Reddy and A. Satake, "A Comparison of Various Finite Element Models of Natural Convection in Enclosures," *Journal of Heat Transfer*, **102**, 659–666 (1980).

5. W. N. R. Stevens, "Finite Element, Stream Function-Vorticity Solution of Steady Laminar Natural Convection," *International Journal for Numerical Methods in Fluids*, **2**, 349–366 (1982).

6. V. Haroutunian, M. S. Engelman, and I. Hasbani, "Segregated Finite Element Algorithms for the Numerical Solution of Large-Scale Incompressible Flow Problems," *International Journal for Numerical Methods in Fluids*, **17**, 323–348 (1993).

7. S. V. Patankar, *Numerical Heat Transfer and Fluid Flow*, McGraw-Hill, New York (1980).

8. D. K. Gartling, "Finite Element Analysis of Convective Heat Transfer Problems with Phase Change," *Computer Methods in Fluids*, K. Morgan, C. Taylor, and C. A. Brebbia (eds.), Pentech Press, London, U.K. (1974).

9. C. Beckermann and R. Viskanta, "Double-Diffusive Convection During Dendritic Solidification of a Binary Mixture," *Physico Chemical Hydrodynamics*, **10**, 195–213 (1988).

10. P. J. Prescott, F. P. Incropera, and W. D. Bennon, "Modeling of Dendritic Solidification Systems: Reassessment of the Continuum Momentum Equation," *International Journal of Heat and Mass Transfer*, **34**, 2351–2359 (1991).

11. D. K. Gartling, C. E. Hickox, and R. C. Givler, "Simulations of Coupled Viscous and Porous Flow Problems," *International Journal of Computational Fluid Dynamics*, **7**, 23–48 (1996).

12. V. R. Voller, A. D. Brent, and C. Prakash, "The Modeling of Heat, Mass and Solute Transport in Solidification Systems," *International Journal of Heat and Mass Transfer*, **32**, 1719–1731 (1989).

13. D. K. Gartling and P. A. Sackinger, "Finite Element Simulation of Vacuum Arc Remelting," *International Journal for Numerical Methods in Fluids*, **24**, 1271–1289 (1997).

14. D. J. Naylor, "Stresses in Nearly Incompressible Materials by Finite Elements with Application to the Calculation of Excess Pore Pressures," *International Journal for Numerical Methods in Engineering*, **8**, 443–460 (1974).

15. E. Hinton, F. C. Scott, and R. E. Ricketts, "Local Least Squares Stress Smoothing for Parabolic Isoparametric Elements," *International Journal for Numerical Methods in Engineering*, **9**, 235–256 (1975).

16. M. Engelman, R. L. Sani, and P. M. Gresho, "The Implementation of Normal and/or Tangential Boundary Conditions in Finite Element Codes for Incompressible Fluid Flow," *International Journal for Numerical Methods in Fluids*, **2**, 225–238 (1982).

17. B. E. Larock and L. R. Herrmann, "Improved Flux Prediction Using Low Order Finite Elements," in *Proceedings of the First International Conference on Finite Elements in Water Resources*, W. G. Gray and G. F. Pinder (eds.), Pentech Press, London, U. K. (1977).

18. R. S. Marshall, J. C. Heinrich, and O. C. Zienkiewicz, "Natural Convection in a Square Enclosure by a Finite Element, Penalty Function Method, Using Primitive Fluid Variables," *Numerical Heat Transfer*, **1**, 315–330 (1978).

19. P. M. Gresho, R. L. Lee, and R. L. Sani, "The Consistent Method for Computing Derived Boundary Quantities When the Galerkin FEM is Used to Solve Thermal and/or Fluids Problems," in *Proceedings of the Second International Conference on Numerical Methods in Thermal Problems*, Venice, Italy (1981).

20. D. K. Gartling, "NACHOS II – A Finite Element Computer Program for Incompressible Flow Problems," Sandia National Laboratories Report, SAND86–1816 and SAND86–1817, Albuquerque, New Mexico (1978).

21. FIDAP Manual, Version 5.0, Fluid Dynamics International, Evanston, Illinois (1990).

22. D. H. Pelletier, J. N. Reddy, and J. A. Schetz, "Some Recent Developments and Trends in Finite Element Computational Natural Convection," in *Annual Review of Numerical Fluid Mechanics and Heat Transfer*, Vol. 2, C. L. Tien and T. C. Chawla (eds.), Hemisphere, New York, 39–85 (1989).

23. D. K. Gartling, "Convective Heat Transfer Analysis by the Finite Element Method," *Computer Methods in Applied Mechanics and Engineering*, **12**, 365–382 (1977).

24. G. de Vahl Davis and I. P. Jones, "Natural Convection in a Square Cavity: A Comparison Exercise," *International Journal for Numerical Methods in Fluids*, **3**, 227–248 (1983).

25. T. D. Blacker and M. B. Stephenson, "Paving – A New Approach to Automated Quadrilateral Mesh Generation," Sandia National Laboratories Report, SAND90–0249, Albuquerque, New Mexico (1990).

26. D. K. Gartling, "A Finite Element Analysis of Volumetrically Heated Fluids in an Axisymmetric Enclosure," in *Finite Elements in Fluids, Vol. 4*, R. H. Gallagher, D. Norrie, J. T. Oden, and O. C. Zienkiewicz (eds.), John Wiley & Sons, New York (1982).

27. D. K. Gartling, "A Finite Element Formulation for Incompressible, Conjugate, Viscous/Porous Flow Problems," in *Proceedings of International Conference on Computational Methods in Flow Analysis, Vol. 1*, H. Niki and M. Kawahara (eds.), Okayama, Japan, 619–626 (1988).

28. D. K. Gartling, "The Numerical Simulation of Viscous Flow with Change of Phase and Chemical Reaction," in *Proceedings of International Conference on Computational Engineering Science, Vol. 2*, S. Atluri and G. Yagawa (eds.), Atlanta, Georgia (1988).

29. B. Launder, "On the Computation of Convective Heat Transfer in Complex Turbulent Flows," *Journal of Heat Transfer*, **110**, 1112–1128 (1988).

30. A. J. Reynolds, "The Prediction of Turbulent Prandtl and Schmidt Numbers," *International Journal of Heat and Mass Transfer*, **18**, 1055–1069 (1975).

31. M. R. Baer, R. E. Benner, R. J. Gross, and J. W. Nunziato, "Modeling and Computation of Deflagration-to-Detonation Transition (DDT) in Reactive Granular Materials," *Lectures in Applied Mathematics*, **24**, American Mathematical Society, Providence, Rhode Island (1986).

Chapter 6

Non-Newtonian Fluids

6.1 Introduction

In Chapters 4 and 5 we studied the finite element models of *Newtonian fluids* (i.e., fluids whose constitutive behavior is linear). Fluids that are not described by the Newtonian constitutive relations are commonly encountered in a wide variety of industrial processes. For example, such materials include motor oils, high molecular weight liquids such as polymers, slurries, pastes, and other complex mixtures. The processing and transport of such fluids are central problems in the chemical, food, plastics, petroleum, and polymer industries.

Non-Newtonian behavior manifests itself in a number of different ways. Most such fluids exhibit a shear rate dependent viscosity, with "shear thinning" (i.e., decreasing viscosity with increasing shear rate) being the most prevalent behavior. Other phenomena associated with the elasticity and memory of the fluid, such as recoil, are also observed in many situations. Differences in the normal stress components occur in many flows and lead to such well-known effects as rod climbing or the Weissenberg effect, and the curvature of the free surface in an open channel flow. A comprehensive list and discussion of these and other non-Newtonian effects is given in the book by Bird et al. [1].

For the present discussion non-Newtonian fluids can conveniently be separated into two distinct categories: (1) inelastic fluids or fluids without memory, and (2) viscoelastic fluids, in which memory effects are significant. The distinction is an important one from both a physical and computational point of view. Basically, inelastic fluids can be viewed as generalizations (in some sense) of the Newtonian fluid. The viscosity function for such materials depends on the rate of deformation of the fluid and thus allows "shear thinning" effects to be modeled. For numerical computations, inelastic fluids can be treated using minor extensions to the standard finite element models developed for Newtonian fluids. Viscoelastic fluids, on the other hand, represent a significant departure from the Newtonian limit in terms of both physical behavior and computational complexity. The primary difficulty here is the "memory" of the fluid; the motion of a material element depends not only on the present stress state, but also on the deformation history of the material element. This history dependence leads to very complex constitutive equations and the need for special computational procedures. It is the purpose of the present chapter to study some aspects of the finite element simulation of both inelastic and viscoelastic fluids.

Following this introduction, a brief outline of the continuum equations, boundary conditions, and pertinent constitutive equations for inelastic fluids will be presented, followed by a description of their finite element models. The emphasis will be on the computational treatment of the material nonlinearities inherent in these types of fluids. The second part of the chapter will be concerned with the so-called simple fluid with fading memory. We will again state the standard balance laws in a convenient form and then describe some of the many different constitutive equations that have been found useful for numerical experimentation and computation. Several numerical examples of both inelastic and viscoelastic flows will also be presented.

The approach taken in the following sections will be very pragmatic with regard to constitutive relations for non-Newtonian fluids. No attempt will be made to theoretically justify the models used or describe in detail their positive or negative aspects, as this topic is well outside the scope of the present text. For readers interested in these questions or other topics in rheology, a number of references can be consulted, including the textbooks by Bird et al. [1], Lodge [2], Walters [3] and Tanner [4]. The book by Crochet et al. [5] covers applications of finite difference and finite element methods to non-Newtonian fluids. The text by Pearson [6] treats a variety of theoretical and practical topics regarding the analysis of industrial flow problems.

6.2 Governing Equations of Inelastic Fluids

For completeness and ready reference, the basic equations of motion are reviewed once again. The section concludes with a discussion of representative inelastic, non-Newtonian constitutive relations.

6.2.1 Conservation Equations

The equations resulting from conservation of mass, momentum, and energy for flows of viscous, incompressible, inelastic fluids consist of the standard continuity equation, the Navier–Stokes equations, and the energy equation. For most flows of interest the assumptions of incompressibility and laminar flow are easily justified. Utilizing standard index notation, where repeated subscripts imply summation, the governing equations in a rectangular Cartesian coordinate system can be summarized as follows:

Conservation of Mass:

$$\frac{\partial u_i}{\partial x_i} = 0 \qquad (6.2.1)$$

Conservation of Momentum:

$$\rho \left(\frac{\partial u_i}{\partial t} + u_j \frac{\partial u_i}{\partial x_j} \right) = \frac{\partial \sigma_{ij}}{\partial x_j} + \rho g_i \qquad (6.2.2)$$

Conservation of Energy:

$$\rho C \left(\frac{\partial T}{\partial t} + u_j \frac{\partial T}{\partial x_j} \right) = -\frac{\partial q_i}{\partial x_i} + Q + \Phi \qquad (6.2.3)$$

The constitutive equations for the total stress, σ_{ij}, and heat flux, q_i, are given by

$$\sigma_{ij} = -P\delta_{ij} + 2\mu D_{ij}, \quad \mu = \mu(D_{ij}) \tag{6.2.4}$$

and

$$q_i = -k\frac{\partial T}{\partial x_i} \tag{6.2.5}$$

where Eq. (6.2.4) is written for an isotropic, inelastic, non-Newtonian or generalized Newtonian fluid [1]. Equation (6.2.5) is the well-known Fourier's heat conduction law. The flow kinematics are given by

$$D_{ij} = \frac{1}{2}(L_{ij} + L_{ji}) ; \quad L_{ij} = \frac{\partial u_i}{\partial x_j} \tag{6.2.6}$$

where D_{ij} and L_{ij} are the rate of deformation and velocity gradient tensors, respectively. The coefficient μ in Eq. (6.2.4) is a scalar viscosity function which, in general, depends on the rate of deformation tensor in a manner that will be made clear in a subsequent section. The function Φ in Eq. (6.2.3) represents the viscous dissipation in the fluid, and it enters the energy equation as a volume source of thermal energy. The dissipation depends on the velocity gradients and is given by

$$\Phi = 2\mu D_{ij}D_{ij} \tag{6.2.7}$$

The remaining terms in Eqs. (6.2.1)–(6.2.5) are the same as defined in earlier chapters for Newtonian flows.

Though the above equation set is quite general and describes most practical inelastic flow problems, several simplifications to the equation system are often considered. Many processing operations involving non-Newtonian fluids occur at very low Reynolds numbers, where $Re = \rho U_{ref}D/\mu$ and U_{ref} is a reference velocity and D is a characteristic dimension in the flow problem. For such cases the nonlinear terms in the momentum equation may be neglected. Despite this simplification, the momentum equations are still nonlinear due to the variation of the viscosity function with the rate of deformation tensor. There are many instances when viscous dissipation can also be neglected, in which case the energy equation is simplified. Various other simplifications involving the energy equation are possible, depending on whether buoyancy effects or temperature-dependent material properties are considered important. Such possibilities are treated in the same manner as for a Newtonian fluid and thus will not be considered further in the present chapter.

6.2.2 Boundary Conditions

Equations (6.2.1)–(6.2.7) represent, in general, a coupled problem that requires boundary conditions on both the fluid motion and the energy transport. The necessary boundary conditions are of the standard type and consist of specified velocities or tractions for the momentum equations and specified temperature or heat flux for the energy equation. Symbolically the boundary conditions for the momentum equation are

$$u_i = f_i^u(s_k, t) \qquad \text{on } \Gamma_u \tag{6.2.8a}$$

$$T_i = \sigma_{ij}(s_k, t) n_j(s_k) = f_i^T(s_k, t) \qquad \text{on } \Gamma_T \tag{6.2.8b}$$

where s_k are the coordinates along the boundary, t is the time, n_i is the outward unit normal to the boundary, and Γ_f is the total boundary enclosing the fluid domain, Ω_f, with $\Gamma_f = \Gamma_u \cup \Gamma_T$. Note that the conditions written in Eq. (6.2.8) are in component form, i.e., there is a condition on each component of the velocity and stress vectors.

The thermal boundary conditions, for the nonisothermal case, are given by

$$T = f^T(s_k, t) \qquad \text{on } \Gamma_T \tag{6.2.9a}$$

$$-\left(k\frac{\partial T}{\partial x_i}\right) n_i = q_i n_i = q_c + q_r + q_a = f_q(s_k, t) \qquad \text{on } \Gamma_q \tag{6.2.9b}$$

where Γ_{ht} is the total boundary enclosing the heat transfer region and $\Gamma_{ht} = \Gamma_T \cup \Gamma_q$. Also, q_a indicates a specified flux and q_c and q_r refer to the convective and radiative components given by

$$q_c = h_c(s_k, T, t)(T - T_c) \tag{6.2.10a}$$
$$q_r = h_r(s_k, T, t)(T - T_r) \tag{6.2.10b}$$

where h_c and h_r are the convective and radiative heat transfer coefficients, and T_c and T_r are the reference (or sink) temperatures for convective and radiative heat transfer.

The above boundary conditions apply to most standard situations where the fluid is contained by fixed boundaries or enters/leaves the domain, Ω_f. One other type of boundary (or interface) condition that requires mention concerns conditions along a free surface between two fluids. Many non-Newtonian (and Newtonian) flows involve situations where the fluid (liquid) is extruded, spun, drawn, or flows in a sheet or jet, such that a free surface interface exists in the problem domain. In most cases, one of the fluids is a gas and its motion relative to the other fluid is neglected. The case of an interface between two immiscible liquids can also be included in this formulation (see Section 1.8.1). The free surface problem was considered in detail in Chapter 4 for the Newtonian fluid and does not differ significantly for the non-Newtonian case. When stress or force balances are considered along the interface, the inelastic formulation is complicated by the nonlinear constitutive behavior defined in (6.2.4). Otherwise, the interface formulation and solution methods are the same as described in Section 4.10 and they will not be repeated here.

6.2.3 Constitutive Equations

The form of the constitutive equation for a generalized Newtonian fluid was given above as

$$\sigma_{ij} = -P\delta_{ij} + 2\mu(D_{ij})D_{ij} \tag{6.2.11}$$

where σ_{ij} are the components of the total stress tensor, P is the pressure, δ_{ij} is the Kronecker delta (or components of the unit tensor), and D_{ij} are the components of the rate of deformation tensor. Of interest in the present section are particular forms for the *deviatoric stress* or *extra stress* components defined by

$$\tau_{ij} = 2\mu(D_{ij})D_{ij} \tag{6.2.12}$$

The viscosity for non-Newtonian fluids is found to depend on the rate of deformation tensor (see [1,4,7]):

$$\mu = \mu(D_{ij}) = \mu(I_1, I_2, I_3) \tag{6.2.13}$$

where the I_i are the *invariants* of D_{ij}, defined by

$$I_1 = tr(\mathbf{D}) = \sum_i D_{ii} \tag{6.2.14a}$$

$$I_2 = \frac{1}{2}tr(\mathbf{D}^2) = \frac{1}{2}\sum_i \sum_j D_{ij}D_{ji} \tag{6.2.14b}$$

$$I_3 = \frac{1}{3}tr(\mathbf{D}^3) = \frac{1}{3}\sum_i \sum_j \sum_k D_{ij}D_{jk}D_{ki} \tag{6.2.14c}$$

where tr denotes the trace. For an incompressible fluid, $I_1 = \nabla \cdot u = 0$. Also, there is no theoretical or experimental evidence to suggest that the viscosity depends on I_3; thus, the dependence on the third invariant is eliminated. Equation (6.2.12) then reduces to

$$\mu = \mu(D_{ij}) = \mu(I_2) \tag{6.2.15}$$

for a generalized Newtonian fluid. The viscosity can also depend on the thermodynamic state of the fluid, which for incompressible fluids usually implies a dependence only on the temperature.

Though Eq. (6.2.15) gives the general functional form for the viscosity function, experimental observation and a limited theoretical base must be used to provide specific models for non-Newtonian viscosities. A variety of models have been proposed and correlated with experimental data (e.g., [1]). Several of the most useful and popular models are cataloged below.

Power-law model

The simplest and most familiar non-Newtonian viscosity model is the power-law model which has the form

$$\mu = KI_2^{(n-1)/2} \tag{6.2.16}$$

where n and K are parameters, which are perhaps functions of temperature, termed the *power law index* and *consistency*, respectively. One of the most common features of many non-Newtonian fluids is the "power law" decrease in the apparent viscosity with increasing shear (deformation) rate as modeled by Eq. (6.2.16). Such fluids, with an index $n < 1$ are termed *shear thinning* or *pseudoplastic*. A few materials are *shear thickening* or *dilatant* and have an index $n > 1$. The Newtonian viscosity function is obtained with $n = 1$. The admissible range of the index is bounded below by zero due to stability considerations.

When considering nonisothermal flows, the following empirical relations for n and K have proved useful

$$n = n_0 + B \left(\frac{T - T_0}{T_0}\right) \tag{6.2.17}$$

$$K = K_0 exp\left(-A[T - T_0]/T_0\right) \tag{6.2.18}$$

where subscript zero indicates a reference condition and A and B are constants for each fluid.

Carreau model

A major deficiency in the power-law model is that it fails to predict upper and lower limiting viscosities for extreme values of the deformation rate, I_2. This problem is alleviated in the multiple parameter Carreau model, which is of the form

$$\mu = \mu_\infty + (\mu_0 - \mu_\infty) \left(1 + [\lambda I_2]^2\right)^{(n-1)/2} \tag{6.2.19}$$

In Eq. (6.2.19), μ_0 and μ_∞ are the initial and infinite shear rate viscosities, respectively, and λ is a time constant. The remaining parameters were defined previously.

To illustrate the differences between the power-law and Carreau models, a plot of $\log \mu$ versus $\log I_2$ is shown in Figure 6.2.1 for several examples of each model. Like the power-law model, the Carreau viscosity is seen to have a "power law" region, which is bounded on either end by regions of constant viscosity.

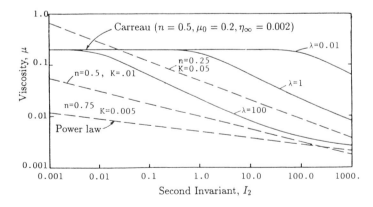

Figure 6.2.1: Viscosity functions for the power-law and Carreau models.

Bingham model

The Bingham fluid differs from most other fluids in that it can sustain an applied stress without fluid motion occurring. The fluid possesses a yield stress, τ_0, such that when the applied stresses are below τ_0 no motion occurs; when the applied stresses exceed τ_0 the material flows, with the viscous stresses being proportional to the excess of the stress over the yield condition. Typically, the constitutive equation after yield is taken to be Newtonian (Bingham model), though other forms such as a power-law equation (Herschel–Buckley model) are possible.

In a general form, the Bingham model is expressed as [8,9]

$$\tau_{ij} = \left(\frac{\tau_0}{\sqrt{I_2}} + 2\mu \right) D_{ij} \quad \text{when } \frac{1}{2} tr(\tau^2) \geq \tau_0^2 \qquad (6.2.20)$$

$$D_{ij} = 0 \qquad \text{when } \frac{1}{2} tr(\tau^2) < \tau_0^2 \qquad (6.2.21)$$

From Eq. (6.2.20) the apparent viscosity of the material beyond the yield point is $(\tau_0/\sqrt{I_2} + 2\mu)$. For a Herschel–Buckley fluid the μ in (6.2.20) is given by Eq. (6.2.16). The inequalities in Eqs. (6.2.20) and (6.2.21) describe a von Mises yield criterion. The implementation of the Bingham model into a computational procedure requires that (6.2.20) and (6.2.21) be modified slightly due to the condition $D_{ij} = 0$. This point will be covered in detail in a later section.

The above models serve to illustrate some of the typical viscosity functions that are available to model inelastic, non-Newtonian fluids. A more extensive list is available in [1].

6.3 Finite Element Models of Inelastic Fluids

6.3.1 Introduction

The finite element formulation of non-Newtonian flows follows very closely the formulations developed for Newtonian flow problems. Therefore, in the present section only a brief overview of the general formulation will be given, with more attention focused on those aspects that are unique to the non-Newtonian problem. For a detailed description of the finite element models of viscous incompressible flow problems, the reader is referred to Chapter 4.

As in the case of Newtonian fluids, there are several different formulations that may be used to construct the finite element models of a non-Newtonian fluid. The equations given in Section 6.2.1 are in terms of the velocity, pressure, and temperature as the dependent variables. However, equations using the stream function and vorticity or the stream function alone could also be employed. The primary argument in favor of the use of primitive variables (i.e., velocity, pressure, and temperature) for non-Newtonian flows comes from the free surface problem. The free surface boundary conditions given previously in Section 4.10 contain the pressure explicitly, and therefore can be more conveniently imposed in a primitive variable formulation. Also, as stated earlier, the use of the stream function-vorticity or stream function formulations present difficulties in the imposition of boundary conditions, and the stream function formulation requires higher-order interpolation (i.e., C^1-continuity).

Here, we develop finite element models based on the primitive variables. The mixed and penalty models described in Chapter 4 are presented here for the nonisothermal flow of inelastic fluids.

6.3.2 Mixed Model

Weak forms of Eqs. (6.2.1)–(6.2.3) can be developed from their Galerkin integrals, as explained in Chapter 5 (see Section 5.2). The velocity components are approximated by Lagrange interpolation functions, one order higher than those used for the pressure. Suppose that the dependent variables (u_i, P, T) are approximated by expansions of the form

$$T(\mathbf{x}, t) = \sum_{n=1}^{N} \psi_n(\mathbf{x}) T_n(t) = \mathbf{\Psi}^T \mathbf{T} \tag{6.3.1a}$$

$$u_i(\mathbf{x}, t) = \sum_{m=1}^{M} \psi_m(\mathbf{x}) u_i^m(t) = \mathbf{\Psi}^T \mathbf{u}_i \tag{6.3.1b}$$

$$P(\mathbf{x}, t) = \sum_{l=1}^{L} \phi_l(\mathbf{x}) P_l(t) = \mathbf{\Phi}^T \mathbf{P} \tag{6.3.1c}$$

where $\mathbf{\Psi}$ and $\mathbf{\Phi}$ are vectors of interpolation (or shape) functions, $\mathbf{T}$, $\mathbf{u}_i$, and $\mathbf{P}$ are (column) vectors of nodal values of temperature, velocity components, and pressure, respectively, and superscript $(\cdot)^T$ denotes a transpose of the enclosed vector or matrix. Note that the standard practice of interpolating the temperature and velocity variables with the same shape functions has been employed here. Substitution of Eqs. (6.3.1a)–(6.3.1c) into the weak forms of Eqs. (6.2.1)–(6.2.4) [i.e., Eq. (5.2.1)] results in the following set of nonlinear algebraic equations [see Eqs. (5.2.4)–(5.2.6) for details]:

Continuity:

$$-\mathbf{Q}^T \mathbf{u} = \mathbf{0} \tag{6.3.2}$$

Momentum:

$$\mathbf{M}\dot{\mathbf{u}} + \mathbf{C}\mathbf{u} + \mathbf{K}(\mu)\mathbf{u} - \mathbf{Q}\mathbf{P} + \mathbf{B}\mathbf{T} = \mathbf{F} \tag{6.3.3}$$

Energy:

$$\mathbf{N}\dot{\mathbf{T}} + \mathbf{D}\mathbf{T} + \mathbf{L}\mathbf{T} = \mathbf{G} \tag{6.3.4}$$

where the superposed dot represents a time derivative and $\mathbf{u}^T = \{\mathbf{u}_1^T, \mathbf{u}_2^T, \mathbf{u}_3^T\}$. This set of equations is virtually identical to those used in Newtonian problems except for the dependence of the viscous diffusion term $\mathbf{K}$ on the velocity, i.e., the rate of deformation (because of viscosity's dependence on the rate of deformation tensor), and possibly temperature.

For the three-dimensional case, Eqs. (6.3.2)–(6.3.4) have the following explicit form [the continuity equation (6.3.2) and momentum equations (6.3.3) are combined into one]:

$$
\begin{bmatrix} \mathbf{M} & 0 & 0 & 0 \\ 0 & \mathbf{M} & 0 & 0 \\ 0 & 0 & \mathbf{M} & 0 \\ 0 & 0 & 0 & 0 \end{bmatrix} \begin{Bmatrix} \dot{\mathbf{u}}_1 \\ \dot{\mathbf{u}}_2 \\ \dot{\mathbf{u}}_3 \\ \dot{\mathbf{P}} \end{Bmatrix} + \begin{bmatrix} \mathbf{C(u)} & 0 & 0 & 0 \\ 0 & \mathbf{C(u)} & 0 & 0 \\ 0 & 0 & \mathbf{C(u)} & 0 \\ 0 & 0 & 0 & 0 \end{bmatrix} \begin{Bmatrix} \mathbf{u}_1 \\ \mathbf{u}_2 \\ \mathbf{u}_3 \\ \mathbf{P} \end{Bmatrix}
$$

$$
+ \begin{bmatrix} \hat{\mathbf{K}}_{11} & \mathbf{K}_{21} & \mathbf{K}_{31} & -\mathbf{Q}_1 \\ \mathbf{K}_{12} & \hat{\mathbf{K}}_{22} & \mathbf{K}_{32} & -\mathbf{Q}_2 \\ \mathbf{K}_{13} & \mathbf{K}_{23} & \hat{\mathbf{K}}_{33} & -\mathbf{Q}_3 \\ -\mathbf{Q}_1^T & -\mathbf{Q}_2^T & -\mathbf{Q}_3^T & 0 \end{bmatrix} \begin{Bmatrix} \mathbf{u}_1 \\ \mathbf{u}_2 \\ \mathbf{u}_3 \\ \mathbf{P} \end{Bmatrix} = \begin{Bmatrix} \mathbf{F}_1(\mathbf{T}) \\ \mathbf{F}_2(\mathbf{T}) \\ \mathbf{F}_3(\mathbf{T}) \\ 0 \end{Bmatrix} \qquad (6.3.5)
$$

$$
[\mathbf{N}]\left\{\dot{\mathbf{T}}\right\} + [\mathbf{D(u)}]\left\{\mathbf{T}\right\} + [\mathbf{L}]\left\{\mathbf{T}\right\} = \left\{\mathbf{G(T)}\right\} \qquad (6.3.6)
$$

The coefficient matrices shown in Eqs. (6.3.5) and (6.3.6) are defined by [from Eq. (6.2.12)]

$$
\hat{\mathbf{K}}_{11} = 2\mathbf{K}_{11} + \mathbf{K}_{22} + \mathbf{K}_{33}
$$
$$
\hat{\mathbf{K}}_{22} = \mathbf{K}_{11} + 2\mathbf{K}_{22} + \mathbf{K}_{33}
$$
$$
\hat{\mathbf{K}}_{33} = \mathbf{K}_{11} + \mathbf{K}_{22} + 2\mathbf{K}_{33} \qquad (6.3.7a)
$$

$$
\mathbf{M} = \int_{\Omega^e} \rho_0 \mathbf{\Psi}\mathbf{\Psi}^T \, dx \; ; \quad \mathbf{C(u)} = \int_{\Omega^e} \rho_0 \mathbf{\Psi}(\mathbf{\Psi}^T \mathbf{u}_j)\frac{\partial \mathbf{\Psi}^T}{\partial x_j} \, dx
$$

$$
\mathbf{K}_{ij} = \int_{\Omega^e} \mu \frac{\partial \mathbf{\Psi}}{\partial x_i} \frac{\partial \mathbf{\Psi}^T}{\partial x_j} \, dx \; ; \quad \mathbf{Q}_i = \int_{\Omega^e} \frac{\partial \mathbf{\Psi}}{\partial x_i} \mathbf{\Phi}^T \, dx
$$

$$
\mathbf{F}_i(\mathbf{T}) = -\int_{\Omega^e} \rho_0 g_i \beta \mathbf{\Psi}\mathbf{\Phi}^T dx + \int_{\Omega^e} \rho_0 g_i \beta T_0 \mathbf{\Psi} \, dx + \oint_{\Gamma^e} \mathbf{\Psi} T_i ds
$$

$$
\mathbf{D(u)} = \int_{\Omega^e} \rho_0 C \mathbf{\Phi}(\mathbf{\Psi}^T \mathbf{u}_j)\frac{\partial \mathbf{\Phi}^T}{\partial x_j} \, dx
$$

$$
\mathbf{N} = \int_{\Omega^e} \rho_0 C \mathbf{\Psi}\mathbf{\Psi}^T \, dx \; ; \quad \mathbf{L} = \int_{\Omega^e} k \frac{\partial \mathbf{\Psi}}{\partial x_i} \frac{\partial \mathbf{\Psi}^T}{\partial x_i} \, dx
$$

$$
\mathbf{G} = \int_{\Omega^e} \mathbf{\Psi} Q \, dx + \int_{\Omega^e} 2\mu \mathbf{\Psi}\mathbf{\Phi} \, dx + \oint_{\Gamma^e} \mathbf{\Psi} q_n \, ds \qquad (6.3.7b)
$$

where summation on repeated indices is implied. Note that the finite element model is nonlinear because of the nonlinearity in the convective terms as well as the viscosity. In addition, the conductivity k can be a function of temperature. Equations (6.3.2)–(6.3.4) can be combined into a single matrix equation

$$
\begin{bmatrix} \mathbf{M} & 0 & 0 \\ 0 & 0 & 0 \\ 0 & 0 & \mathbf{N} \end{bmatrix} \begin{Bmatrix} \dot{\mathbf{u}} \\ \dot{\mathbf{P}} \\ \dot{\mathbf{T}} \end{Bmatrix} + \begin{bmatrix} \mathbf{C(u)} + \mathbf{K(u,T,\mu)} & -\mathbf{Q} & \mathbf{B(T)} \\ -\mathbf{Q}^T & 0 & 0 \\ 0 & 0 & \mathbf{D(u)} + \mathbf{L(T)} \end{bmatrix} \begin{Bmatrix} \mathbf{u} \\ \mathbf{P} \\ \mathbf{T} \end{Bmatrix}
$$

$$= \left\{ \begin{array}{c} \mathbf{F(T)} \\ 0 \\ \mathbf{G(T, u)} \end{array} \right\} \tag{6.3.8}$$

or in a more symbolic format as

$$\bar{\mathbf{M}}\dot{\mathbf{U}} + \bar{\mathbf{K}}(\mathbf{u}, \mathbf{T}, \mu)\mathbf{U} = \bar{\mathbf{F}} \tag{6.3.9}$$

where

$$\mathbf{U}^T = \{\mathbf{u}_1^T, \mathbf{u}_2^T, \mathbf{u}_3^T, \mathbf{P}^T, \mathbf{T}^T\} \tag{6.3.10}$$

This completes the development of the mixed finite element model for the inelastic case.

6.3.3 Penalty Model

In the penalty function method, the continuity equation is treated as a constraint (see Section 5.3) and the problem is reformulated as an unconstrained problem. The pressure, which is the Lagrange multiplier, does not appear explicitly as a dependent variable in the formulation, although it is a part of the boundary stresses [see Eq. (4.3.5)]. In two dimensions, an approximation for the pressure can be post-computed from the relation

$$P = -\gamma_e \left(\frac{\partial u_1}{\partial x_1} + \frac{\partial u_2}{\partial x_2} \right) \tag{6.3.11}$$

where γ_e is the penalty parameter. The penalty finite element model is given by

$$\begin{bmatrix} \mathbf{M} & 0 \\ 0 & \mathbf{M} \end{bmatrix} \left\{ \begin{array}{c} \dot{\mathbf{u}}_1 \\ \dot{\mathbf{u}}_2 \end{array} \right\} + \begin{bmatrix} \mathbf{C(u)} & 0 \\ 0 & \mathbf{C(u)} \end{bmatrix} \left\{ \begin{array}{c} \mathbf{u}_1 \\ \mathbf{u}_2 \end{array} \right\} +$$

$$\left(\begin{bmatrix} 2\mathbf{K}_{11} + \mathbf{K}_{22} & \mathbf{K}_{21} \\ \mathbf{K}_{12} & \mathbf{K}_{11} + 2\mathbf{K}_{22} \end{bmatrix} + \begin{bmatrix} \hat{\mathbf{K}}_{11} & \hat{\mathbf{K}}_{21} \\ \hat{\mathbf{K}}_{12} & \hat{\mathbf{K}}_{22} \end{bmatrix} \right) \left\{ \begin{array}{c} \mathbf{u}_1 \\ \mathbf{u}_2 \end{array} \right\} = \left\{ \begin{array}{c} \mathbf{F}_1 \\ \mathbf{F}_2 \end{array} \right\} \tag{6.3.12}$$

where $\mathbf{M}, \mathbf{C(u)}, \mathbf{K}_{ij}$ (which depend on the viscosity) and $\mathbf{F}_i$ are the same as those defined in Eq. (6.3.7), and

$$\hat{\mathbf{K}}_{ij} = \int_{\Omega^e} \gamma_e \frac{\partial \mathbf{\Psi}}{\partial x_i} \frac{\partial \mathbf{\Psi}^T}{\partial x_j} d\mathbf{x} \tag{6.3.13}$$

The energy equation remains unchanged as in Eq. (6.3.6). In matrix form, Eqs. (6.3.13) and (6.3.6) can be expressed as

$$\bar{\mathbf{M}}\dot{\mathbf{U}} + \bar{\mathbf{K}}(\mathbf{u}, \mathbf{T}, \mu)\mathbf{U} = \bar{\mathbf{F}} \tag{6.3.14}$$

where

$$\mathbf{U}^T = \{\mathbf{u}_1^T, \mathbf{u}_2^T, \mathbf{u}_3^T, \mathbf{T}^T\} \tag{6.3.15}$$

This completes the discussion of the penalty finite element model for the inelastic case.

6.3.4 Matrix Evaluations

The appearance of a nonlinear diffusion term, $\mathbf{K}$, in Eqs. (6.3.9) and (6.3.14) influences two basic aspects of the finite element analysis, namely, the matrix evaluation and solution algorithms. Matrix evaluation methods will be discussed first.

The techniques used for the construction of element level matrices that contain variable coefficients have generally fallen into one of two categories — the reconstruction methods and the hypermatrix methods. The differences in these methods and their application to non-Newtonian formulations are best explained via a specific example. Consider the construction of a particular component of the $\mathbf{K}$ matrix in Eq. (6.3.9), for example,

$$\mathbf{K}_{11} = \mathbf{K}_{\mathbf{xx}} = \int_{\Omega_e} \frac{\partial \mathbf{\Psi}}{\partial x} \mu(I_2, T) \frac{\partial \mathbf{\Psi}^T}{\partial x} \, d\mathbf{x} \qquad (6.3.16)$$

where $\mathbf{\Psi}$ is the vector of shape functions used for the velocity field. The viscosity function is shown in its most general form with a dependence on I_2 and the temperature; the invariant I_2 depends on velocity gradients [see Eq. (6.2.17)].

For most finite element applications the integration in Eqs. (6.3.7) and (6.3.16) is performed via a numerical quadrature. Such a computation requires that each function in the integrand be evaluated at each integration point within the element. Therefore, the evaluation of $\mathbf{K}$ for a non-Newtonian fluid requires that the velocity gradients and perhaps temperature be evaluated at integration points, using the values from the latest available solution $(\bar{u}_m, \bar{T}_n)$ at time t_s:

$$T(\mathbf{x}, t_s) = \sum_{n=1}^{N} \theta_n(\mathbf{x}) \bar{T}_n(t_s) \qquad (6.3.17a)$$

$$u_i(\mathbf{x}, t_s) = \sum_{m=1}^{M} \psi_m(\mathbf{x}) \bar{u}_i^m(t_s) \qquad (6.3.17b)$$

In many standard finite element programs the above quadrature procedure is carried out each time the element matrix is required. As a result of the iterative methods used to solve the nonlinear equations, the matrix formation operations may be required many times, especially for time-dependent problem, resulting in a large computational cost.

An alternative to the matrix reconstruction method, the hypermatrix approach may be used. This method reduces some of the computational cost but at the cost of some additional storage and I/O in the computer program. Since the viscosity is a function of position in the element (due to its functional dependence on the velocity field and temperature), it is natural to interpolate the viscosity in the same way as a dependent variable. Thus, let the viscosity be represented by

$$\mu = \mathbf{\Psi}^T \hat{\mu} \qquad (6.3.18)$$

where $\boldsymbol{\Psi}$ is a vector of shape functions and $\hat{\mu}$ is a vector of nodal point viscosity values. Substitution of Eq. (6.3.18) in Eq. (6.3.16) yields

$$\mathbf{K_{xx}} = \int_{\Omega_e} \frac{\partial \boldsymbol{\Psi}}{\partial x} (\boldsymbol{\Psi}^T \hat{\mu}) \frac{\partial \boldsymbol{\Psi}^T}{\partial x} \, dx \qquad (6.3.19)$$

Since the $\hat{\mu}$ are the nodal values, the integral in Eq. (6.3.19) can be constructed once and stored as a three-dimensional array $\mathbf{K}(NI, NJ, NK)$, where each index corresponds to one of the shape functions in (6.3.19). This matrix is called the *hypermatrix*. A product of the hypermatrix with a known vector of nodal point viscosities produces the required element matrix without repeated numerical quadrature. This technique has been used successfully (see [10,11]) for the nonlinear advection terms in the momentum equations [i.e., the $\mathbf{C(u)u}$ term in Eq. (6.3.7)] as well as for material property variations (see [12]). It is most effective when the variable coefficient is a nodal point quantity or depends on a nodal point quantity, since this permits shape function evaluations to be avoided. Unfortunately, this is not the case for the non-Newtonian viscosity which depends on velocity gradients. It is well-known that the most accurate points within a quadrilateral element at which to evaluate derivatives are the 2×2 Gauss integration points (see [13]). Thus, for maximum accuracy in the viscosity evaluation the invariant I_2 should be evaluated at the Gauss points. However, the extrapolation of Gauss point values of I_2 to the nodes by standard methods (see [14,15]) in order to use Eq. (6.3.19) has not proven to be a viable technique for most non-Newtonian viscosity models. In general, predictions of I_2 at the nodes via extrapolation are very inaccurate and lead to poor overall accuracy or nonconvergence of the solution method. As an alternative to extrapolation an averaging method is recommended in which the Gauss point values of I_2 are averaged over the element and a single value used to evaluate the viscosity at the nodes.

Figure 6.3.1 shows velocity profiles obtained with both the extrapolation and averaging procedures described above. The results are for a power-law fluid in a cylindrical tube. The inaccuracies generated by extrapolation are on the order of 30% while the averaging procedure yields results within a few percent of the analytical result. The averaging procedure in conjunction with Eq. (6.3.19) has been used successfully in other problems with a variety of viscosity models (see [16,17]).

To conclude this section it should be pointed out that the choice of a reconstruction or hypermatrix procedure is dependent on the type of computing resources available and the structure of the finite element software. Both methods are equivalent and effective. The relative costs of CPU and I/O time will heavily influence the selection.

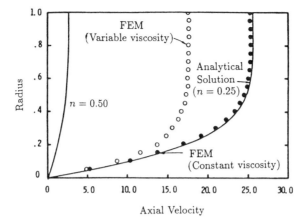

Figure 6.3.1: Velocity profile for a power fluid in a tube; various methods for viscosity computation.

6.4 Solution Methods for Inelastic Fluids

The solution procedure selected for the inelastic, non-Newtonian problem given by the finite element models in Section 6.3 must be capable of treating several different types of nonlinearities. For nonzero Reynolds number flows the nonlinear advection terms are significant and in many cases may dominate the problem. The non-Newtonian fluid introduces a second type of nonlinear behavior while the presence of unknown free surface boundaries can introduce a geometric nonlinearity. In the present section, only the first two types of phenomena will be considered; free surface solutions were covered in a previous chapter. Also, the discussion will focus on steady isothermal problems since many of the procedures and results obtained for this type of problem can be extended in a straightforward manner to more complex situations. The solution methods described in Chapters 4 and 5 are also applicable here with only minor alterations.

For the steady-state case, the mixed method Eqs. (6.3.2) and (6.3.3) take the form

$$\mathbf{C}(\mathbf{u})\mathbf{u} + \mathbf{K}(\mu, \mathbf{u}) - \mathbf{Q}\mathbf{P} = \mathbf{F} \qquad (6.4.1a)$$

$$-\mathbf{Q}^T\mathbf{u} = 0 \qquad (6.4.1b)$$

or, as a single matrix equation we have

$$\hat{\mathbf{K}}(\mathbf{U})\mathbf{U} = \hat{\mathbf{F}} \qquad (6.4.2)$$

where the vector $\mathbf{U}$ now contains the velocity components $(\mathbf{u}_1, \mathbf{u}_2, \mathbf{u}_3)$ and the pressure $\mathbf{P}$. The dependence of μ on $\mathbf{u}$ has been absorbed into $\hat{\mathbf{K}}$. There are a wide variety of iterative algorithms that can be applied to Eq. (6.4.2), and we will consider only two such schemes in detail here. Similar comments apply to the penalty method model from Eq. (6.3.14).

The simplest method is Picard's method (also known as successive substitution or functional iteration), which can be written in the following form

$$\hat{\mathbf{K}}(\mathbf{U}^n)\mathbf{U}^{n+1} = \hat{\mathbf{F}}^n \qquad (6.4.3)$$

where the superscript n indicates the iteration level. Equation (6.4.3) describes a particularly simple procedure in which the nonlinear coefficients in the problem are evaluated using dependent variable data from the previous iteration. The method has a linear rate of convergence but works for a relatively large range of problems. For shear thinning materials it is observed that as the power law index, n [see Eq. (6.2.16) for example] decreases, the rate of convergence decreases markedly (see [5]).

A more sophisticated algorithm with a higher rate of convergence is Newton's method. For the nonlinear equation in (6.4.2) Newton's method may be written as

$$\mathbf{J}(\mathbf{U}^n)\left[\mathbf{U}^{n+1} - \mathbf{U}^n\right] = -\hat{\mathbf{K}}(\mathbf{U}^n)\mathbf{U}^n + \hat{\mathbf{F}}^n \qquad (6.4.4)$$

where $\mathbf{J}$ is the Jacobian matrix defined by

$$\mathbf{J}(\mathbf{U}^n) = \frac{\partial}{\partial \mathbf{U}}\left[\hat{\mathbf{K}}(\mathbf{U})\mathbf{U} - \hat{\mathbf{F}}\right]\Big|_{\mathbf{U}^n} \qquad (6.4.5)$$

Newton's method is the standard solution procedure for Newtonian problems since it is well-suited to the quadratic nonlinearity occurring in the advection terms of the momentum equations. However, experience has shown that this procedure, as written in Eq. (6.3.29), does not perform well for many types of generalized Newtonian fluids (see [18]). In particular, Newton's method does not work well for viscosity models with shear thinning. It is therefore recommended that the non-Newtonian behavior in Eq. (6.4.4) be treated using a Picard method; the advection terms, if present, should be treated using Newton's method.

To illustrate the behavior of the Picard and Newton algorithms, a simple creeping flow problem ($Re = 0$) was solved using a power-law viscosity model. The problem consists of the ubiquitous driven cavity flow in which a fluid is contained in a square cavity three sides of which are stationary; the fourth side of the cavity moves at unit velocity in its own plane. Shown in Figure 6.4.1 are plots of the convergence measure (relative norm on the change in the solution between iterations) versus iteration number for both algorithms and several values of the power-law index. The Picard scheme converges for all values of the index below unity (shear thinning) but diverges for shear thickening fluids. The Newton method performs in the opposite way, with rapid convergence for shear thickening and divergence for smaller values of the index.

Other methods of solution of Eq. (6.4.4) are possible though the Picard and Newton methods represent the behavioral limits for most algorithms. Tanner and Milthorpe [18] have used a combination of a Picard and Newton method successfully for several nonlinear viscosity models. Also, Engelman [19] has found the quasi-Newton or variable metric method to be very cost effective for non-Newtonian flows.

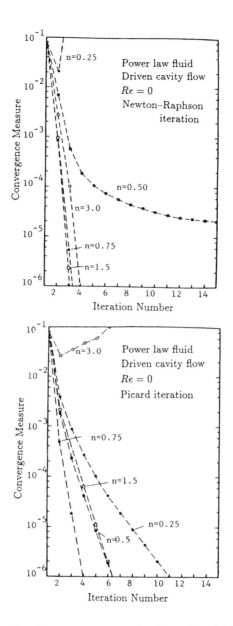

Figure 6.4.1: Iterative convergence for Picard and Newton methods.

The solution procedures outlined above are applicable to all of the common inelastic, non-Newtonian models. However, the Bingham model [Eqs. (6.2.20) and (6.2.21)] deserves some further comments due to its unique method of implementation. For a typical velocity-based finite element method, the exact Bingham constitutive equation is quite difficult to use, mainly as a result of the requirement that no motion take place below the yield condition. To circumvent this difficulty it has proved useful to employ an approximate Bingham equation given by

$$\tau_{ij} = \left(\frac{\tau_0(1 - \mu/\mu_r)}{\sqrt{I_2}} + 2\mu\right) D_{ij} \quad \text{when} \quad \frac{1}{2}tr(\tau^2) \geq \tau_0^2 \qquad (6.4.6a)$$

$$\tau_{ij} = 2\mu_r D_{ij} \qquad\qquad\qquad \text{when} \quad \frac{1}{2}tr(\tau^2) < \tau_0^2 \qquad (6.4.6b)$$

where μ_r is a pre-yield viscosity and $\mu/\mu_r \ll 1$. A plot of the constitutive model given by (6.4.6) is shown in Figure 6.4.2. Unlike the true Bingham material, the fluid described by Eq. (6.4.6) can undergo deformation below the yield point though the magnitude of the motion can be made arbitrarily small (i.e., approach the Bingham model) by increasing μ_r relative to μ. Experience has shown that when $\mu/\mu_r \sim .01$ or less then the solution is virtually independent of μ_r and thus approximates the Bingham model with excellent accuracy [17,18].

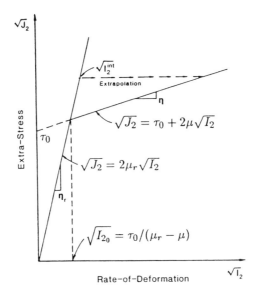

Figure 6.4.2: Bi-viscosity model for computations involving Bingham fluids.

Iterative convergence of Bingham fluids using a Picard method [see Eq. (6.4.3)] can be very slow since the apparent viscosity after yield is essentially like that of a power-law fluid with index $n = 0$. To accelerate the convergence a particular form of extrapolation can be used. Note first that for the present model the yield stress, τ_0, corresponds to a value of the rate of deformation (invariant) given by $\sqrt{I_{2_0}} = \tau_0/(\mu_r - \mu)$ as seen in Figure 6.4.2. At the start of the iterative procedure the fluid is usually assumed to have a viscosity, μ_r, from which an initial velocity field and values of I_2 may be computed. For those elements which have yielded ($\sqrt{I_2} \geq \sqrt{I_{2_0}}$) an extrapolation at constant stress should be used to estimate a new value of the apparent viscosity. Failure to perform this extrapolation can double or triple the number of iterations needed to reach convergence. Based on Figure 6.4.2 (dashed arrow) it is possible to provide an analytical expression for the projected value of I_2 needed for the viscosity evaluation. That is,

$$\sqrt{I_2} = \frac{2\mu_r\sqrt{I_2^{int}} - \tau_0}{2\mu} \tag{6.4.7}$$

where the superscript *int* indicates the value of I_2 computed initially using a viscosity of μ_r. The above procedure has been used in a variety of steady-state problems with good success [16,17] and is readily extended to time-dependent flows.

6.5 Governing Equations of Viscoelastic Fluids

As noted in the introduction, non-Newtonian behavior has many facets. In the previous sections we dealt with the modeling of "shear rate" dependent viscosity behavior but ignored the problems associated with fluid elasticity and memory. In the following several sections we will focus on viscoelastic fluids and their numerical simulation by the finite element method.

6.5.1 Conservation Equations

The analytical description of the motion of a continuous medium is based on conservation of mass, momentum, and energy, and the associated equations of state and constitutive relations. The present development will be limited to laminar flows of incompressible, viscous isotropic fluids. The fluid motion is assumed to be isothermal to avoid the discussion of the energy equation, as it was discussed in detail in Chapter 5. We also assume that, for simplicity, the flows are two-dimensional. Of course, the extension of the present discussion to three dimensions is conceptually straightforward, although in practice the subject is sufficiently complex and computationally taxing to have received relatively little attention. The discussion will also be limited to simple fluids with fading memory [4–7,20,21], a material description that has received the most widespread attention and development.

The equations of interest for two-dimensional flows are written in the Cartesian coordinate system (x_1, x_2) using the Eulerian description:

Conservation of Mass:

$$\frac{\partial u_i}{\partial x_i} = 0 \tag{6.5.1}$$

Conservation of Momentum:

$$\rho\left(\frac{\partial u_i}{\partial t} + u_j\frac{\partial u_i}{\partial x_j}\right) = \frac{\partial \sigma_{ij}}{\partial x_j} + \rho g_i \tag{6.5.2}$$

where u_i denotes the *i*th component of velocity, σ_{ij} the components of the total (Cauchy) stress tensor, ρ the density, g_i the component of body force, and t denotes time. The standard index notation is used, with summation on repeated indices.

The field equations in (6.5.1) and (6.5.2) are to be solved in conjunction with a constitutive equation for the fluid and boundary and initial conditions of the flow problem. The various constitutive equations of viscoelastic fluid behavior are a major focus of this chapter, and will be discussed in the next section. Since the boundary conditions depend on the dependent variables employed in the problem, they are discussed subsequent to the discussion of the constitutive relations.

6.5.2 Constitutive Equations

For an incompressible fluid the total stress is given by

$$\sigma_{ij} = -P\delta_{ij} + \tau_{ij} \tag{6.5.3}$$

where P is a pressure and τ_{ij} is the extra-stress. It should be noted that due to the particular forms that will be used to describe the extra-stress, the pressure defined in Eq. (6.5.3) will not, in general, be the same as a Newtonian pressure or hydrostatic pressure. That is, the extra-stress may not be traceless (i.e., $\tau_{ii} \neq 0$), in which case P is not the usual mean normal stress found in Newtonian fluid mechanics [22,23].

The central issue for the constitutive description of a viscoelastic fluid is the choice of an equation that relates the extra-stress in (6.5.3) to the flow kinematics. For the general class of materials called simple fluids, such a relationship can be established in functional form where the current extra-stress is related to the history of deformation in the fluid. Typically, such a relation is expressed as

$$\tau_{ij} = \mathcal{F}_{ij}[G_{kl}(s), \ 0 < s < \infty] \tag{6.5.4}$$

where $\mathcal{F}_{ij}$ is a tensor-valued functional, G_{kl} is a finite deformation tensor (related to the Cauchy–Green tensor), and $s = t - t'$ is the time lapse from time t' to the present time, t. Unfortunately, the generality of a functional form, such as (6.5.4), does not produce usable constitutive equations for general flow problems; the solution of practical problems requires that approximate forms of (6.5.4) be developed.

Since there are numerous ways to approximate the general functional given in (6.5.4), a wide variety of constitutive equations have developed out of the simple fluid theory. To date, none of these approximate constitutive equations are universally applicable; that is, no single constitutive equation is capable of predicting all of the observed behavior in elastic fluids. In essence, the present state-of-the-art mandates that a spectrum of constitutive equations be considered for use, with the specific choice being dictated by the ability of a given model to predict the dominant, non-Newtonian effects expected in a particular application.

To further complicate the situation, approximate constitutive equations can be of several different forms. For certain limiting flow conditions, such as "small" strains, "slow" flows, or "slightly" elastic fluids, the functional in (6.5.4) can be approximated by an expansion in terms of a small parameter. Such a formulation leads to a set of "hierarchical" equations for the extra-stress in terms of the rate of deformation tensor and its various derivatives [5,6]. In these models the extra-stress is given explicitly in terms of the flow kinematics, which in principle, allows Eqs. (6.5.1)–(6.5.3) to be written in terms of only the velocity and pressure, as in a Newtonian flow. Though convenient in form, these hierarchical formulas are of very limited value in general flows due to the "smallness" assumptions used in their derivation. Some computational work has been done using these types of models [24–26]; however, due to their limited applicability they will not be given further consideration here.

The two remaining major categories of approximate constitutive relations include the integral and differential models. The integral model represents the extra-stress in terms of an integral over past time of the fluid deformation history. For a differential model the extra-stress is determined from a differential (evolution) equation that relates the stress and stress rate to the flow kinematics. The choice between a differential and integral formulation is a crucial one with regard to numerical simulation since the computational algorithms for each type are very different. In the present book we will emphasize the differential form since the majority of the current numerical work involves this formulation. References to work with integral models will also be given.

Differential models

The differential constitutive equations to be considered here are implicit, rate-type models, generally associated with the names of Oldroyd, Maxwell, and Jeffrey. Before proceeding to the description of specific models it is important to list a number of definitions. For an Eulerian reference frame the material time derivative (or convected derivative) of a symmetric, second-order tensor can be defined in several ways, all of which are frame invariant. Let φ_{ij} denote the Cartesian components of a second-order tensor. Then the *upper-convected* (or codeformational) derivative is defined by

$$\overset{\nabla}{\varphi}_{ij} = \frac{\partial \varphi_{ij}}{\partial t} + u_k \frac{\partial \varphi_{ij}}{\partial x_k} - L_{ik}\varphi_{kj} - L_{jk}\varphi_{ki} \qquad (6.5.5a)$$

and the *lower-convected* derivative is defined as

$$\overset{\Delta}{\varphi}_{ij} = \frac{\partial \varphi_{ij}}{\partial t} + u_k \frac{\partial \varphi_{ij}}{\partial x_k} + L_{ki}\varphi_{kj} + L_{kj}\varphi_{ki} \qquad (6.5.5b)$$

where u_k are the components of the velocity vector and L_{ij} are the components of the velocity gradient tensor $\mathbf{L}$, defined by

$$\mathbf{L} = \nabla \mathbf{u}, \quad \text{or} \quad L_{ij} = \frac{\partial u_j}{\partial x_i} \qquad (6.5.6)$$

Since both Eqs. (6.5.5a) and (6.5.5b) are admissible convected derivatives, their linear combination is also admissible:

$$\overset{\circ}{\varphi}_{ij} = \left(1 - \frac{a}{2}\right)\overset{\nabla}{\varphi}_{ij} + \left(\frac{a}{2}\right)\overset{\Delta}{\varphi}_{ij} \qquad (6.5.7)$$

Equation (6.5.7) is a general convected derivative which reduces to (6.5.5a) for $a = 0$ and (6.5.5b) for $a = 2$. When $a = 1$ [average of (6.5.5a) and (6.5.5b)] the convected derivative in (6.5.7) is termed a *corotational* or *Jaumann derivative*.

Using τ as the extra-stress tensor with the Cartesian components denoted by τ_{ij}, the upper-convected derivative of the stress components is given by

$$\overset{\triangledown}{\tau}_{ij} = \frac{\partial \tau_{ij}}{\partial t} + u_k \frac{\partial \tau_{ij}}{\partial x_k} - L_{ik}\tau_{kj} - L_{jk}\tau_{ki} \qquad (6.5.8a)$$

and the lower-convected derivative is

$$\overset{\triangle}{\tau}_{ij} = \frac{\partial \tau_{ij}}{\partial t} + u_k \frac{\partial \tau_{ij}}{\partial x_k} + L_{ki}\tau_{kj} + L_{kj}\tau_{ki} \qquad (6.5.8b)$$

The corotational derivative is

$$\overset{\circ}{\tau}_{ij} = \left(1 - \frac{a}{2}\right)\overset{\triangledown}{\tau}_{ij} + \left(\frac{a}{2}\right)\overset{\triangle}{\tau}_{ij} \qquad (6.5.9)$$

All of these derivatives have been used in various differential constitutive equations. The selection of one type of derivative over another is usually based on the physical plausibility of the resulting constitutive equation and the matching of experimental data to the model for simple (viscometric) flows.

The simplest differential constitutive models are the upper- and lower-convected Maxwell fluids, which are defined by the following equations:

Upper-Convected Maxwell Fluid:

$$\tau_{ij} + \lambda \overset{\triangledown}{\tau}_{ij} = 2\mu^P D_{ij} \qquad (6.5.10a)$$

Lower-Convected Maxwell Fluid:

$$\tau_{ij} + \lambda \overset{\triangle}{\tau}_{ij} = 2\mu^P D_{ij} \qquad (6.5.10b)$$

where λ is a relaxation time for the fluid, μ^P is a viscosity, and D_{ij} are the components of the rate of deformation (or strain rate) tensor defined by

$$\mathbf{D} = \frac{1}{2}\left[\nabla\mathbf{u} + (\nabla\mathbf{u})^T\right], \text{ or } D_{ij} = \frac{1}{2}(L_{ij} + L_{ji}) \qquad (6.5.11)$$

The upper-convected Maxwell model (6.5.10a) has been used extensively in testing numerical algorithms; the lower-convected and corotational forms of the Maxwell fluid predict physically unrealistic behavior and are not generally used.

By employing the general convected derivative (6.5.9) in a Maxwell-like model the following constitutive equation is produced:

Johnson–Segalman Model:

$$\tau_{ij} + \lambda \overset{\circ}{\tau}_{ij} = 2\mu^P D_{ij} \qquad (6.5.12)$$

which is the Johnson–Segalman fluid [27]. By slightly modifying Eq. (6.5.12) to include a variable coefficient for τ_{ij}, the Phan Thien–Tanner model [28] is derived.

Phan Thien–Tanner Model:

$$Y(\tau)\tau_{ij} + \lambda\overset{\circ}{\tau}_{ij} = 2\mu^p D_{ij} \qquad (6.5.13a)$$

where

$$Y(\tau) = 1 + \frac{\epsilon\lambda}{\mu^p}\ tr(\tau) \qquad (6.5.13b)$$

and ϵ is a constant. This equation is somewhat better than (6.5.12) in representing actual material behavior.

The Johnson–Segalman and Phan Thien–Tanner models suffer a common defect in that, for a monotonically increasing shear rate, there is a region where the shear stress decreases — a physically unrealistic behavior. To correct this anomaly the constitutive equations can be altered using the following procedure. Let the extra-stress be decomposed into two partial stresses, τ_{ij}^s and τ_{ij}^p such that

$$\tau_{ij} = \tau_{ij}^s + \tau_{ij}^p \qquad (6.5.14)$$

where τ_{ij}^s is a purely viscous and τ_{ij}^p is a viscoelastic stress component. The decomposition used here is sometimes thought of in terms of a combination of a Newtonian solvent (superscript s) and a polymer additive (superscript p). Using the Johnson–Segalman fluid as an example, then

$$\tau_{ij}^s = 2\mu^s\ D_{ij}$$

$$\tau_{ij}^p + \lambda\overset{\circ}{\tau}_{ij}^p = 2\mu^p\ D_{ij} \qquad (6.5.15)$$

It is possible to eliminate the partial stresses in (6.5.14) and (6.5.15) to produce a new constitutive relation of the following form

$$\tau_{ij} + \lambda\overset{\circ}{\tau}_{ij} = 2\bar{\mu}\left(D_{ij} + \lambda'\overset{\circ}{D}_{ij}\right) \qquad (6.5.16)$$

with $\bar{\mu} = (\mu^s + \mu^p)$ and $\lambda' = \lambda\mu^s/\bar{\mu}$, where λ' is a retardation time. The constitutive equation in (6.5.16) is recognized as a type of Oldroyd fluid – it is, in fact, a reduced form of the eight-constant Oldroyd model [1,5]. For particular choices of the convected derivative in Eq. (6.5.16), specific models can be generated. When $a = 0$ ($\overset{\circ}{\tau}_{ij}$ becomes $\overset{\triangledown}{\tau}_{ij}$) then (6.5.14) becomes the Oldroyd B fluid; the case $a = 2$ ($\overset{\circ}{\tau}_{ij}$ becomes $\overset{\triangle}{\tau}_{ij}$) produces the Oldroyd A fluid. In order to ensure a monotonically increasing shear stress the inequality $\mu^s \geq \mu^p/8$ must be satisfied. The stress decomposition employed above can also be used with the Phan Thien–Tanner model to produce a correct shear stress behavior.

In all of the above constitutive equations the material parameters, λ and μ^p, were assumed to be constants. For some constitutive equations the constancy of these parameters leads to material (or viscometric) functions that do not accurately represent the behavior of real elastic fluids. For example, the shear viscosity predicted by a Maxwell fluid is a constant, when in fact viscoelastic fluids normally exhibit a shear thinning behavior. This situation can be remedied to some degree by allowing the parameters λ and μ^p to be functions of the invariants of the rate of deformation tensor as was done for the generalized Newtonian fluid (see Section 6.2.3). Using the upper-convected Maxwell fluid as an example, then

$$\tau_{ij} + \lambda(I_2)\overset{\triangledown}{\tau}_{ij} = 2\mu^p(I_2)D_{ij} \tag{6.5.17}$$

where I_2 is the second invariant of the strain rate tensor $\mathbf{D}$ [see Eq. (6.5.11)], $I_2 = 1/2(D_{ij}D_{ij})$. The constitutive equation in (6.5.17) is usually termed a White–Metzner model [29]. White–Metzner forms of other differential models, such as the Oldroyd fluids, have also been developed and used in various situations.

The above list of differential models is by no means exhaustive though it does include many of the constitutive equations that have been used in computational work. Though the equations have been stated without derivation or motivation, they all rest on a reasonable theoretical basis. Some of the equations (e.g., Oldroyd and Maxwell) were developed as generalizations of simple, linear viscoelastic models. Linear viscoelasticity can be formally developed from the previously cited "hierarchical" equations or informally from mechanical analogies and heuristic arguments [1]. Other differential models, such as the Johnson–Segalman and Phan Thien–Tanner fluids, were derived using statistical mechanics ideas and a conceptual (network) model for the microstructure of a viscoelastic fluid. Due to their approximate nature, all of the above models are limited in their ability to represent true viscoelastic behavior; the more complex models provide a reasonable qualitative description of many observed phenomena. A catalog of the strengths, weaknesses, and limitations of these and other differential models can be found in [1,4,5,23].

Integral models

An approximate integral model for a viscoelastic fluid represents the extra-stress in terms of an integral over the past history of the fluid deformation. A general form for a single integral model can be expressed as (see [30])

$$\tau_{ij} = \int_{-\infty}^{t} 2m(t-t')H_{ij}(t,t')dt' \tag{6.5.18}$$

where t is the current time, m is a scalar memory function (or relaxation kernel), and H_{ij} is a nonlinear deformation measure (tensor) between the past time, t', and t.

There are many possible forms for both the memory function and the deformation measure. Normally the memory function is a decreasing function of the time lapse $s = t - t'$. Typical of such a function is the exponential given by

$$m(t - t') = m(s) = \frac{\mu_0}{\lambda^2} \, e^{-s/\lambda} \qquad (6.5.19)$$

where the parameters μ_0, λ, and s were defined previously. Like the choice of a convected derivative in a differential model, the selection of a deformation measure for use in Eq. (6.5.18) is somewhat arbitrary. One particular form that has received some attention is given by

$$H_{ij} = \phi_1(I, II)C_{ij}^{-1} + \phi_2(I, II)C_{ij} \qquad (6.5.20)$$

In Eq. (6.5.20) C_{ij} is the Cauchy–Green deformation tensor, C_{ij}^{-1} is its inverse, called the Finger tensor, and the ϕ_i are scalar functions of the invariants of the deformation tensors, i.e., $I = tr(C_{ij}^{-1})$ and $II = tr(C_{ij})$. The indicated deformation tensor is defined by (see Malvern [31])

$$C_{ij} = \frac{\partial \chi_m}{\partial x_i} \frac{\partial \chi_m}{\partial x_j} \qquad (6.5.21)$$

where χ_m is the location at time t' of a fluid particle that is now at position x_i and time t. The form of the deformation measure in Eq. (6.5.20) is still quite general, though specific choices for the functions ϕ_i and the memory function m lead to several well-known constitutive models. Among these are the Kaye–BKZ fluid [1,4,23] and the Lodge rubber-like liquid [32].

As a specific example of an integral model of the type given by the previous equations we consider the Maxwell fluid. Setting $\phi_1 = 1$ and $\phi_2 = 0$ in Eq. (6.5.20) and using the memory function shown in (6.5.19) allow a constitutive equation of the following form

$$\tau_{ij} = \frac{\mu_0}{\lambda^2} \int_{-\infty}^{t} exp\left[-(t - t')/\lambda\right] \left[C_{ij}^{-1}(t') - \delta_{ij}\right] dt' \qquad (6.5.22)$$

In writing Eq. (6.5.22), the Finger deformation tensor in (6.5.20) has been replaced with the more usual Finger strain tensor (e.g., [4,30,31]). The above constitutive equation is an integral equivalent to the upper-convected Maxwell model shown in differential form in Eq. (6.5.10a). Note that in this case the extra-stress is given in an explicit form, though its evaluation requires that the strain history be known for each fluid particle. Also, it is important to emphasize that though the Maxwell fluid has both a differential and integral form, this is not generally true for other constitutive equations.

Since the emphasis here is on differential models we will not dwell on other specific forms of integral constitutive equations. A good source for further discussion of such models is the book by Bird et al. [1]. It is appropriate to note in closing this section that some of the integral models are very good at reproducing realistic viscoelastic behavior [4,23]. However, this improvement in modeling accuracy is offset to a large degree by the difficulties in using such models in general computational schemes. Further comments on computational procedures for integral models will be reserved for a later section.

6.5.3 Boundary Conditions

The equations in (6.5.1) and (6.5.2) are to be solved in conjunction with an appropriate constitutive relation as discussed in the previous section. The set of boundary conditions for the problem at hand consists of either specified velocities or tractions on the boundary of the fluid domain, with one condition needed for each component of velocity. However, there is also an additional set of conditions for the viscoelastic problem due to the "memory" of the fluid. The extra condition, in reality, is an initial condition on the components of the fluid extra-stress, though it is often implemented as a boundary condition, especially for flow-through type boundaries. For time-dependent problems the initial fluid stresses must be given at all points in the problem domain. All problems that contain an inflow boundary require the specification of a strain (stress) history of the fluid crossing the boundary. Completely confined flows do not require any additional data beyond the initial stress-state. Additional comments on boundary conditions can be found in [33].

6.6 Finite Element Model of Differential Form

6.6.1 Preliminary Comments

In this section we will outline typical finite element procedures for the solution of viscoelastic flow problems. At the outset it is important to realize that a standard, well-established computational procedure has not yet evolved for this particular class of flow problems. Numerous different formulations have been proposed [21], but none has proved adequate for the solution of flows with highly elastic fluids. Further comments on the unresolved general problems in this area will be given in a later section.

The wide variety of possible constitutive equations and finite element formulations makes detailed explanation of many specific algorithms impossible in this limited text. Therefore, the detailed discussion will be focused on a single, mixed finite element formulation for a particular differential constitutive equation. Other models and formulations follow easily from this basic outline. A brief description of typical procedures for integral models will also be presented.

6.6.2 Summary of Governing Equations

The governing equations of interest are summarized below in rectangular Cartesian form.

Conservation of Mass:

$$\frac{\partial u_i}{\partial x_i} = 0 \qquad (6.6.1)$$

Conservation of Momentum:

$$\rho \left(\frac{\partial u_i}{\partial t} + u_j \frac{\partial u_i}{\partial x_j} \right) = \frac{\partial \sigma_{ij}}{\partial x_j} \qquad (6.6.2)$$

Constitutive Equations:

$$\sigma_{ij} = -P\delta_{ij} + \tau_{ij} \qquad (6.6.3)$$

$$\tau_{ij} = \tau_{ij}^s + \tau_{ij}^p \qquad (6.6.4)$$

$$\tau_{ij}^s = 2\mu^e D_{ij} \qquad (6.6.5)$$

$$\tau_{ij}^p + \lambda \overset{\nabla}{\tau}_{ij}^p = 2\mu^p D_{ij} \qquad (6.6.6)$$

where in Eq. (6.6.2) the body force term has been neglected for simplicity. The constitutive equation employed is the Oldroyd B model. Other models could be included by adding a generic function of the stress to the right-hand side of the constitutive equation. However, the basic algorithm remains unchanged. Also, note that the solvent or viscous stress in (6.6.5) has been expressed in terms of an effective viscosity. In many cases, this is taken as the solvent viscosity ($\mu^e = \mu^s$) as illustrated in equation (6.5.15). In other situations, the stress split is made arbitrary and μ^e represents only a part of the total viscosity. The permissible variation in this definition has been used to advantage in developing solution algorithms.

The above set of equations can be combined in various ways, each of which leads to a slightly different finite element procedure. For example, combining Eqs.(6.6.3)–(6.6.5) and substituting for σ_{ij} in the momentum balance (6.6.2) leads to the following set of equations:

$$\frac{\partial u_i}{\partial x_i} = 0 \qquad (6.6.7)$$

$$\rho \left(\frac{\partial u_i}{\partial t} + u_j \frac{\partial u_i}{\partial x_j} \right) = -\frac{\partial P}{\partial x_i} + \frac{\partial}{\partial x_j} (2\mu^e D_{ij}) + \frac{\partial}{\partial x_j} \left(\tau_{ij}^p \right) \qquad (6.6.8)$$

$$\tau_{ij}^p + \lambda \overset{\nabla}{\tau}_{ij}^p = 2\mu^p D_{ij} \qquad (6.6.9)$$

This equation set is a natural extension of work on a Newtonian fluid, and was the first one considered by workers in viscoelastic flows [34–36]. Much of the early work centered on the Upper Convected Maxwell (UCM) fluid ($\mu^e = 0$) in the creeping or Stokes flow limit. Under these circumstances, as pointed out by Crochet et al. [5], the mixed finite element formulation associated with (6.6.7)–(6.6.9) has some restrictions with regard to the choice of approximating functions for the extra-stress variables.

Another possible formulation was proposed by Chang et al. [37] and is the one that will be described here, in some detail. The implicit constitutive equation (6.6.6) can be rearranged and substituted into (6.6.8). The result is a set of equations of the following form

$$\frac{\partial u_i}{\partial x_i} = 0 \tag{6.6.10}$$

$$\rho\left(\frac{\partial u_i}{\partial t} + u_j\frac{\partial u_i}{\partial x_j}\right) = -\frac{\partial P}{\partial x_i} + \frac{\partial}{\partial x_j}(2\bar{\mu}D_{ij}) - \frac{\partial}{\partial x_j}\left(\lambda\overset{\nabla}{\tau}{}^p_{ij}\right) \tag{6.6.11}$$

$$\tau^p_{ij} + \lambda\overset{\nabla}{\tau}{}^p_{ij} = 2\mu^p D_{ij} \tag{6.6.12}$$

where $\bar{\mu} = (\mu^e + \mu^p)$, D_{ij} is defined in (6.5.11), and the stress rate, $\overset{\nabla}{\tau}_{ij}$, is given in (6.5.8a). The inclusion of the stress rate in the momentum equation appears to be of some benefit for numerical computations [5]. The finite element forms of the above two formulations are often termed MIX (mixed method) formulations and form the basis for many of the later advanced developments.

For two-dimensional geometries the equations in (6.6.10)–(6.6.12) provide six relations for the six unknowns, u_1, u_2, P, τ^p_{11}, τ^p_{22}, and τ^p_{12}. In an axisymmetric geometry the system would increase to seven equations with the circumferential (hoop) stress, τ^p_{33}, as the seventh unknown. The distinguishing feature of this system, as compared to a Newtonian problem, is the implicit nature of the constitutive equation that forces the extra-stresses to remain as dependent variables.

6.6.3 Finite Element Model

Application of the finite element procedure to the partial differential equations in (6.6.10)–(6.6.12) follows the standard format. The velocity components, pressure, and extra-stress components are approximated by expansions of the form

$$u_i = \sum_{m=1}^{M}\Psi_m u_i^m = \boldsymbol{\Psi}^T\mathbf{u}_i \tag{6.6.13a}$$

$$P = \sum_{n=1}^{N}\Phi_n P_n = \boldsymbol{\Phi}^T\mathbf{P} \tag{6.6.13b}$$

$$\tau_{ij} = \sum_{k=1}^{K}\Pi^k\tau_{ij}^k = \boldsymbol{\Pi}^T\boldsymbol{\tau}_{ij} \tag{6.6.13c}$$

where $\boldsymbol{\Phi}$, $\boldsymbol{\Psi}$, and $\boldsymbol{\Pi}$ are vectors of basis functions, $\mathbf{u}_i$, $\mathbf{P}$, and $\boldsymbol{\tau}_{ij}$ are vectors of nodal unknowns, and M, N, and K indicate the number of nodes at which the various unknowns are defined. The superscript p for the extra-stress has been omitted for clarity. Using the finite element approximations (6.6.13) in standard weighted-integral statements (i.e., weak forms) of the system in (6.6.10)–(6.6.12) (see Chapter 5), the following system of finite element equations can be obtained:

Momentum:

$$\left[\int_{\Omega^e} \rho_0 \boldsymbol{\Phi}\boldsymbol{\Phi}^T \mathbf{u}_j \frac{\partial \boldsymbol{\Phi}^T}{\partial x_j}\, d\mathbf{x}\right]\mathbf{u}_i + \left[\int_{\Omega^e} \mu \frac{\partial \boldsymbol{\Phi}}{\partial x_j}\frac{\partial \boldsymbol{\Phi}^T}{\partial x_j}\, d\mathbf{x}\right]\mathbf{u}_i + \left[\int_{\Omega^e} \mu \frac{\partial \boldsymbol{\Phi}}{\partial x_j}\frac{\partial \boldsymbol{\Phi}^T}{\partial x_i}\, d\mathbf{x}\right]\mathbf{u}_j$$

$$+ \left[\int_{\Omega^e} -\frac{\partial \boldsymbol{\Phi}}{\partial x_i}\boldsymbol{\Psi}^T\, d\mathbf{x}\right]\mathbf{P} - \left[\int_{\Omega^e} \lambda \frac{\partial \boldsymbol{\Phi}}{\partial x_j}\overset{\triangledown}{\tau}_{ij}\, d\mathbf{x}\right] = \left[\oint_{\Gamma^e} \boldsymbol{\Phi}\tau_{ij}n_j\, ds\right] \qquad (6.6.14)$$

Continuity:

$$\left[\int_{\Omega^e} -\boldsymbol{\Psi}\frac{\partial \boldsymbol{\Phi}^T}{\partial x_i}\, d\mathbf{x}\right]\mathbf{u}_i = 0 \qquad (6.6.15)$$

Constitutive:

$$\left[\int_{\Omega^e} \boldsymbol{\Pi}\ \boldsymbol{\Pi}^T\, d\mathbf{x}\right]\tau_{ij} + \left[\int_{\Omega^e} \boldsymbol{\Pi}\lambda\overset{\triangledown}{\tau}_{ij}\, d\mathbf{x}\right] - \left[\int_{\Omega^e} \mu^p \boldsymbol{\Pi}\frac{\partial \boldsymbol{\Phi}^T}{\partial x_j}\, d\mathbf{x}\right]\mathbf{u}_i$$

$$- \left[\int_{\Omega^e} \mu^p \boldsymbol{\Pi}\frac{\partial \boldsymbol{\Phi}^T}{\partial x_i}\, d\mathbf{x}\right]\mathbf{u}_j = 0 \qquad (6.6.16)$$

where the steady-state form of the equations has been considered. In arriving at the above equations, the Green–Gauss or divergence theorem has been used to reduce the second-order diffusion terms to first-order terms plus a boundary integral. The appearance of the boundary integral corresponds to the "natural" boundary conditions for the problem. Also, the stress rate has not been written out due to its complexity; the two-dimensional components of the stress rate are written below in terms of the finite element functions,

$$\overset{\triangledown}{\tau}_{11} = \boldsymbol{\Phi}^T\mathbf{u}_1\frac{\partial \boldsymbol{\Pi}^T}{\partial x_1}\tau_{11} + \boldsymbol{\Phi}^T\mathbf{u}_2\frac{\partial \boldsymbol{\Pi}^T}{\partial x_2}\tau_{11} - 2\frac{\partial \boldsymbol{\Phi}^T}{\partial x_1}\mathbf{u}_1\boldsymbol{\Pi}^T\tau_{11} - 2\frac{\partial \boldsymbol{\Phi}^T}{\partial x_2}\mathbf{u}_1\boldsymbol{\Pi}^T\tau_{12}$$

$$(6.6.17)$$

$$\overset{\triangledown}{\tau}_{22} = \boldsymbol{\Phi}^T\mathbf{u}_1\frac{\partial \boldsymbol{\Pi}^T}{\partial x_1}\tau_{22} + \boldsymbol{\Phi}^T\mathbf{u}_2\frac{\partial \boldsymbol{\Pi}^T}{\partial x_2}\tau_{22} - 2\frac{\partial \boldsymbol{\Phi}^T}{\partial x_1}\mathbf{u}_2\boldsymbol{\Pi}^T\tau_{12} - 2\frac{\partial \boldsymbol{\Phi}^T}{\partial x_2}\mathbf{u}_2\boldsymbol{\Pi}^T\tau_{22}$$

$$(6.6.18)$$

$$\overset{\triangledown}{\tau}_{12} = \boldsymbol{\Phi}^T\mathbf{u}_1\frac{\partial \boldsymbol{\Pi}^T}{\partial x_1}\tau_{12} + \boldsymbol{\Phi}^T\mathbf{u}_2\frac{\partial \boldsymbol{\Pi}^T}{\partial x_2}\tau_{12} - \frac{\partial \boldsymbol{\Phi}^T}{\partial x_1}\mathbf{u}_2\boldsymbol{\Pi}^T\tau_{11} - \frac{\partial \boldsymbol{\Phi}^T}{\partial x_2}\mathbf{u}_1\boldsymbol{\Pi}^T\tau_{22}$$

$$- \left(\frac{\partial \boldsymbol{\Phi}^T}{\partial x_1}\mathbf{u}_1 + \frac{\partial \boldsymbol{\Phi}^T}{\partial x_2}\mathbf{u}_2\right)\boldsymbol{\Pi}^T\tau_{12} \qquad (6.6.19)$$

Equations (6.6.17)–(6.6.19) can be substituted directly into (6.6.14) and (6.6.16) to produce the complete form of the finite element equations.

Once the form of the interpolation functions $\boldsymbol{\Phi}$, $\boldsymbol{\Psi}$, and $\boldsymbol{\Pi}$ is specified (i.e., a particular element is selected), and the geometry of the element is known (i.e., x_i) then the integrals in (6.6.14)–(6.6.16) may be evaluated to produce the required coefficient matrices. The integrals are evaluated via a numerical quadrature procedure, as discussed in Chapters 2 and 3. The discrete system is given by the following matrix equations (it is common to write the momentum and continuity equations as a single system):

$$
\begin{bmatrix} \mathbf{C}_i(\mathbf{u}_i) & 0 & 0 \\ 0 & \mathbf{C}_i(\mathbf{u}_i) & 0 \\ 0 & 0 & 0 \end{bmatrix} \begin{Bmatrix} \mathbf{u}_1 \\ \mathbf{u}_2 \\ \mathbf{P} \end{Bmatrix} + \begin{bmatrix} 2\mathbf{K}_{11}+\mathbf{K}_{22} & \mathbf{K}_{21} & \mathbf{Q}_1 \\ \mathbf{K}_{12} & \mathbf{K}_{11}+2\mathbf{K}_{22} & \mathbf{Q}_2 \\ \mathbf{Q}_1^T & \mathbf{Q}_2^T & 0 \end{bmatrix} \begin{Bmatrix} \mathbf{u}_1 \\ \mathbf{u}_2 \\ \mathbf{P} \end{Bmatrix}
$$

$$
+ \begin{bmatrix} \mathbf{R}_1^{11}(\mathbf{u}_1,\mathbf{u}_2) & \mathbf{R}_1^{22}(\mathbf{u}_1,\mathbf{u}_2) & \mathbf{R}_1^{12}(\mathbf{u}_1,\mathbf{u}_2) \\ \mathbf{R}_2^{11}(\mathbf{u}_1,\mathbf{u}_2) & \mathbf{R}_2^{22}(\mathbf{u}_1,\mathbf{u}_2) & \mathbf{R}_2^{12}(\mathbf{u}_1,\mathbf{u}_2) \\ 0 & 0 & 0 \end{bmatrix} \begin{Bmatrix} \tau_{11} \\ \tau_{22} \\ \tau_{12} \end{Bmatrix} = \begin{Bmatrix} \mathbf{F}_1 \\ \mathbf{F}_2 \\ 0 \end{Bmatrix} \quad (6.6.20)
$$

and for the constitutive equation

$$
\begin{bmatrix} \mathbf{N} & 0 & 0 \\ 0 & \mathbf{N} & 0 \\ 0 & 0 & \mathbf{N} \end{bmatrix} \begin{Bmatrix} \tau_{11} \\ \tau_{22} \\ \tau_{12} \end{Bmatrix} - \begin{bmatrix} 2\mathbf{D}_1(\mathbf{u}_1) & 0 & 2\mathbf{D}_2(\mathbf{u}_1) \\ 0 & 2\mathbf{D}_2(\mathbf{u}_2) & 2\mathbf{D}_1(\mathbf{u}_2) \\ \mathbf{D}_1(\mathbf{u}_2) & \mathbf{D}_2(\mathbf{u}_1) & \mathbf{D}_i(\mathbf{u}_i) \end{bmatrix} \begin{Bmatrix} \tau_{11} \\ \tau_{22} \\ \tau_{12} \end{Bmatrix}
$$

$$
+ \begin{bmatrix} \mathbf{C}_i^*(\mathbf{u}_i) & 0 & 0 \\ 0 & \mathbf{C}_i^*(\mathbf{u}_i) & 0 \\ 0 & 0 & \mathbf{C}_i^*(\mathbf{u}_i) \end{bmatrix} \begin{Bmatrix} \tau_{11} \\ \tau_{22} \\ \tau_{12} \end{Bmatrix} - \begin{bmatrix} 2\mathbf{L}_1 & 0 & 0 \\ 0 & 2\mathbf{L}_2 & 0 \\ \mathbf{L}_1 & \mathbf{L}_2 & 0 \end{bmatrix} \begin{Bmatrix} \mathbf{u}_1 \\ \mathbf{u}_2 \\ \mathbf{P} \end{Bmatrix} = \begin{Bmatrix} 0 \\ 0 \\ 0 \end{Bmatrix}
$$

$$(6.6.21)$$

where sum on repeated indices is implied. The coefficient matrices shown in Eqs. (6.6.20) and (6.6.21) are defined by

$$
\mathbf{C}_i(\mathbf{u}_j) = \int_{\Omega^e} \rho_0 \boldsymbol{\Phi}\boldsymbol{\Phi}^T \mathbf{u}_j \frac{\partial \boldsymbol{\Phi}^T}{\partial x_i}\, d\mathbf{x}, \quad \mathbf{K}_{ij} = \int_{\Omega^e} \bar{\mu} \frac{\partial \boldsymbol{\Phi}}{\partial x_i}\frac{\partial \boldsymbol{\Phi}^T}{\partial x_j}\, d\mathbf{x}
$$

$$
\mathbf{Q}_i = -\int_{\Omega^e} \frac{\partial \boldsymbol{\Phi}}{\partial x_i} \boldsymbol{\Psi}^T\, d\mathbf{x}, \qquad \mathbf{F}_i = \oint_{\Gamma^e} \boldsymbol{\Phi}\tau_{ij}n_j\, ds
$$

$$
\mathbf{N} = \int_{\Omega^e} \boldsymbol{\Pi}\,\boldsymbol{\Pi}^T\, d\mathbf{x}, \qquad \mathbf{C}_i^*(\mathbf{u}_j) = \int_{\Omega^e} \lambda\boldsymbol{\Pi}\boldsymbol{\Phi}^T \mathbf{u}_j \frac{\partial \boldsymbol{\Pi}^T}{\partial x_i}\, d\mathbf{x}
$$

$$
\mathbf{D}_i(\mathbf{u}_j) = \int_{\Omega^e} \lambda\boldsymbol{\Pi}\frac{\partial \boldsymbol{\Phi}^T}{\partial x_i}\mathbf{u}_j\boldsymbol{\Pi}^T\, d\mathbf{x}, \qquad \mathbf{L}_i = \int_{\Omega^e} \mu^p\boldsymbol{\Pi}\frac{\partial \boldsymbol{\Phi}^T}{\partial x_i}\, d\mathbf{x} \quad (6.6.22)
$$

and the matrices $\mathbf{R}_k^{ij}$ are defined by

$$
\begin{aligned}
\mathbf{R}_1^{11} &= -\mathbf{S}_{11}(\mathbf{u}_1) - \mathbf{S}_{12}(\mathbf{u}_2) + 2\mathbf{T}_{11}(\mathbf{u}_2) + \mathbf{T}_{21}(\mathbf{u}_2) \\
\mathbf{R}_1^{22} &= \mathbf{T}_{22}(\mathbf{u}_1) \\
\mathbf{R}_1^{12} &= -\mathbf{S}_{21}(\mathbf{u}_1) - \mathbf{S}_{22}(\mathbf{u}_2) + 2\mathbf{T}_{12}(\mathbf{u}_1) + \mathbf{T}_{21}(\mathbf{u}_1) + \mathbf{T}_{22}(\mathbf{u}_2) \\
\mathbf{R}_2^{11} &= \mathbf{T}_{11}(\mathbf{u}_2) \\
\mathbf{R}_2^{22} &= -\mathbf{S}_{21}(\mathbf{u}_1) - \mathbf{S}_{22}(\mathbf{u}_2) + \mathbf{T}_{12}(\mathbf{u}_1) + 2\mathbf{T}_{22}(\mathbf{u}_2) \\
\mathbf{R}_2^{12} &= -\mathbf{S}_{12}(\mathbf{u}_2) - \mathbf{S}_{11}(\mathbf{u}_1) + 2\mathbf{T}_{21}(\mathbf{u}_2) + \mathbf{T}_{12}(\mathbf{u}_2) + \mathbf{T}_{11}(\mathbf{u}_1) \quad (6.6.23)
\end{aligned}
$$

The $\mathbf{S}$ and $\mathbf{T}$ matrices are defined as

$$
\mathbf{S}_{ij}(\mathbf{u}_k) = \int_{\Omega^e} \lambda\frac{\partial \boldsymbol{\Phi}}{\partial x_i}\boldsymbol{\Phi}^T \mathbf{u}_k \frac{\partial \boldsymbol{\Pi}^T}{\partial x_j}\, d\mathbf{x}
$$

$$
\mathbf{T}_{ij}(\mathbf{u}_k) = \int_{\Omega^e} \lambda\frac{\partial \boldsymbol{\Phi}}{\partial x_i}\frac{\partial \boldsymbol{\Phi}^T}{\partial x_j}\mathbf{u}_k\boldsymbol{\Pi}^T\, d\mathbf{x} \quad (6.6.24)
$$

Finally, Eqs. (6.6.20) and (6.6.21) can be expressed symbolically as

$$\mathbf{C}(\mathbf{u})\mathbf{U} + \mathbf{K}\mathbf{U} + \mathbf{R}(\mathbf{u})\tau = \mathbf{F} \qquad (6.6.25)$$

$$\mathbf{N}\tau - \mathbf{D}(\mathbf{u})\tau + \mathbf{C}^*(\mathbf{u})\tau - \mathbf{L}\mathbf{U} = 0 \qquad (6.6.26)$$

where the vector of unknowns is $\mathbf{U}^T = \{\mathbf{u}_1^T, \mathbf{u}_2^T, \mathbf{P}^T\}$. Equations (6.6.25) and (6.6.26) represent the finite element equations for a viscoelastic (Oldroyd B) fluid for the steady-state case; very little consideration has been given to the time-dependent case. The displayed matrix system is highly nonlinear and fully coupled. Equation (6.6.25) is recognized as the standard form for a Newtonian problem with the addition of the $\mathbf{R}(\mathbf{u})\sigma$ term which comes from the stress rate. Equation (6.6.26) is the finite element analog of the implicit constitutive equation and is seen to be an equation of the advection type for the extra-stress.

Before discussing solution procedures for the matrix system, some comments on the approximations used in (6.6.13) are required. In defining the approximations to the dependent variables all of the basis functions were taken to be different. From experience with mixed methods for Newtonian problems it is well-known that the pressure approximation should be of lower order than the velocity approximation. For most applications in viscoelastic simulations this condition has been met by using quadratic functions for the velocity and linear approximations for the pressure. The choice of an appropriate approximation for the extra-stress is less clearly defined. As mentioned earlier, the problem formulation based on Eqs. (6.6.7)–(6.6.9) has a restriction on the permissible extra-stress basis functions. A solvability condition for the resulting matrix problem requires that the extra-stresses be approximated to the same order as the velocity components [5]. In the present formulation there are no such restrictions and the choice of the interpolation functions used for stress components is quite arbitrary. Heuristic arguments were made for having the extra-stress and pressure be of the same order since they are both "stress-like" variables. However, many studies using both linear and quadratic stress approximations have been produced and in virtually all cases the mixed method has been limited to flows with small fluid elasticity. In the mid 1980s Marchal and Crochet [38,39] demonstrated that some improvement in behavior of a mixed method could be achieved by using higher-order approximations for the extra-stress. Their general criterion was that the stress approximation should be of sufficient order to represent velocity gradients. They achieved this approximation in practical computations by subdividing each velocity element into 3×3 or 4×4 bilinear subelements on which the extra-stresses were interpolated. This procedure, along with some upwinding techniques, allowed computations to be made at large values of fluid elasticity. The benefits of higher-order stresses were also examined by Rao and Finlayson [40], where a cubic stress approximation was employed. Despite this and more recent work, the problem of stress approximation is not fully understood nor resolved. Further comments will be made in Section 6.9.

6.6.4 Solution Methods

To discuss possible solution procedures for the discretized viscoelastic problem in (6.6.25) and (6.6.26), it is convenient to write the matrix system in a more compact form. Combining the incompressibility constraint with the momentum equation allows the following system to be derived

$$\mathbf{K_{uu}(u)U + K_{u\tau}(u)}\tau = \mathbf{F} \qquad (6.6.27a)$$

$$\mathbf{K_{\tau\tau}(u)}\tau + \mathbf{K_{\tau u}U} = 0 \qquad (6.6.27b)$$

where the vector $\mathbf{U}$ now represents the velocity and pressure variables. The $\mathbf{K_{uu}}$ term contains the advection, diffusion, pressure gradient, and incompressibility terms from the momentum equation. The second equation in (6.6.27) is a rearranged version of the constitutive equation in (6.6.26).

As with any strongly coupled system of equations, there are two basic approaches to the solution of the system given in (6.6.27) – artificially decouple (split) the system and employ a cyclic solution process on the decoupled equations or directly solve the combined equation set. Though split equation algorithms are generally less effective than combined schemes, they have the potential for some significant computational benefits in the present system.

Consider the simplest cyclic Picard method available for (6.6.27), which is written here for cycle $n + 1$ as

Step 1:

$$\mathbf{K_{uu}(u}^n)\mathbf{U}^{n+1} = \mathbf{F} - \mathbf{K_{u\tau}(u}^n)\tau^n \qquad (6.6.28a)$$

Step 2:

$$\mathbf{K_{\tau\tau}(u}^{n+1})\tau^{n+1} = -\mathbf{K_{\tau u}U}^{n+1} \qquad (6.6.28b)$$

At each step of (6.6.28a) a matrix solution is required. However, the matrices in (6.6.28a) are significantly smaller than the matrix associated with the combined system in (6.6.27). Reduction of the size of the matrix problem is especially important for the viscoelastic problem since the number of unknowns per nodal point is very large (e.g., the number of unknowns in a viscoelastic flow can be more than double those of a Newtonian problem for the same finite element mesh). Unfortunately, little work has been done to improve the relatively poor iterative (convergence) performance of methods such as (6.6.28); virtually all viscoelastic problems to date have been solved using a combined equation method.

If the solution to the coupled matrix system is to be obtained by simultaneous solution of the equations in (6.6.26), then it is appropriate to rewrite the system in the operator form

$$\hat{\mathbf{K}}(\hat{\mathbf{U}})\hat{\mathbf{U}} = \hat{\mathbf{F}} \qquad (6.6.29)$$

where $\hat{\mathbf{U}}$ now contains all of the problem unknowns. The nonlinear problem shown in (6.6.19) may be solved using any of the fixed point iteration schemes. The Picard method

$$\hat{\mathbf{K}}(\hat{\mathbf{U}}^n)\hat{\mathbf{U}}^{n+1} = \hat{\mathbf{F}} \qquad (6.6.30)$$

was used by Coleman [36] for several viscoelastic flow problems. The most commonly used method, however, is Newton's method [26,41–44], which for (6.6.19) can be written as

$$\mathbf{J}(\hat{\mathbf{U}}^n)\left[\hat{\mathbf{U}}^{n+1} - \hat{\mathbf{U}}^n\right] = -\hat{\mathbf{K}}(\hat{\mathbf{U}}^n)\hat{\mathbf{U}}^n + \hat{\mathbf{F}} \qquad (6.6.31a)$$

where

$$\mathbf{J}(\hat{\mathbf{U}}^n) = \frac{\partial}{\partial \hat{\mathbf{U}}}\left[\hat{\mathbf{K}}(\hat{\mathbf{U}})\hat{\mathbf{U}} - \hat{\mathbf{F}}\right]\Bigg|_{\hat{\mathbf{U}}^n} \qquad (6.6.31b)$$

The Jacobian, $\mathbf{J}$, for the system given in (6.6.25) and (6.6.26) is very complex but can, in fact, be evaluated analytically. Newton's method converges rapidly and, in general, works very well for flows involving materials with low to moderate elasticity. For highly elastic fluids none of the above algorithms allow convergence, as will be discussed subsequently.

There has been no discussion of solution methods for transient flows, mainly due to the lack of previous work in this area. It is certainly possible to extend the steady-state formulation of Section 6.5 into the time-dependent regime. Following the ideas developed for Newtonian problems [45], the resulting ordinary differential equation system could be integrated in time using an implicit method, such as the trapezoid rule. However, there are two basic difficulties in such an approach. First, since the matrix system for a viscoelastic formulation is so large, the repeated solution of an implicit integration algorithm may not be economically justifiable. Splitting the system during the integration procedure [as in (18)] could alleviate some of the computational burden but would increase concerns regarding stability. The second concern is with the constitutive equation and the fact that the time scale for the material response can be significantly different than the characteristic time for the flow process [46]. Such a situation suggests that a single integration time scale may be inappropriate or not economical; the momentum and constitutive equations may require quite different integration procedures. A method developed by Gartling [47], which is similar to the work of Fortin and Fortin [48], addresses some of the questions regarding transient flows.

6.7 Additional Models of Differential Form

The finite element models developed in Section 6.6 formed the basis for viscoelastic simulations until the later part of the 1980s. Various perturbations on the basic mixed method were attempted to improve computational performance, improve convergence, and most importantly, provide solutions at increasing levels of fluid elasticity. This last problem, the "high Weissenberg number problem" (see Section 6.9), was particularly troublesome as the MIX type algorithms all failed at moderate levels of fluid elasticity. The cause of this failure is complex; some of the issues associated with the phenomena are discussed in a later section.

A number of other formulations have been developed to try to alleviate the high Weissenberg difficulties. In each case, the main objective has been to maintain the momentum and continuity equations as a fully elliptic equation system with well-understood mathematical and numerical properties (e.g., saddle point problem, LBB condition, dominant advection terms, etc.). Also, the constitutive equation treatment has been altered to better account for the hyperbolic nature of the equation set. The following subsections describe the main algorithms developed from these basic ideas. The finite element form of these algorithms will not be detailed as the computational implementation follows directly from the previous section.

6.7.1 Explicitly Elliptic Momentum Equation Method

The first significant reformulation of the viscoelastic flow problem is due to King et al. [49] and is termed the explicitly elliptic momentum equation (EEME) method. It was proven by Joseph et al. [50] that under certain conditions the equation set for the UCM model formulated via a MIX type method could undergo a change of type (change in the characteristics of the equations from imaginary to real) and lose the dominant elliptic character of the system. The essence of the analysis was a criterion on the dependent variables that must be met in order to ensure a consistent and stable numerical solution and prevent any change in the basic characteristics of the equation set. The EEME method was the result of a search for a method that would explicitly satisfy the change of type criterion.

The formulation begins with the basic continuity, momentum, and UCM constitutive models

$$\frac{\partial u_i}{\partial x_i} = 0 \tag{6.7.1}$$

$$-\frac{\partial P}{\partial x_i} + \frac{\partial}{\partial x_j}\left(\tau_{ij}^p\right) = 0 \tag{6.7.2}$$

$$\tau_{ij}^p + \lambda \overset{\nabla p}{\tau_{ij}} = 2\mu^p D_{ij} \tag{6.7.3}$$

where the momentum equation is simplified by considering only steady Stokes flow and no body forces. Taking the divergence of the stress in (6.7.3) and substituting into (6.7.2) produces a momentum equation of the form

$$-\frac{\partial P}{\partial x_i} + \frac{\partial}{\partial x_j}\left(2\mu^p D_{ij}\right) - \frac{\partial}{\partial x_j}\left(\lambda \overset{\nabla p}{\tau_{ij}}\right) = 0 \tag{6.7.4}$$

This equation is similar in form to (6.6.11) though for a different constitutive model. Equation (6.7.4) may be rewritten using the definition of the upper convected derivative as

$$-\frac{\partial P}{\partial x_i} + \frac{\partial}{\partial x_j}\left(\mu^p L_{ij} + \mu^p L_{ji}\right) - \frac{\partial}{\partial x_j}\left(\lambda u_k \frac{\partial \tau_{ij}^p}{\partial x_k} - L_{ik}\tau_{kj}^p - L_{jk}\tau_{ki}^p\right) = 0 \tag{6.7.5}$$

Defining a new tensor variable, χ_{ij}, based on the change of type criterion, the momentum equation may be reformed as

$$\frac{\partial}{\partial x_k}(\chi_{ij}L_{jk}) + L_{ij}\frac{\partial \chi_{jk}}{\partial x_k} = \frac{\partial q}{\partial x_k} \qquad (6.7.6)$$

where

$$\chi_{ij} = \delta_{ij} - \lambda\tau_{ij}^p \qquad (6.7.7)$$

$$q = P + \lambda u_j\frac{\partial P}{\partial x_j} \qquad (6.7.8)$$

The EEME method consists of Eqs. (6.7.1), (6.7.3), and (6.7.6) with the definitions in (6.7.7) and (6.7.8). This formulation is similar in development to the MIX method of Section 6.6. The major difference between the methods is that the EEME was conceived for an UCM model (no solvent or viscous stress component) while the MIX methods were usually employed with constitutive models having a viscous component. A MIX formulation for an UCM has been shown by a number of investigators to be very limited in terms of maximum attainable Weissenberg number. The advantage of the EEME method is the enforced elliptic nature of the momentum equation in (6.7.6), which is maintained as the fluid elasticity is increased. The tensor variable χ_{ij}, is analogous to an anisotropic diffusivity; the variable q is a modified pressure. The lower-order derivative terms are similar to advection terms. The momentum equation is thus similar to a Navier Stokes problem. In a finite element procedure the modified pressure q is used (and interpolated) in the same way as in a Newtonian formulation.

The EEME formulation was used with a finite element implementation, primarily by the MIT group [49,51], to solve a number of viscoelastic problems. A variety of stress approximations were tested and found to produce consistent and convergent results at reasonably large values of fluid elasticity. The failure of the EEME method at higher values of elasticity was attributed to the poor resolution of elastic boundary layers. The low-order derivative terms in (6.7.6) become dominant at high elasticity, much like the advection terms in a high Reynolds number viscous flow. Note that the EEME method can be extended to constitutive models with a viscous stress contribution, such as the Oldroyd models. However, its inherent design makes it most relevant to situations where the viscous stress is much less than the viscoelastic stress.

6.7.2 Elastic Viscous Stress Splitting Method

Despite the improvements seen with the EEME algorithm, its focus on UCM type models reduced its universality. However, the philosophy of the EEME formulation, involving a strongly elliptic momentum/continuity system and a properly represented hyperbolic constitutive equation, was accepted as a way to improve viscoelastic simulation methods. A number of investigators had also demonstrated that the addition of a viscous contribution to the stress model was computationally helpful. The viscous contribution regularizes the mathematical problem and eliminates the deleterious change of equation type problem that plagues the UCM model.

The elastic viscous split stress (EVSS) method was developed [51] to generalize the EEME method and focus on constitutive models containing a purely viscous component. The formulation followed from previous work on stress splitting, but changed significantly the method for handling higher-order derivatives in the resulting equations. The development of the method again begins with the steady Stokes form of the continuity and momentum equations. In this illustration the Oldroyd model will be used as the constitutive equation with a viscous component.

The basic equations are

$$\frac{\partial u_i}{\partial x_i} = 0 \tag{6.7.9}$$

$$-\frac{\partial P}{\partial x_i} + \frac{\partial \tau_{ij}}{\partial x_j} = 0 \tag{6.7.10}$$

$$\tau_{ij}^p + \lambda \overset{\nabla p}{\tau_{ij}} = 2\mu^p D_{ij} \tag{6.7.11}$$

$$\tau_{ij} = 2\mu^e D_{ij} + \tau_{ij}^p \tag{6.7.12}$$

For the EVSS scheme, a change of variable is introduced with

$$S_{ij} = \tau_{ij}^p - 2\mu^p D_{ij} \tag{6.7.13}$$

Substituting (6.7.13) and (6.7.12) into the momentum equation and (6.7.13) into the constitutive relation yields the following momentum and stress rate equations

$$-\frac{\partial P}{\partial x_i} + \frac{\partial}{\partial x_j}\left(2\bar{\mu}^e D_{ij}\right) + \frac{\partial S_{ij}}{\partial x_j} = 0 \tag{6.7.14}$$

$$S_{ij}^p + \lambda \overset{\nabla}{S_{ij}^p} = 2\lambda\mu^p \overset{\nabla}{D}_{ij} \tag{6.7.15}$$

where the viscosity $\bar{\mu}^e$ is the sum of μ^e and μ^p.

The standard EVSS formulation thus consists of equations (6.7.9), (6.7.14), and (6.7.15) with the definitions for S_{ij} and D_{ij}. This system maintains the elliptic nature of the continuity and momentum equation set but introduces a complication for the finite element implementation. The upper convected derivative of the rate of deformation tensor D_{ij} in (6.7.15) contains second derivatives of the velocity field. A number of approaches have been demonstrated to circumvent this problem and include an integration by parts [52] leading to an inconvenient boundary term, the retention of D_{ij} as an unknown and use of a projection method [51], and a discrete version of retaining D_{ij} as an unknown variable [52]. This last method is termed a Discrete EVSS (DEVSS) and has found considerable favor within the viscoelastic community. Other perturbations and refinements of the EVSS formulation have recently been demonstrated. An adaptive procedure has been employed in [53] and [54] to split the stress leading to methods termed Adaptive Viscous Split Stress (AVSS) and Discrete AVSS (DAVSS). The effective viscosity μ^e in (6.7.12) is allowed to vary such that a balance between the viscous and viscoelastic stress levels is maintained and thereby keep the solution procedure under control.

It must be emphasized that the stress splitting methods are beneficial in maintaining the mathematical properties of the momentum/continuity problem; the proper treatment of the hyperbolic constitutive equation remains an important part of the overall algorithm. In most of the EEME and EVSS methods, some type of upwind method is employed in the formulation of the constitutive equations. The Streamline Upwind Petrov-Galerkin (SUPG) method mentioned in Chapter 4 is one popular approach to the convection operator. The discontinuous Galerkin (DG) method, first outlined by Lesaint and Raviart [55], is another method for computationally treating first-order hyperbolic equations. This method was first applied to the viscoelastic problem in [48].

The additional formulations outlined above have improved considerably the finite element computational capabilities for viscoelastic simulations. However, substantial work remains to be done, since most of these methods are not affordable in three-dimensional, transient, and/or multiple relaxation time simulations.

6.8 Finite Element Model of Integral Form

Finite element procedures for viscoelastic fluids of the integral type have generally received less attention than those for differential models. This can be attributed primarily to the difficulties in implementation of an integral constitutive equation. In this section we will outline the steps needed to use such a model and provide references to work that describes the details of several algorithms.

For consideration of a specific example the Maxwell fluid described by Eq. (6.5.22) will be used. The basic equations are rewritten here for reference

$$\frac{\partial u_i}{\partial x_i} = 0 \tag{6.8.1}$$

$$\rho \left(\frac{\partial u_i}{\partial t} + u_j \frac{\partial u_i}{\partial x_j} \right) = -\frac{\partial P}{\partial x_i} + \frac{\partial}{\partial x_j} (\tau_{ij}) \tag{6.8.2}$$

$$\tau_{ij} = \frac{\mu_0}{\lambda^2} \int_{-\infty}^{t} exp\left[-(t - t')/\lambda\right] \left[C_{ij}^{-1}(t') - \delta_{ij}\right] dt' \tag{6.8.3}$$

Note that the stress decomposition procedure of Eq. (6.5.13) can also be used with an integral model; its use in the present case would lead to an integral form of the Oldroyd B fluid.

The balance laws, (6.8.1) and (6.8.2), can be discretized in the standard way, using a mixed (velocity, pressure) finite element procedure. The result of such a process is the following familiar set of matrix equations

$$\mathbf{Q}^T \mathbf{u} = 0 \tag{6.8.4}$$

$$\mathbf{C}_i(\mathbf{u})\mathbf{u} + \mathbf{Q}_i \mathbf{P} = \mathbf{F}_i - \int_{\Omega^e} \frac{\partial \mathbf{\Phi}}{\partial x_j} \tau_{ij} d\mathbf{x} \tag{6.8.5}$$

which are written here for the steady-state case. Note that in writing Eq. (6.8.5), $\mathbf{\Phi}$ is assumed to be the velocity basis (weighting) function; an integration by parts is responsible for the form of the stress integral. The arrangement of terms in (6.8.5) anticipates the use of an iterative solution procedure. In the first part of the iteration scheme, Eqs. (6.8.4) and (6.8.5) can be solved for the velocity and pressure given a known (or predicted) stress field. The task then is to develop an algorithm for the prediction of the extra-stress given a velocity field from (6.8.5).

From Eq. (6.8.3), it is apparent that to predict the extra-stress at a particular location, say, a nodal point or quadrature point, the strain history for that point must be evaluated over past time. This computation is usually done in three stages. In the first stage, the particle path (streamline for steady flow) through the point in question is constructed. Procedures for this construction range from the Lagrangian-like computation of displacement fields for each finite element [56] to the direct computation of the streamlines [57]. The construction of the trajectory allows the previous locations of the particle (i.e., node or quadrature point) to be established. The second stage requires the evaluation of the strain measure (e.g., C_{ij}^{-1}) along the particle trajectory. The final stage uses a numerical (Laguerre) quadrature to evaluate the history integral (6.8.3) and produce the extra-stress at a particular location.

The details of the particle tracking and strain evaluation depend directly on the particular type of finite element approximation used for the velocity field. A discussion of several different approaches may be found in the work of Viriyayuthakorn and Caswell [56], Dupont et al. [57] and Bernstein et al. [58]. Once a prediction of the stress field has been made the stress forcing function in (6.8.5) can be evaluated and a new velocity obtained. This cyclic procedure is continued until convergence of the dependent variables is achieved. Details regarding the implementation of the various iteration procedures may also be found in [56,57].

There are several drawbacks to the integral methods as currently implemented. The basic structure of the equations in (6.8.2)–(6.8.4) does not permit the easy use of higher-order iterative methods, such as Newton's method. Convergence is thus limited to a linear rate, which can be a severe computational burden. The particle tracking algorithms appear to be the weak point of the overall procedure. For recirculating flows the methods must be of sufficient accuracy to ensure the closure of streamlines. This is a rather difficult task when dealing with approximate numerical schemes and an incompressible fluid. Also, the extension of the particle tracking ideas to time-dependent flows would appear to be very difficult. Despite these disadvantages the work on integral constitutive equations should continue to increase, since these models are somewhat better at predicting realistic viscoelastic behavior than their differential counterparts.

6.9 Unresolved Problems

6.9.1 General Comments

To conclude this brief treatment of finite element methods for viscoelastic flows, it is appropriate to outline some of the unresolved problems and areas of research in this field. Numerical difficulties in the simulation of Newtonian fluids are generally associated with the occurrence of large values of some relevant nondimensional parameter, e.g., the Reynolds number, Peclet number, or Rayleigh number. The situation is quite similar for the viscoelastic problem though the nondimensional parameters are not as familiar.

As noted in a previous section, there are two time scales involved in a viscoelastic flow – the time constant for the material (relaxation time, λ) and the characteristic time for the flow process, t_f. The ratio of these time parameters is called the Deborah number, $De = \lambda/t_f$. Very often the characteristic time for the flow is taken as the reciprocal of a typical (wall) shear rate, so that $De = \lambda \dot{\gamma}_w$. The nondimensionalization of simple constitutive equations, such as the Maxwell model, can lead to the definition of a second characteristic parameter. The Weissenberg number is typically defined as $W = \lambda U/L$ where U and L are a characteristic velocity and length scale for the flow. In many cases the Weissenberg and Deborah numbers can have the same definition; they also tend to be used interchangeably, which leads to some confusion in the literature.

The Deborah and Weissenberg numbers both indicate the relative importance of fluid elasticity for a given flow problem. High values of W (or De) indicate a material response that is very solid-like, while low values of the parameter represent "small" departures from normal viscous fluid behavior.

The continuing outstanding problem in the numerical simulation of viscoelastic fluids is the "high Weissenberg (or Deborah) number problem." In essence, this phrase refers to the observed behavior that, as the Weissenberg number is increased for a given problem, a critical value is reached (W_{cr}) beyond which the numerical algorithm fails. Failure in the present context normally consists of an initial degradation in the solution (e.g., spatial oscillations in the dependent variables, especially the stresses) followed by nonconvergence of the solution algorithm. The high Weissenberg limit is a universal phenomenon in that it has been observed in all numerical methods (finite element and finite difference) and for all types of constitutive equations. The critical value of the Weissenberg number is very sensitive to the particular flow geometry, numerical algorithm, constitutive equation, and computational mesh. Unfortunately, from an engineering analysis viewpoint the limiting value W_{cr} is usually small enough to make the obtainable solutions uninteresting (i.e., they are perturbations to the Newtonian solution), while the important simulations at high Weissenberg number remain unsolved.

The failure of numerical methods to provide solutions for highly elastic fluids has spawned a great deal of research activity as evidenced by the numerous publications and several international workshops on the subject [59]. Though the basic causes of the problem have yet to be fully identified, a number of issues have been explored that contribute to the basic understanding of several parts of this complex problem. A summary of some of these topics is given below. For the reader interested in further details regarding the many facets of the "high Weissenberg number problem" the text by Crochet et al. [5], the reviews by Crochet [60] and Keunings [61], and recent issues of the *Journal of Non-Newtonian Fluid Mechanics* are recommended. An excellent review of the current computational state-of-the-art has been compiled by Baaijens [62] and a review of the more mathematical issues associated with modeling viscoelastic fluids can be found in the work of Renardy [63].

6.9.2 Choice of Constitutive Equation

Much of the early numerical work on viscoelastic flows was carried out using the Maxwell constitutive model. This was mainly due to the relative simplicity of the equation and the belief that it was prototypical of the more complex constitutive relations. Work by Phan Thien [64–66] demonstrated that for some flows there are inherent instabilities in the Maxwell constitutive model that occur above a critical Weissenberg number. Such a proof demonstrates one of the subtle dangers of this class of problems — the use of approximate constitutive equations that are difficult or impossible to test analytically for non-physical behavior in complex flows.

Many recent investigations have abandoned the Maxwell/Oldroyd family of fluid models in favor of more realistic (and complex) equations such as those due to Johnson–Segalman and Phan Thien–Tanner. There is of course no proof that the more complex models do not also contain built-in instabilities that may or may not be physically realistic. Despite the results shown in [48,56,57] most investigators do not feel that the high Weissenberg limit is due solely to the choice of the constitutive model, though it is certainly regarded as a contributing factor. It should also be noted that virtually all numerical work has considered only single relaxation time models, while it is clear that most constitutive models require multiple relaxation times to approximate realistic fluid behavior. The large computational burden for single relaxation time models has prevented more realistic relaxation spectra from being investigated.

6.9.3 Solution Uniqueness and Existence

The complexity of the equations describing viscoelastic flows precludes any proof of solution uniqueness or existence. Indeed, from the behavior of the simple Newtonian fluid it is to be expected that bifurcations and multiple solutions will exist for flows of these types of fluids. Various investigators [26,41,46] working with several different geometries and constitutive relations have demonstrated the occurrence of limit points (loss of solution with increasing Weissenberg number) and bifurcation points (multiple solutions) in numerical simulations. Typically such features of the flow are computed using a continuation (or imbedding) procedure and Newton's method. Corner singularities routinely cause numerical difficulties and mathematical uncertainty in viscoelastic flows.

The question remains as to whether such behavior is an attribute of the physical model (differential constitutive equation) or is an artifact of the numerical discretization process. Keunings [67] and Debbaut and Crochet [68] have provided arguments for such phenomena being of a numerical nature in at least two particular simulations. Their findings, however, do not eliminate the concern over encountering true bifurcations or limit points under other circumstances.

Various experimental investigations of viscoelastic flows, such as [69], have demonstrated the occurrence of time-dependence and three-dimensional flow structure in relatively simple two-dimensional geometries. With the numerical emphasis on two-dimensional, steady simulations, this type of behavior is clearly not predictable. Issues such as these emphasize the difficulty in discriminating between behavior of the discretized equations, the continuum equations, and the physical system.

6.9.4 Numerical Algorithm Problems

A particularly disturbing behavior that was observed in many numerical simulations is a pathology associated with mesh refinement studies. Typically, a solution on one mesh discretization could be obtained up to a critical Weissenberg number of, say, W_{cr}^{coarse}; a refined mesh solution would produce a solution up to W_{cr}^{fine}, where $W_{cr}^{fine} < W_{cr}^{coarse}$. Stated differently, the finite element procedure did not appear to converge as the mesh was refined — a fundamental property of a stable algorithm. Such behavior was reported for several two-dimensional simulations [5,44] and studied in detail in a one-dimensional flow [42].

In [42] it was conjectured that this behavior was due to an improper choice of the approximating functions for the extra-stress. Little theoretical guidance is available for the choice of stress approximation relative to say the velocity or pressure; the LBB condition must usually still be satisfied for the momentum and continuity equations. As noted previously, the work of Marchal and Crochet [38,39] and others [40] produced some support for the need for higher-order stress approximations. Such anomalous convergence behavior could also stem from the use of an inappropriate solution algorithm or formulation for the constitutive equation. Again, improvements in the treatment of the hyperbolic nature of the constitutive equation seem to yield improved computations.

Extensive work in this area by Dupret et al. [70] eventually led to a better understanding of some aspects of the problem. Viscoelastic flows are particularly difficult to model due to: (a) the presence of normal stresses, which lead to very high stress gradients and thin elastic boundary layers, (b) the occurrence of singularities and stress concentrations, and (c) the strong hyperbolic character of the constitutive equation. Large stress gradients and singularities place a high demand on mesh resolution for the stress variables. The hyperbolic nature of the constitutive equation requires that an appropriate numerical method be used for the equation set; upwind techniques or Petrov–Galerkin methods are one possible approach to this problem. When any of the above demands are not met, as was the case in most standard, mixed finite element methods, the discrete equation systems for the viscoelastic problem could produce improper evolutionary behavior. Dupret et al. [70] tracked this behavior by monitoring a characteristic tensor and correlating its loss of positive definiteness with the occurrence of numerical limit points in the solution. The combination of higher-order stress approximations and upwind techniques for the constitutive equation proved to stabilize the method, and allow solutions at very high Weissenberg numbers with appropriate mesh convergence behavior. However, upwinding and Petrov–Galarkin methods are not without their own problems. The addition of artificial diffusion with these methods must be carefully evaluated to ensure that a true solution is not masked/altered by the numerical technique. The use of streamline upwind methods or discontinuous Galerkin methods has been beneficial in this regard.

A proof by Fortin and Pierre [71] showed under what conditions a higher-order stress approximation was required and confirmed the method developed by Marchal and Crochet [39]. This proof, though not applicable to all formulations, also pointed to the need to properly formulate the momentum/continuity equations as a strongly elliptic system. The EVSS formulation [51] eventually developed out of these considerations and led to substantial improvements in the numerical formulation. It should be noted that a general proof for an LBB type relation for all constitutive equations and system formulations is still lacking. In essence, guidance is still often lacking in what type of stress interpolation is admissible.

6.9.5 Equation Change of Type

The work of Joseph et al. [50] demonstrated that the partial differential equations describing viscoelastic flows can sometimes undergo a formal change of type from elliptic to hyperbolic as the Weissenberg number increases. This is analogous to the transonic flow problem where local supercritical flow regions can be embedded in an otherwise subcritical flow. The proof was for a constitutive model without a viscous component, such as the Upper Convected Maxwell model. This work caused a great deal of activity since so much of the early computational work was based on this model. In combination with the work of Dupret et al. [70] on loss of evolutionary behavior, the change of type analysis spurred development of the EEME method [49]. The proper formulation of the elliptic part of the viscoelastic problem was reinforced by this work.

The change of type was found to occur for the equations describing the fluid vorticity; the occurrence of real characteristics in such an equation admits to the possibility of discontinuities in the vorticity. The occurrence of shear waves has been observed in several transient solutions of viscoelastic flows [47,72–74]. Standard velocity based finite element methods are capable of resolving such flows, as will be demonstrated in the next section.

6.9.6 Closure

With the development of formulations such as the EVSS and its derivative methods, the high Weissenberg number problem is not as formidable as it once was. Solutions at high values of fluid elasticity have been produced and verified through mesh convergence studies. Not all problems have been eliminated however. At very high values of elasticity, numerical methods still have convergence problems. Most investigators believe this is a problem of mesh resolution and thin elastic boundary layers. Until higher resolution simulations are produced, this remains conjecture. It is clear that many simulations, when compared with experiment, fall short of a predictive capability. The constitutive relation is blamed in this case as being not representative of the viscoelastic material. The trend here is to use multimode models (multiple relaxation times) or more sophisticated micro models coupled with the macro flow equations. Finally, though some of the issues described above have been worked around computationally, a clear mathematical understanding of many of these areas is still not available.

6.10 Numerical Examples
6.10.1 Preliminary Comments

In this section a few example problems are described to demonstrate some of the procedures discussed in previous sections. The first examples consider inelastic, non-Newtonian flows and the second group illustrates a couple of viscoelastic applications. Most of the inelastic examples presented here were obtained using the mixed finite element model contained in the NACHOS II code [10], unless otherwise stated. Use of the penalty finite element models for power-law fluids can be found in [75–77]. For further examples, the literature cited in the text should be consulted.

The computations in the viscoelastic section will be limited to a few problems that illustrate some of the non-Newtonian behavior of a viscoelastic fluid. A variety of simulations have been reported in the literature. However, as example problems they suffer from a common shortcoming in that the reported solutions are often little different from those of a Newtonian fluid, mainly due to the "high Weissenberg limit" described above. One exception to the last comment is the work on die swell by Tanner [78], Crochet and Keunings [43,79,80], and others [81], which does show some departure from the Newtonian case for even slightly elastic fluids.

6.10.2 Buoyancy Driven Flow in a Cavity

The first example is concerned with the flow of a power-law fluid inside a rectangular cavity (see Figure 6.10.1). The vertical walls of the cavity are held at uniform but different temperatures, while the horizontal boundaries are insulated. The differentially heated walls produce a buoyancy driven flow in the cavity. The computational boundary conditions for this problem consist of specified temperatures on the vertical walls, zero fluxes on the horizontal walls, and zero velocities on all four boundaries. The pressure level was set by specifying a pressure value at one point on the boundary of the cavity.

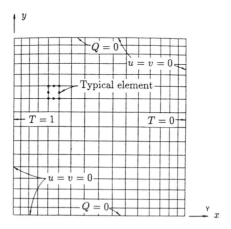

Figure 6.10.1: Schematic of natural convection in a cavity.

Figures 6.10.2 and 6.10.3 contain contour plots of the temperature and stream function obtained for three values of the power-law index. The $n = 1.0$ case corresponds to a Newtonian fluid at a Rayleigh number of $Ra = 10^4$ while the other two cases are for shear thinning fluids at the same nominal Rayleigh number. The effect of the shear thinning viscosity is evident as the fluid motion becomes more vigorous and complex for the same basic driving force. The illustrated cases were solved using a Picard iteration scheme starting from a zero initial solution field and required approximately 10 to 12 iterations to reach convergence.

6.10.3 Driven Cavity Flow

An isothermal driven cavity flow is considered as a second example. A Bingham fluid is contained in a planar cavity, three sides of which are fixed, with the fourth side moving at unit velocity in its own plane. For purposes of computation, the Bingham parameters are taken to be $\mu = 1.0$ and $\mu_r = 1000.0$ with τ_0 being varied over the range $0 \leq \tau_0 \leq 0.50$. The nominal Reynolds number for the problem is $Re = 100$. The computational mesh for this problem is the same as shown in Figure 6.10.1. Again, a Picard scheme was used to solve the problem with the extrapolation procedure of Section 6.4 also being employed.

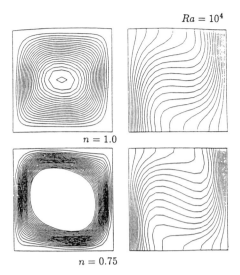

$Ra = 10^4$

$n = 1.0$

$n = 0.75$

Figure 6.10.2: Streamlines and isotherms, natural convection in a cavity, power law fluid, $n = 0.75$.

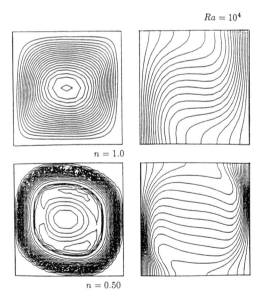

$Ra = 10^4$

$n = 1.0$

$n = 0.50$

Figure 6.10.3: Streamlines and isotherms, natural convection in a cavity, power law fluid, $n = 0.50$.

Figure 6.10.4 contains contour plots of the stream function at four different values of the yield stress. The $\tau_0 = 0$ case is a Newtonian fluid. With increasing yield stress, the motion in the cavity is reduced and relegated to a small area near the moving lid of the cavity. This effect is seen more clearly in the velocity profiles shown in Figure 6.10.5. Note also that, as shown in Figure 6.10.5, the velocity profiles show a "break" near the vortex center. This is due to the relative magnitude of the deformation rate above and below the vortex center. Actually, the break point should occur at zero velocity but is offset slightly due to the use of a constant viscosity within each element and a relatively coarse mesh.

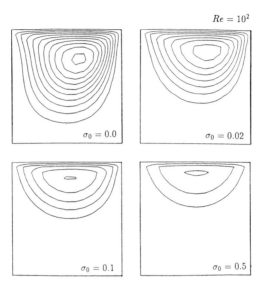

Figure 6.10.4: Streamlines for driven cavity flow; Bingham fluid; various yield stresses.

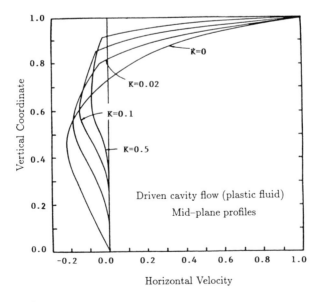

Figure 6.10.5: Midplane velocity profiles for driven cavity flow; Bingham fluid.

6.10.4 Squeeze Film Flow

The third example considers the flow of a plastic (Bingham) fluid in a squeeze film apparatus. This type of device is often used to measure the yield parameters for lubricating greases and other "stiff" materials. The geometry and a typical mesh are shown in Figure 6.10.6. Two circular plates are separated by a fluid layer of initial height h. At time zero the upper plate moves downward with a fixed velocity. The motion of the plate is slow enough that a creeping flow assumption may be used and a series of steady-state problems examined with h as a parameter.

Boundary conditions for this simulation consisted of a specified nonzero vertical velocity and a zero horizontal velocity along the top surface of the domain. Along the vertical centerline the horizontal velocity is set to zero and the shear stress (natural boundary condition) vanishes due to symmetry. The horizontal symmetry plane has a zero vertical velocity and a zero shear stress. The outflow boundary assumes a zero vertical velocity (parallel flow assumption) and a constant normal traction. A Picard iteration scheme with extrapolation was used as a solution procedure; up to 12 iterations were required to obtain convergence.

Radial velocity profiles for two plate separations are shown in Figure 6.10.7, where they are compared to an analytical solution developed in [17]. The fluid near the center plane of the device flows as an unyielded plug while near the plate a more viscous boundary layer-like flow occurs. The demarcation line between yielded and unyielded material is illustrated in Figure 6.10.8. The agreement with the analytical solution is quite good.

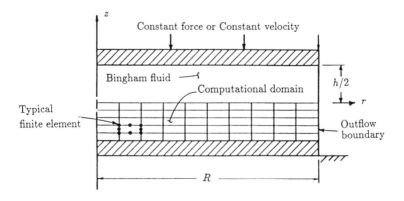

Figure 6.10.6: Schematic of squeeze film flow.

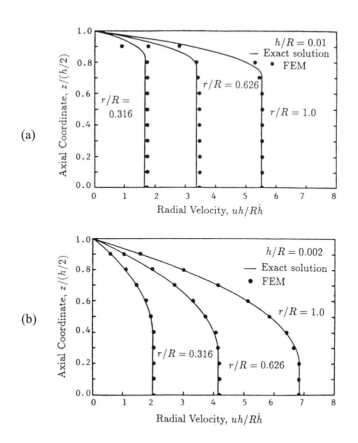

Figure 6.10.7: Radial velocity profiles for the squeeze film flow problem (Bingham fluid). (a) $h/R = 0.01$. (b) $h/R = 0.002$.

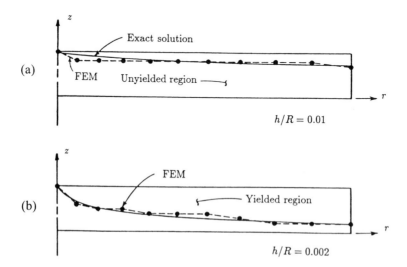

Figure 6.10.8: Plots of the yield surface for the squeeze film flow problem (Bingham fluid). (a) $h/R = 0.01$. (b) $h/R = 0.002$.

6.10.5 Time-Dependent Poiseuille Flow

The first viscoelastic example is a geometrically simple flow with a complex physical response. Consider the problem of an initially quiescent viscoelastic fluid contained between infinite parallel plates. At time zero a specified, constant pressure gradient is imposed. The objective is to compute the time-dependent response of the fluid until it reaches a steady state (Poiseuille flow). A schematic of the problem is shown in Figure 6.10.9.

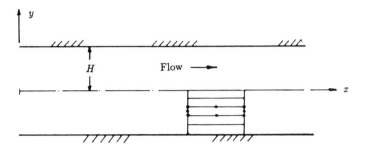

Figure 6.10.9: Schematic of Poiseuille flow.

The above problem was solved analytically by Rivlin [74] for both a Newtonian and an upper-convected Maxwell fluid. The numerical solutions presented here were obtained using the finite element scheme described in [47,48]; results are reported for Newtonian and Maxwell fluids.

Shown in Figures 6.10.10 and 6.10.11 are representative time histories and velocity profiles for the response of a Newtonian fluid. The evolution from the rest state to a fully developed parabolic profile is seen to be a smooth, asymptotic process. Comparison of the numerical results with the analytical work of Rivlin shows excellent agreement (less than 1% difference).

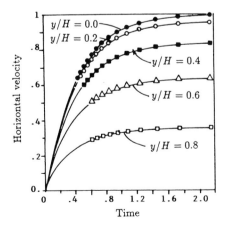

Figure 6.10.10: Velocity histories for Poiseuille flow, Newtonian fluid.

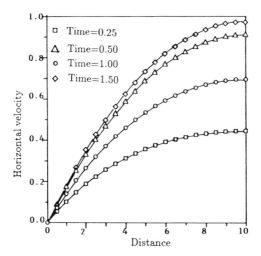

Figure 6.10.11: Velocity profiles for Poiseuille flow, Newtonian fluid.

The response of a viscoelastic fluid, such as a Maxwell material, is quite different. Figure 6.10.12 shows a typical time history for the Maxwell fluid. The damped oscillatory behavior is quite evident and represents a streamwise oscillation of the fluid. This is seen more clearly in Figure 6.10.13, where velocity profiles at various times are illustrated. Early in the start-up process the velocity field shows the propagation of a shear wave from the wall toward the centerline of the channel. The subsequent reflection of the wave causes the oscillatory behavior seen in Figure 6.10.12. At long times the velocity field approaches the expected steady-state profile. A comparison of this solution with the analytical result again shows excellent agreement.

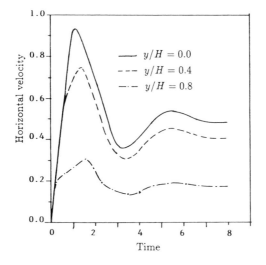

Figure 6.10.12: Velocity histories for Poiseuille flow, Maxwell fluid.

Figure 6.10.14 shows a series of shear stress profiles that illustrates the early time behavior of the shear wave as it propagates across the channel. The normal stress component has a similar behavior. Though the present example is quite simple, it does illustrate the complex behavior that can occur in fluids with memory.

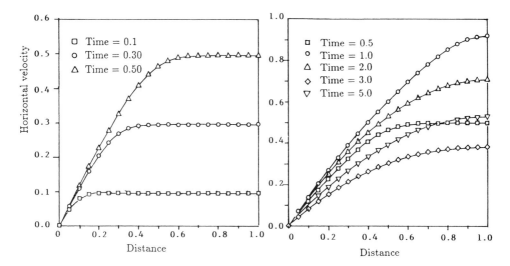

Figure 6.10.13: Velocity profiles for Poiseuille flow, Maxwell fluid.

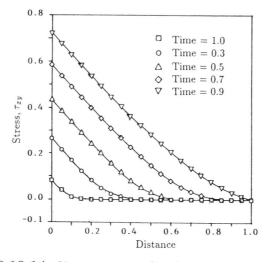

Figure 6.10.14: Shear stress profiles for Poiseuille flow, Maxwell fluid.

6.10.6 Four-to-One Contraction Problem

A standard test problem for the study of viscoelastic flow behavior is the entry flow in axisymmetric, four-to-one contraction. This geometry is popular since it is of technological importance for the polymer processing industry and also exhibits significant flow variation with changes in fluid elasticity. In particular, for several different polymers, as the fluid elasticity is increased (i.e., Weissenberg number is increased) the vortex structure in the largest entry tube undergoes a significant increase in length. The objective of many numerical simulations has been to predict this trend in vortex growth with a suitable constitutive model.

To illustrate typical results for the four-to-one contraction the method proposed by Rao and Finlayson [40] is employed. Figure 6.10.15 contains a schematic of the problem with the velocity boundary conditions shown; a typical quadrilateral mesh is shown in Figure 6.10.16. An Oldroyd B constitutive model was used with a typical mixed method; a cubic approximation for the stress components was used in conjunction with quadratic velocities and a linear pressure. A Newton iteration method with continuation was used to solve the four-to-one contraction for a series of Weissenberg numbers. It was found that to achieve significant values of the Weissenberg number an upwind procedure for the constitutive equation was required. Full details of all the combinations of formulations and algorithms that were tried on this problem are available in [82].

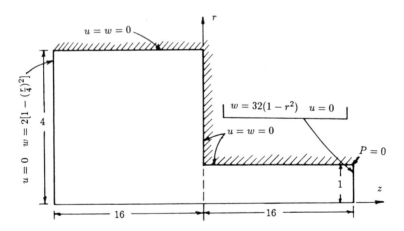

Figure 6.10.15: Schematic for axisymmetric 4:1 contraction problem.

Computed results for the four-to-one contraction in the form of stream function contours are shown in Figure 6.10.17. The plots correspond to increasing values of the recoverable shear, S_R, which is directly proportional to the Weissenberg number. The simulations predict the correct trend in that the vortex length increases as the fluid elasticity increases. Quantitative comparisons with experimental data show that numerical simulations underpredict the vortex growth. This discrepancy is at least partly the result of using approximate constitutive models.

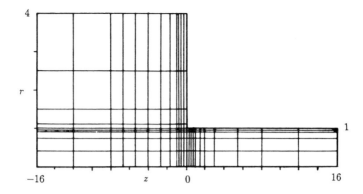

Figure 6.10.16: Quadrilateral mesh for 4:1 contraction. The stress field is approximated using cubic polynomials on each element.

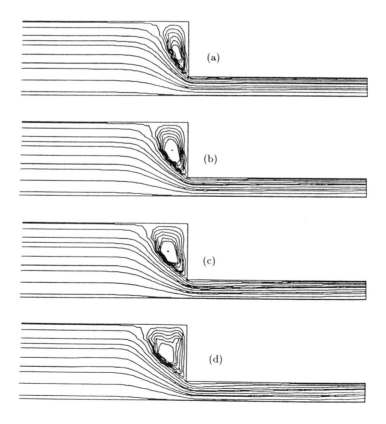

Figure 6.10.17: Streamline for flow in an axisymmetric, 4:1 contraction using an Oldroyd B fluid with recoverable shear values of (a) $S_R = 10$, (b) $S_R = 20$, (c) $S_R = 30$, and (d) $S_R = 40$ (results are taken from [82]).

346 THE FINITE ELEMENT METHOD IN HEAT TRANSFER AND FLUID DYNAMICS

Problems

6.1 The general Oldroyd model is given by (see [5])

$$\tau_{ik} + \lambda \overset{\circ}{\tau}_{ik} + \mu \tau_{jj} D_{ik} - \mu_1 \left(\tau_{ij} D_{jk} + \tau_{kj} D_{ji} \right)$$

$$= 2\mu_0 \left(D_{ik} + \lambda_1 \overset{\circ}{D}_{ik} - \mu_2 D_{ij} D_{jk} \right)$$

where μ_0 is a constant viscosity coefficient, and $(\lambda, \lambda_1, \mu, \mu_1, \mu_2)$ are material constants. Deduce the upper-convected, lower-convected, and corotational Maxwell models by assigning values to the material constants in the Oldroyd model.

6.2 The White–Metzner model [29] is defined by

$$\tau_{ik} + \lambda(I_2) \overset{\triangledown}{\tau}_{ik} = 2\mu_0(I_2) D_{ik}$$

which describes viscoelastic effects in flow problems that are dominated by the shear viscosity. Develop the finite element model governing viscoelastic flows in two-dimensional Cartesian geometries using the White–Metzner constitutive equation (see [81]).

6.3 Repeat Problem 3 for flows in axisymmetric two-dimensional geometries.

References for Additional Reading

1. R. B. Bird, R. C. Armstrong, and O. Hassager, *Dynamics of Polymeric Liquids, Vol. 1: Fluid Mechanics*, Second Edition, John Wiley & Sons, New York (1971).

2. A. S. Lodge, *Elastic Liquids*, Academic Press, New York (1964).

3. K. Walters, *Rheometry*, Chapman and Hall, London, U.K. (1975).

4. R. I. Tanner, *Engineering Rheology*, Clarendon Press, Oxford, U.K. (1985).

5. M. J. Crochet, A. R. Davies, and K. Walters, *Numerical Simulation of Non-Newtonian Flow*, Elsevier, Amsterdam, The Netherlands (1984).

6. J. R. A. Pearson, *Mechanics of Polymer Processing*, Elsevier, London, U.K. (1985).

7. R. B. Bird, W. E. Stewart, and E. N. Lightfoot, *Transport Phenomena*, John Wiley & Sons, New York (1960).

8. J. G. Oldroyd, "A Rational Formulation of the Equations of Plastic Flow for a Bingham Solid," *Proceedings of the Cambridge Philosophical Society*, **43**, 100–105 (1947).

9. W. Prager, *Introduction to the Mechanics of Continua*, Ginn and Company, Boston, Massachusetts (1961).

10. D. K. Gartling, "NACHOS II–A Finite Element Computer Program for Incompressible Flow Problems," Sandia National Laboratories Report, SAND86–1816, Albuquerque, New Mexico (1987).

11. R. L. Lee and P. M. Gresho, "Development of a Three-Dimensional Model of the Atmospheric Boundary Layer Using the Finite Element Method," *Lawrence Livermore National Laboratory Report, UCRL–52366*, Livermore, California (1977).

12. D. K. Gartling, "Finite Element Analysis of Convective Heat Transfer Problems with Change of Phase," in *Computer Methods in Fluids*, K. Morgan, C. Taylor, and C. A. Brebbia (eds.), Pentech Press, London, U.K., 257–284 (1980).

13. D. J. Naylor, "Stresses in Nearly Incompressible Materials by Finite Elements with Application to the Calculation of Excess Pore Pressures," *International Journal for Numerical Methods in Engineering*, **8**, 443–460 (1974).

14. E. Hinton and J. S. Campbell, "Local and Global Smoothing of Discontinuous Finite Element Functions Using a Least Squares Method," *International Journal for Numerical Methods in Engineering*, **8**, 461–480 (1974).

15. E. Hinton, F. C. Scott, and R. E. Ricketts, "Local Least Squares Stress Smoothing for Parabolic Isoparametric Elements," *International Journal for Numerical Methods in Engineering*, **9**, 235–238 (1975).

16. D. K. Gartling, "The Numerical Simulation of Plastic Fluids," in *Proceedings of Third International Conference on Numerical Methods in Laminar and Turbulent Flow*, C. Taylor, C. Johnson, and I. Smith (eds.), Pineridge Press, Swansea, U.K., 669–679 (1983).

17. D. K. Gartling and N. Phan Thien, "A Numerical Simulation of a Plastic Fluid in a Parallel-Plate Plastometer," *Journal of Non-Newtonian Fluid Mechanics*, **14**, 347–360 (1984).

18. R. I. Tanner and J. F. Milthorpe, "Numerical Simulation of the Flow of Fluids with Yield Stresses," in *Proceedings of Third International Conference on Numerical Methods in Laminar and Turbulent Flow*, C. Taylor, C. Johnson, and I. Smith (eds.), Pineridge Press, Swansea, U.K., 680–690 (1983).

19. M. S. Engelman, "Quasi-Newton Methods in Fluid Dynamics," in *The Mathematics of Finite Elements and Applications, IV*, J. R. Whiteman (ed.), Academic Press, London, U.K., 479–487 (1982).

20. C. Truesdell and W. Noll, "The Non-Linear Field Theories of Mechanics," in *Encyclopedia of Physics, Vol III*, S. Flügge (ed.), Springer-Verlag, Berlin, Germany (1965).

21. M. J. Crochet and K. Walters, "Numerical Methods in Non-Newtonian Fluid Mechanics," *Annual Review of Fluid Mechanics*, **15**, 241–260 (1983).

22. A. C. Pipkin and R. I. Tanner, "A Survey of Theory and Experiment in Viscometric Flows of Viscoelastic Liquids," *Mechanics Today, Vol. 1*, S. Nemat-Nasser (ed.), 262–321 (1972).

23. R. I. Tanner, "Recent Progress in Rheology," *Journal of Applied Mechanics*, **105**, 1181–1190 (1983).

24. M. J. Crochet and G. Pilate, "Numerical Study of the Flow of a Fluid of Second Grade in a Square Cavity," *Computers and Fluids*, **3**, 283–291 (1975).

25. K. R. Reddy and R. I. Tanner, "Finite Element Approach to Die-Swell Problems of Non-Newtonian Fluids," *Proceedings of the Sixth Australian Hydraulics and Fluid Mechanics Conference*, Adelaide, Australia, 431–434 (1977).

26. M. A. Mendelson, P. W. Yeh, R. A. Brown, and R. C. Armstrong, "Approximation Error in Finite Element Calculation of Viscoelastic Flow," *Journal of Non-Newtonian Fluid Mechanics*, **10**, 31–54 (1982).

27. M. W. Johnson and D. Segalman, "A Model for Viscoelastic Fluid Behavior Which Allows Non-Affine Deformation," *Journal of Non-Newtonian Fluid Mechanics*, **2**, 255–270 (1977).

28. N. Phan Thien and R. I. Tanner, "A New Constitutive Equation Derived From Network Theory," *Journal of Non-Newtonian Fluid Mechanics*, **2**, 353–365 (1977).

29. J. L. White and A. B. Metzner, "Development of Constitutive Equations for Polymeric Melts and Solutions," *Journal of Applied Polymer Science*, **7**, 1867–1889 (1963).

30. C. J. S. Petrie, "Measures of Deformation and Convected Derivatives," *Journal of Non-Newtonian Fluid Mechanics*, **5**, 147–176 (1979).

31. L. E. Malvern, *Introduction to the Mechanics of a Continuous Medium*, Prentice-Hall, Englewood Cliffs, New Jersey (1969).

32. A. S. Lodge, *Elastic Liquids*, Academic Press, New York (1964).

33. Y. Shimazaki and E. G. Thompson, "Elasto Visco-Plastic Flow With Special Attention to Boundary Conditions," *International Journal for Numerical Methods in Engineering*, **17**, 97–112 (1981).

34. M. Kawahara and N. Takeuchi, "Mixed Finite Element Method for Analysis of Viscoelastic Fluid Flow," *Computers and Fluids*, **5**, 33–45 (1977).

35. M. J. Crochet and M. Bezy, "Numerical Solution for the Flow of Viscoelastic Fluids," *Journal of Non-Newtonian Fluid Mechanics*, **5**, 201–218 (1979).

36. C. J. Coleman, "A Finite Element Routine for Analyzing Non-Newtonian Fluids, Part I: Basic Method and Preliminary Results," *Journal of Non-Newtonian Fluid Mechanics*, **7**, 289–301 (1980).

37. P. W. Chang, T. W. Patten, and B. A. Finlayson, "Collocation and Galerkin Finite Element Methods for Viscoelastic Fluid Flow, I. Description of Method and Problems with Fixed Geometries," *Computers and Fluids*, **7**, 267–283 (1979).

38. J. M. Marchal and M. J. Crochet, "Hermitian Finite Elements for Calculating Viscoelastic Flow," *Journal of Non-Newtonian Fluid Mechanics*, **20**, 187–207 (1986).

39. J. M. Marchal and M. J. Crochet, "A New Mixed Finite Element for Calculating Viscoelastic Flow," *Journal of Non-Newtonian Fluid Mechanics*, **26**, 77–114 (1987).

40. R. R. Rao and B. A. Finlayson, "Viscoelastic Flow Simulation Using Cubic Stress Finite Elements," *Journal of Non-Newtonian Fluid Mechanics*, **43**, 61–82 (1992).

41. P. W. Yeh, M. E. Kim, R. C. Armstrong, and R. A. Brown, "Multiple Solutions in the Calculation of Axisymmetric Contraction Flow of an Upper Convected Maxwell Fluid," *Journal of Non-Newtonian Fluid Mechanics*, **16**, 173–194 (1984).

42. D. K. Gartling, "One Dimensional Finite Element Solutions for a Maxwell Fluid," *Journal of Non-Newtonian Fluid Mechanics*, **17**, 203–231 (1985).

43. M. J. Crochet and R. Keunings, "Die Swell of a Maxwell Fluid: Numerical Predictions," *Journal of Non-Newtonian Fluid Mechanics*, **7**, 199–212 (1980).

44. R. Keunings and M. J. Crochet, "Numerical Simulation of the Flow of a Viscoelastic Fluid Through an Abrupt Contraction," *Journal of Non-Newtonian Fluid Mechanics*, **14**, 279–299 (1984).

45. P. Gresho, R. Lee, R. Sani, and T. Stullich, "On the Time-Dependent FEM Solution of the Incompressible Navier–Stokes Equations in Two- and Three-Dimensions," in *Recent Advances in Numerical Methods in Fluids, Vol. 1*, Pineridge Press, Swansea, U.K. (1980).

46. S. L. Josse and B. A. Finlayson, "Reflections on the Numerical Viscoelastic Flow Problem," *Journal of Non-Newtonian Fluid Mechanics*, **16**, 13–36 (1984).

47. D. K. Gartling, "A Finite Element Algorithm for Time-Dependent Viscoelastic Flows," Sandia National Laboratories Report, Albuquerque, New Mexico (1988).

48. M. Fortin and A. Fortin, "A New Approach for the FEM Simulation of Viscoelastic Flows," *Journal of Non-Newtonian Fluid Mechanics*, **32**, 295–

310 (1989).

49. R. C. King, M. R. Apelian, R. C. Armstrong, and R. A. Brown, "Numerically Stable Finite Element Techniques for Viscoelastic Calculations in Smooth and Singular Geometries," *Journal of Non–Newtonian Fluid Mechanics*, **29**, 147–216 (1988).

50. D. D. Joseph, M. Renardy, and J. C. Saut, "Hyperbolicity and Change of Type in the Flow of Viscoelastic Fluids," *Archive of Rational Mechanics and Analysis*, **87**, 213-251 (1985).

51. D. Rajagopalan, R. C. Armstrong, and R. A. Brown, "Finite Element Methods for Calculation of Steady, Viscoelastic Flow Using Constitutive Equations with a Newtonian Viscosity," *Journal of Non-Newtonian Fluid Mechanics*, **36**, 159–192 (1990).

52. R. Guenette and M. Fortin, "A New Mixed Finite Element Method for Computing Viscoelastic Flows," *Journal of Non-Newtonian Fluid Mechanics*, **60**, 27–52 (1995).

53. J. Sun, N. Phan Thien, and R. I. Tanner, "An Adaptive Viscoelastic Stress Splitting Scheme and Its Applications: AVSS/SI and AVSS/SUPG," *Journal of Non-Newtonian Fluid Mechanics*, **65**, 75–91 (1996).

54. J. Sun, M. D. Smith, R. C. Armstrong, and R. A. Brown, "Finitre Element Method for Viscoelastic Flows Based on the Discrete Adaptive Viscoelastic Stress Splitting and the Discontinuous Galerkin Method: DAVSS-G/DG," *Journal of Non-Newtonian Fluid Mechanics*, **86**, 281–307 (1999).

55. P. Lesaint and P. A. Raviart, "On a Finite Element Method for Solving the Neutron Transport Equation," in *Mathematical Aspects of Finite Elements in Partial Differential Equations*, C. de Borr (Ed.), Academic Press, New York (1974).

56. M. Viriyayuthakorn and B. Caswell, "Finite Element Simulation of Viscoelastic Flow," *Journal of Non-Newtonian Fluid Mechanics*, **6**, 245–267 (1980).

57. S. Dupont, J. M. Marchal, and M. J. Crochet, "Finite Element Simulation of Viscoelastic Fluids of the Integral Type," *Journal of Non-Newtonian Fluid Mechanics*, **17**, 157–183 (1985).

58. B. Bernstein, M. K. Kadivar, and D. S. Malkus, "Steady Flow of Memory Fluids with Finite Elements: Two Test Problems," *Computer Methods in Applied Mechanics and Engineering*, **27**, 279–302 (1981).

59. R. C. Armstrong, R. A Brown, and B. Caswell (eds.), "Papers From the Third International Workshop on Numerical Simulation of Viscoelastic Flows," *Journal of Non-Newtonian Fluid Mechanics*, **16** (1984).

60. M. J. Crochet, "Numerical Simulation of Viscoelastic Flow: A Review," *Rubber Reviews*, **62**, 426–455 (1989).

61. R. Keunings, "Progress and Challenges in Computational Rheology," *Rheologica Acta*, **29**, 556–570 (1990).

62. F. P. T. Baaijens, "Mixed Finite Element Methods for Viscoelastic Flow Analysis: A Review," *Journal of Non-Newtonian Fluid Mechanics*, **79**, 361–385 (1998).

63. M. Renardy, "Current Issues in Non-Newtonian Flows: A Mathematical Perspective", *Journal of Non-Newtonian Fluid Mechanics*, **90**, 243–259 (2000).

64. N. Phan Thien, "Coaxial-Disk Flow of an Oldroyd-B Fluid: Exact Solution and Stability," *Journal of Non-Newtonian Fluid Mechanics*, **13**, 325–340 (1983).

65. N. Phan Thien and W. Walsh, "Squeeze-Film Flow of an Oldroyd-B Fluid: Similarity Solution and Limiting Weissenberg Number," *Journal of Applied Mathematics and Physics (ZAMP)*, **35**, 747–759 (1984).

66. N. Phan Thien, "Cone-and-Plate Flow of the Odlroyd-B Fluid is Unstable," *Journal of Non-Newtonian Fluid Mechanics*, **17**, 37–44 (1985).

67. R. Keunings, "Mesh Refinement Analysis of the Flow of a Maxwell Fluid Through an Abrupt Contraction," *Proceedings of Fourth International Conference on Numerical Methods in Laminar and Turbulent Flow*, C. Taylor et al. (eds.), Pineridge Press, Swansea, U.K. (1985).

68. B. Debbaut and M. J. Crochet, "Further Results on the Flow of a Viscoelastic Fluid Through an Abrupt Contraction," *Journal of Non-Newtonian Fluid Mechanics*, **20**, 173–185 (1986).

69. J. V. Lawler, S. J. Muller, R. A. Brown, and R. C. Armstrong, "Laser Doppler Velocimetry, Measurements of Velocity Fields and Transitions in Viscoelastic Fluids," *Journal of Non-Newtonian Fluid Mechanics*, **20**, 51 (1986)

70. F. Dupret, J. M. Marchal, and M. J. Crochet, "On the Consequence of Discretization Errors in the Numerical Calculation of Viscoelastic Flow," *Journal of Non-Newtonian Fluid Mechanics*, **18**, 173–186 (1985).

71. M. Fortin and R. Pierre, "On the Convergence of the Mixed Method of Crochet and Marchal for Viscoelastic Flows," *Computer Methods in Applied Mechanics and Engineering*, **7**, 1035–1052 (1987).

72. N. Phan Thien, "Coaxial–Disk Flow and Flow About a Rotating Disk of a Maxwell Fluid," *Journal of Fluid Mechanics*, **128**, 427–442 (1983).

73. N. Phan Thien and R. I. Tanner, "Viscoelastic Squeeze-Film Flows – Maxwell Fluids," *Journal of Fluid Mechanics*, **129**, 265–281 (1983).

74. R. S. Rivlin, "Run-Up and Decay of Plane Poiseuille Flow," *Journal of Non-Newtonian Fluid Mechanics*, **14**, 203–217 (1984).

75. M. Iga and J. N. Reddy, "Penalty Finite Element Analysis of Free Surface Flows of Power-Law Fluids," *International Journal of Non-Linear Mechanics*, **24**(5), 383–399 (1989).

76. M. P. Reddy and J. N. Reddy, "Finite Element Analysis of Flows of Non-Newtonian Fluids in Three-Dimensional Enclosures," *International Journal of Non-Linear Mechanics*, **27**(1), 9–26 (1992).

77. M. P. Reddy and J. N. Reddy, "Numerical Simulation of Forming Processes Using a Coupled Fluid Flow and Heat Transfer Model," *International Journal for Numerical Methods in Engineering*, **35**, 807–833 (1992).

78. R. I. Tanner, "Extrudate Swell," in *Computational Analysis of Polymer Processing*, J. R. A. Pearson and S. M. Richardson (eds.), Elsevier, London, U.K., 63–91 (1983).

79. M. J. Crochet and R. Keunings, "On Numerical Die Swell," *Journal of Non-Newtonian Fluid Mechanics*, **10**, 85–94 (1982).

80. M. J. Crochet and R. Keunings, "Finite Element Analysis of Die-Swell of a Highly Elastic Fluid," *Journal of Non-Newtonian Fluid Mechanics*, **10**, 339–356 (1982).

81. J. N. Reddy and V. A. Padhye, "A Penalty Finite Element Model for Axisymmetric Flows of Non-Newtonian Fluids," *Numerical Methods in Partial Differential Equations*, **4**, 33–56 (1988).

82. R. R. Rao, "Adaptive Finite Element Analysis of Non-Newtonian Flow," Ph.D. Dissertation, Department of Chemical Engineering, University of Washington, Seattle, Washington (1990).

Chapter 7

Coupled Problems

7.1 Introduction

Coupled problems in applied mechanics are generally defined as those analyses that require the solution of more than one physical process or physical phenomenon for adequate representation of the overall system. In reality, most engineering problems fall into this category, though it is still common practice to perform single physics studies or sequential, but disjoint, analyses for many applications. Coupled phenomena are very prevalent in the areas of fluid mechanics and heat transfer as is evident from some of the topics from previous chapters. In Chapter 3, coupled problems involving heat conduction and radiation and heat conduction and chemical reaction were considered. At the beginning of Chapter 5 it was noted that convective heat transfer was a coupled problem since it involved two different physical phenomena, namely, fluid mechanics and heat transfer within a fluid. Likewise, the conjugate problem of convective heat transfer in a fluid adjacent to a heat conducting solid is a type of coupled problem. In the present chapter we are going to revisit coupled problems, though the emphasis will be on problems involving more than one discipline or branch of mechanics. Specifically, the finite element solution of solid mechanics and electromagnetics problems will first be presented. Subsequent sections will describe how these types of solutions may be coupled to fluid mechanics and heat transfer simulations to provide a more complete analysis.

7.2 Coupled Boundary Value Problems

Boundary value problems describing different types of mechanics may be coupled through a variety of mechanisms and with varying degrees of interaction. Both of these characteristics are difficult to generalize and quantify and lead to a certain vagueness when discussing coupled problems in generic terms. Before getting to the specific cases of interest here, it is worthwhile to set some terminology and outline some of the complexities that may occur.

The partial differential equations describing different phenomena are coupled when any terms in either equation are functions of the dependent variable or its derivatives from the other equation. This functional dependence may occur directly in source or volume terms, material coefficients, and/or boundary conditions. The dependent variables from one equation may also act more indirectly on the second equation by causing alterations in the geometry of the problem and changes to the temporal behavior of the problem. The degree to which one equation is coupled to a second equation is particularly difficult to define. The terms "weak" and "strong" coupling are often used without precise definitions. In the present case, a "weakly" coupled problem is defined as one in which the transfer of dependent variable data between equations need not occur at every solution step. In essence, the influence of one physical process on the other is sufficiently mild that only periodic updating is required for an accurate representation of the overall system response. For a weakly coupled problem the data transfer may be bidirectional or unidirectional, i.e., there is no mandate that the processes be equally influential. The definition of a "strongly" coupled problem is obvious from the previous definition. If data must be transferred or shared between equations at each step of the solution to maintain accuracy of overall simulation, the problem is defined as strongly coupled. These definitions are not rigorous but do allow some general classifications for coupled problems.

The degree of coupling will also influence the style and choice of computational algorithms used to solve each equation set and the interaction between discretized equations. In general, our preference is to solve strongly coupled physical processes as fully coupled equation sets as this usually provides the most robust and strongly convergent algorithm. The natural convection problem of Chapter 5 is an example of this type of coupled problem. The fully coupled approach is not, however, always feasible. Both physical and numerical characteristics of the particular coupled problem may render full coupling computationally inefficient or impractical. Large disparities in length and time scales between physical processes are two characteristics that may influence the strong coupling algorithm. The solution procedure for weakly coupled physical processes follows from its definition and is obviously some type of cyclic algorithm with separate solution methods for each equations set intertwined with periodic exchanges of data. The frequency and timing of data updates are very problem dependent and one of the characteristics that make general coupling methods and software difficult to develop.

7.3 Fluid Mechanics and Heat Transfer

7.3.1 Introduction

To proceed with the coupled problem discussion it is necessary to become specific with regard to physical processes and coupling mechanisms. In the present section the fluid/thermal problem will again be outlined with particular attention paid to possible dependencies on the solid mechanics and electromagnetics problems to be considered subsequently.

7.3.2 Continuum Equations

The boundary value problem for the nonisothermal flow of a viscous, incompressible fluid is described by the standard conservation relations for mass, momentum, and energy. For a Cartesian coordinate system using the Eulerian description, these relations are

Conservation of Mass:

$$\frac{\partial u_i}{\partial x_i} = 0 \tag{7.3.1}$$

Conservation of Momentum:

$$\rho_0 \left(\frac{\partial u_i}{\partial t} + u_j \frac{\partial u_i}{\partial x_j} \right) - \frac{\partial}{\partial x_j} \left[-P\delta_{ij} + \mu \left(\frac{\partial u_i}{\partial x_j} + \frac{\partial u_j}{\partial x_i} \right) \right] + \rho_0 g_i \beta (T - T_0) = 0 \tag{7.3.2}$$

Conservation of Energy:

$$\rho_0 C_v \left(\frac{\partial T}{\partial t} + u_j \frac{\partial T}{\partial x_j} \right) - \frac{\partial}{\partial x_i} \left(k_{ij} \frac{\partial T}{\partial x_j} \right) - Q - \Phi = 0 \tag{7.3.3}$$

The parameters and symbols used in (7.3.1)–(7.3.3) are standard and the same as used in Chapter 5. Note that the porous flow problem could also be considered here as part of the coupling problem with little variation in the discussion. The boundary conditions for the nonisothermal flow are

$$u_i = f_i^u(s_k, t) \qquad \text{on } \Gamma_u \tag{7.3.4a}$$

$$T_i \equiv \sigma_{ij}(s_k, t) n_j(s_k) = f_i^T(s_k, t) \quad \text{on } \Gamma_T \tag{7.3.4b}$$

for the fluid mechanics part of the problem, and

$$T = f^T(s_k, t) \quad \text{on } \Gamma_T \tag{7.3.5a}$$

$$-\left(k_{ij} \frac{\partial T}{\partial x_j} \right) n_i \equiv q_i n_i = q_c + q_r + q_a = f^q(s_k, t) \quad \text{on } \Gamma_q \tag{7.3.5b}$$

for the heat transfer part of the problem.

When discussing coupling, solid body thermal analysis will be considered distinct from the fluid mechanics problem since the equation system is substantially different as are the coupling interactions. In the no flow case, Eqs. (7.3.1) and (7.3.2) are neglected and Eq. (7.3.3) reduces to the simple heat conduction problem

$$\rho C \frac{\partial T}{\partial t} = \frac{\partial}{\partial x_i} \left(k_{ij} \frac{\partial T}{\partial x_j} \right) + Q \tag{7.3.6}$$

with boundary conditions as given in Eq. (7.3.5). The addition of radiation and/or chemical reaction to the problem in Eq. (7.3.6) can be anticipated but will not be described in equation form. The radiation and chemical reaction problems were described in detail in Chapter 3.

The general thermal problem described above can be coupled to a solid mechanics problem via several mechanisms. Deformation of the material leads to a new spatial orientation (new coordinates) and new density distribution. The deformation may also produce new surface orientations that influence radiation and/or surface exposures that lead to changes in thermal boundary conditions, e.g., contact. Mechanical failure can also lead to new surface definitions. Dissipation mechanisms during deformation may produce a significant volumetric heat source for the energy balance. Surface tractions in the mechanical problem will influence thermal contact at material interfaces and frictional heating models. Finally, reaction rates may be influenced by the material stress state (pressure).

The usual sources of coupling for a thermal problem to an electromagnetics problem are somewhat fewer in number. These include volumetric heat sources due to Joule heating and the possibility of thermal property dependence on electric and magnetic field strength.

7.3.3 Finite Element Models

The finite element method for the boundary value problems described above have been presented in great detail in previous chapters. For completeness and ease of reference the standard matrix form of the discretized equations will be presented here. For the heat conduction problem, the finite element equations are given by Eqs. (3.6.3) and (3.6.4) as

$$\mathbf{M}\dot{\mathbf{T}} + \hat{\mathbf{K}}\mathbf{T} = \hat{\mathbf{F}} \tag{7.3.7}$$

with

$$\hat{\mathbf{K}} = \mathbf{K} + \mathbf{C} + \mathbf{R} \tag{7.3.8a}$$

$$\hat{\mathbf{F}} = \mathbf{Q} - \mathbf{q}_a + \mathbf{F}_{hc} + \mathbf{F}_{hr} \tag{7.3.8b}$$

In the standard uncoupled problem, the matrices $\mathbf{M}, \hat{\mathbf{K}}$, and $\hat{\mathbf{F}}$ are functions of temperature and/or time. The coupled problem will introduce other dependencies which will be discussed in a later section.

In the case of the nonisothermal, viscous flow problem, the finite element equations are given by Eqs. (5.2.10)–(5.2.12), which are

$$-\mathbf{Q}^T\mathbf{u} = 0 \tag{7.3.9}$$

$$\mathbf{M}\dot{\mathbf{u}} + \mathbf{C}\mathbf{u} + \mathbf{K}\mathbf{u} - \mathbf{Q}\mathbf{P} + \mathbf{B}\mathbf{T} = \mathbf{F} \tag{7.3.10}$$

$$\mathbf{N}\dot{\mathbf{T}} + \mathbf{D}\mathbf{T} + \mathbf{L}\mathbf{T} = \mathbf{G} \tag{7.3.11}$$

and represent the continuity, momentum, and energy equations. Like the heat conduction problem the dependencies of the various terms in (7.3.9)–(7.3.11) will be described in a later section for the various types of coupling.

7.4 Solid Mechanics

7.4.1 Introduction

Computational solid mechanics is a very broad discipline with a well-developed finite element experience base and widespread use in engineering analyses. The solid mechanics applications that will be of concern in the context of coupled problems are limited to the regime of quasi-statics, large deformation, and inelastic response in solid materials. Problems with a significant dynamic content are also of interest but are not considered here because of the limited coupling with the diffusion dominated thermal problems and viscous, incompressible flows that are the main focus of the text. The coverage of even this limited area in solid mechanics must, by necessity, be rather superficial with little explanation of important details, such as constitutive behavior and finite element solution methods. The interested reader may consult any of several comprehensive texts dedicated to finite elements in solid mechanics [1–5].

7.4.2 Continuum Equations

The starting point for describing the equations of solid mechanics is usually a series of definitions for the kinematics, deformation, strain, and stress. The majority of these definitions will be presented here without proof or derivation; additional details may be found in a solid mechanics or continuum mechanics book [6,7]. In solid mechanics it is usual to start with an initial configuration of a body and describe its subsequent motion and deformation. A Lagrangian or material description is standard in which a material point is located by

$$\mathbf{x} = \chi(\mathbf{X}, t) = \mathbf{X} + \mathbf{u}(\mathbf{X}, t) \qquad (7.4.1a)$$
$$x_i = \chi(X_i, t) = X_i + u_i(X_i, t) \qquad (7.4.1b)$$

where $\mathbf{X}$ are the *material coordinates* of the material point in the reference or initial configuration, $\mathbf{x}$ are the *spatial coordinates* of the material point at the current time t, and $\mathbf{u}$ is the displacement vector from the reference configuration to the current configuration. Both vector and Cartesian component forms of the relations are written in (7.4.1a,b) and will be used throughout this section. The mapping expressed in (7.4.1a,b) is assumed to be invertible. The function χ is sometimes used to make the relations (7.4.1a,b) more general but will not be used here. Though the problems of interest are quasi-static, time is a convenient variable with which to follow the motion and deformation of the body.

The motion of the material point occupying the position $\mathbf{X}$ in the reference configuration is expressed in differential form as

$$dx = \frac{\partial \mathbf{x}}{\partial \mathbf{X}} d\mathbf{X} = \mathbf{F} \cdot d\mathbf{X} = \frac{\partial (\mathbf{X} + \mathbf{u})}{\partial \mathbf{X}} d\mathbf{X} = [\mathbf{I} + \mathbf{D}] d\mathbf{X} \qquad (7.4.2a)$$

$$dx_i = \frac{\partial x_i}{\partial X_j} dX_j = F_{ij}\, dX_j = \frac{\partial (x_i + u_i)}{\partial X_j} dX_j = [\delta_{ij} + D_{ij}] dX_j \quad (7.4.2b)$$

where $\mathbf{F}$ is the deformation gradient tensor, $\mathbf{I}$ is the unit tensor, and $\mathbf{D}$ is the displacement gradient tensor with components

$$\mathbf{D} = \frac{\partial \mathbf{u}}{\partial \mathbf{X}}, \quad D_{ij} = \frac{\partial u_i}{\partial X_j} \qquad (7.4.3)$$

A strain measure is defined as the difference in the squared length of a line segment in the initial and current configurations

$$dx^2 - d\mathbf{X}^2 = d\mathbf{X}^T \mathbf{F}^T \mathbf{F} d\mathbf{X} - d\mathbf{X}^2 = \left[\mathbf{F}^T \mathbf{F} - \mathbf{I} \right] d\mathbf{X}^2 = 2\,\mathbf{e}\, d\mathbf{X}^2 \quad (7.4.4a)$$

$$dx_i dx_i - dX_i dX_i = dX_j F_{ij} F_{ik} dX_k - \delta_{jk} dX_j dX_k$$
$$= [F_{ij} F_{ik} - \delta_{jk}] dX_j dX_k = 2 e_{jk} dX_j dX_k \qquad (7.4.4b)$$

where the factor 2 is inserted for convenience. This strain measure is called the Green-Lagrange or Green strain tensor that is useful for large deformation analysis. Using the definition $\mathbf{F} = [\mathbf{I} + \mathbf{D}]$ from (7.4.2), the strain tensor $\mathbf{e}$ can be expressed as

$$\mathbf{e} = \mathbf{e}^1 + \mathbf{e}^2 = \frac{1}{2} [\mathbf{I} + \mathbf{D}]^T [\mathbf{I} + \mathbf{D}] - \frac{1}{2} \mathbf{I} = \frac{1}{2} \left[\mathbf{D} + \mathbf{D}^T \right] + \frac{1}{2} \mathbf{D}^T \mathbf{D} \qquad (7.4.5a)$$

and component form

$$e_{ij} = e_{ij}^1 + e_{ij}^2 = \frac{1}{2} [D_{ij} + D_{ji}] + \frac{1}{2} D_{ki} D_{kj} = \frac{1}{2} \left[\frac{\partial u_i}{\partial X_j} + \frac{\partial u_j}{\partial X_i} + \left(\frac{\partial u_k}{\partial X_i} \right) \left(\frac{\partial u_k}{\partial X_j} \right) \right]$$
$$(7.4.5b)$$

The superscripts 1 and 2 refer to the linear and quadratic parts of the strain tensor. If the displacement derivatives are small, their products are negligible compared to the linear terms and the strain tensor becomes

$$\mathbf{e} = \mathbf{e}^1 = \frac{1}{2} \left[\mathbf{D} + \mathbf{D}^T \right] \qquad (7.4.6a)$$

$$e_{ij} = e_{ij}^1 = \frac{1}{2} \left[\frac{\partial u_i}{\partial X_j} + \frac{\partial u_j}{\partial X_i} \right] \qquad (7.4.6b)$$

which defines the familiar infinitesimal or small strain tensor.

Other strain measures are derivable and useful in finite element analysis; the choice is mainly one of convenience for the type of solution method used. When path-dependent material response is important, it is usual to formulate the equations in terms of strain rates and velocities rather than strains and displacements. These types of relations can be derived from the above definitions. Because we will not be exploring constitutive relations in detail, the rate-dependent definitions will not be discussed here. References such as 5–7 should be consulted for additional information.

Because of the desire to accommodate large deformations, a suitable stress tensor must be defined that is compatible with the Green strain tensor. The standard Cauchy stress tensor, σ, is defined as a function of the spatial coordinates $\mathbf{x}$, and is referenced to the current configuration. Since the Green strain tensor in (7.4.5) was referenced to the initial configuration, a stress tensor with the same reference is needed. The second (2nd) Piola–Kirchhoff (or Lagrangian) stress tensor fits this requirement and is related to the Cauchy stress by

$$\tilde{\sigma} = \frac{\rho_0}{\rho}\mathbf{F}^{-1}\sigma\mathbf{F}^{-T} \tag{7.4.7a}$$

$$\tilde{\sigma}_{ij} = \frac{\rho_0}{\rho}\frac{\partial X_i}{\partial x_k}\sigma_{k\ell}\frac{\partial X_j}{\partial x_\ell} \tag{7.4.7b}$$

An inverse relation exists where the Cauchy stress is a function of the Piola–Kirchhoff stress

$$\sigma = \frac{\rho}{\rho_0}\mathbf{F}\tilde{\sigma}\mathbf{F}^{T} \tag{7.4.8a}$$

$$\sigma_{ij} = \frac{\rho}{\rho_0}\frac{\partial x_i}{\partial X_k}\tilde{\sigma}_{k\ell}\frac{\partial x_j}{\partial X_\ell} \tag{7.4.8b}$$

The Piola–Kirchhoff stress tensor is compatible with or energetically conjugate to the Green strain measure and can be used in the construction of constitutive relations. The Cauchy stress (or true stress, traction per unit area) is important as a solution quantity from a finite element model.

With the previous assumption of quasi-static behavior and specifying a Lagrangian reference frame , the equilibrium equations for a solid are provided by the principle of conservation of linear momentum

$$\nabla \cdot \sigma = \rho\mathbf{b}, \quad \frac{\partial \sigma_{ij}}{\partial x_j} = \rho b_i \tag{7.4.9}$$

where σ is the Cauchy stress tensor, ρ is the density and $\mathbf{b}$ is the body force vector. A similar momentum relation could be written in terms of the initial configuration and the Piola–Kirchhoff stress. The form in (7.4.9) will be used here as it is the most familiar. Though many of the symbols used in the present discussion will be the same as used for the fluid mechanics description, context should make the definitions precise.

A statement of mass conservation is also required and in material coordinates is given by

$$\rho\mathbf{J} = \rho\left|\frac{\partial \mathbf{x}}{\partial \mathbf{X}}\right| = \rho_0 \tag{7.4.10}$$

which is valid for each material point. In Eq. (7.4.10) the current density is ρ, the initial density is ρ_0 and J is the determinant of the Jacobian and specifies the ratio of the current configuration to the initial state. The mass conservation equation need not be enforced explicitly in most Lagrangian methods but is used to account for density changes in the material.

Coupling of the solid mechanics problem to a thermal analysis may occur through several phenomena and requires the addition of the energy equation to the above equations. Temperature- and/or species-dependent material properties (constitutive parameters) are a common occurrence. Also, the material state, such as the extent of reaction, decomposition or gas fraction, and material addition or removal due to a thermal process will influence the mechanical response. Changes in mechanical boundary conditions, e.g., pressure loading, may occur due to heat transfer. Mechanical response may be influenced by electromagnetics primarily through Lorentz body forces (magnetic pressures) and field-dependent constitutive response, as illustrated by piezoelectric materials.

7.4.3 Constitutive Relations

Constitutive relations for the stress-strain response of a solid material are quite numerous and range in complexity from simple, linear models to multi-variable, path-dependent, nonlinear models. Here, we will limit the outline to linear and nonlinear elastic constitutive relations and leave the complexities of elastic-plastic, creep, and viscoelastic behavior to the more specialized texts [3,6,7].

A standard elastic constitutive relation (generalized Hooke's law) that is valid for large deformation can be written as

$$\tilde{\sigma} = \mathbf{C} \, \mathbf{e} \tag{7.4.11a}$$
$$\tilde{\sigma}_{ij} = C_{ijk\ell} \, e_{k\ell} \tag{7.4.11b}$$

where $\tilde{\sigma}$ is the second Piola–Kirchhoff stress and $\mathbf{e}$ is the Green strain tensor. For an isotropic, linear elastic material, the elasticity tensor $\mathbf{C}$ is a constant and can be written in its usual form involving the Lamé constants λ and μ

$$C_{ijk\ell} = \lambda \delta_{ij} \delta_{k\ell} + \mu \left(\delta_{ik} \delta_{j\ell} + \delta_{i\ell} \delta_{jk} \right) \tag{7.4.12}$$

$$\lambda = \frac{E\nu}{(1+\nu)(1-2\nu)} \quad , \qquad \mu = \frac{E}{2(1+\nu)} \tag{7.4.13}$$

where E is the modulus of elasticity and ν is Poisson's ratio. This type of model is generally valid when strains are small, though displacements may still be large. For small strains and small displacements, the linear elastic model is recovered when the Cauchy stress is related to the infinitesimal strain tensor as

$$\sigma = \mathbf{C} \, \mathbf{e}^l \quad \text{or} \quad \sigma_{ij} = C_{ijk\ell} \, e^l_{k\ell} = \lambda e_{kk} \delta_{ij} + 2\mu e^l_{ij} \tag{7.4.14}$$

and the elasticity tensor is still defined by Eq. (7.4.12).

When strains are not small, the relation in (7.4.11) can still be used though the elasticity tensor is no longer a constant. If the tensor $\mathbf{C}$ depends on the strains, the functional dependence must be in terms of the invariants of the strain tensor. This type of material is usually termed hyperelastic and the elasticity tensor is most often derived from a strain energy functional. Another nonlinear elastic material model defines the elasticity tensor to be a function of a number of variables including the strain, stress, load, damage, and so on. This type of model is termed hypoelastic and is usually written in an incremental form. Note that the above constitutive relations can be augmented with a term that produces a thermal stress. This additional effect is quite common and leads directly to coupling with an energy equation.

7.4.4 Boundary Conditions

The boundary conditions for the momentum equation in (7.4.9) are standard and consist of specified displacements on part of the boundary

$$u_i = f_i^u(s_k, t) \quad \text{on} \quad \Gamma_u \tag{7.4.15}$$

and specified tractions on the remainder of the boundary

$$\tau_{ij} n_j = f_i^\tau(s_k, t) \quad \text{on} \quad \Gamma_\tau \tag{7.4.16}$$

The displacement conditions are often homogeneous and may be used to eliminate the rigid body motions from the deforming geometry. Specified displacements or velocities are also used to define the boundary motion while the region undergoes deformation. Applied tractions, often in the form of normal pressures, provide another method for loading the structure. The contact boundary condition, either frictionless or with friction, is in widespread use but is difficult to implement. In quasi-static problems it is often difficult to uniquely define the geometric aspects of contact and to know how to apportion the forces and deformation between contacting bodies of comparable strength.

7.4.5 Finite Element Models

The development of a finite element model for solid mechanics is normally accomplished through the principles of virtual work or the principle of minimum total potential energy [8]. These are equivalent to the weak forms or variational problems, and they form the basis of most finite element models in solid mechanics. The method of weighted residuals is also equivalent to these two principles and results in an appropriate weak form of the boundary value problem. In this section the principle of minimum total potential energy is used to derive a finite element model [5–8].

The total potential energy, Π, for a solid body consists of the strain energy U and the potential energy V of the volume and surface forces. The strain energy for the solid region is expressed in terms of the Piola–Kirchhoff stress components $\tilde{\sigma}_{ij}$ and the Green strain components e_{ij} as

$$U = \int_{\Omega_0} \frac{1}{2} \tilde{\sigma}_{ij} \, e_{ij} \, d\Omega \tag{7.4.17}$$

where Ω_0 is the volume of the body in the reference state. Using the constitutive relation (7.4.11) and the decomposition of the strain into linear and quadratic terms, the strain energy can be written as

$$U = \int_{\Omega_0} \frac{1}{2} C_{ijk\ell}(e_{k\ell}^1 + e_{k\ell}^2)(e_{ij}^1 + e_{ij}^2) \, d\Omega \qquad (7.4.18a)$$

or

$$U = \int_{\Omega_0} \frac{1}{2} \left[C_{ijk\ell} e_{k\ell}^1 e_{ij}^1 + C_{ijk\ell}(e_{k\ell}^1 + e_{k\ell}^2)e_{ij}^2 + C_{ijk\ell} e_{k\ell}^2 e_{ij}^1 \right] d\Omega \qquad (7.4.18b)$$

where the terms have been grouped to facilitate a later interpretation. The strain components appearing in Eq. (7.4.18b) can be expressed in terms of the displacement derivatives using the definitions in (7.4.5). The potential energy of the applied forces is given by

$$V = -\int_{\Omega_0} b_i u_i \, d\Omega - \int_{\Gamma_0} f_i^T u_i \, d\Gamma \qquad (7.4.19)$$

where Γ_0 is the boundary of the region Ω_0 and u_i are the displacement components. Note that the total potential energy $\Pi = U + V$ is only a function of the displacements. The principle of minimum total potential energy, $\delta \Pi = 0$, yields the equilibrium equations and force boundary conditions for the problem.

A finite element approximation is defined for the nodal values of displacement components u_i in the usual way

$$u_i(\mathbf{x}, t) = \mathbf{\Phi}^T \mathbf{u}_i \qquad (7.4.20)$$

where $\mathbf{\Phi}$ are the shape functions. The strain-displacement relations can then be expressed in matrix form as

$$\mathbf{e}^1 = \mathbf{D}^1 \mathbf{u} \;, \quad \mathbf{e}^2 = \mathbf{u}^T \mathbf{D}^2 \mathbf{u} \qquad (7.4.21)$$

where the $\mathbf{D}^i$ operators contain the various derivatives found in (7.4.5). Substituting (7.4.20) and (7.4.21) into the expressions for $\Pi = U + V$ and minimizing Π with respect to the nodal displacements, we obtain a finite element model of the problem in the following form:

$$\left[\mathbf{K}^0(\mathbf{u}) + \mathbf{K}^1(\mathbf{u}) + \mathbf{K}^2(\mathbf{u}) \right] \mathbf{u} = \mathbf{F} \qquad (7.4.22)$$

where each of the stiffness terms corresponds to a grouping in (7.4.18b), and the possible dependency of the elasticity tensor on the strain has been indicated. The linear elastic, small strain and small deformation problem is embodied in the first matrix $\mathbf{K}^0$; material nonlinearity may also be present in this term. The second matrix $\mathbf{K}^1$ is usually identified with geometric stiffness since it contains the quadratic strain terms. The matrix $\mathbf{K}^2$ is a nonlinear term that contains the quadratic strains and possible material nonlinearities.

7.4.6 Solution Methods - Quasi-Static Solid Mechanics

The finite element equations developed in the previous section for the quasi-static response of a solid, generally represent a system of highly nonlinear algebraic equations. Two basic methods of solution may be considered. In the first case, the full matrix system in (7.4.22) is constructed and solved using any of the fixed point iterative methods, such as Newton's method. There are generally numerous difficulties with this approach, though when the problem is suitable the method may be very effective. The biggest drawbacks to direct iteration are nonconvergence of the iterative process and construction of the Jacobian for Newton's method. With any nonlinear problem, the starting point for the iterative procedure is crucial in achieving convergence (see Appendix C). The starting guess for most problems involving geometric and/or material nonlinearities will be far enough from the final, equilibrium solution that convergence will not be possible. The construction of the Jacobian is also difficult for many material models that are not analytic; the geometric nonlinearities can be incorporated analytically into the Jacobian through a tangent stiffness matrix.

The second approach to solving (7.4.22) involves an incrementation or imbedding process, in which the load is applied in a series of increments and a sequence of equilibrium problems is solved. This method has the distinct advantage of keeping the next solution in the sequence "close" to the previously converged solution. Iterative methods, like Newton's method, are more robust in this incremental algorithm and convergence is more rapid. Though not formulated here, path- or history-dependent material models must be approached with an incrementation strategy. The variety of incrementation procedures is quite large and outside the scope of this discussion. Additional details on solution algorithms for solid mechanics problems may be found in references [3,5,7].

7.5 Electromagnetics

7.5.1 Introduction

Problems involving the coupling of electromagnetic (EM) fields with fluid and thermal transport have a broad spectrum of applications ranging from astrophysics to manufacturing and to electromechanical devices and sensors. Here we will limit the discussion to the interaction of electromagnetic fields with solid bodies or incompressible fluids that are good electrical conductors. This eliminates many of the interesting problems involving plasmas, and concentrates the applications in the area of metals and liquid metal flows. In many of these problems an applied or induced magnetic field provides an additional body force to the fluid, which results in a convective motion. For high current applications, resistive heating may also be important as a volumetric energy source.

In the following section an outline of the field equations for electromagnetics is given along with their coupling to the nonisothermal, viscous flow problem and the solid body heat conduction problem. Subsequent sections show how the EM problem is redefined in terms of potential functions that are more suitable for finite element model development, and then describe some of the numerical issues associated with EM field simulation. A good introduction to coupled fluid-EM problems is available in [9]; general EM field theory is available in texts such as the one by Jackson [10]. Finite element models for EM applications are well covered in the texts by Jin [11], Sadiku [12], Binns et al. [13] and Silvester [14]. The coverage of the electromagnetics problem is substantially more detailed than the solid mechanics problem mainly because it is less familiar to practitioners of thermal sciences.

7.5.2 Maxwell's Equations

The appropriate mathematical description of electromagnetic phenomena in a conducting material region, Ω_C, is given by Maxwell's equations. In rational MKSA notation these equations are expressed as (see [9,10])

$$\nabla \times \mathbf{E} = -\frac{\partial \mathbf{B}}{\partial t} \tag{7.5.1}$$

$$\nabla \times \mathbf{H} = \mathbf{J} + \frac{\partial \mathbf{D}}{\partial t} \tag{7.5.2}$$

$$\nabla \cdot \mathbf{B} = 0 \tag{7.5.3}$$

$$\nabla \cdot \mathbf{D} = \rho \tag{7.5.4}$$

where the field variables are the electric field intensity, $\mathbf{E}$, the magnetic field intensity, $\mathbf{H}$, the magnetic flux density, $\mathbf{B}$, the electric flux (displacement) density, $\mathbf{D}$, the conduction current density, $\mathbf{J}$, and the source charge density, ρ. Typically, Eq. (7.5.1) is referred to as Faraday's law, Eq. (7.5.2) as Ampere's law (as modified by Maxwell), and Eq. (7.5.4) as Gauss' law. A continuity condition on the current density is also defined by

$$\nabla \cdot \mathbf{J} = \frac{\partial \rho}{\partial t} \tag{7.5.5}$$

Note that only three of the above five equations are independent; the combinations (7.5.1), (7.5.2) and (7.5.4) or (7.5.1), (7.5.2) and (7.5.5) form valid descriptions of the fields.

Constitutive relations

To complete the formulation, the constitutive relations for the material are required. The fluxes are functionally related to the field variables by

$$\mathbf{D} = f_D(\mathbf{E}, \mathbf{B}) \tag{7.5.6}$$

$$\mathbf{H} = f_H(\mathbf{E}, \mathbf{B}) \tag{7.5.7}$$

$$\mathbf{J} = f_J(\mathbf{E}, \mathbf{B}) \tag{7.5.8}$$

where the functions (f_D, f_H, f_J) may also depend on external variables such as temperature or mechanical stress. The form of the material response to applied

E or **B** fields can vary strongly depending on the state of the material, its microstructure and the strength, and time-dependent behavior of the applied field.

Conductive and dielectric materials

For conducting materials, the standard f_J relation is Ohm's law which relates the current density **J** to the electric field intensity **E**

$$\mathbf{J} = \sigma \cdot \mathbf{E} \tag{7.5.9}$$

where σ is the conductivity tensor. For isotropic materials σ is a scalar. In general, the conductivity may be a function of **E** or an external variable such as temperature. This form of Ohm's law applies to stationary conductors. If the conductive material is moving in a magnetic field, then Eq. (7.5.9) is modified to read

$$\mathbf{J} = \sigma \cdot \mathbf{E} + \sigma \cdot (\mathbf{u} \times \mathbf{B}) \tag{7.5.10}$$

where **u** is the velocity vector describing the motion of the conductor and **B** is the magnetic flux vector.

For dielectric materials, the standard f_D function relates the electric flux density **D** to the electric field **E** and polarization vector **P**

$$\mathbf{D} = \epsilon_0 \cdot \mathbf{E} + \mathbf{P} \tag{7.5.11}$$

where ϵ_0 is the permittivity of free space. The polarization is generally related to the electric field through

$$\mathbf{P} = \epsilon_0 \chi_e \mathbf{E} + \mathbf{P}_0 \tag{7.5.12}$$

where χ_e is the electric susceptibility tensor that accounts for the different types of polarization, and $\mathbf{P}_0$ is the remnant polarization that may be present in some materials. The electric susceptibility is always positive and may be a simple scalar (linearly polarizable material) or a field-dependent scalar or tensor. In some situations the polarization exhibits a hysteretic behavior with respect to the electric field (ferroelectric material) and the susceptibility is defined as the local slope of the **P** versus **E** curve. Combining Eqs. (7.5.11) and (7.5.12), we obtain

$$\mathbf{D} = (\epsilon_0 \mathbf{I} + \epsilon_0 \chi_e) \cdot \mathbf{E} + \mathbf{P}_0 = \epsilon_0 (\mathbf{I} + \chi_e) \cdot \mathbf{E} + \mathbf{P}_0 = \epsilon \cdot \mathbf{E} + \mathbf{P}_0 \tag{7.5.13}$$

where **I** is the unit tensor and $\epsilon = \epsilon_0 (\mathbf{I} + \chi_e)$. Note that if $\mathbf{P}_0$ exists, it only needs to be explicitly defined for linear materials (i.e., when ϵ is independent of **E**); for nonlinear constitutive behavior, $\mathbf{P}_0$, is usually absorbed in the functional form for ϵ.

Magnetic materials

For magnetic materials, the standard f_H function relates the magnetic field intensity **H** to the magnetic flux **B**

$$\mathbf{H} = \frac{1}{\mu_0} \mathbf{B} - \mathbf{M} \tag{7.5.14}$$

where μ_0 is the permeability of free space and $\mathbf{M}$ is the magnetization vector. The magnetization can be related to either the magnetic flux or magnetic intensity

$$\mathbf{M} = \frac{\chi_m}{(\mathbf{I} + \chi_m)} \frac{1}{\mu_0} \cdot \mathbf{B} + \mathbf{M}_0 \qquad (7.5.15a)$$

or

$$\mathbf{M} = \chi_m \cdot \mathbf{H} + (\mathbf{I} + \chi_m) \cdot \mathbf{M}_0 \qquad (7.5.15b)$$

where χ_m is the magnetic susceptibility for the material and $\mathbf{M}_0$ is the remnant magnetization. If the susceptibility is negative, the material is diamagnetic; while a positive susceptibility defines a paramagnetic material. Generally, these susceptibilities are quite small and are often neglected. Ferromagnetic materials have large positive susceptibilities and produce a nonlinear (hysteretic) relationship between $\mathbf{B}$ and $\mathbf{H}$. These materials may also exhibit spontaneous and remnant magnetization. Combining Eq. (7.5.14) with either Eq. (7.5.15a) or Eq. (7.5.15b) leads to

$$\mathbf{H} = \frac{1}{\mu_0(\mathbf{I} + \chi_m)} \cdot \mathbf{B} - \mathbf{M}_0 = \frac{1}{\mu} \cdot \mathbf{B} - \mathbf{M}_0 = \nu \cdot \mathbf{B} - \mathbf{M}_0 \qquad (7.5.16)$$

Here $\mu = \mu_0(\mathbf{I} + \chi_m)$ is the permeability tensor for the material. The relative permeability is $\mu_r = \mu/\mu_0 = (\mathbf{I} + \chi_m)$ and the reluctivity ν, is defined as the inverse of the permeability. The remnant magnetization, $\mathbf{M}_0$, need only be explicitly defined for linear magnetic materials (e.g., permanent magnets); in the nonlinear case $\mathbf{M}_0$ is usually absorbed into the $\mu(\mathbf{B})$ function.

Electromagnetic forces and volume heating

The coupling of electromagnetic fields with a fluid or thermal problem occurs through the dependence of material properties on EM field quantities and the production of EM-induced body forces and volumetric energy production. The Lorentz body force in a conductor due to the presence of electric currents and magnetic fields is given by

$$\mathbf{F}_B = \rho\mathbf{E} + \mathbf{J} \times \mathbf{B} \qquad (7.5.17)$$

where, in the general case, the current is defined by Eq. (7.5.10). The first term on the right-hand side of Eq. (7.5.17) is the electric field contribution to the Lorentz force; the magnetic term $\mathbf{J} \times \mathbf{B}$ is usually of more interest in applied mechanics problems. The energy generation or Joule heating in a conductor is described by

$$Q_J = \mathbf{J} \cdot \mathbf{E} \qquad (7.5.18)$$

which takes on a more familiar form if the simplified ($\mathbf{u} = 0$) form of Eq. (7.5.10) is used to produce

$$Q_J = \sigma^{-1}(\mathbf{J} \cdot \mathbf{J}) \qquad (7.5.19)$$

The above forces and heat source occur in the fluid momentum and energy equations, respectively.

Quasi-static approximation

For good conductors, the conduction current $\mathbf{J}$ is large compared to the displacement current $\mathbf{D}$ for most frequencies of interest. Neglecting the displacement current allows Ampere's law (7.5.2) to be simplified and Coulomb's law (7.5.4) to be omitted. Also, the continuity relation is simplified since the divergence of $\nabla \times \mathbf{H}$ is zero by a vector identity. The omission of the displacement current is termed a quasi-static approximation (pre-Maxwell equations) since the propagation of electromagnetic waves is precluded. Under this assumption Maxwell's equations for a conducting region become

$$\nabla \times \mathbf{E} = -\frac{\partial \mathbf{B}}{\partial t} \tag{7.5.20}$$

$$\nabla \times \mathbf{H} = \mathbf{J} \tag{7.5.21}$$

$$\nabla \cdot \mathbf{B} = 0 \tag{7.5.22}$$

The continuity condition is

$$\nabla \cdot \mathbf{J} = 0 \tag{7.5.23}$$

The simplified forms of the required constitutive relations are

$$\mathbf{B} = \mu \cdot \mathbf{H} \quad \text{or} \quad \mathbf{H} = \nu \cdot \mathbf{B} \tag{7.5.24}$$

$$\mathbf{J} = \sigma \cdot \mathbf{E} + \sigma \cdot (\mathbf{u} \times \mathbf{B}) \quad \text{or} \quad \mathbf{J} = \sigma \cdot \mathbf{E} \tag{7.5.25}$$

7.5.3 Electromagnetic Potentials

For many static and quasi-static applications it is usual to introduce a set of potential functions to represent the electric and magnetic field variables and reduce the number of partial differential equations requiring solution. Two basic systems of potentials may be considered: (a) the electric scalar potential V and a magnetic vector potential $\mathbf{A}$, and (b) the electric vector potential $\mathbf{T}$ and the scalar magnetic potential ψ. The $\mathbf{T}$-ψ formulation is of limited value for general analyses since there are significant difficulties in representing multiply-connected domains. Though the $\mathbf{A} - V$ formulation generally leads to a larger number of differential equations, it is preferred for numerical computation due to its complete generality.

From Gauss' relation $\nabla \cdot \mathbf{B} = 0$, it follows that $\mathbf{B}$ is derivable from a vector potential. By definition, then

$$\mathbf{B} = \nabla \times \mathbf{A} \tag{7.5.26}$$

where $\mathbf{A}$ is the *magnetic vector potential*. In addition, from Faraday's law (7.5.20), one has

$$\nabla \times \mathbf{E} = -\frac{\partial \mathbf{B}}{\partial t} = -\frac{\partial (\nabla \times \mathbf{A})}{\partial t} \tag{7.5.27a}$$

Rearranging the above equation, we obtain

$$\nabla \times \left(\mathbf{E} + \frac{\partial \mathbf{A}}{\partial t} \right) = 0 \tag{7.5.27b}$$

For a scalar V, the vector identity $\nabla \times \nabla V = \mathbf{0}$ holds and allows the definition

$$-\nabla V \equiv \mathbf{E} + \frac{\partial \mathbf{A}}{\partial t}$$

or

$$\mathbf{E} = -\nabla V - \frac{\partial \mathbf{A}}{\partial t} \tag{7.5.28}$$

The scalar function V is known as the *electric potential*. The definitions in Eqs. (7.5.26) and (7.5.28) may be used in the appropriate forms of Ampere's law and the current continuity equation to produce the needed field equations for $\mathbf{A}$ and V in conductors and in free-space regions. Using Eqs. (7.5.24)–(7.5.26) and (7.5.28) in Eq. (7.5.21), one obtains

$$\nabla \times [\nu \cdot (\nabla \times \mathbf{A})] = -\sigma \cdot \frac{\partial \mathbf{A}}{\partial t} - \sigma \cdot \nabla V \tag{7.5.29}$$

Also, in the conduction regions the electric field is described by the continuity equation (7.5.23), which is rewritten in terms of $\mathbf{A}$ and V by the use of Eqs. (7.5.25b) and (7.5.28)

$$\nabla \cdot \left(-\sigma \cdot \nabla V - \sigma \cdot \frac{\partial \mathbf{A}}{\partial t} \right) = 0 \tag{7.5.30}$$

Equations (7.5.29)–(7.5.30) describe the general, quasi-static electromagnetic field problem for free space and conductors in terms of the magnetic vector potential $\mathbf{A}$ and electric scalar potential V. The above derivation has been executed for stationary conductors with simple material behavior; the formulation is altered slightly if conductors are moving and/or constitutive behavior is more complex. These equations are appropriate for solution by the finite element method.

The general problem outlined above may now be specialized to particular material regions and types of current ($\mathbf{J}$) specifications. For conduction regions without source currents, both Eqs. (7.5.29) and (7.5.30) are generally required to describe both the electric and magnetic fields. Conduction regions that have specified currents may require some alteration of Eqs. (7.5.29)–(7.5.30), depending on the form of the current source (7.5.8). Electric currents described by distributions of the electric potential require no alteration to the equation set since this specification would appear as a boundary condition on the variable V. However, when current densities are assumed known and are specified directly, the total current density in Ohm's law must be rewritten as the combination of two parts. Define the total current density as

$$\mathbf{J} = \mathbf{J}_s + \mathbf{J}_e \tag{7.5.31}$$

where the known source current $\mathbf{J}_s$ is associated with the electric potential term $-\sigma \cdot \nabla V$, and the induced or self-induced current is defined by the time-dependent magnetic potential term $-\sigma \cdot (\partial \mathbf{A}/\partial t)$. Therefore, in a conductor with a known current density $\mathbf{J}_s$, Eq. (7.5.29) is rewritten as

$$\nabla \times (\nu \cdot \nabla \times \mathbf{A}) + \sigma \cdot \frac{\partial \mathbf{A}}{\partial t} = \mathbf{J}_s \tag{7.5.32}$$

Also, Eq. (7.5.30) is no longer required in the source region since the imposed current density is assumed to be divergence free; the induced currents in the source conductor are also divergence free due to the gauge condition that will be discussed in a subsequent section. Note that the total current through the source conductor is given by

$$I = \int_\Gamma \mathbf{J} \cdot \hat{\mathbf{n}} \, d\Gamma = \int_\Gamma \mathbf{J}_s \cdot \hat{\mathbf{n}} \, d\Gamma - \int_\Gamma \sigma \cdot \frac{\partial \mathbf{A}}{\partial t} \cdot \hat{\mathbf{n}} \, d\Gamma \qquad (7.5.33)$$

where $\hat{\mathbf{n}}$ is the vector normal to the cross-sectional area Γ of the conductor. For the general time-dependent problem, the total current, $\mathbf{J}$, cannot be specified a priori, since the self-induced portion of the current is obtained as a part of the solution. This implies that an iterative procedure is needed if specified current problems are defined.

The equations needed for free space or dielectric regions are simply those of (7.5.29) with σ set to zero; Eq. (7.5.30) is not required as no electric fields are considered. Finally, note that simplification of the equations for all regions is possible for problems of reduced dimensionality. If the geometry is two-dimensional (planar or axisymmetric) and the currents and potential gradients are oriented orthogonal to the plane, then Eq. (7.5.30) is not required and Eq. (7.5.29) reduces to a single equation for the remaining (axial or circumferential) component of the magnetic potential.

The general time-dependent equations given in (7.5.29)–(7.5.30) are applicable to any type of time varying field problem. However, in the often encountered special case of a single frequency, time-harmonic excitation (e.g., alternating current), the equations may be simplified through use of a phasor representation. Let any specified current densities be represented as a time harmonic excitation and assume that the electric scalar and magnetic vector potentials have a time-harmonic form that can be represented as

$$\mathbf{J}_s = \mathbf{J}_{s0} \, e^{i\omega t} = (\mathbf{J}_s^R + i \mathbf{J}_s^I) \, e^{i\omega t}$$
$$V = V_0 \, e^{i\omega t} = (V^R + i V^I) \, e^{i\omega t}$$
$$\mathbf{A} = \mathbf{A}_0 e^{i\omega t} = (\mathbf{A}^R + i \mathbf{A}^I) \, e^{i\omega t} \qquad (7.5.34)$$

where ω is the circular frequency $(= 2\pi f, \; f$ is the imposed AC frequency in Hz), $i = \sqrt{-1}$ and t is the time. The superscripts R and I denote the real and imaginary components of a variable. Then substituting (7.5.34) into (7.5.29) and (7.5.30) and eliminating the common exponential factor produces

$$\nabla \times (\nu \cdot \nabla \times \mathbf{A}_0) + i\omega\sigma\mathbf{A}_0 + \sigma \cdot \nabla V_0 = 0 \qquad (7.5.35)$$
$$\nabla \cdot (\sigma \cdot \nabla V_0 + i\omega\sigma \cdot \mathbf{A}_0) = 0 \qquad (7.5.36)$$

and for conduction regions with source currents (7.5.32) produces

$$\nabla \times (\nu \cdot \nabla \times \mathbf{A}_0) + i\omega\sigma\mathbf{A}_0 = \mathbf{J}_{s0} \qquad (7.5.37)$$

These complex equations describe the amplitudes for the potentials. Note that the boundary conditions for V and $\mathbf{A}$ must also be expressed in terms of the harmonic approximation given in (7.5.34). Implicit in the use of the phasor representation is the assumption that material properties are independent of the temporal behavior of the electromagnetic fields. This is a particularly stringent requirement that is violated when considering high field applications such as induction heating or most types of ferromagnetic materials.

7.5.4 Boundary and Interface Conditions

Boundary and interface conditions for the quasi-static, electromagnetic field problem are most easily described by reference to the generic domain shown in Figure 7.5.1. The region Ω is composed of a number of different materials $(\Omega = \Omega_C \cup \Omega_J \cup \Omega_D)$, several of which are illustrated in the figure. The boundary or interface between two conductors is denoted by Γ_{CC} while the boundary between a dielectric or free space region and a conductor is labeled Γ_{DC}. Note that since the equations for Ω_J and Ω_D are the same except for a source function, no specific designation for an interface between these regions is required. The external boundary of the entire domain Ω is defined by Γ which may be composed of one or more well-defined segments. A two-dimensional representation of the region is used for simplicity.

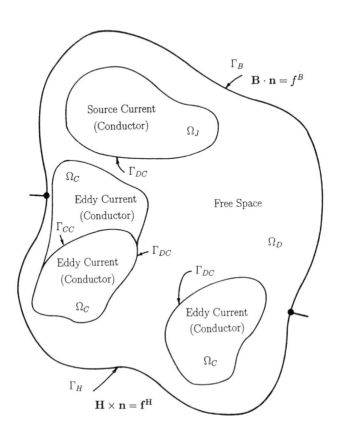

Figure 7.5.1: Schematic of regions for general electromagnetic field problem.

The electromagnetic problem requires that on the exterior free space boundary either the magnetic flux or the magnetic field be specified at all points of the boundary, Γ. In equation form these conditions are given by

$$\mathbf{B} \cdot \hat{\mathbf{n}} = (\nabla \times \mathbf{A}) \cdot \hat{\mathbf{n}} = f^B(s_k, t) \quad \text{on} \quad \Gamma_B \qquad (7.5.38)$$

$$\mathbf{H} \times \hat{\mathbf{n}} = (\nu \cdot \mathbf{B}) \times \hat{\mathbf{n}} = (\nu \cdot \nabla \times \mathbf{A}) \times \hat{\mathbf{n}} = \mathbf{f^H}(s_k, t) \quad \text{on} \quad \Gamma_H \quad (7.5.39)$$

In Eqs. (7.5.38) and (7.5.39) the f^B and $\mathbf{f^H}$ functions are specified values of the known boundary magnetic flux and magnetic field. Also, $\hat{\mathbf{n}}$ is the outward unit normal to the boundary Γ, s_k are coordinates defined on the boundary, t is the time, and $\Gamma = \Gamma_B \cup \Gamma_H$. The functions f^B and $\mathbf{f^H}$ are generally simple expressions for most boundaries of practical interest. When a conductor forms part of the external boundary the above conditions are augmented with a condition on the current flux or the electric field. That is,

$$\mathbf{J} \cdot \hat{\mathbf{n}} = \sigma \cdot \mathbf{E} \cdot \hat{\mathbf{n}} = \left(-\sigma \cdot \frac{\partial \mathbf{A}}{\partial t} - \sigma \cdot \nabla V \right) \cdot \hat{\mathbf{n}} = f^J(s_k, t) \quad \text{on} \quad \Gamma_J \qquad (7.5.40)$$

or

$$\mathbf{E} \times \hat{\mathbf{n}} = -\nabla V - \frac{\partial \mathbf{A}}{\partial t} \times \hat{\mathbf{n}} = \mathbf{f^E}(s_k, t) \quad \text{on} \quad \Gamma_E \qquad (7.5.41)$$

Along a material interface, such as Γ_{CC} or Γ_{DC}, the usual assumption is that the normal component of the magnetic flux is continuous and the tangential component of the magnetic field is discontinuous by an amount equal to the surface current, $\mathbf{J}_{surf}$. These conditions are specified by

$$(\mathbf{B}_2 - \mathbf{B}_1) \cdot \hat{\mathbf{n}} = (\nabla \times \mathbf{A}_1 - \nabla \times \mathbf{A}_2) \cdot \hat{\mathbf{n}} = 0 \quad \text{on} \quad \Gamma_{CC}, \Gamma_{DC} \qquad (7.5.42)$$

$$(\mathbf{H}_2 - \mathbf{H}_1) \times \hat{\mathbf{n}} = (\nu_1 \cdot \nabla \times \mathbf{A}_1 - \nu_2 \cdot \nabla \times \mathbf{A}_2) \times \hat{\mathbf{n}} = \mathbf{J}_{surf} \quad \text{on} \quad \Gamma_{CC}, \Gamma_{DC}$$
$$(7.5.43)$$

where the subscripts 1 and 2 indicate variables evaluated on either side of the interface. In many cases the surface current is not important and may be neglected.

The divergence and curl relations for the electric field also provide two conditions at a material interface. In this case the normal component of the current density is continuous and the tangential components of the electric field are continuous. That is,

$$(\mathbf{J}_2 - \mathbf{J}_1) \cdot \hat{\mathbf{n}} = (\sigma_2 \cdot \mathbf{E}_2 - \sigma_1 \cdot \mathbf{E}_1) \cdot \hat{\mathbf{n}} = 0 \quad \text{on} \quad \Gamma_{CC} \qquad (7.5.44)$$

$$(\mathbf{E}_2 - \mathbf{E}_1) \times \hat{\mathbf{n}} = \mathbf{0} \quad \text{on} \quad \Gamma_{CC} \qquad (7.5.45)$$

for the boundary between two conductors and

$$\mathbf{J}_2 \cdot \hat{\mathbf{n}} = \sigma_2 \cdot \mathbf{E}_2 \cdot \hat{\mathbf{n}} = 0 \quad \text{on} \quad \Gamma_{DC} \qquad (7.5.46)$$

$$\mathbf{E}_2 \times \hat{\mathbf{n}} = \mathbf{0} \quad \text{on} \quad \Gamma_{DC} \qquad (7.5.47)$$

for the boundary between a very good conductor and a dielectric where the subscript 2 refers to the conducting region.

7.5.5 Gauge Conditions

The quasi-static form of Maxwell's equations is given by Eqs. (7.5.29)–(7.5.30) in terms of the magnetic and electric potentials. The original or primitive variable form of Maxwell's Eqs. (7.5.20)–(7.5.22) can be shown to provide unique solutions for the $\mathbf{B}$ and $\mathbf{E}$ fields when appropriate boundary conditions are specified. However, with the introduction of the potential variables, uniqueness of the solution is not retained – Eqs. (7.5.29) and (7.5.30) define the curl of $\mathbf{A}$, but $\mathbf{A}$ itself is only defined up to the gradient of an arbitrary scalar function. Typically this arbitrariness in $\mathbf{A}$ is resolved by defining the divergence of $\mathbf{A}$ and supplying appropriate boundary conditions for $\mathbf{A}$ (rather than boundary conditions for the curl of $\mathbf{A}$). The incorporation of a $\nabla \cdot \mathbf{A}$ constraint, termed gauging, may be accomplished in any of several ways. The Coulomb gauge is one particular choice that has found extensive use in numerical simulation methods. In this case the magnetic vector potential is made unique by the constraint

$$\nabla \cdot \mathbf{A} = 0 \qquad\qquad (7.5.48)$$

Other choices for the gauge condition, such as the Lorentz gauge, select a nonhomogeneous form for Eq. (7.5.48). Note that in some cases of reduced dimensionality, the explicit use of Eq. (7.5.48) is not required since vector $\mathbf{A}$ is automatically divergence free.

The implementation of Eq. (7.5.48) in a numerical method may take any of several forms, including modification of the field equations, penalty methods, projection methods, and the construction of divergence free basis functions. All of these techniques incur a computational penalty in terms of either additional work or the modification of the equation system to a less desirable form.

For some applications a unique value of the magnetic potential is not required and the above gauge condition may be neglected. In particular, if the magnetic field is time independent, then all field quantities are related to the curl of $\mathbf{A}$ and a unique value of $\mathbf{A}$ is not required. However, for the time-dependent case, the electric field $\mathbf{E}$ is related to the time derivative of $\mathbf{A}$ in Eq. (7.5.28) and the $\mathbf{A}$ field must be unique if $\mathbf{E}$ is required.

7.5.6 Static Field Problems

Within the general framework established above, a number of simpler static problems may also be defined. Each of these problem classes may have importance in the context of coupling with other mechanics problems or as stand alone analyses in electromagnetics. As subclasses of the general formulation they may be solved with many of the same numerical techniques as the general quasi-static problem.

Electrostatics

The electrostatic problem is described by Coulomb's law (7.5.4) and the definition of the electric flux (displacement current). Combining Eq. (7.5.4) with the simple conductive form of the constitutive relation in (7.5.11) leads to

$$\nabla \cdot \mathbf{D} = \nabla \cdot (\epsilon \cdot \mathbf{E}) = \rho \qquad (7.5.49)$$

where ρ is the spatial distribution of (free) electric charge. Faraday's law (7.5.1) under steady conditions ($\nabla \times \mathbf{E} = 0$) implies that $\mathbf{E}$ is derivable from a scalar potential, $\mathbf{E} = -\nabla V$, and thus (7.5.49) becomes

$$\nabla \cdot (\epsilon \cdot \nabla V) = -\rho \qquad (7.5.50)$$

which is valid for electrically conductive materials. Note that constant potential regions (conductors without a specified charge distribution) may be removed from the analysis domain or approximated as a high permittivity material. For dielectrics, the free charge is zero and the more general form of the constitutive relation in (7.5.11) can be used to produce

$$\nabla \cdot (\epsilon \cdot \nabla V) = \nabla \cdot \mathbf{P}_0 \qquad (7.5.51)$$

Nonlinear dielectrics would ignore the source term in (7.5.51) and use $\epsilon(\mathbf{E})$. Boundary conditions for electrostatics generally involve specification of the scalar potential, V, or the definition of the electric flux normal to the boundary (i.e., the normal derivative of the potential). When the spatial variation of the potential has been determined, the electric field $\mathbf{E}$ and the electric flux may be found from the relevant definitions.

Steady current flow

For time-independent problems the system in (7.5.29) and (7.5.30) becomes decoupled and the current continuity condition (7.5.30) may be written as

$$\nabla \cdot (-\sigma \cdot \nabla V) = 0 \qquad (7.5.52)$$

with

$$\mathbf{J} = -\sigma \cdot \nabla V \qquad (7.5.53)$$

Equations (7.5.52) and (7.5.53) describe steady electric currents within a conductor. Boundary conditions on the system would normally include specification of the electric potential (voltage) over part of the boundary and/or the current density normal to the boundary, i.e., $\frac{\partial V}{\partial n}$. Once the electric potential and current distributions have been found then the Joule heating could be recovered from the definition in Eq. (7.5.19).

Magnetostatics

Ampere's law (7.5.30) or (7.5.32) for the time-independent case becomes

$$\nabla \times (\nu \cdot \nabla \times \mathbf{A}) = -\sigma \cdot \nabla V = \mathbf{J}_s \qquad (7.5.54)$$

This describes the magnetic field due to specified current distributions $\mathbf{J}_s$. Note that the conduction currents could be specified directly or computed from the steady current flow Eq. (7.5.52) for the electric potential V. When the magnetic potential is known, then the magnetic field $\mathbf{B}$ may be computed from its definition in (7.5.26). In addition, the Lorentz forces can be found from Eq. (7.5.17).

7.5.7 Finite Element Models for EM Fields

The potential equations in (7.5.29) and (7.5.30) represent the three components of the vector magnetic potential and the scalar electric potential for the general quasi-static field problem. Using the standard method of weighted residual techniques these equations can be converted to appropriate weak forms for subsequent use with a finite element approximation. This process will be outlined here for the case of a conducting material; simplifications for free space or source current regions are obvious.

Quasi-static potential equations

The weak or weighted integral forms corresponding to (7.5.29) and (7.5.30) are obtained by defining a vector weighting function $\mathbf{W}$ and a scalar weighting function W, multiplying (7.5.29) and (7.5.30) by the appropriate function and integrating each equation over the conducting region. That is,

$$\int_{\Omega_C} \mathbf{W}\cdot\nabla\times(\nu\cdot\nabla\times\mathbf{A})\,d\Omega + \int_{\Omega_C} \mathbf{W}\cdot\sigma\cdot\frac{\partial\mathbf{A}}{\partial t}\,d\Omega + \int_{\Omega_C} \mathbf{W}\cdot\sigma\cdot\nabla V\,d\Omega = 0 \quad (7.5.55)$$

and

$$\int_{\Omega_C} W\nabla\cdot(-\sigma\cdot\nabla V)\,d\Omega + \int_{\Omega_C} W\nabla\cdot(-\sigma\cdot\frac{\partial\mathbf{A}}{\partial t})\,d\Omega = 0 \quad (7.5.56)$$

The weak forms in (7.5.55) and (7.5.56) may be further transformed by utilizing Gauss' theorem to reduce the highest order derivative terms. Proceeding first with the magnetic potential equation, use the definition $\mathbf{H} = \nu\cdot\nabla\times\mathbf{A}$ and rewrite (7.5.55) as

$$\int_{\Omega_C} \mathbf{W}\cdot\nabla\times\mathbf{H}\,d\Omega + \int_{\Omega_C} \mathbf{W}\cdot\sigma\cdot\frac{\partial\mathbf{A}}{\partial t}\,d\Omega + \int_{\Omega_C} \mathbf{W}\cdot\sigma\cdot\nabla V\,d\Omega = 0 \quad (7.5.57)$$

Using the vector identity

$$\nabla\cdot(\mathbf{W}\times\mathbf{H}) = \mathbf{H}\cdot\nabla\times\mathbf{W} - \mathbf{W}\cdot\nabla\times\mathbf{H} \quad (7.5.58)$$

Eq. (7.5.57) is written as

$$\int_{\Omega_C} \mathbf{H}\cdot\nabla\times\mathbf{W}\,d\Omega - \int_{\Omega_C}\nabla\cdot(\mathbf{W}\times\mathbf{H})\,d\Omega + \int_{\Omega_C}\mathbf{W}\cdot\sigma\cdot\frac{\partial\mathbf{A}}{\partial t}\,d\Omega$$
$$+ \int_{\Omega_C}\mathbf{W}\cdot\sigma\cdot\nabla V\,d\Omega = 0 \quad (7.5.59)$$

Introducing the divergence theorem for the second integral and the vector identity $(\mathbf{W} \times \mathbf{H}) \cdot \mathbf{n} = \mathbf{W} \cdot (\mathbf{H} \times \mathbf{n})$ then (7.5.59) becomes

$$
\int_{\Omega_C} \mathbf{H} \cdot \nabla \times \mathbf{W} \, d\Omega - \int_{\Gamma_C} \mathbf{W} \cdot (\mathbf{H} \times \mathbf{n}) \, d\Gamma + \int_{\Omega_C} \mathbf{W} \cdot \sigma \cdot \frac{\partial \mathbf{A}}{\partial t} \, d\Omega
$$
$$
+ \int_{\Omega_C} \mathbf{W} \cdot \sigma \cdot \nabla V \, d\Omega = 0 \qquad (7.5.60)
$$

where $\mathbf{n}$ is the outward normal to the boundary Γ_C. Finally, reintroducing the definition of $\mathbf{H}$, rearranging and introducing the natural boundary condition for the magnetic potential, lead to the required form of the integral statement

$$
\int_{\Omega_C} \mathbf{W} \cdot \sigma \cdot \frac{\partial \mathbf{A}}{\partial t} \, d\Omega + \int_{\Omega_C} \nabla \times \mathbf{W} \cdot (\nu \cdot \nabla \times \mathbf{A}) \, d\Omega
$$
$$
= - \int_{\Omega_C} \mathbf{W} \cdot \sigma \cdot \nabla V \, d\Omega + \int_{\Gamma_C} \mathbf{W} \cdot \mathbf{f}^{\mathbf{H}} \, d\Gamma \qquad (7.5.61)
$$

The electric potential equation (7.5.56) is transformed in a similar manner. Using the divergence theorem on the first two integrals in (7.5.56) leads to

$$
- \int_{\Omega_C} \nabla W \cdot (-\sigma \cdot \nabla V) \, d\Omega - \int_{\Omega_C} \nabla W \cdot \left(-\sigma \cdot \frac{\partial \mathbf{A}}{\partial t} \right) d\Omega
$$
$$
= - \int_{\Gamma_C} W \left(-\sigma \cdot \frac{\partial \mathbf{A}}{\partial t} - \sigma \cdot \nabla V \right) \cdot \mathbf{n} \, d\Gamma \qquad (7.5.62)
$$

This may be rearranged into a standard form, with the natural boundary conditions for the potential included to produce

$$
\int_{\Omega_C} \nabla W \cdot \sigma \cdot \nabla V \, d\Omega + \int_{\Omega_C} \nabla W \cdot \sigma \cdot \frac{\partial \mathbf{A}}{\partial t} \, d\Omega = \int_{\Gamma_C} W f^J \, d\Gamma \qquad (7.5.63)
$$

Equations (7.5.61) and (7.5.63) are weak forms of the potential equations for electromagnetics and are suitable for use with a finite element approximation. Though not shown explicitly, the above formulation can be easily extended to include the case of motion of the conductor and/or more complex material behavior, e.g., remnant magnetization. To proceed with the finite element model, the region Ω_C is discretized into an assemblage of finite elements and the weighted integral statements are applied to each element. Within each element the vector magnetic potential and scalar electric potential are approximated by expansions of the form

$$
\mathbf{A}(\mathbf{x}, t) = \mathbf{\Phi}^T(\mathbf{x}) \mathbf{A_x}(t) \mathbf{e_x} + \mathbf{\Phi}^T(\mathbf{x}) \mathbf{A_y}(t) \mathbf{e_y} + \mathbf{\Phi}^T(\mathbf{x}) \mathbf{A_z}(t) \mathbf{e_z} \quad (7.5.64\text{a})
$$
$$
V(\mathbf{x}, t) = \mathbf{\Psi}^T(\mathbf{x}) \mathbf{V}(t) \qquad (7.5.64\text{b})
$$

which are written here for the three-dimensional Cartesian case with the $\mathbf{e_i}$ being unit vectors for the coordinate system; similar expressions can be constructed for the axisymmetric and two-dimensional cases. In Eqs.

(7.5.64a,b) $\mathbf{\Phi}$ and $\mathbf{\Psi}$ represent vectors of interpolation functions, $\mathbf{A_i}$ and $\mathbf{V}$ are vectors of nodal point unknowns and superscript T indicates a vector transpose. Note that the assumed approximations for the dependent variables are, as usual, semi-discrete with the spatial dependence being discretized through interpolation and the temporal dependence remaining continuous. For the Galerkin method the weight functions $\mathbf{W}$ and W are selected to be the same functions as used to represent the variables. That is,

$$\mathbf{W}(\mathbf{x}) = \mathbf{\Phi}(\mathbf{x})\mathbf{e_x} + \mathbf{\Phi}(\mathbf{x})\mathbf{e_y} + \mathbf{\Phi}(\mathbf{x})\mathbf{e_z} \qquad (7.5.65a)$$
$$W(\mathbf{x}) = \mathbf{\Psi}(\mathbf{x}) \qquad (7.5.65b)$$

Substituting the definitions in (7.5.64) and (7.5.65) into the weighted residual equations in (7.5.61) and (7.5.63) produces the following set of discrete equations for each element

$$\mathbf{M\dot{A}} + \mathbf{KA} + \mathbf{NV} = \mathbf{F}_A \qquad (7.5.66)$$
$$\mathbf{N}^T\dot{\mathbf{A}} + \mathbf{LV} = \mathbf{F}_V \qquad (7.5.67)$$

where the superposed dot indicates a time derivative. Equation (7.5.66) represents the three component equations for the magnetic potential.

The matrix system shown above is unsymmetric and is of an undesirable form from the standpoint of time integration. To restore symmetry, the following definition proposed by Chari et al. [15] may be employed

$$\mathbf{V} \equiv \frac{\partial \mathbf{v}}{\partial t} = \dot{\mathbf{v}} \qquad (7.5.68)$$

Using this definition Eqs. (7.5.66) and (7.5.67) can be rewritten as

$$\mathbf{M\dot{A}} + \mathbf{KA} + \mathbf{N}\dot{\mathbf{v}} = \mathbf{F}_A \qquad (7.5.69)$$
$$\mathbf{N}^T\dot{\mathbf{A}} + \mathbf{L}\dot{\mathbf{v}} = \mathbf{F}_V \qquad (7.5.70)$$

and in a completely assembled form

$$\begin{bmatrix} \mathbf{M} & \mathbf{N} \\ \mathbf{N}^T & \mathbf{L} \end{bmatrix} \begin{Bmatrix} \dot{\mathbf{A}} \\ \dot{\mathbf{v}} \end{Bmatrix} + \begin{bmatrix} \mathbf{K} & 0 \\ 0 & 0 \end{bmatrix} \begin{Bmatrix} \mathbf{A} \\ \mathbf{v} \end{Bmatrix} = \begin{Bmatrix} \mathbf{F}_A \\ \mathbf{F}_V \end{Bmatrix} \qquad (7.5.71)$$

The component matrices defined in (7.5.71) are defined by the following integrals that arise from the weighted residual statements. The matrices are written here in terms of vector notation; the explicit forms involving derivatives on the element and Jacobians for the transformation to the master element can be obtained by use of obvious substitutions from previous chapters

$$\mathbf{M} = \int_{\Omega_e} \mathbf{\Phi} \cdot \sigma \cdot \mathbf{\Phi}^T \, d\Omega, \qquad \mathbf{K} = \int_{\Omega_e} \nabla \times \mathbf{\Phi} \cdot \nu \cdot \nabla \times \mathbf{\Phi}^T \, d\Omega$$

$$\mathbf{N} = \int_{\Omega_e} \mathbf{\Phi} \cdot \sigma \cdot \nabla \mathbf{\Psi}^T \, d\Omega, \qquad \mathbf{N}^T = \int_{\Omega_e} \nabla \mathbf{\Psi} \cdot \sigma \cdot \mathbf{\Phi}^T \, d\Omega$$

$$\mathbf{L} = \int_{\Omega_e} \nabla \mathbf{\Psi} \cdot \sigma \cdot \nabla \mathbf{\Psi}^T \, d\Omega, \qquad \mathbf{F}_A = \int_{\Gamma_e} \mathbf{\Phi} \cdot \mathbf{f}^H \, d\Gamma$$

$$\mathbf{F_M} = \int_{\Omega_e} \nabla \times \mathbf{\Phi} \cdot \mathbf{M_0} \, d\Omega, \qquad \mathbf{F}_V = \int_{\Gamma_e} \mathbf{\Psi} f^J \, d\Gamma \qquad (7.5.72)$$

Gauge condition

When the magnetic potential must be made unique, the Coulomb gauge defined in (7.5.48) is added to the field equations as a constraint. From a computational standpoint there are several methods available for developing a discrete form of (7.5.48). One of the simplest implementations is the penalty method. The penalty method is most easily invoked by casting the original quasi-static field problem in terms of a functional to be minimized. To simplify the derivation the specified current form of the magnetostatic problem will be considered; the more general quasi-static or eddy current problem is developed in an analogous manner. Consider the static form of (7.5.29) or more precisely, Eq. (7.5.54)

$$\nabla \times (\nu \cdot \nabla \times \mathbf{A}) = \mathbf{J_s} \qquad (7.5.73)$$

The functional associated with (7.5.73) is

$$I(\mathbf{A}) = \frac{1}{2} \int_\Omega \nabla \times \mathbf{A} \cdot \nu \cdot \nabla \times \mathbf{A} \, d\Omega - \int_\Omega \mathbf{J_s} \cdot \mathbf{A} \, d\Omega \qquad (7.5.74)$$

A finite element approximation for $\mathbf{A}$ may be used in (7.5.74), which when minimized with respect to the dependent variables produces a discrete system that is the same as the appropriate magnetostatic subset of the Galerkin equations in (7.5.71). To invoke the Coulomb gauge on this system, the functional in (7.5.74) is augmented with a least squares penalty term

$$I_C(\mathbf{A}) = I(\mathbf{A}) + \frac{\lambda}{2} \int_\Omega (\nabla \cdot \mathbf{A})^2 \, d\Omega \qquad (7.5.75)$$

where λ is the penalty parameter. When a finite element approximation, such as (7.5.64) is used in the augmented functional in (7.5.75), and variations are taken with respect to the components of $\mathbf{A}$, the result is a discrete system of the following form

$$(\mathbf{K} + \lambda \mathbf{K_{div}})\mathbf{A} = \mathbf{F}_A + \mathbf{F}_J \qquad (7.5.75)$$

where most of the matrices and vectors are defined in (72) and the penalty term is defined by

$$\mathbf{K_{div}} = \int_{\Omega_e} \nabla \cdot \mathbf{\Phi} \nabla \cdot \mathbf{\Phi}^T \, d\Omega \qquad (7.5.76)$$

This process is very similar to the penalty method used to enforce incompressibility in the viscous flow problem. The penalty enforcement of the Coulomb gauge requires the construction of an additional matrix that is added to the appropriate terms in the magnetic vector potential equation. The penalty parameter λ is set to a large number with the free space reluctivity being a good choice ($\lambda = 1/\mu_0 = \nu_0$). In order to avoid an over-constrained system, the penalty matrix must be singular. The standard method for achieving this situation is to reduce the order of the quadrature rule used to evaluate the integral in (7.5.76).

Static field equations

The finite element equations for the simplified static field problems are developed in the same manner as outlined in the previous sections. For some situations, the equations are merely subsets of the more complex quasi-static problem.

The basic equation for electrostatics is either (7.5.50) or (7.5.51) depending on the type of material. If combined, the relevant equation is

$$\nabla \cdot (\epsilon \cdot \nabla V) = -\rho \nabla \cdot \mathbf{P}_0 \tag{7.5.77}$$

This has a weak form that is given by

$$\int_\Omega \nabla W \cdot \epsilon \cdot \nabla V \, d\Omega = -\int_\Omega W \rho \, d\Omega + \int_\Omega \nabla W \cdot \mathbf{P}_0 \, d\Omega + \int_\Gamma W f^D \, d\Gamma \tag{7.5.78}$$

where the natural boundary condition is a specification of the electric flux (gradient of the potential) normal to the boundary. Let the electric potential be approximated by the steady form of the finite element representation given in (7.5.64b)

$$V(\mathbf{x}) = \mathbf{\Psi}^T(\mathbf{x})\mathbf{V} \tag{7.5.79}$$

Using the Galerkin definition and the scalar weight function in (7.5.65b), Eq. (7.5.78) leads to the matrix equation

$$\mathbf{L}_\epsilon \mathbf{V} = \mathbf{F}_\rho + \mathbf{F}_P + \mathbf{F}_D \tag{7.5.80}$$

where

$$\mathbf{L}_\epsilon = \int_{\Omega_e} \nabla \mathbf{\Psi} \cdot \epsilon \cdot \nabla \mathbf{\Psi}^T \, d\Omega, \quad \mathbf{F}_\rho = \int_{\Omega_e} \mathbf{\Psi} \rho \, d\Omega$$

$$\mathbf{F}_P = \int_{\Omega_e} \nabla \mathbf{\Psi} \cdot \mathbf{P}_0 \, d\Omega, \quad \mathbf{F}_D = \int_{\Gamma_e} \mathbf{\Psi} f^D \, d\Gamma \tag{7.5.81}$$

The steady current flow problem is a subset of the general quasi-static formulation and can be defined immediately as

$$\mathbf{L}\mathbf{V} = \mathbf{F}_\mathbf{V} \tag{7.5.82}$$

where the matrix and vector are defined in (7.5.72).

The magnetostatics problem is also a subset of the general quasi-static problem and corresponds to the equations for a free space region with specified currents. The relevant finite element equations are derived from (7.5.69)

$$\mathbf{K}\mathbf{A} = \mathbf{F}_A + \mathbf{F}_J \tag{7.5.83}$$

where the vector $\mathbf{F}_J$ is a weighted integral over the element volume of the specified current density.

7.5.8 Solution Methods - EM Fields

The finite element equations for the general quasi-static electromagnetics problem, as given in Eq. (7.5.71), represent a set of ordinary differential equations not unlike the equation sets that were encountered in the heat conduction and viscous flow formulations. Because of this similarity, virtually all of the solution methods used in the transport problems can be used in the electromagnetics problem without modification.

For static problems, the equations in (7.5.71) reduce to nonlinear algebraic equations that can be linearized and solved via iteration using Picard or Newton methods. The general time-dependent case is well-suited to implicit time integration methods such as backward Euler or the trapezoid rule. Both constant time step and adaptive time step, predictor/corrector methods, have been used successfully with these types of integrators.

The only problem type that differs significantly from the transport equations is the time harmonic field problem. If the nodal point variables are defined through a phasor representation with frequency ω [as was done for the continuum case in Eq. (7.5.34)], the equations in (7.5.69) and (7.5.70) become

$$i\omega \mathbf{M} \mathbf{A}_0 + \mathbf{K} \mathbf{A}_0 + i\omega \mathbf{N} \mathbf{v}_0 = \mathbf{F}_{A_0} \qquad (7.5.84)$$

$$i\omega \mathbf{N}^T \mathbf{A}_0 + i\omega \mathbf{L} \mathbf{v}_0 = \mathbf{F}_{v_0} \qquad (7.5.85)$$

or in the combined matrix form

$$\begin{bmatrix} i\omega \mathbf{M} & i\omega \mathbf{N} \\ i\omega \mathbf{N}^T & i\omega \mathbf{L} \end{bmatrix} \left\{ \begin{array}{c} \mathbf{A}_0 \\ \mathbf{v}_0 \end{array} \right\} + \begin{bmatrix} \mathbf{K} & 0 \\ 0 & 0 \end{bmatrix} \left\{ \begin{array}{c} \mathbf{A}_0 \\ \mathbf{v}_0 \end{array} \right\} = \left\{ \begin{array}{c} \mathbf{F}_A \\ \mathbf{F}_V \end{array} \right\} \qquad (7.5.86)$$

where the unknowns are now complex and defined by $V_0 = (V^R + iV^I)$ and $\mathbf{A}_0 = (\mathbf{A}^R + i\mathbf{A}^I)$. Recall also that $V_0 = \dot{v}_0 = i\omega v_0$.

Through the use of the phasor representation, the time harmonic problem has been reduced to a steady, linear matrix problem; the matrix coefficients and nodal point variables are complex.

The linear algebra problem associated with each of the steady, time-dependent or harmonic solution methods may be solved using either direct matrix solvers or iterative solvers of the conjugate gradient or Krylov type. Note that the matrix systems are usually unsymmetric. The gauge condition alters the matrix structure and may cause convergence difficulties for iterative methods. The complex coefficients in the time harmonic case can be used directly in solvers set up for complex arithmetic. An alternative is to separate the complex matrix into real and imaginary parts and solve twice the number of equations using only real coefficients.

7.6 Coupled Problems in Mechanics

7.6.1 Introduction

The equations from the three previous sections may be combined in a variety of ways to describe a large class of problems in mechanics. The fact that there is such a large spectrum of problems implies that meaningful discussions of coupling will be difficult to accomplish in a general way. Therefore, we will revert to a simple and expedite method of description that considers the pairwise combinations of five mechanics areas: heat conduction (HC), nonisothermal viscous flow (VF), quasi-static solid mechanics (SM), electric fields (EF), and magnetic fields (MF). The symmetric coupling matrix shown in Figure 7.6.1 is used to organize and focus the discussion. Note that at the intersection of each pair of mechanics areas, one or more of the usual types of coupled problems are listed. No attempt is made to be thorough in this listing and only the most common interactions are indicated. In the matrix and in the following sections, some of the possible data dependencies for each interaction will be described, the typical strength of the interaction will be cited, and the types of algorithms that may be used will be suggested. Since this is a book on fluid mechanics and heat transfer, the focus will be on the first two rows of the matrix.

	1 Heat Conduction/ Radiation	2 Viscous/ Porous Flow	3 Quasi- Static Solids	4 Electric Fields	5 Magnetic Fields
1 Heat Conduction/ Radiation		1-2 Conjugate Heat Transfer B (S)	1-3 Thermal Stress V.P.G. (W-S)	1-4 Resistive Heating V.P. (W-S)	1-5 Inductive Heating V.P. (W-S)
2 Viscous/ Porous Flow	2-1		2-3 Fluid/Solid Interaction B.G. (S)	2-4 EHD Electro- Rheology V.P. (S-W)	2-5 MHD Inductive Stirring V.P. (W-S)
3 Quasi-Static Solids	3-1	3-2		3-4 Electro- Striction Piezoelectric B.P.G.(S-W)	3-5 Magnetic Stress V.G. (W-S)
4 Electric Fields	4-1	4-2	4-3		4-5 Eddy Currents V.P. (S)
5 Magnetic Fields	5-1	5-2	5-3	5-4	

Types of Coupling: B – Boundary conditions/surface flux
V – Volume terms; G – Geometry
P – Property/constitutive dependence

Strength of Coupling: S – Strong
W – Weak

Figure 7.6.1: Matrix of possible mechanics interactions.

7.6.2 Heat Conduction - Viscous Fluid Interactions 1 & 2

For completeness of the exposition, some of the coupled problem situations from previous chapters are reviewed here in summary form. Consider the interaction of heat conduction and radiation and conduction and chemical reaction from Chapter 3. These are strongly coupled problems with conduction/radiation being a boundary condition interaction and possibly a thermophysical property interaction if the emissivity is a function of temperature. The conduction/chemical reaction problem is primarily a volume source interaction on the conduction side of the interaction and a (kinetics) property interaction for the reaction equations. Because of the strong coupling the preferred method of solution would be a simultaneous solution of each pair of equations.

The completely coupled solution process is often not practical in these two cases because of the characteristics of the radiation and chemical reaction equations. In the conduction/radiation case, the matrix structure of the radiation equation is full due to the view factor (surface-to-surface) coupling while the conduction matrix is sparse. The size of the radiation matrix and its precipitous growth with increase in mesh resolution often make the combined matrix problem not tractable. The combined conduction/reaction problem suffers mainly from a problem of disparate time scales and stiffness in the reaction equations. The excessively small time steps required in the chemistry solution are an unnecessary computational penalty for the thermal diffusion problem. Cyclic or operator splitting methods may not be optimal from a coupling perspective, but are practical from an overall algorithm point of view as detailed in Chapter 3.

The conjugate problem that couples solid body conduction with a nonisothermal flow is another strongly coupled problem with an interaction at the boundary. This class of problems is typically solved as a fully coupled matrix problem. The addition of conducting regions (scalar equation) to a primarily viscous flow problem (vector equation) presents no real increase in computational burden. Time constants for the two equation sets are usually of the same order of magnitude so that integration methods may follow the smaller time scale without too much of a penalty and coupling is straightforward. Note that it is possible to treat the conjugate problem in a decoupled or cyclic manner. In this case, the temperature and flux at the common boundary are used as the coupling variables. One variable is used as a boundary condition while the second variable is used to measure the convergence or agreement with the other equation set. Though feasible, this is not a recommended solution method for this type of problem.

7.6.3 Heat Conduction - Quasi-Static Solid Interactions 1 & 3

Thermal stress problems are one of the most heavily studied and commonly encountered types of coupled mechanics problems. In most cases the coupling is relatively weak with the thermal field providing a thermal strain through the mechanical constitutive relation; in some cases a temperature dependence of the parameters in the constitutive equation adds to the interaction. When the interaction is limited to these effects, the problem can be easily solved via a one-way or decoupled procedure. With Eq. (7.3.7) representing the finite element heat conduction problem

$$\mathbf{M\dot{T}} + \mathbf{\hat{K}T} = \mathbf{\hat{F}} \qquad (7.6.1)$$

and Eq. (7.4.22) defining the mechanical equilibrium

$$\mathbf{K(u,T)u} = \mathbf{F} \qquad (7.6.2)$$

the solution procedure is fairly evident. The conduction problem, being independent of any mechanical variables, is solved first over the time interval of interest. Temperature data are passed to the solid mechanics equations and these are solved for the displacement, strain, and stress fields over the temperature history. Since (7.6.2) is a quasi-static model, incrementation of the structural loading is substituted for a time integration procedure. The temperature field at any given load step can be found by interpolation of the finite element solution. This type of problem could also be solved through a fully coupled procedure though this is not a computationally attractive alternative. Since the thermal diffusion process has the smaller time constant (compared to the infinite time constant for the solid mechanics problem), the thermal problem must be solved a fairly large number of times over the time interval. A fully coupled algorithm would thus force the solid mechanics problem to be solved many more times than would normally be required to achieve an accurate stress solution. The additional solutions of (7.6.2) make this approach ill-advised.

The coupling between Eqs. (7.6.1) and (7.6.2) increases in strength when mechanical dissipation is important and/or very large deformations occur in the structure such that the geometric definition of the heat conduction problem is altered. Mechanical dissipation appears as a volumetric source term in the conduction equation and is directly dependent on the stress and strain rate fields. That is, the force vector in (7.6.1) is $\mathbf{\hat{F}}(\sigma,\dot{\epsilon})$. Depending on the dissipation rate the problem may be either weakly or strongly coupled. If the heat generation rate is slow and temperature effects in the solid negligible, the solid mechanics problem could be solved over the loading history of interest to produce a dissipation history. The dissipation history could then be used to solve the conduction problem for the temperature response of the region. As the dissipation rate increases and temperature dependence increases, the coupling is more pronounced and Eqs. (7.6.1) and (7.6.2) must be solved more implicitly. A fully coupled method can be used, though a cyclic algorithm is more cost effective. Again, the conduction problem sets the time scale (time step) for the integration; the solid mechanics problem is solved after each thermal time step or whenever the fields change sufficiently to require an update in the dissipation. This decision on the frequency of solution for a subordinate process is one of the more difficult judgments to make in the use of cyclic or staggered solution strategies.

The coupling between (7.6.1) and (7.6.2) is very strong when large changes in geometry occur in the solid and the boundary conditions in the thermal problem are effected. If radiation is part of the thermal problem geometric changes imply changes in the radiation view factors and a redistribution of surface flux. Also, if deformation leads to new contacts between surfaces or the separation of previously contacting surfaces, the thermal conditions may be significantly altered. This type of coupling may be effectively treated with a cyclic algorithm, though the structural solution and geometrical updates would usually have to be computed at every thermal time step or whenever deformation was significant. The recomputing of radiation view factors for a dynamic geometry is a significant computational burden and makes this type of problem quite challenging.

7.6.4 Heat Conduction - Electric Field Interactions 1 & 4

Resistive or Joule heating occurs when energy is dissipated by an electric current flowing through a conductor. This type of problem would be represented by the conduction equation in (7.6.1)

$$\mathbf{M}\dot{\mathbf{T}} + \hat{\mathbf{K}}\mathbf{T} = \hat{\mathbf{F}}(\mathbf{J}) \tag{7.6.3}$$

and the steady current flow equation (7.5.52) and its finite element model (7.5.82)

$$\mathbf{L}(\mathbf{T})\,\mathbf{V} = \mathbf{F}_V \tag{7.6.4}$$

The variable dependencies are shown in (7.6.3) and (7.6.4). The volume heating in (7.6.3) is given by $Q_J = \sigma^{-1}\mathbf{J}^2$ and $\mathbf{J}$ is related to the gradient of the electric potential, $\mathbf{V}$, which is the unknown in (4). The electrical conductivity is usually a function of temperature, thus making the diffusion operator in (7.6.4) dependent on $\mathbf{T}$.

If these dependencies are both present, Eqs. (7.6.3) and (7.6.4) can be most easily solved as a fully coupled system. An implicit time integration procedure applied to (7.6.3) results in a generally nonlinear matrix problem as described in Chapter 3. This matrix can be combined with (7.6.4) to represent the two scalar equations. At each time step (or iteration, if the problem is steady), the combined nonlinear problem can be solved using either a Picard iteration if the nonlinearities are fairly mild or Newton's method for stronger nonlinear behavior. This problem can also be solved via a decoupled or cyclic procedure. Because (7.6.3) is time dependent, the conduction problem is the master process and sets the time scale for the problem. At a given time step, the known temperature field can be passed to the current flow equation, the temperature-dependent conductivity evaluated and the current density and Joule heating computed. The Joule heating is transferred back to the conduction equation to allow the next update on the temperature field. This process may be repeated until convergence occurs at the current time; in many cases the changes that occur over a time step are sufficiently small that only one iteration through the data exchange is necessary to maintain acceptable accuracy.

7.6.5 Heat Conduction - Electromagnetic Field Interactions 1 & 4 & 5

Isolated magnetic fields do not generate any significant interactions with a thermal field, though the full coupling of both electric and magnetic fields with temperature effects describe a number of engineering processes of importance. Induction heating for melting and surface treatment (hardening) are commonly used processes that depend on this type of coupling. Eddy current analysis for the performance of electromagnetic devices, such as motors and generators, is another source of coupled problems. In the present description the focus will be on induction heating where extremes in the interaction are observed; the less demanding thermal performance problems will be subsets of this type of coupling.

To describe the nonisothermal eddy current problem, the conduction equation (7.3.7) is required

$$\mathbf{M}\dot{\mathbf{T}} + \hat{\mathbf{K}}\mathbf{T} = \hat{\mathbf{F}}(\mathbf{J}) \tag{7.6.5}$$

as is the quasi-static, electromagnetic system defined by the field problem in Eqs. (7.5.29) and (7.5.30) and the corresponding finite element model in equations (7.5.69) and (7.5.70)

$$\mathbf{M}(\mathbf{T})\dot{\mathbf{A}} + \mathbf{K}(\mathbf{T}, \mathbf{B})\mathbf{A} + \mathbf{N}(\mathbf{T})\dot{\mathbf{v}} = \mathbf{F}_A \tag{7.6.6}$$

$$\mathbf{N}^T(\mathbf{T})\dot{\mathbf{A}} + \mathbf{L}(\mathbf{T})\dot{\mathbf{v}} = \mathbf{F}_V \tag{7.6.7}$$

The variable dependencies are shown in Eqs. (7.6.5)–(7.6.7), which are written for the general case where the electrical problem is potential or voltage driven. When the problem is current driven, Eq. (7.6.7) is no longer needed and the $\dot{\mathbf{v}}$ term in (7.6.6) is known and associated with the known current density. These subcases were covered previously in Section 7.5.3. The volume heating in (7.6.5) is again given by the Joule heating relation

$$Q_J = \sigma^{-1}\mathbf{J}^2$$

and $\mathbf{J}$ is related to the combination of the gradient of the electric potential, $\mathbf{V}$, if it is present, and the time derivative of the magnetic vector potential [see Eq. (7.5.28) and following]. The time derivative term is the induced or eddy current due to the fluctuating magnetic field. The dependencies in Eqs. (7.6.6) and (7.6.7) correspond to the possible variation of the electrical conductivity with temperature and the magnetic permeability with both temperature and the magnetic flux density.

In ferromagnetic materials the variation of the magnetic permeability is a major complicating factor in the analysis since it is a hysteretic function of $\mathbf{B}$ and $\mathbf{H}$ which is scaled by the temperature. The magnetic permeability may vary by three orders of magnitude over typical magnetic field strengths (at moderate temperatures) while above the Curie temperature the magnetic permeability reverts to a constant equal to free space permeability.   A final complication in this type of coupled problem stems from the limited penetration into the material of the induced current fields and Joule heating when the applied fields are moderate to high frequencies. This skin depth effect implies that the simulation regions that are critical to the heat transfer part of the problem do not necessarily overlap all of the electromagnetic regions of importance.

In many applications the electromagnetic fields are time harmonic and the phasor representation of (7.6.6) and (7.6.7) [Eqs. (7.5.84) and (7.5.85)] are the appropriate finite element equations with similar functional dependencies. Even though the electromagnetic problem is nonlinear, a phasor representation can be used with single frequency driving fields if material properties are treated appropriately. This method is illustrated in [14]. Though the coupling is very strong, the appropriate solution strategy for Eqs. (7.6.5)–(7.6.7), or their time harmonic equivalents, is a cyclic procedure due to the vastly different time scales for most applications. The diffusion time constant for the heat conduction problem is significantly longer than equilibration of the induced electromagnetic fields; the electromagnetic fields are virtually constant over the diffusion time interval even if they are periodic. The heat conduction problem, therefore, is the master process and sets the time integration procedure. At a given time step, the current temperature field is passed to the electromagnetic problem. If the time-dependent forms of (7.6.6) and (7.6.7) are used, the fields are integrated forward to the end of the time interval using an appropriately small time step. The Joule heating history is computed from the time-dependent fields and integrated to provide an energy rate over the time step for each resistively heated finite element. When the time harmonic forms of (7.6.6) and (7.6.7) are used, the procedure is similar except that a time independent, often nonlinear field problem must be solved and the resulting complex variables must be manipulated to generate the Joule heating. After the electromagnetic fields are updated, the new Joule heating data are used as a source term for the next conduction step. Note that in some applications, the geometry of the problem may change with time. This is not usually a difficulty because both the conduction and electromagnetics problems can be solved in a Lagrangian reference frame making geometric updates straightforward in the coupled situation. When radiation is a consideration, changing geometries will increase the computational work for the heat transfer solution step.

Even though the electromagnetic problem is the subordinate process and is simply providing the "loads" for the thermal problem it is, in this case, the more difficult problem. The inclusion of the gauge condition (7.5.48) makes the solution of the magnetic vector potential equation very difficult in much the same way as the incompressibility constraint hinders the solution of the viscous flow problem. In addition, the large changes in magnetic permeability with both field strength and temperature, plus the hysteretic nature of the property variation, generate strong material nonlinearities that inhibit convergence. Finally, though the time harmonic case is simplified in the time domain, the use of complex variables increases significantly the computational burden due to the matrix and variable vector sizes.

7.6.6 Viscous Flow - Quasi-Static Solid Interactions 2 & 3

Fluid structure interaction problems are quite common and this type of coupling has been well studied. In many simulations the fluid model can be simplified to a potential flow where only pressure loads on the structure are of interest and fluid separation is not an important feature of the flow. The coupling of the fluid potential equation with the solid mechanics problem will not be considered here as the main interest remains the viscous flow problem. The major difficulties in coupled viscous fluid and solid mechanics problems stem from the inherently different coordinate descriptions used in the two fields and the possible variations in defining a dominant physical process.

Because the preferred description for solid mechanics is Lagrangian and the usual description for fluids is an Eulerian coordinate system, some compromise must be found for a useful coupled problem description. The standard solution is to work in the usual Lagrangian reference frame for the solid and specify the fluid motion in an Arbitrary Lagrangian Eulerian (ALE) framework, as was outlined in Section 4.10.3. This approach has the advantage of maintaining an accurate definition of the fluid/solid interface which is the primary coupling mechanism in this type of problem. The possible large deformation of the solid is well captured by this approach and the use of appropriate strain measures in the solid mechanics formulation. Also, the ALE method with a mesh moving scheme coupled to the interface motion (which is determined from the solid mechanics problem), provides a robust and accurate algorithm for the fluid motion. Many of the details associated with ALE methods were described in Section 4.10.3 and will not be repeated here. The second area of difficulty is to determine the primary physical process so that a solution algorithm for the coupled system can be constructed. Unfortunately, the fluid motion can drive the solid deformation or the solid motion and deformation can drive the fluid motion. Since neither process is always dominant a fully coupled algorithm is suggested.

To describe the fluid structure interaction problem the viscous flow model from equations (7.3.9) and (7.3.10) is required,

$$-\mathbf{Q}^T \mathbf{u} = 0 \tag{7.6.8}$$

$$\mathbf{M}\dot{\mathbf{u}} + \mathbf{C}(\mathbf{u}^m)\mathbf{u} + \mathbf{K}\mathbf{u} - \mathbf{Q}\mathbf{P} = \mathbf{F}(\mathbf{u}^m) \tag{7.6.9}$$

which is written here in its isothermal form. Coupling with a nonisothermal flow is certainly possible, in which case the energy equation from (7.3.11) would be added to the above and a heat conduction equation would be required for the solid region. This complication adds little to the overall algorithm and therefore will not be explicitly considered. The quasi-static solid mechanics problem is again given by Eq. (7.4.22)

$$\mathbf{K}(\mathbf{u}, \mathbf{u}^m, \mathbf{T})\mathbf{u} = \mathbf{F}(\mathbf{P}, \mathbf{u}) \tag{7.6.10}$$

For ease of notation it is assumed that the solid mechanics problem is written in terms of velocities; the use of the same matrix and vector symbols in (7.6.9) and (7.6.10) does not imply that these are the same matrices. The variable dependencies are indicated in (7.6.8)–(7.6.10) and are seen to be a strong function of the mesh movement $\mathbf{u}^m$ (interface geometry) and the fluid loads on the structure.

As noted above, the most general solution procedure for this type of problem is a fully coupled method in which Eqs. (7.6.8)–(7.6.10) are solved in a single matrix problem. A time integration scheme would be applied to the fluid equations while the static solid equations would have the appropriate loads applied for the current time. The Picard or Newton's method could be used to converge the solution at the current time. The fluid and solid mechanics regions do not overlap; the entire domain is meshed by a single finite element discretization. If a pseudo-structural mesh movement scheme is employed as part of the ALE method, then solid mechanics equations are solved over the entire domain. The mesh movement equations may be fully coupled with the fluid and solid equations or may be decoupled and used to update the fluid mesh region in a subcycle. For efficiency, this fully coupled method assumes that the time scales for the fluid and solid regions are not too different, which they probably cannot be, due to the strong coupling at the boundary. This algorithm does pay a penalty in the matrix solution because the fluid equations are nonsymmetric while the solid equations are symmetric and could be solved with less computational effort.

7.6.7 Viscous Flow - Electric Field Interactions 2 & 4

There are a number of problems that involve a viscous fluid and an electric field, most of which are unfamiliar to the applied mechanics community. Electrohydrodynamics involves the motion of a fluid due to an electric pressure which is proportional to the square of the applied electric field. The fluid may be either conducting or a dielectric. The process of flow electrification involves a dielectric fluid and the convection of charge from the double layers that form near boundaries. This type of problem is common in electrochemical devices such as batteries. A more familiar problem, and one that will be considered here, is the flow of an electrorheological fluid. In this case an applied electric field causes a large increase in the viscosity of the fluid, which can act as a control mechanism for the flow. The electrorheological application is not a particularly difficult type of coupled problem since the coupling is essentially in one direction with the electric field influencing the flow.

The equations for this type of problem include the isothermal, viscous flow equations from (7.3.9) and (7.3.10) or

$$-\mathbf{Q}^T\mathbf{u} = 0 \tag{7.6.11}$$

$$\mathbf{M}\dot{\mathbf{u}} + \mathbf{C}\mathbf{u} + \mathbf{K}(\mathbf{E})\mathbf{u} - \mathbf{Q}\mathbf{P} = \mathbf{F} \tag{7.6.12}$$

and the potential equation for the electric field from Eq. (7.5.82)

$$\mathbf{L}\,\mathbf{V} = \mathbf{F}_V \tag{7.6.13}$$

The viscous term in Eq. (7.6.12) is a function of the electric field, which is computed from the gradient of the electric potential, the unknown in (7.6.13). The master process is the viscous flow problem which is integrated in time by standard methods. When a voltage is applied in some region of the problem, the potential equation must be solved for the instantaneous electric field. These field data are then transferred to the flow equation to allow the viscosity function to be evaluated for the next time step. The process is continued as long as a voltage is applied to the problem. The subordinate process in this case involves a series of steady solutions with the boundary condition (voltage) having a time variation.

7.6.8 Viscous Flow - Electromagnetic Field Interactions 2 & 4 & 5

The topical area of magnetohydrodynamics (MHD) brings together the descriptions of nonisothermal, fluid mechanics and the electromagnetics of conducting materials. The MHD field includes a variety of problem types many of which are quite complex especially with regard to coupling and property variations. Rather than attempt a superficial coverage of general MHD problems, we will narrow the focus here to a class of problems that are important in material processing and manufacturing applications. Processes such as high current melting and inductive stirring fall into this problem category. The major assumption involved in this type of problem is that the magnetic Reynolds number is small. The magnetic Reynolds number is usually defined as $Re_m = UL\sigma\mu_m$, where σ is the electrical conductivity, μ_m is the magnetic permeability, and U and L are a representative flow velocity and length scale, respectively. The magnetic Reynolds number represents a ratio of magnetic convection to magnetic diffusion. For many common processes involving liquid metals, the magnetic Reynolds number is small, which implies that the magnetic field lines are unaffected by the flow field. The opposite extreme of a high magnetic Reynolds number ensures that the magnetic field lines are strongly convected by the flow field. This limit is typical of fusion applications and problems in astrophysics.

With the small magnetic Reynolds number assumption, the equations required for the MHD problem include the nonisothermal, viscous flow equations

$$-\mathbf{Q}^T\mathbf{u} = 0 \qquad (7.6.13)$$
$$\mathbf{M\dot{u}} + \mathbf{Cu} + \mathbf{Ku} - \mathbf{QP} + \mathbf{BT} = \mathbf{F(J,B)} \qquad (7.6.14)$$
$$\mathbf{N\dot{T}} + \mathbf{DT} + \mathbf{LT} = \mathbf{G(J)} \qquad (7.6.15)$$

and the potential forms of the electromagnetic equations

$$\mathbf{M(T)\dot{A}} + \mathbf{K(T)A} + \mathbf{N(T)\dot{v}} = \mathbf{F}_A \qquad (7.6.16)$$
$$\mathbf{N}^T(\mathbf{T})\dot{\mathbf{A}} + \mathbf{L(T)\dot{v}} = \mathbf{F}_V \qquad (7.6.17)$$

In the flow equations (7.6.13)–(7.6.15), the main electromagnetic coupling is through the resistive or Joule heating in the energy equation (which is proportional to the square of the current density) and the Lorentz body force in the momentum equation (which is proportional to the current density and the magnetic flux density). The electromagnetic equations are coupled to the flow problem primarily through temperature-dependent properties; a change of phase in the flow region may require some adjustment in geometry and boundary conditions to account for expansion or contraction of the fluid region. The small magnetic Reynolds number assumption produces the reduced coupling of the electromagnetics to the flow problem. If this assumption is not invoked, the magnetic equation (7.6.16) would have an additional term that depends on the fluid velocity.

Equations (7.6.13)–(7.6.17) form a strongly coupled set when all of the dependencies are present. Though a fully coupled algorithm is attractive from the convergence point of view, the very large size of the system and the generally differing time scales make such an approach impractical. In many applications, the electromagnetic fields are slowly varying or steady and the flow process will set the time scale and be the dominant process. The fact that the fluid and thermal "loads" come from the electromagnetics, while only properties vary in the other coupling direction, also argues for this choice of master and slave process. The standard staggered algorithm would thus proceed with the current temperature field being transferred to the electromagnetics equations (7.6.16) and (7.6.17). A time integration or steady solution for the electric and magnetic fields would produce the needed Joule heating and Lorentz forces. Transferring these data back to the nonisothermal flow equations would allow the next fluid and thermal fields to be computed. In the limit of temperature-independent electromagnetic properties and steady EM fields, Eqs. (7.6.16) and (7.6.17) reduce to a magnetostatics description, which only needs to be solved once for resistive heating and body forces. Slowly varying electric and magnetic fields could also be treated as a series of steady states that may be precomputed and made available to the flow problem as a loading history.

7.6.9 Quasi-Static Solid - Electromagnetic Field Interactions 3 & 4 & 5

The interaction of solid mechanics with electromagnetics will not be treated here in detail because it is outside the scope of the text. It is sufficient to note that these types of interactions can produce coupled problems that are similar in type and complexity to those described in the previous sections. Electric fields may cause stresses and deformations in a solid (electrostriction) and deformation may induce an electrical response (piezoelectric effect). These types of interactions are primarily through the constitutive relations for the material. Also, a strong magnetic field may produce a magnetic stress (magnetostriction) as described by the Maxwell stress tensor. The algorithms for coupling these types of interactions would certainly be similar to the fluid and thermal procedures outlined above.

7.7 Implementation of Coupled Algorithms

A primary feature of coupled field problems in mechanics is the almost endless variety of possible interactions. This diversity plus the accompanying distribution of workable numerical algorithms, as illustrated in the previous sections, complicates the construction of finite element software. Though this text is not oriented toward the descriptions of code architectures and finite element implementation issues, it is important to discuss some general aspects of coupled problem implementation.

The first choice in implementing a coupled or multiphysics solution method is whether the algorithm will be in a single code or multiple software packages. For strongly coupled problems that will be solved in a fully coupled method, the choice is obviously a single code. In many ways this is the simplest type of implementation since only one matrix problem needs to be considered and all the equations are solved at once. Finite element interpolation order for the various fields, equation ordering within the matrix, type of integration and/or iterative method demand consideration but are not major impediments to implementation. The major drawback to this approach is that the code does one and only one type of coupled problem.

A second approach to coupling involves a single code or software package that is composed of modules for the different types of physical phenomena that are to be modeled. The coupling or required data exchanges are then orchestrated from a driver routine. This type of design would work for either strongly or weakly coupled problems since any strongly coupled interaction can be solved iteratively. The interactions in this case are predefined in terms of what data will be transferred from one process to another; new types of interactions require code modifications that may be quite extensive if the alterations are in constitutive behavior rather than boundary or source terms. An advantage in this design is the sharing of finite element infrastructure between modules. An element library, solver package, and data management routines can be made common between physics modules and reduce code maintenance. Since the individual modules may be run independently, the overall utility of the software is increased significantly over the previous, fully coupled approach. The disadvantage of the multiple module implementation is the size of the overall code, the number of modules that may be required for general applications, and designing enough flexibility into the driver to anticipate the numerous types and strengths of interactions that may be needed.

A final methodology that strongly resembles the multiple module concept introduces the idea of coupling separately maintained mechanics codes into an interacting package. No driver routine is needed though the code containing the dominant mechanics process will act as the master routine. The secondary mechanics codes will act as slave processes and provide data on the demand of the master. For such a design to be feasible, the individual codes must be sufficiently general to make proper use of any externally supplied data and return needed quantities to the master. Also, since no assumptions about which physical process will be dominant in any particular application, all codes should be constructed to function in either the master or slave role. The communication between individual codes could be achieved through data file transfers. Another alternative is the use of message passing software developed for parallel computing environments. Message passing utilities, such as the PVM (Parallel Virtual Machine) libraries [16] or the MPI (Message Passing Interface) libraries [17], allow the memory to memory transfer of data, while each code is running on its own processor. The advantage of the independent code design is that the very substantial investment in stand alone mechanics code development can be readily applied to coupled problems with relatively minor software modifications. Stand alone mechanics codes also tend to contain more extensive capabilities and more robustness for an application than a module in a more monolithic code. The seamless coupling of such specialized codes provides an increase in overall capability. The drawbacks to this design are the need to set some standards for data transfer protocols and develop some general interface routines that use either file reads and writes or the message passing utilities.

In addition to the overall architecture, there are a number of algorithm details that must be resolved and implemented during the development of a coupled problem or multiphysics capability. Many of these items have been mentioned throughout this chapter but the major issues will be collected here as a series of questions to summarize the subject.

1. How many physical processes will be allowed in any simulation? Though coupling of only two processes have been described above, there are a variety of important problems in which three physical phenomena can be coupled.

2. What types of coupling will be allowed? Will the data dependencies be in source terms, constitutive relations, boundary conditions, geometric changes, or all of these areas?

3. What types of data will be exchanged between processes? Will the coupling be limited to nodal variables, or will element and global data also be permitted in the exchange?

4. How will the dominant or master process be selected? Will this be predetermined so that only certain types of physics may drive the simulation or will any process be selectable as the master?

5. What assumptions will be made regarding the domains for the different physics? Will these areas be forced to coincide or will they be completely arbitrary in extent and overlap? Will the mesh discretizations be the same for each physical process or will data interpolation be required?

6. How will the solution procedures for each process be synchronized? Will only similar time dependencies, both steady or time dependent, be allowed in the coupling or will any useful combination be permitted? How will the time integration processes in each process be allowed to influence each other? How will the frequency of data exchange be determined?

As was outlined in the individual discussions of the various types of coupled heat transfer and fluid mechanics problems, the answers to the above questions can vary considerably. In any given situation the answers are fairly obvious but when planning a general capability the decisions are more complicated. Though complete generality in multiphysics simulations has not as yet been achieved, considerable progress has been made in linking the most important types of interactions together to obtain more realistic finite element solutions.

7.8 Numerical Examples

7.8.1 Introduction

The coupled problems illustrated in this section are relatively complex engineering analyses that are intended to demonstrate some of the multiphysics areas described in this chapter. Due to space limitations, each problem is simply outlined and some representative results described. The finite element codes used in these computations and the method of code coupling vary from example to example. In the first example of a thermal stress problem, the COYOTE [18] heat transfer code was connected via subroutine calls to the quasi-static solid mechanics code, JAS [19]. The second and fourth examples involving electromagnetics employed individual codes linked by the PVM [16] message passing utilities. The second problem of induction heating used the thermal code COYOTE [18] and the electromagnetics code TORO [20], while the remelting problem was solved with the flow code NACHOS II [21] connected to TORO. The third example was solved with the flow code GOMA [22], which has a solid mechanics and ALE capability included as part of a fully coupled algorithm.

7.8.2 Thermal-Stress Example

The thermal-structural interaction example is a geometrically simple problem with a fairly complex material response. A cylindrical canister is filled with a foam-like material; located within the foam-like material is a solid of rectangular cross section. The canister is subjected to a thermal boundary condition that is representative of a heat source at one end of the canister. As the temperature of the container rises, the foam-like material begins to decompose, producing an off-gas and the eventual disappearance of the foam-like material. The container is pressurized by the gas production causing deformation of the canister wall and changes in the geometry for the thermal analysis. Figure 7.8.1 shows a meshed, cut away view of the canister and its contents at the initial time.

A full three-dimensional simulation of this problem was produced using a thermal analysis code as a master process and coupled to the quasi-static, solid mechanics code through a subroutine interface. A chemical kinetic description was used to detail the heat and mass transfer due to the decomposition of the foam-like material. The chemistry was solved in conjunction with the temperature field via an operator splitting technique. As the foam-like material completed the decomposition reaction in each element, the element was removed from the simulation (element death). Mass transfer from the decomposition reaction was accounted for in the expanding clear space of the container through use of a bulk node (see Section 3.11.4). Radiation inside the container was included in the analysis with view factors being constantly updated due to element death and changes in the shape of the container walls. The pressure computed in the bulk node was passed to the solid mechanics code for use as a boundary condition on the container walls. The temperature field was also transferred to the solid mechanics code for use in the temperature-dependent, elastic-plastic constitutive model of the container wall. The foam-like material was not modeled in the structural computation because of its very limited load carrying capacity. An updated, deformed geometry was transferred back to the thermal analysis code at the completion of each solution step.

Figure 7.8.1: Schematic and mesh for heated canister.

The deformed geometries for the canister at several times during the heating process are shown in Figure 7.8.2. The foam-like material begins to disappear (element death) at 250 seconds after heating begins and creates a clear space at the bottom of the canister. Visible deformation begins at approximately 300 seconds and continues to a time of 450 seconds after which additional deformations are relatively small. The deformations in Figure 7.8.2 are scaled by a factor of 1.5. The state of the foam within the canister is illustrated in the cut away plots in Figure 7.8.3 which are viewed from the heated end of the container. The recession of the foam-like material is easily seen in this figure, as only the currently active elements are plotted; an undeformed configuration of the canister is used for reference. At the last time shown, the block buried in the foam-like material is just starting to be exposed. Though the mesh discretization is not excessively refined, the complexity of the physical processes modeled required that a parallel computation be utilized to reduce the analysis time. This analysis is due to R. G. Schmitt (personal communication).

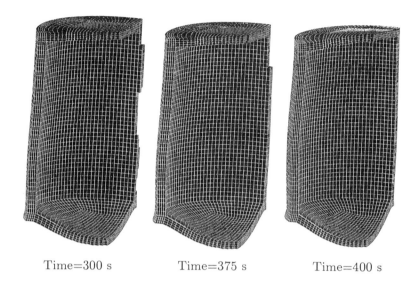

Time=300 s   Time=375 s   Time=400 s

Figure 7.8.2: Deformed geometry for heated/pressurized canister.

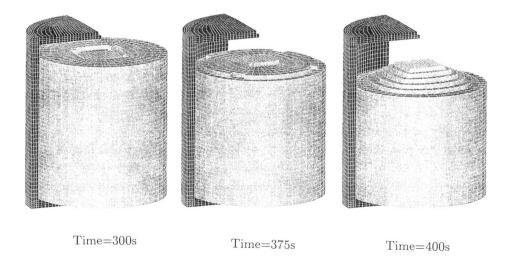

Time=300s   Time=375s   Time=400s

Figure 7.8.3: Material decomposition and removal for heated/pressurized canister; currently active elements are shown.

7.8.3 Thermal-Electromagnetic Example

Induction heating is used extensively in manufacturing processes, especially in the surface treatment (hardening) of ferromagnetic materials. In general terms, induction hardening consists of a part placed in the vicinity of a alternating current coil. The time varying coil current induces an eddy current in the part which in turn produces a resistive heating and temperature rise in the part. The high frequency of the applied coil current and shallow penetration of the magnetic field (skin depth effect) in the part, limit the thermal effect to the part surface. Complexities in this process include strongly temperature-dependent electromagnetic properties, a strongly nonlinear magnetic permeability and the relative motion between the part and the coil when a realistic industrial process is simulated.

Shown in Figure 7.8.4 is a schematic of a simplified induction hardening process in which a cylindrical part is heated within an annular coil of square cross section. The finite element mesh used for the axisymmetric analysis is shown in Figure 7.8.5. The time harmonic electromagnetics problem was solved over the domain shown in the figure where the current density was specified for the coil. The thermal problem was solved over the cylindrical part and the coil, each of which was subjected to Joule heating; the air space was not included in the thermal analysis. Radiation between the part and coil could be included in the analysis but was omitted for this demonstration. For this application the thermal problem is the master process since the time constant for thermal diffusion is significantly longer than the time constant for induction. The thermal problem was integrated over the heating cycle of 5.4 seconds using an implicit integration method. At every time step, the current temperature field was transferred to the electromagnetics code via PVM and the (nonlinear) time harmonic eddy current problem solved for the Joule heating in the part and coil. For the present example, the coil current was 8 kA and was run at a frequency of 7.6 kHz. These data were returned to the thermal code via PVM and used in the next temperature solution. At the conclusion of the heating cycle the part was subjected to a spray quench which was applied via a convective heat transfer boundary condition on the vertical surface of the part.

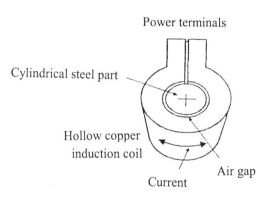

Figure 7.8.4: Schematic of induction hardening problem.

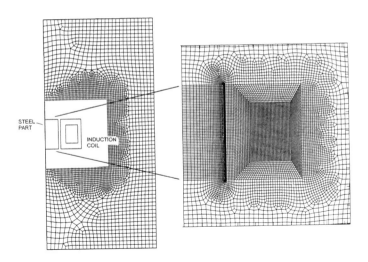

Figure 7.8.5: Finite element mesh for axisymmetric induction hardening problem.

Figure 7.8.6 shows computed and measured temperature histories for various locations within the midplane of the cylinder. The agreement between model and experiment is reasonably good and shows the rapid rise in surface temperature. Note that the surface temperature exceeds the Curie temperature at which point the magnetic permeability falls to a free space value. At quench, the surface temperature drops immediately while the internal temperatures continue to rise due to conduction.

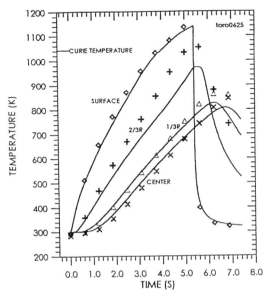

Figure 7.8.6: Predicted and measured temperature histories for induction heated cylindrical part.

In the real analysis setting, the thermal problem is augmented by a reaction kinetics solution that describes the phase transformation of the material and the hardness of the processed part. Ultimately, the coupled solution described here is also passed to a quasi-statics solid mechanics code that evaluates residual stresses and deformation due to the heat treatment. Further details on this problem are available in [23].

The same type of problem with a longer cylinder that translates along its axis, through the coil, was solved. A sliding surface was placed within the (air space) mesh between the coil and the cylinder. Along the slide surface a multipoint constraint was employed to enforce continuity of the magnetic vector potential. The mesh inside the slide surface moved with the translating cylinder while the mesh outside the slide surface remained stationary with the coil. The time-dependent motion of the cylinder was specified. The thermal problem required no constraint since the air was not included in the analysis. If radiation had been included in the analysis, continual updating of the view factors would be necessary to account for the motion of the cylinder.

Figure 7.8.7 contains a series of contour plots which illustrate the temperature field at several times during the motion of the rod through the coil. Motion of the cylinder begins 0.5 second after the coil current is applied and continues for 1 second after heating is terminated; the coil is energized for a period of 5 seconds. The process is sufficiently rapid so that the heat loss from the cylinder is minimal and no thermal boundary conditions were applied; no quench process was modeled in this example.

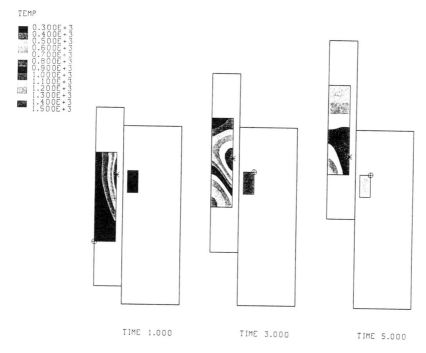

Figure 7.8.7: Temperature contours for a cylinder moving through an alternating current coil.

7.8.4 Fluid-Solid Interaction Example

As an example of the interaction between a viscous fluid and a solid, the problem of creating a drop by pushing fluid through an orifice is considered. A cross section of a typical axisymmetric geometry is shown in Figure 7.8.8 along with an initial mesh for the simulation. A piston is located in a small fluid reservoir and is moved toward the orifice at an almost constant rate. Fluid is pushed through the orifice forming an axisymmetric drop (no gravity). The boundary behind the piston is an inflow boundary that simulates the connection to a large reservoir. During the piston motion, the piston and reservoir wall are slightly deformed due to the relative thinness of the solid sections and the fluid pressure.

The drop formation problem was simulated using a fully coupled ALE technique to track the motion of the fluid free surface (with surface tension) and motion/deformation of the solid regions. Due to the large geometric change in the fluid region, mesh distortion became unacceptable at several times during the transient and remeshing of the fluid region was required. The solid was treated as an elastic material.

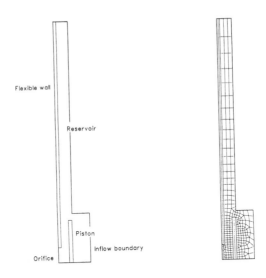

Figure 7.8.8: Schematic and initial finite element mesh for drop formation problem.

Shown in Figure 7.8.9 is a series of deformed mesh plots that illustrate the motion of the piston and the formation of a fluid drop. Necking of the drop is apparent in the second plot and reaches incipient pinch-off by the last plot. No mechanism for pinch-off is available in this method and the simulation was terminated at the last time shown. The deflection of the reservoir wall is not visible in the unmagnified plots of Figure 7.8.9; the deformation is shown at one time in Figure 7.8.10 where the displacements have been magnified by a factor of five. The deflection of the wall is obvious since the wall was initially parallel with the piston. Small deflections of the piston were observed early in the transient as the piston started to accelerate the fluid; these deflections were not as large as the bending of the reservoir wall.

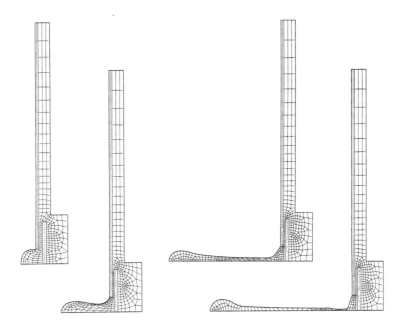

Figure 7.8.9: Finite element solution for drop formation problem. Remeshing occurs at three times during the simulation.

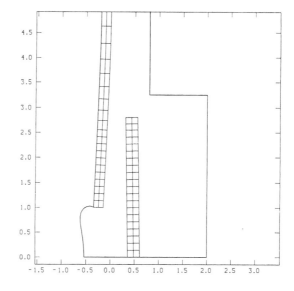

Figure 7.8.10: Reservoir wall deformation during drop formation; the displacements are magnified by a factor of five.

7.8.5 Fluid-Electromagnetic Example

The last example in this section is a simulation of the vacuum arc remelt (VAR) process that involves nonisothermal viscous fluid flow, flow in a porous medium, solidification and Joule heating, and Lorentz forces from a magnetostatic field. The problem and solution techniques are described in detail in [24] and will only be summarized here. Figure 7.8.11 shows a schematic of the VAR process and a typical finite element mesh for the axisymmetric analysis of the solidifying ingot. In the VAR process, a high current is passed through a consumable electrode, creating a metal vapor plasma arc between the electrode and the melt pool contained within the crucible. The arc provides the energy for melting the small electrode.

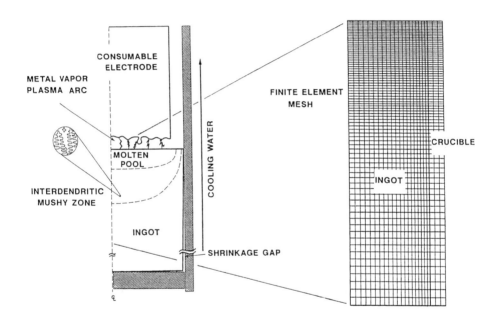

Figure 7.8.11: Schematic for vacuum arc remelt process and typical finite element mesh for crucible.

The objective is to create a larger ingot with better and more uniform metallurgical properties. In the finite element simulation of the solidifying ingot, the melt pool is represented as a viscous incompressible fluid with significant buoyancy forces; the mushy zone, which is typically present in alloy solidification, is modeled as a porous media using the Darcy–Brinkman equations of Section 4.4. Solid body conduction describes heat flow in the solidified portion of the ingot. The melt pool fluid is a conducting medium and the interaction of the current and magnetic field produces body forces and resistive heating within the crucible. A low magnetic Reynolds number is typical for this flow and the magnetostatic equations from Section 7.5 provide the appropriate description of the fields. The electromagnetic problem is weakly coupled to the flow problem since the properties are temperature and phase dependent. Boundary conditions are fairly complex and represent mass, energy, and current input from the arc and interface conditions with the crucible wall. As the ingot solidifies it shrinks in diameter and pulls away from the wall. This effect is modeled by increasing the resistance to heat transfer and electrically insulating the wall below the shrinkage point. A coupled solid mechanics solution would be required to eliminate this model parameter.

The algorithm used to solve this type of problem is a cyclic procedure with the fluid mechanics and heat transfer code being the master process; the magnetostatic code was invoked whenever an update on the volume heating or body force was required. Because time-independent solutions were of primary interest, a solution strategy involving zeroeth order continuation and combinations of the Picard and Newton iteration for the flow problem was adopted. The solution process was delicate with small increments in surface heat flux from the arc being used to "advance" the solution while buoyancy and Lorentz forces were scaled to maintain a balance between these opposing effects. Data transfer between the two finite element solutions occurred at user-specified intervals and was processed by the PVM utilities. Contour plots for the main variables in a typical VAR simulation are shown in Figure 7.8.12.

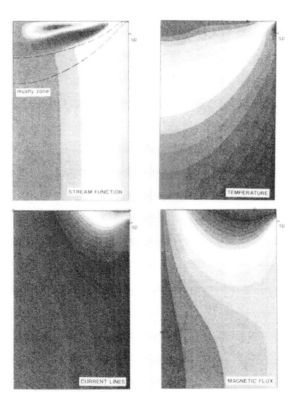

Figure 7.8.12: Contour plots for a typical VAR solution with buoyancy and Lorentz effects included.

Of primary interest is the stream function plot that illustrates the motion of the liquid metal from its small inflow at the top surface of the melt pool, through the clockwise rotation of the electromagnetically driven, Lorentz cell, the counterclockwise rotation of the thermally driven cell, through the porous layer, and finally as a solid body translation with the solid ingot. The temperature field shows some influence of the cell structure and the separation of the ingot from the wall. The electromagnetic fields are also strongly influenced by the wall boundary conditions. These results are for a fairly large current applied to the ingot. As the current is reduced, the thermal cell grows in size and limits the Lorentz cell to a thin region at the top of the melt pool. Without the Lorentz force, very large thermal cell solutions with unrealistically large melt pools are predicted. Additional details on modeling of the VAR process can be found in [24,25].

References for Additional Reading

1. O. C. Zienkiewicz and R. L. Taylor, *The Finite Element Method, Volume 2: Solid and Fluid Mechanics, Dynamics and Non-Linearity*, McGraw-Hill, New York (1991).

2. R. D. Cook, D. S. Malkus, and M. E. Plesha, *Concepts and Applications of Finite Element Analysis*, John Wiley & Sons, New York (1989).

3. K. J. Bathe, *Finite Element Procedures in Engineering Analysis*, Prentice-Hall, Englewood Cliffs, New Jersey (1982).

4. J. T. Oden, *Finite Elements of Nonlinear Continua*, McGraw-Hill, New York (1972).

5. M. A. Crisfield, *Non-linear Finite Element Analysis of Solids and Structures, Volume 1 - Essentials*, John Wiley & Sons, Chichester, U.K. (1991).

6. L. E. Malvern, *Introduction to the Mechanics of a Continuous Medium*, Prentice-Hall, Englewood Cliffs, New Jersey (1969).

7. ABAQUS Theory Manual, Version 5.8, Hibbitt, Karlsson & Sorenson, Inc., Pawtucket, Rhode Island (1998).

8. J. N. Reddy, *Energy and Variational Methods in Applied Mechanics*, John Wiley & Sons, New York (1984).

9. W. F. Hughes and F.J. Young, *The Electromagnetodynamics of Fluids*, John Wiley & Sons, New York (1966).

10. J. D. Jackson, *Classical Electrodynamics*, Second Edition, John Wiley & Sons, New York (1975).

11. J. Jin, *The Finite Element Method in Electromagnetics*, John Wiley & Sons, New York (1993).

12. M. N. O. Sadiku, *Numerical Techniques in Electromagnetics*, CRC Press, Boca Raton, Florida (1992).

13. K. J. Binns, P. J. Lawrence, and C. W. Trowbridge, *The Analytical and Numerical Solution of Electric and Magnetic Fields*, John Wiley & Sons, New York (1992).

14. P. P. Silvester and P. L. Ferrari, *Finite Elements for Electrical Engineers*, Third Edition, Cambridge University Press, Cambridge, U.K. (1996).

15. M. V. K. Chari, A. Konrad, M. A. Palmo, and J. DÁngelo, "Three-Dimensional Vector Potential Analysis for Machine Field Problems," *IEEE Transactions on Magnetics*, **MAG-18**, 436–446 (1982).

16. A. Geist, A. Beguelin, J. Dongarra, W. Jiang, R. Manchek, and V. Sunderam, *PVM: Parallel Virtual Machine, A User's Guide and Tutorial for Network Parallel Computing*, MIT Press, Cambridge, Massachusetts (1996).

17. W. Gropp, E. Lusk, and A. Skjellum, *Using MPI, Portable Parallel Programming with the Message Passing Interface*, MIT Press, Cambridge, Massachusetts (1995).

18. D. K. Gartling, R. E. Hogan, and M. W. Glass, "COYOTE - A Finite Element Computer Program for Nonlinear Heat Conduction Problems," Sandia National Laboratories Report, SAND94-1173 and SAND94-1179, Albuquerque, New Mexico (2000).

19. M. L. Blanford, "JAS3D A Multi-Strategy Iterative Code fro Solid Mechanics Analysis, User's Instructions, Sandia National Laboratories Report, in progress, Albuquerque, New Mexico (2000).

20. D. K. Gartling, "TORO - A Finite Element Computer Program for Nonlinear Quasi-Static Problems in Electromagnetics," Sandia National Laboratories Report, SAND95-2472 and SAND96-0903, Albuquerque, New Mexico (2000).

21. D. K. Gartling, "NACHOS II - A Finite Element Computer Program for Incompressible Flow Problems," Sandia National Laboratories Report, SAND86-1816 and SAND86-1817, Albuquerque, New Mexico (1986).

22. P. R. Schunk, P. A. Sackinger, R. R. Rao, K. S. Chen, R. A. Cairncross, T. A. Baer, and D. A. Labreche, "GOMA 2.0 - A Full-Newton Finite Element Program for Free and Moving Boundary Problems with Coupled Fluid/Solid Momentum, Energy, Mass, and Chemical Species Transport: User's Guide," Sandia National Laboratories Report, SAND97-2404, Albuquerque, New Mexico (1998).

23. D. R. Adkins, D. K. Gartling, J. B. Kelly, and P. M. Kahle, "TORO II Simulations of Induction Heating in Ferromagnetic Materials," in *Proceedings of the 1st International Induction Heat Treating Symposium*, Indianapolis, Indiana (1997).

24. D. K. Gartling and P. A. Sackinger, "Finite Element Simulation of Vacuum Arc Remelting," *International Journal for Numerical Methods in Fluids*, **24**, 1271-1289 (1997).

25. F. J. Zanner and L. A. Bertram, "Vacuum Arc Remelting - An Overview", in *Proceedings 8th International Conference on Vacuum Metallurgy*, **1**, 512–552 (1985).

Chapter 8

Advanced Topics: Parallel Processing

8.1 Introduction

Parallel processing is still a relatively new facet of computing with a significant history in computational mechanics of less than ten years. However, the technology is a certainty and will continue to have a major impact on computational mechanics for the foreseeable future. Many of the currently popular computing systems are based on the sequential processing, von Neumann architecture, in which instructions are executed in sequence, one at a time, on a single data element. For computational algorithms that have a sequential character, this type of processing is optimal. However, most computational problems are not strictly sequential, but have processes that are independent of each other, and could therefore be executed simultaneously. For these parallel tasks or algorithms, sequential processing is obviously not optimal. In the past, increases in computer performance have been primarily achieved through increases in speed of the central processing unit (CPU). However, as the rate of increase in CPU speed slows down, and costs for complex CPUs increase, computer developers have turned to alternative architectures to achieve high computing performance. A logical choice is to use multiple processors and exploit the parallelism inherent in most numerical algorithms. This hardware development has led to the rapidly growing field of parallel computing.

This chapter provides a brief introduction to the topic of parallel processing and its relation to finite element algorithms. In previous chapters we have mostly avoided discussions of finite element implementation since there are many comparable ways to achieve the required algorithmic result. Some texts [1–4] have illustrated the finite element implementation process by providing detailed descriptions and listings of source code for a variety of applications. Appendix A provides a minimal introduction of this type for some two-dimensional problems of interest in this text. The implementation of finite element procedures in a parallel computing environment involves a substantial increase in complexity with many algorithms requiring a heavy dependence on areas in computer science. A good introduction to parallel computing is found in [5].

In the next section an introductory description of parallel systems is provided. We will then outline some of the major algorithmic areas that must be addressed if a finite element procedure is going to be adapted to or designed for a parallel computer. It is not our intention to be extremely detailed in this chapter but it is important to call attention to methods and code structures that have been tested and put into use in engineering simulations. Likewise, it is important to point out where finite element methods and parallel processing are the most conflicted and where a substantial code development effort may be required.

8.2 Parallel Systems

8.2.1 Classification

The standard terminology often used for classifying parallel computers is the scheme proposed by Flynn [6]. Even though this scheme does not include many new developments in classification [7], it still forms a good basis. Following are the four classes in the scheme proposed by Flynn [6]:

1. Single Instruction Single Data (SISD)

2. Single Instruction Multiple Data (SIMD)

3. Multiple Instruction Single Data (MISD)

4. Multiple Instruction Multiple Data (MIMD).

An instruction stream is the sequence of instructions executed by the processing element (PE) and the data stream is the sequence of data on which instructions are performed.

The traditional sequential computers (Micro and Mini computers) fall under the class of the SISD system. They have a single processing element executing instructions in sequence on a single stream of data. The SIMD systems have many processing elements. They execute the same set of instructions in each of these processing elements but on different data streams. Vector computers fall under this class. On the other hand, MISD computers execute a different set of instructions in each processing element but on the same data stream. These systems are useful in application areas like signal processing and are often termed pipeline processors. The fourth category, MIMD machines, is the most general type among the parallel processing systems. They have many processing elements executing their own sets of instructions on different data streams.

Parallel processing systems can be further classified based on the characteristics listed below. Most of the parallel machines of interest in computational mechanics come under the MIMD or SIMD class of computers. The classification discussed here is relevant primarily to these systems.

Granularity of the processing elements

Granularity of the system can be either fine or coarse, where a coarse grain system has fewer processors than the fine grain system. A coarse grain system typically operates with a few (say, 100) powerful processing elements, while a fine grain system would usually have many (say, 1000) ordinary processing elements. The boundary between fine and coarse grain computers is not rigidly defined. Systems with more than a 1000 processors are usually termed massively parallel computers.

Topology of interconnections

The topology of interconnections is the manner in which the individual processing elements are connected. In terms of hardware, the connections may be arranged as a ring, tree, pipeline, or a hypercube. With the advent of new systems and with a virtual communication facility, the programmer can assume that each processor is connected to every other processor. The system will take care of the data communication through the intermediate processors based on the actual physical topology. Efficiency of such a system will be high if the actual layout and virtual layout are identical. Figure 8.1.1 shows the commonly used interconnect topologies. These types of processor topologies are common to multiprocessors, i.e., computers with all of the processors located in a single system.

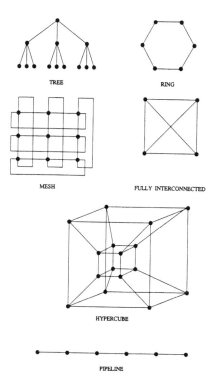

Figure 8.1.1: Topology of interconnections.

Networks of single processor computers can be arranged to provide a parallel computing system; a heterogeneous network of multiprocessor machines can be used to form a virtual parallel machine. These types of networked architectures are usually referred to as *distributed computing*. The major difference between a parallel or massively parallel computer and a distributed computing system is the variability of the processing unit. On a multiprocessor every processor is essentially the same, whereas in a distributed (virtual) machine each processor on the network could be very different in terms of speed and capability. Likewise, data formats may vary significantly between processors on a network. These differences lead to software portability issues and concerns with the basic parallel paradigm, language selection, and communication protocols.

Distribution of control across the processing elements

The control of the system may be with a single processor, which directs the system, as is often the case in SIMD machines. Also, in MIMD machines, even though a master (or main) processor is identified, each processor operates on its own, independently. However, interprocessor communication demands synchronous operation.

Memory access

Each processing element can have its own local memory and/or share the global memory of the system. Both types of architectures have advantages and disadvantages. In a local memory system, synchronous communication becomes vital and deadlocks/delays in communication may cause drastic deterioration of the system efficiency. In a shared memory system, processing elements compete with each other in addressing the memory locations. Bottlenecks in communication remain one of the major problems in a shared memory system. Again, the most common architectures used in computational mechanics are the shared memory MIMD machines and the distributed memory MIMD computers with the latter being the most popular. A detailed discussion of these characteristics is available in [5–8].

8.2.2 Languages and Communication Utilities

There are many types of parallelism in computing. For some applications, special computer languages or language extensions have been developed to simplify program development. A *data parallel* model seeks to exploit parallel capabilities in the performance of even simple arithmetic tasks such as a vector-matrix product. In this case, corresponding components of the matrix and vector are distributed to different processors for multiplication and returned for summation. Microtasking or multitasking of this type originally required compiler directives to indicate sections of the code that could be spread across multiple processors. Currently, languages such as High Performance (HP) Fortran [8] and Fortran 90 [10] implicitly support data parallel constructs. For systems based on transputers [11], the language of choice is Occam, which was developed specifically for this type of processor [11,12].

The usual approach to parallel computing of mechanics problems is a domain decomposition or partitioning paradigm with a single program, multiple data (SPMD) software implementation. In this situation, the problem data are divided or partitioned equally among the available processors. A copy of the finite element mechanics code runs on each processor and computes its locally assigned part of the problem. Crucial to the success of this paradigm is the requirement to efficiently communicate (transfer) data between processors such that the global problem can be solved from its distributed pieces. The two traditional methods for processor to processor communication on standard MIMD machines are the use of shared memory or the use of an explicit message passing utility. Message passing has become something of a standard because of its generality, especially considering the difficulties associated with distributed computing.

In the standard SPMD implementation, no special languages or extensions are required. The use of ANSI standard Fortran 77 and C is widespread. Finite element programs require a mixture of data processing procedures, some of which are best suited to C or Fortran 90 constructs. Other parts of the code are optimally constructed using the unit stride facilities of Fortran. Most current codes are a mixture of the two languages. The use of Object Oriented designs and the C++ language [13] is gaining in popularity.

Utilization of a distributed memory parallel system requires the ability to perform two basic functions: create tasks and set communication channels between processors. Portable, system independent libraries with standardized interfaces have been developed to accomplish these functions. The two most popular packages of this type are the Parallel Virtual Machine (PVM) library [14] and the Message Passing Interface (MPI) software [15]. Other systems, such as Linda [16] and P4 [17] exist but do not have the overall popularity of PVM and MPI. PVM is the older of the two communication packages and is best suited for systems with a few processors and an architecture that operates with a master processor and a number of slave processors. Distributed computing architectures with heterogeneous network processors are well aligned with PVM. The MPI software has found the widest utility and is heavily used on multiprocessor machines and networks, as well as massively parallel machines. MPI is especially convenient for the SPMD models found in most mechanics applications. Some language extensions that support task and channel functionality are also available, though not widely used. The CC++ [18] language is an extension of C++ and supports parallelism from within the language; Fortran M [19] is an extension of Fortran with similar functionality.

8.2.3 Performance

The performance of a parallel code is generally estimated by its parallel efficiency, which indicates how much is gained by solving the problem in parallel. The best way to estimate this is to compare it with a sequential version of the same code. This can be done in two different ways [20].

Algorithmic efficiency

This is defined as the ratio of CPU time taken for running the parallel code on N_p processing elements to the time taken when the code is run on a single processing element. The sequential code used here is essentially a parallel code run on a single processor. Hence the sequential code will also have the delays in communication reflected in its timing. This will indicate the parallel efficiency of the algorithm rather than the problem. The formula for algorithmic efficiency is

$$\eta_{alg} = \frac{t_{par}}{t_{seq1} N_p} 100 \qquad (8.2.1)$$

where t_{par} is the time taken to execute the program on a parallel machine with N_p processors and t_{seq1} is the time taken to run the same code on a single processor.

Actual/beneficial efficiency

Another measure, the actual or beneficial efficiency is defined as the ratio of CPU time taken for running the parallel code on N_p processing elements to the time taken by the fastest sequential code run on one processing element (t_{seq2}). This efficiency will reflect the parallel efficiency of the problem, or in other words it will reflect both the level of parallelism in the problem under consideration and the efficiency of the parallel algorithm used. It indicates the actual benefit derived by resorting to parallel processing. It is given by the following equation:

$$\eta_{act} = \frac{t_{par}}{t_{seq2} N_p} 100 \qquad (8.2.2)$$

The actual efficiency will be always be less than or equal to the algorithmic efficiency. If these efficiencies are less than 50%, then it is not usually worth considering the case. Sometimes a speed-up ratio is also used to rate a program. This is just the ratio of time taken by the parallel program to the time taken by the sequential program. The speed-up ratio should be greater than one. The speed-up factor can be calculated by multiplying the efficiency by $N_p/100$.

Scalability

Another important issue for performance is scalability. It is expected that systems with increasing numbers of processors will continue to be developed. Software that maintains performance with an increasing number of processors is therefore important for effective use of a parallel system. Early definitions of scalability sought to decrease computational time by increasing the number of processors on a fixed work load. It was believed that the serial portions of the algorithm and communication costs would ultimately limit performance as the number of processors increased. This is a statement of Amdahl's law which has as its focus a decreasing work load per processor. A current view of scalability asks for a constant work load per processor, in which case the overall work load would increase with the number of processors. In this situation, any degradation in performance can be assigned to communication cost.

8.3 FEM and Parallel Processing

8.3.1 Preliminary Comments

Previous chapters have outlined the steps/procedures for finite element algorithms for a number of specific problem types. The implementation of all of these finite element methods follows a series of generic steps. In the first part of this section, these generic finite element procedures will be discussed in the context of a parallel computing implementation. A later section will describe some of the issues that arise due to specific computational and modeling techniques such as multipoint constraints and code coupling.

As noted previously, the most common parallel architecture of interest in the computational mechanics community is the MIMD architecture with a shared memory, or more preferably, with a distributed memory. The MIMD architecture argues for a decomposition or partitioning strategy when considering approaches to finite element solutions in a parallel environment. Note that here we are considering only the solution to a single finite element model; other, more coarse grained levels of parallelism are possible when considering multiphysics applications. If the computer has N processors, then the finite element mesh is decomposed into N roughly equal groups of (contiguous) nodes or elements with each group assigned to a processor. Each processor computes the finite element algorithm for its group of nodes or elements with interprocessor communication required for completion of the algorithm at processor boundaries where nodes or elements are shared. Input and output from the process may be handled through a special, dedicated front end processor, by one of the N processors that is assigned the duty, or through all the processors and a parallel I/O hardware system. Often the finite element software is structured such that a copy of the executable code is running on each processor; at appropriate points in the execution, the processors are synchronized and data are exchanged between processors. This is the SPMD model mentioned previously.

As sketched here, the domain decomposition or partitioning approach to parallel finite element solutions is the most common paradigm currently in use. The discussion to follow will be oriented toward this type of procedure. Note that the term domain decomposition is also used to describe parallel solution methods for large systems of equations. The two areas are obviously related but not necessarily coincident. We use the term here only to describe an approach to parallel implementation of finite element methods.

8.3.2 Generic FEM Steps

A finite element algorithm or program is naturally divided into the following major steps or units:

- External preprocessing

- Internal preprocessing

- Solution processing

  - Element matrix building

  - Matrix solving

  - Solution control

- Internal postprocessing

- External postprocessing

The first and last processing steps usually occur outside of the basic finite element program in specialized mesh generators and graphics packages. However, these external processes must be included when discussing a parallel application because they play a significant role in the problem setup. Each of these processing steps is considered individually in the next several sections.

8.3.3 External Preprocessing

Preprocessing for a finite element model usually refers to the generation of the element mesh from some type of geometric description of the problem region. In many cases the problem geometry resides in a computer-aided design (CAD) file. Mesh generation sets the location of the nodal points (global coordinates) and defines the elements in terms of the nodes through construction of the nodal connectivity. The connectivity is an ordered list relating the nodes in an element to the global node numbering in the assembled model. Elements in the mesh are normally assigned material i.d.s at this point and element surfaces and nodes are flagged for boundary conditions. The output from this step is a file, in some standard format, that provides all the required data for each element in the mesh.

For parallel applications, preprocessing involves several additional steps. The distribution of the finite element algorithm to a number of processors is accomplished through domain decomposition or mesh partitioning algorithm. Within this procedure, the overall problem is divided into N groups of elements for execution on N processors of the parallel platform. Efficient execution of the parallel problem demands that the computational work on each processor be as equal as possible and that the communication between processors be minimized. This part of the process is referred to as load-balancing. Various software packages, such as Chaco [21] and Parmetis [22], have been developed to perform this type of mesh decomposition and processor load-balancing. The mesh decomposition algorithm is often based on viewing the mesh as a graph and utilizing sophisticated graph partitioning methods. Recursive bisection and octree methods have also been used as partitioning procedures. In each case, the partitioning method must consider the general case where the number of unknowns per node and per element may vary between elements. This circumstance is usually handled by weighting the vertices of the graph to reflect the work associated with the node or element. Also, partitioning may be nodal or element based. In a node-based decomposition, nodes are assigned to processors and elements may be split between processors. An element-based decomposition assigns elements to a processor with nodes then being shared between processors. The type of needed decomposition and load balance is thus dependent on the architecture of the finite element code.

The domain decomposition and load-balancing software typically functions by reading the mesh generation file, decomposing the mesh for the stated number of processors and producing its own output file or files. The form of the partitioned output again depends on the structure of the finite element code. If the finite element code is designed to have a master process and a number of slave processes, the domain decomposition data can be stored in a single scalar file. This file would contain the load balance data needed by the master process to distribute the node, element, and communication data to the individual processors. Finite element codes designed to run with a copy of the code on each processor (the SPMD model) require the decomposition data in a set of parallel files. Each of the parallel files would contain some global information plus the nodal, element, and communication data specific to a processor.

Figure 8.3.1 shows the finite element mesh of a casting with a partial gating system that was partitioned for use in a heat conduction/radiation simulation using eight processors. The original geometry was meshed using a standard mesh generator with the output being processed by a graph partitioning method within the Chaco [21] code. Additional utilities were used to reformat the output file into a series of eight parallel files that were subsequently read by a heat transfer code. Note that the original mesh contained 3297 elements (6661 nodes) and the element-based partitioning yielded 412 elements on seven processors and 413 elements on one processor. The number of nodes on each processor varied from a low of 790 to a high of 944.

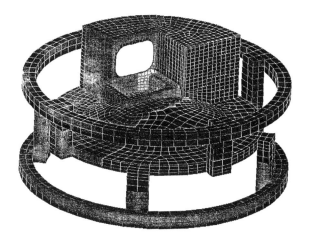

Figure 8.3.1: A finite element mesh showing a partitioning for use on a parallel computer.

In the course of this discussion, it has been implied that the mesh generation occurs on a serial computer and the decomposition and load-balancing follow as a utility process. In many situations this type of procedure is adequate. For extremely large and/or complex finite element models, it may not be possible to generate the mesh on a serial computer. Parallel mesh generation is an area of current investigation and will be required as simulations become more sophisticated. Also, static domain decompositions are adequate for static element topologies; adaptive meshing or remeshing will require dynamic load-balancing capabilities [23].

8.3.4 Internal Preprocessing

The first steps within the finite element code usually entail the reading of input data, reading of mesh data, the allocation of memory for the problem and the setting of execution and control pointers and flags. These same steps are required in a parallel code though the source of the data may be different. As noted in the previous section, mesh data may be available from a parallel input file for some types of codes or may be passed from the master processor and a serial file. Input data that specify material properties, boundary condition values, solution control, and postprocessing requirements usually come from a single, serial file and must be broadcast to the appropriate processors. Processor communication utilities, such as MPI, must be initialized. Assuming a distributed memory machine, local memory on a processor is allocated based on the number of elements or nodes assigned to the processor. Print or text output from the code is generally written from a single processor, even if the code design does not specifically include a master or front-end processor. At the conclusion of initialization and data checking, the processors are synchronized and the first step of the solution is ready for processing.

8.3.5 Solution Processing

A solution step for a finite element method consists of two major parts - element matrix building and solution of the global matrix system. For time-dependent simulations or nonlinear steady problems this step will be repeated for each time step and/or each iteration. The solution of the equations is usually the most time-consuming part of the analysis and can benefit substantially from a parallel implementation. For time-dependent solutions obtained via explicit methods, no matrix solution is required though a parallel implementation is still very effective.

Element matrices

The finite element method lends itself naturally to parallel computing especially in the equation building part of the algorithm. The usual structure for a serial finite element code contains a loop over the elements during which element matrices are computed (via numerical quadrature) and added to the global matrix. In most cases, the evaluation of the matrices for an individual element is independent of other elements. Hence, generation of element matrices can be carried out simultaneously, i.e., in parallel. The parallel finite element code, therefore, has a loop over all the elements on the processor, where element matrices are computed and added into the part of the global matrix represented on the processor. This architecture presumes that an element-based decomposition has been used in the mesh partitioning so that all the data for an element reside on the processor. The part of the global matrix that is assembled on each processor represents a combination of both fully summed equations and partially summed equations for nodes shared between processors. At the completion of the element loop, this submatrix is passed to the parallel matrix solver for further processing and solution. Note that the element-based decomposition is the natural choice for element construction since an element is the primary data entity.

In the case where the partitioning is node based, equation construction is slightly more complex. The looping is over the nodes on the processor. An inverse connectivity is used to determine which elements (and nodes) are connected to the node. The integration of the shape functions for the nodal equations will require coordinate data from other processors and therefore communication between processors. At the completion of the loop, a submatrix will be available to the parallel solution routine. This type of decomposition is well suited for most parallel, iterative matrix solution methods but is not well aligned with finite element data processing.

Matrix solvers

The matrix solver used for a parallel code implementation can be either iterative or direct; the serial versions of these two types of solvers are discussed in Appendix B. Direct solvers obtain the solution to the system of simultaneous equations in a fixed number of steps. On the other hand, iterative solvers do not converge to the solution in a fixed number of iterations. The number of iterations required depends on the mathematical structure of the system of equations being solved. Nonlinearity of the problem and the diagonal dominance of the system of equations affect the convergence rate. In some cases, it may not even be possible to obtain a solution using an iterative solver—especially, without a preconditioning of the coefficient matrix. The major advantage of most iterative solvers is that they are amenable to the element-by-element (EBE) or block EBE solution approach. The system of equations is solved without direct assembly of the element matrices. These algorithms have very high parallel efficiency when compared with the direct solvers.

Most direct solvers use a Gauss elimination scheme as their base algorithm. This algorithm consists of two steps, the forward elimination and the backward substitution. The forward elimination can be programmed in parallel with a good parallel efficiency, but with a significant increase in solver complexity. The parallel forward elimination is much like a substructure process where the interior nodes (fully summed equations) on a processor are eliminated, leaving the border nodes (partially summed equations). For efficiency, the summation and subsequent elimination of groups of border nodes should be redistributed among all the processors. The continuation of this process to its conclusion involves substantial data tracking and communication. Finally, the backward substitution is sequential in nature, and this will decrease the overall parallel efficiency of the solver. Also, the forward elimination will be less efficient for banded systems.

The Gauss–Jordan solver is a less efficient solver in a sequential mode, and it is rarely used in sequential programming. If N is the number of unknowns, Gauss elimination performs $O(N^3/3)$ multiplications and the same number of additions during the elimination operation (multiplication and division and addition and subtraction are taken as identical operations). The back substitution scheme involves $O(N^2/2)$ multiplications and additions. On the other hand, the Gauss–Jordan solver performs $O(N^3/2)$ multiplications and additions during the elimination step, and there is no back substitution. Since there are more operations in the Gauss–Jordan method than the Gauss elimination scheme, it is not preferred; but the issues take different priorities in parallel processing. The Gauss–Jordan algorithm is devoid of the back substitution step, which is highly sequential. This is a major advantage. Also more computation is performed for a unit of communication in the Gauss–Jordan method than the Gauss elimination scheme. This will result in an improved parallel efficiency [23]. For a complete system, Gauss–Jordan will break even with Gauss elimination, even in terms of CPU time, with only a few processors. However, for a banded system, a greater number of processors will be required to achieve parity.

Among the many iterative solvers, like the Jacobi iteration method and the Gauss–Seidel method, the Conjugate Gradient Method (CGM) is very efficient and effective for many finite element equation systems. In theory, it will yield convergent solution in less than N iterations, where N is the number of unknowns. In practical computations, round-off errors will affect convergence and it often requires more than N iterations to achieve a solution. Hence, restarting procedures have to be used. Even with this penalty, it is faster than the Gauss–Seidel iteration method by an order of magnitude.

The conjugate gradient method is often used for minimizing a function. It uses the gradient information of the function to minimize it, without evaluating the function itself. This method was originally developed for solving systems of simultaneous equations. The original method can solve only symmetric, positive-definite systems. Using an appropriate preconditioning matrix, other systems can also be solved. Details of the conjugate gradient method algorithm with preconditioning can be found in [24] as well as Appendix B.

In developing a parallel implementation for an iterative matrix solution method, two key features are exploited. Algorithms like the conjugate gradient method involve matrix–vector products and vector dot products. These steps can be easily split into parallel operations by expressing the product of a matrix with a vector as the sum of the products of their components. The key idea is

$$[A]\{x\} = \left[\sum_{i=1}^{N_{elem}} [A_i]\right] \{x\} \tag{8.3.1}$$

$$[A]\{x\} = \sum_{i=1}^{N_p} \left[\sum_{j=M_{i1}}^{M_{i2}} [A_j]\right] \{x\} \tag{8.3.2}$$

where N_p is the number of processors and N_{elem} is the number of elements. The values of M_{i1} and M_{i2} can be calculated in different ways. Only the matrix $[A]$ is expressed as a sum and not the load vector or right-hand side. Hence the load vector generated by the processors has to be added to get the final load vector, which involves some communication. If the load vector is also expressed as a sum, the process will be even more complicated. Also, it will be computationally intensive with a heavy communication load. This can be readily seen from the following equations:

$$A \cdot x = (A_1 + A_2)(x_1 + x_2) = A_1 x_1 + A_1 x_2 + A_2 x_1 + A_2 x_2 \tag{8.3.3}$$

$$A \cdot x = \left[\sum_{i=1}^{N} A_i\right] \left[\sum_{j=1}^{N} x_j\right] = \sum_{i=1}^{N} \sum_{j=1}^{N} A_i x_j \tag{8.3.4}$$

If there are N values of A and x, each value of A is associated with all the N components of x. This places a requirement on the processors to know all the values, and there is a larger number multiplications as well. It is therefore not advisable to express the vectors involved in the matrix-vector product as a sum of its element contributions.

The other product frequently encountered is a vector dot product. Consider two vectors $\{x\}$ and $\{y\}$ of dimension N. The dot product can be written as

$$\{x\} \cdot \{y\} = \sum_{i=1}^{N} x_i y_i \tag{8.3.5}$$

$$\{x\} \cdot \{y\} = \sum_{i=1}^{N_p} \left[\sum_{j=M_{i3}}^{M_{i4}} x_j y_j\right] \tag{8.3.6}$$

The values of M_{i3} and M_{i4} are calculated from the element connectivity of the elements present in the ith processor. Care should be taken to make sure that a particular node is identified with only one processor in calculating the dot product, even though it may be associated with many processors. This is due to the fact that the vectors are assembled and they are global. Depending upon the particular algorithm under consideration, the methodology can be restructured.

Although the matrix solution process occupies a large majority of the execution time in a finite element code, it is no longer necessary to understand the implementation details of the various matrix solvers. A number of software projects have produced solver libraries with convenient, standardized interfaces that can be used easily with many finite element formulations. Many of these libraries have both direct and iterative matrix solvers and usually have available a number of preconditioning options. Typical of these solver packages are LAPACK [25] for shared memory machines and ScaLAPACK [26], PETSC [27] and Aztec [28] for distributed memory architectures. The use of well-documented solver libraries is an effective method for quickly getting a finite element algorithm running in parallel. As noted above, much of a finite element computation is naturally parallel and easy to code. The matrix solution is the difficult part of the parallel implementation and involves most of the "computer science" aspects of a finite element code. Though solver libraries may be conveniently used for expediting program development, it is important to understand the relationship between various matrix solver methods and characteristics of the finite element equations (matrices).

Solution control

At the conclusion of a matrix solution, various computations must be completed that influence the overall progression of the algorithm. In particular, norms on the solution field must be computed and checked to determine if convergence has occurred. Also, for time-dependent simulations tests for reaching steady state are computed and a new time step may be required. Each of these computations requires some communication between processors though these tasks are minor compared to the data exchange that occurs within the matrix solver.

8.3.6 Internal Postprocessing

At the completion of the finite element solution, some postprocessing of the data may be required. The computation of auxiliary quantities, such as flux data or globally integrate values, may occur at the end of each time step or nonlinear iteration or at the completion of the steady state solution. In a serial algorithm these data are again processed within a loop on the elements. The parallel implementation of this type of computation is similar though some communication may be required.

If fluxes are computed at the integration points of an element, then the computation is completely parallel if the decomposition was element based. When continuous flux data are required, the usual process is to extrapolate the integration point fluxes to the nodes and average the nodal data between elements attached to the node. This type of averaging requires some interprocessor communication. Using communication utilities such as MPI make the computation straightforward. Quantities that are integrated over surfaces or volumes likewise require data exchange between processors but nothing outside the scope of MPI-like capabilities.

Output of postprocessing data may be accommodated by either of the two methods used to input data. For many systems, data would be collected on the master processor or a designated processor and written to a printed output file and/or a postprocessing file. Codes using the SPMD model would more likely collect data on a single processor for printing while using parallel output files for postprocessing data. The parallel files are usually concatenated into a single file prior to use with a graphics package.

8.3.7 External Postprocessing

External postprocessing refers to the use of specialized graphics programs to investigate the solution fields generated by the finite element code. These software packages have a variety of methods for graphically displaying data, manipulating data, and probing the computed fields. For applications generated by a parallel algorithm the concerns related to postprocessing stem mainly from the size of the computed data sets. It is a significant challenge to graphically render nodal or element data from time-dependent finite element meshes containing several million elements. Not only is rendering time an issue but temporary and permanent storage of the data sets is a formidable problem. Work on parallel graphics software is in progress and will be mandatory as finite element simulations continue to increase in size and complexity.

8.3.8 Other Parallel Issues

It is clear from the previous section that most of the parallel issues in a standard finite element algorithm are contained within the matrix or linear algebra solution procedure. However, there are some finite element modeling capabilities that complicate the normal parallel implementation based on mesh partitioning. As described in Chapter 3, contact and multipoint constraint boundary conditions require data external to the current element. Algorithms that require the staggered or cyclic solution of multiple equation sets also add some complexity to a parallel implementation. These topics are considered briefly in the following sections.

Nonlocal data

The efficient use of a domain decomposition method requires that data used by a processor be local to the processor. This is the usual situation when an element-based decomposition is used and the looping is over the elements on a processor. Nonlocal or off-processor data can occur in the implementation of certain types of boundary conditions, such as node-to-surface contact and multipoint constraints (see Section 3.11.3). Implementation of these types of conditions require two steps: the location of the slave node on the master surface and the construction of an equation relating the slave node to the nodes on the master surface. When the slave node is on a different processor from the master surface, additional interprocessor communication is required during both steps of the procedure. Note that preprocessing data is not available for establishing these communication paths since the boundary condition may be dynamic and change location during the course of the solution.

The search for the location of one or more (slave) nodes within a list of master elements is a procedure that occurs in several computational situations. The transfer (interpolation) of data from one mesh to another requires this type of procedure as does the implementation of the boundary conditions mentioned above. In a parallel setting the local search process is most effectively done after the slave nodes and master elements that are geometrically close are redistributed to common processors. This amounts to a slave node-based repartitioning of the data. An effective method for load balancing this problem is a recursive bisection algorithm. After completion of the local, on-processor search procedure, the location data for each slave must be returned to its original processor.

The actual implementation of the multipoint constraint requires a second step in which a constraint matrix is constructed between the slave node and the nodes on the master surface. The contact boundary condition requires this same type of construction if the implementation is fully implicit; an explicit formulation for contact can be constructed without this second step. After the constraint equations are constructed on the processor containing the slave node, the constraint coefficients must be distributed to the processors containing the master surface nodes and added to their portion of the global matrix. These interprocessor communication paths are flagged during the search process since they may change with time and are not included as part of the original mesh partition.

Another type of nonlocal data occurs for modeling features such as a bulk node (see Section 3.11.4). Here, the equation for a bulk node is not part of the original mesh partition and in fact is owned by all the processors. Because the bulk node may be connected to a large number of elements (and nodes) located on different processors, updating of the bulk node equation occurs on every processor. This local update is followed by a global accumulation of data and integration of the bulk node equation on each processor. The bulk node parameters are then available to each processor for use in boundary conditions on the processor.

Multiphysics simulations

The use of cyclic or staggered solution methods for multiphysics or multiple systems of equations, may introduce additional complexities into the parallel implementation. The two issues of main concern are load balancing and data transfer between equation sets. When the equation sets are solved on the same element topology (and same domain decomposition), these problems do not occur.

As a specific example of the load-balancing problem, consider the fairly common situation of solving the coupled heat conduction, enclosure radiation problem or the heat conduction, chemical kinetics problem outlined in Chapter 3. These types of problems would normally be solved with a single software package, so that code coupling is not a complicating factor. A recommended method in both cases is the staggered or alternating solution of the two equation sets as a function of time. In a parallel implementation, the finite element conduction problem is partitioned across the N processors and solved by the standard method. However, when the second equation set is solved it is probable that a number of processors will be idle because its computational domain does not necessarily coincide with the conduction domain. In the radiation case, the view factors and radiosity problem are computed for a subset of element surfaces. In the chemical kinetics problem it is usual for only one material in the problem to be reactive, while the other materials require no chemistry solution. Note that it is quite possible to have both radiation and chemistry in the same problem in which case none of the equation domains are coincident.

The solution to this dilemma is of course to have individual decompositions for each equation set in the problem. The individual partitions may be generated externally if the equation sets are complex, e.g., two or more partial differential equation models. For less complex secondary equation systems, the partitioning may be done internally with a simple algorithm. In the case of radiation view factors or chemistry on an element, the number of equations or elements divided by the number of processors provides an adequate partition. This simple method works in these cases because there is no interaction between the equations within the set. Regardless of the method used for partitioning the second equation set, some additional interprocessor communication is required. In essence, the communication relationship between the partitions must be established to allow the exchange of dependent variables between the sets of equations.

The previous load-balancing discussion centered on the situation encountered when a single finite element code was required to handle multiple equation sets. It was assumed that the variables were defined at the same spatial locations (nodes and elements) in each equation, though the spatial extent of the domains did not have to coincide. The coupling of independent finite element codes, as described in Chapter 7, can present another difficulty for parallel implementation. When the coupled codes use independent finite element meshes and different domain decompositions, the transfer of variable data from one mesh to the other involves added complexity. The basic problem is similar to the multipoint constraint problem described above. To transfer nodal data from mesh A to mesh B, the locations of the mesh B nodes within mesh A must be found. Once located, the nodal data for mesh B can be found from shape function interpolation on mesh A. The search procedure is most effectively done in parallel by redistributing the data across the processors such that groups of mesh B nodes and mesh A elements are geometrically close. A recursive bisection procedure is fairly optimal for this process. Once the interpolated nodal data are computed, the variables are transferred back to the original processor so that the next solution step in the algorithm can be completed. In general the search and interpolation process must be completed in each direction if the two solution fields are mutually dependent. When the geometries are static the search need only be done once, though the interpolation would have to be carried out at each data exchange step. The complexity of this process obviously increases substantially for each coupled code added to the simulation.

8.4 Summary

The future of computational mechanics will be influenced to a great extent by the developments in parallel processing. Computational fluid dynamics, in particular, can and will benefit significantly from the developments in parallel computing. Currently, most of the parallel algorithms are the parallel versions of existing sequential algorithms. It may be more useful to conceive and develop algorithms which are fundamentally parallel. Other than CFD, problems involving artificial vision, signal processing, neural networks, optimization, and general finite element analyses will benefit with the advances in parallel computing. The scope is much broader than what is mentioned here.

References for Additional Reading

1. C. Taylor and T. G. Hughes, *Finite Element Programming of the Navier-Stokes Equations*, Pineridge Press, Swansea, U.K. (1981).

2. J. E. Akin, *Application and Implementation of Finite Element Methods*, Academic Press, Orlando, Florida (1982).

3. T. J. R. Hughes, *The Finite Element Method, Linear Static and Dynamic Finite Element Analysis*, Prentice Hall, Englewood Cliffs, New Jersey (1987).

4. J. N. Reddy, *An Introduction to the Finite Element Method*, Second Edition, McGraw-Hill, New York (1992).

5. I. Foster, *Designing and Building Parallel Programs*, Addison-Wesley, Reading, Massachusetts (1995).

6. M. J. Flynn, "Very High Speed Computing Systems," *Proceedings of IEEE*, **54**, 1901–1909 (1966).

7. B. Davidson, "A Parallel Processing Tutorial," *IEEE Antennas and Propagation Society Magazine*, **32**, 6–19 (1990).

8. M. R. Hord, *Parallel Supercomputing in SIMD Architecture*, CRC Press, Boca Raton, Florida (1990).

9. C. Koebel, D. Loveman, R. Schreiber, G. Steele, and M. Zosel, *The High Performance Fortran Handbook*, MIT Press, Cambridge, Massachusetts (1994).

10. J. Adams, W. Brainerd, J. Martin, B. Smith, and J. Wagener, *The Fortran 90 Handbook*, McGraw-Hill, New York (1992).

11. A. Carling, *Parallel Processing: the Transputer and Occam*, Sigma Press, London, U.K. (1992).

12. J. Galletly, *Occam 2*, Krieger Publishing, Melbourne, Florida (1990).

13. J. Barton and L. Nackman, *Scientific and Engineering C++*, Addison-Wesley, Reading, Massachusetts (1994).

14. A. Geist, A. Beguelin, J. Dongarra, W. Jiang, R. Manchek, and V. Sunderam, *PVM: Parallel Virtual Machine, A User's Guide and Tutorial for Network Parallel Computing*, MIT Press, Cambridge, Massachusetts (1996).

15. W. Gropp, E. Lusk, and A. Skjellum, *Using MPI, Portable Parallel Programming with the Message Passing Interface*, MIT Press, Cambridge, Massachusetts (1995).

16. N. Carriero and D. Gelernter, "Linda in Context," *Communications of the ACM*, **32**, 444–458 (1989).

17. R. Butler and E. Lusk, "Monitors, message, and clusters: The *p*4 parallel programming system," *Parallel Computing*, **20**, 547–564 (1994).

18. P. Sivilotti and P. Carlin, "A Tutorial for CC++," California Institute of Technology Technical Report, CS-TR-94, Pasadena, California (1994).

19. I. Foster, R. Olson, and S. Tuecke, "Programming in Fortran M," Argonne National Laboratory, Mathematics and Computer Sciences Division Technical Report, ANL-93/26, Argonne, Illinois (1993).

20. T. L. Freeman, *Parallel Numerical Algorithms*, Prentice Hall, Englewood Cliffs, New Jersey (1992).

21. B. Hendrickson and R. Leland, "The Chaco User's Guide - Version 2.0," Sandia National Laboratories Report, SAND94-2692, Albuquerque, New Mexico (1994).

22. G. Karypis and V. Kumar, "Parmetis: Parallel Graph Partitioning and Sparse Matrix Ordering Library," University of Minnesota, Department of Computer Science Technical Report, 97-060, Minneapolis, Minnesota (1997).

23. J. J. Modi, *Parallel Algorithms and Matrix Computations*, Oxford University Press, Oxford, U.K. (1988).

24. O. Axelsson, *Iterative Solution Methods*, Cambridge University Press, Cambridge, U.K. (1996).

25. E. Anderson, Z. Bai, C. Bischof, J. Demmel, J. Dongarra, J. DuCroz, A. Greenbau, S. Hammarling, A. McKenney, S. Ostrouchov, and D. Sorensen, "LAPACK Users' Guide, Second Edition," Society for Industrial and Applied Mathematics, Philadelphia, Pennsylvania (1994).

26. L. S. Blackford, J. Choi, A. Cleary, E. D'azevedo, J. Demmel, I. Dhillon, J. Dongarra, S. Hammarling, G. Henry, A. Petitet, K. Stanley, D. Walker, and R. C. Whaley, "ScaLAPACK Users' Guide," Society for Industrial and Applied Mathematics, Philadelphia, Pennsylvania (1997).

27. S. Balay, W. D. Gropp, L. C. McInnes, and B. F. Smith, "PETSc 2.0 Users Manual," Argonne National Laboratory Report, ANL-95/11 - Revision 2.0.28, Argonne, Illinois (2000).

28. S. Hutchinson, L. Prevost, J. Shadid, C. Tong, and R. Tuminaro, "Official Aztec User's Guide, Version 2.1," Sandia National Laboratories Report, SAND99-8801, Albuquerque, New Mexico (1999).

Computer Program *FEM2DHT*

A.1 Introduction

The computer program *FEM2DHT* is a finite element analysis program for the solution of two-dimensional heat transfer and viscous, incompressible fluid flow problems. The program is a modification of the computer program FEM2DV2 from the book by Reddy [1]. The program is discussed here to aid the readers with the basic computer implementation aspects of the finite element method. The program is educational in nature and it does not contain powerful pre- or post-processors found in a commercial finite element software.

A.2 Heat Transfer and Related Problems

The heat transfer part of the program is based on a slightly general form of equation (2.2.1) on page 32 of this book ($a_{11} = k_{xx}$ and $a_{22} = k_{yy}$):

$$-\left[\frac{\partial}{\partial x}\left(a_{11}\frac{\partial u}{\partial x}\right) + \frac{\partial}{\partial y}\left(a_{22}\frac{\partial u}{\partial y}\right)\right] + a_{00}u = f \quad \text{in } \Omega \tag{A.2.1}$$

with the boundary conditions [c.f., equation (2.2.5a,b)]

$$u = \hat{u}(s) \quad \text{on } \Gamma_u \tag{A.2.2a}$$

$$\left(a_{11}\frac{\partial u}{\partial x}n_x + a_{22}\frac{\partial u}{\partial y}n_y\right) + \beta(u - u_\infty) = q \quad \text{on } \Gamma_q \tag{A.2.2b}$$

where Γ_u and Γ_q are disjoint portions of the boundary Γ such that $\Gamma = \Gamma_u \cup \Gamma_q$. For a heat transfer problem, $a_{11} = k_{xx}$ and $a_{22} = k_{yy}$ denote conductivities in the x and y directions, respectively, $f(x, y)$ denotes the known internal heat generation per unit volume, (n_x, n_y) denote the direction cosines of the unit normal vector on the boundary, β denotes the convective heat transfer coefficient, and u_∞ is the ambient temperature.

Of course, Eq. (A.2.1) arises in a variety of other fields and the program *FEM2DHT* can be used to analyze them as long as the governing equation of the problem solved is a special case of Eq. (A.2.1). For example, slow flows of inviscid fluids are often described in terms of either the *velocity potential* ϕ

$$u \equiv -\frac{\partial \phi}{\partial x}, \quad v \equiv -\frac{\partial \phi}{\partial y} \tag{A.2.3}$$

or the *stream function* ψ

$$v \equiv -\frac{\partial \psi}{\partial x}, \quad u \equiv \frac{\partial \psi}{\partial y} \tag{A.2.4}$$

Here (u, v) denote the velocity components. Thus, the program *FEM2DHT* can be used to solve the inviscid flow problems in terms of the velocity potential or the stream function, and then post-compute the velocity field as per Eqs. (A.2.3) and (A.2.4).

A.3 Flows of Viscous Incompressible Fluids

The finite element model of viscous, incompressible flows used in the *FEM2DHT* program is the *reduced integration penalty model* discussed in Section 4.3.3. The flows considered here have no inertia and are therefore termed Stokes flows. The governing equations of the model are a combination of the continuity equation (4.3.1) and momentum equations (4.3.2):

$$-\frac{\partial}{\partial x}(2\mu\frac{\partial u}{\partial x}) - \frac{\partial}{\partial y}[\mu(\frac{\partial u}{\partial y} + \frac{\partial v}{\partial x})] - \gamma\frac{\partial}{\partial x}(\frac{\partial u}{\partial x} + \frac{\partial v}{\partial y}) = 0 \tag{A.3.1a}$$

$$-\frac{\partial}{\partial x}[\mu(\frac{\partial u}{\partial y} + \frac{\partial v}{\partial x})] - \frac{\partial}{\partial y}(2\mu\frac{\partial v}{\partial y}) - \gamma\frac{\partial}{\partial y}(\frac{\partial u}{\partial x} + \frac{\partial v}{\partial y}) = 0 \tag{A.3.1b}$$

with the boundary conditions

$$u = \hat{u} \text{ on } \Gamma_u \text{ or } T_1 = \hat{T}_1 \text{ on } \Gamma_{T_1} \tag{A.3.3a}$$

$$v = \hat{v} \text{ on } \Gamma_v \text{ or } T_2 = \hat{T}_2 \text{ on } \Gamma_{T_2} \tag{A.3.3b}$$

where Γ_u and Γ_{T_1}, and Γ_v and Γ_{T_2} are disjoint portions of the boundary Γ such that $\Gamma = \Gamma_u \cup \Gamma_{T_1}$ and $\Gamma = \Gamma_v \cup \Gamma_{T_2}$. The boundary stress components T_1 and T_2 are given by

$$T_1 = (2\mu\frac{\partial u}{\partial x} - P)n_x + \mu(\frac{\partial u}{\partial y} + \frac{\partial v}{\partial x})n_y \tag{A.3.4a}$$

$$T_2 = \mu(\frac{\partial u}{\partial y} + \frac{\partial v}{\partial x})n_x + (2\mu\frac{\partial v}{\partial y} - P)n_y \tag{A.3.4b}$$

In equations (A.3.3a) and (A.3.3b) γ denotes the *penalty parameter*, and in equations (A.3.4a) and (A.3.4b) P denotes the pressure.

A.4 Description of the Input Data

The following table describes the input data required for the solution of heat transfer and related problems and viscous fluid flow problems using program *FEM2DHT*. Sample input data files are included. For additional examples, the reader may consult the book by Reddy [1, Chapter 13].

Table A.4.1: Description of the input variables to the computer program *FEM2DHT*.

---

* Data Card 1:
     TITLE   - Title of the problem being solved (80 characters)

* Data Card 2:
     ITYPE   - Problem type
               ITYPE = 0, Heat transfer (and like) problems
               ITYPE = 1, Viscous incompressible flow problems
     IGRAD   - Flag for computing the gradient of the solution
               IGRAD = 0, No postprocessing is required
               IGRAD > 0, Postprocessing is required
               When ITYPE=0 and IGRAD=1, the gradient is computed
               as in Eq. (A.3); for ITYPE = 0 and IGRAD > 1, the
               gradient is computed as in Eq. (A.4).
     ITEM   - Indicator for transient analysis:
               ITEM = 0, Steady--state analysis is required
               ITEM > 0, Transient analysis is required
* Data Card 3:
     IELTYP - Element type used in the analysis:
               IELTYP= 0, Triangular elements
               IELTYP> 0, Quadrilateral elements
     NPE   - Nodes per element:
               NPE   = 3, Linear triangle (IELTYP=0)
               NPE   = 4, Linear quadrilateral (IELTYP>0)
               NPE   = 6, Quadratic triangle (IELTYP=0)
               NPE   = 8 or 9, Quadratic quadrilateral (IELTYP>0)
     MESH   - Indicator for mesh generation by the program:
               MESH = 0, Mesh is not generated by the program
               MESH = 1, Mesh is generated by the program for
                          rectangular domains (by MSH2DR)
               MESH > 1, Mesh is generated by the program for
                          general domains (by MSH2DG)
     NPRNT   - Indicator for printing certain output:
               NPRNT = 0, Not print array NOD, element matrices,
                          or global matrices
               NPRNT = 1, Print array NOD and Element 1 matrices:
                          [ELK] and {ELF}
               NPRNT = 2, Print array NOD and assembled matrices,
                          [GLK] and {GLF}
               NPRNT > 2, Combination of NPRNT=1 and NPRNT=2

* Data Card 4: SKIP the card if MESH.EQ.1
     NEM   - Number of elements in the mesh when the user inputs
               the mesh or the mesh is generated by MSH2DG
     NNM   - Number of nodes   in the mesh when the user inputs
               the mesh or the mesh is generated by MSH2DG

* Data Card 5: SKIP the card if MESH.NE.0; otherwise, the card is
                read in a loop on the number of elements (N=1, NEM)
     NOD(N,I)-Connectivity for the N-th element (I=1,NPE)

* Data Card 6: SKIP the card if MESH.NE.0
     GLXY(I,J)-Global x and y coordinates of I-th global node in
               the mesh (J=1, x-coordinate, J=2, y-coordinate)
               Loops on I and J are: ((J=1,2), I=1,NNM); the NNM
               pairs of (x,y)--coordinates are read sequentially

_____ The next FOUR data cards are read in subroutine MSH2DG _____
* * * * *   SKIP Cards 7, 8, 9, and 10 unless MESH.GT.1 * * * * *

* Data Card 7:
     NRECL   - Number of line records to be read in the mesh

```
* Data Card 8: Read the following variables NRECL times
        NOD1    - First global node number of the line segment
        NODL    - Last global node number of the line segment
        NODINC  - Node increment on the line

        X1      - The global x-coordinate of NOD1
        Y1      - The global y-coordinate of NOD1
        XL      - The global x-coordinate of NODL
        YL      - The global y-coordinate of NODL
        RATIO   - Ratio of the first to the last element lengths

* Data Card 9:
        NRECEL  - Number of rows of elements to be read in the mesh

* Data Card 10: Read the following variables NRECEL times
        NEL1    - First element number of the row
        NELL    - Last element  number of the row
        IELINC  - Increment of element number in the row
        NODINC  - Increment of global node number in the row
        NPE     - Number of nodes in each element
        NODE(I) - Connectivity array of the first element in the row

* Data Card 11:  SKIP the card if  MESH.NE.1
        NX      - Number of elements in the x-direction
        NY      - Number of elements in the y-direction

* Data Card 12:  SKIP the card if  MESH.NE.1
        X0      - The x-coordinate of global node 1
        DX(I)   - The x-dimension of the I-th element (I=1,NX)

* Data Card 13:  SKIP the card if  MESH.NE.1
        Y0      - The y-coordinate of global node 1
        DY(I)   - The y-dimension of the I-th element (I=1,NY)

* Data Card 14:
        NSPV    - The number of specified primary variables

* Data Card 15:  SKIP the card if NSPV.EQ.0
        ISPV(I,J)-Node number and LOCAL degree of freedom number of
                 the I-th specified primary variable:
                 ISPV(I,1)=Node number; ISPV(I,2)=Local DOF number
                 The loops on I and J are: ((J=1,2),I=1,NSPV)

* Data Card 16:  SKIP the card if NSPV.EQ.0
        VSPV(I) - Specified value of the I-th primary variable

* Data Card 17:
        NSSV    - Number of (nonzero) specified secondary variables

* Data Card 18:  SKIP the card if NSSV.EQ.0
        ISSV(I,J)-Node number and LOCAL degree of freedom number of
                 the I-th specified secondary variable:
                 ISSV(I,1)=Node number; ISSV(I,2)=Local DOF number
                 The loops on I and J are: ((J=1,2),I=1,NSSV)

* Data Card 19:  SKIP the card if NSSV.EQ.0
        VSSV(I) - Specified value of the I-th secondary variable
                 (I=1,NSSV)

    Data Cards 20 - 26 are for HEAT TRANSFER PROBLEMS (ITYPE = 0) only

* Data Card 20:  SKIP the card if ITYPE.NE.0
        A10     |
        A1X     |  Coefficients of the differential equation
        A1Y     |  a11 = A10 +A1X*X + A1Y*Y

* Data Card 21:  SKIP the card if ITYPE.NE.0
        A20     |
        A2X     |  Coefficients of the differential equation
        A2Y     |  a22 = A20 +A2X*X + A2Y*Y
```

* Data Card 22:   SKIP the card if ITYPE.NE.0
        A00   - Coefficient of the differential equation

* Data Card 23:   SKIP the card if ITYPE.NE.0
        ICONV   - Indicator for convection boundary conditions
                     ICONV = 0, No convection boundary conditions
                     ICONV > 0, Convection boundary conditions present

* Data Card 24:   SKIP the card if ITYPE.NE.0 or ICONV.EQ.0
        NBE   - Number elements with convection

* Data Card 25:   SKIP the card if ITYPE.NE.0 or ICONV.EQ.0
                  The following cards are read for each I, I=1,NBE
        IBN(I)   - I-th element number with convection
        BETA(I)   - Film coefficient for convection on I-th element
        TINF(I)   - Ambient temperature of the I-th element

* Data Card 26:   SKIP the card if ITYPE.NE.0 or ICONV.EQ.0
        INOD(I,J)- Local node numbers of the side with convection
                     (J=1,2; for quadratic elements, give end nodes)
                     Loops on I and J are: ((J=1,2), I=1,NBE)

_____ Data Card 27 is for VISCOUS FLUID FLOWS (ITYPE = 1) only _____

* Data Card 27:   SKIP the card if ITYPE.EQ.0
        AMU   - Viscosity of the fluid
        PENLTY   - Penalty parameter used

_____ Remaining data cards are for ALL problem types _____

* Data Card 28:
        F0 |
        FX | Coefficients to define the source term:
        FY | f = F0 + FX*x + FY*y

_____ Cards 29 thru 32 are for TRANSIENT ANALYSIS only _____

* Data Card 29:   SKIP the card if ITEM.EQ.0
        C0 | Coefficients defining the temporal parts of the
        CX | differential equations:
        CY | CT = C0 + CX*x + CY*y

* Data Card 30:   SKIP the card if ITEM.EQ.0
        NTIME   - Number of time steps for the transient solution
        NSTP   - Time step number at which the source is removed
        INTVL   - Time step interval at which to print the solution
        INTIAL   - Indicator for nature of initial conditions:
                     INTIAL=0, Zero initial conditions are used
                     INTIAL>0, Non-zero initial conditions are used

* Data Card 31:   SKIP the card if ITEM.EQ.0
        DT   - Time step used for the transient solution
        ALFA   - Parameter in the alfa-family of time approximation
                     used:
                     ALFA=0,   The forward difference scheme (C.S.)@
                     ALFA=0.5, The Crank-Nicolson scheme (stable)
                     ALFA=2/3, The Galerkin scheme (stable)
                     ALFA=1,   The backward difference scheme (stable)
                     @C.S.=Conditionally Stable; For all schemes with
                     ALFA < 0.5, the time step DT is restricted to:
                     DT < 2/[MAXEGN*(1-2*ALFA)], where MAXEGN is the
                     maximum eigenvalue of the discrete problem.
        EPSLN   - A small parameter to check if the solution has
                     reached a steady state

* Data Card 32:   SKIP the card if ITEM or INTIAL.EQ.0
        GLU(I)   - Vector of initial values of the primary variables
                     (I=1,NEQ, NEQ=Number of nodal values in the mesh)

Example 1: Convective heat transfer in a square region (see Figure A.4.1) is solved using 2×2 mesh of nine-node rectangular elements. The conductivities are taken to be $a_{11} = a_{22} = 10$ W/(m °C). Echo of the input data along with the output from program *FEM2DHT* is given below and on the next couple of pages.

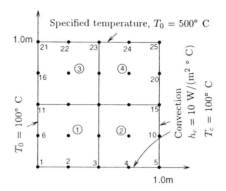

Figure A.4.1: Domain, boundary conditions, and mesh used for the convective heat transfer problem in Example 1.

Input data and output for Example 1: ————————————

```
Example 1: CONVECTIVE heat transfer in a square region
  0    1    0    0                              ITYPE,IGRAD,ITEM,NEIGN
  2    9    1    0                              IEL, NPE, MESH, NPRNT
  2    2                                        NX, NY
  0.0   0.5    0.5                              X0, DX(1), DX(2)
  0.0   0.5    0.5                              Y0, DY(I), DY(2)
  9                                             NSPV
  1 1   6 1   11 1   16 1   21 1   22 1   23 1   24 1   25 1 -ISPV(I,J)
  100.0 100.0 100.0 100.0 500.0 500.0 500.0 500.0 500.0-VSPV(I)
  0                                             NSSV
  10.0   0.0   0.0                              A10, A1X, A1Y
  10.0   0.0   0.0                              A20, A2X, A2Y
   0.0                                          A00
  1                                             ICONV
  4                                             NBE
  1 10.0 100.0 2 10.0 100.0 2 10.0 100.0 4 10.0 100.0-IBN,BETA,TINF
  1 2   1 2   2 3   2 3                         INOD(I,J)
  0.0       0.0   0.0                           F0,   FX,   FY

Example 1: CONVECTIVE heat transfer in a square region
```

```
              OUTPUT   FROM   PROGRAM   *FEM2DHT*   BY   J. N. REDDY


                 ANALYSIS   OF   AN   HEAT   TRANSFER   PROBLEM

      COEFFICIENTS OF THE DIFFERENTIAL EQUATION:
            Coefficient, A10 ........................=   0.1000E+02
            Coefficient, A1X ........................=   0.0000E+00
            Coefficient, A1Y ........................=   0.0000E+00
            Coefficient, A20 ........................=   0.1000E+02
            Coefficient, A2X ........................=   0.0000E+00
            Coefficient, A2Y ........................=   0.0000E+00
            Coefficient, A00 ........................=   0.0000E+00
```

CONVECTIVE HEAT TRANSFER DATA:
    Number of elements with convection, NBE .=   4
    Elements, their LOCAL nodes and convective
    heat transfer data:
    Ele. No.   End Nodes   Film Coeff.   T-Infinity

        1   1   2   0.10000E+02   0.10000E+03
        2   1   2   0.10000E+02   0.10000E+03
        2   2   3   0.10000E+02   0.10000E+03
        4   2   3   0.10000E+02   0.10000E+03

CONTINUOUS SOURCE COEFFICIENTS:
    Coefficient, F0 ....= 0.0000E+00
    Coefficient, FX ....= 0.0000E+00
    Coefficient, FY ....= 0.0000E+00

    ******* A STEADY-STATE PROBLEM is analyzed *******

    *** A mesh of QUADRILATERALS is chosen by user ***

FINITE ELEMENT MESH INFORMATION:
    Element type: 0 = Triangle; >0 = Quad.)..=   2
    Number of nodes per element, NPE ....=   9
    No. of primary deg. of freedom/node, NDF =   1
    Number of elements in the mesh, NEM .....=   4
    Number of nodes in the mesh, NNM ....=   25
    Number of equations to be solved, NEQ ...=   25
    Half bandwidth of the matrix GLK, NHBW ..=   13
    Mesh subdivisions, NX and NY ....=   2   2
    No. of specified PRIMARY variables, NSPV =   9

Node	x-coord.	y-coord.	Speci. primary & secondary variables (0, unspecified; >0, specified)	
			Primary DOF	Secondary DOF
1	0.0000E+00	0.0000E+00	1	0
2	0.2500E+00	0.0000E+00	0	0
3	0.5000E+00	0.0000E+00	0	0
4	0.7500E+00	0.0000E+00	0	0
5	0.1000E+01	0.0000E+00	0	0
6	0.0000E+00	0.2500E+00	1	0
7	0.2500E+00	0.2500E+00	0	0
8	0.5000E+00	0.2500E+00	0	0
9	0.7500E+00	0.2500E+00	0	0
10	0.1000E+01	0.2500E+00	0	0
11	0.0000E+00	0.5000E+00	1	0
12	0.2500E+00	0.5000E+00	0	0
13	0.5000E+00	0.5000E+00	0	0
14	0.7500E+00	0.5000E+00	0	0
15	0.1000E+01	0.5000E+00	0	0
16	0.0000E+00	0.7500E+00	1	0
17	0.2500E+00	0.7500E+00	0	0
18	0.5000E+00	0.7500E+00	0	0
19	0.7500E+00	0.7500E+00	0	0
20	0.1000E+01	0.7500E+00	0	0
21	0.0000E+00	0.1000E+01	1	0
22	0.2500E+00	0.1000E+01	1	0
23	0.5000E+00	0.1000E+01	1	0
24	0.7500E+00	0.1000E+01	1	0
25	0.1000E+01	0.1000E+01	1	0

NUMERICAL INTEGRATION DATA:
    Full quadrature (IPDF x IPDF) rule, IPDF =   3
    Reduced quadrature (IPDR x IPDR),   IPDR =   2
    Quadrature rule used in postproc., ISTR =   2

S O L U T I O N :

Node	x-coord.	y-coord.	Temperature
1	0.00000E+00	0.00000E+00	0.10000E+03
2	0.25000E+00	0.00000E+00	0.11380E+03
3	0.50000E+00	0.00000E+00	0.13273E+03
4	0.75000E+00	0.00000E+00	0.13350E+03
5	0.10000E+01	0.00000E+00	0.11697E+03
6	0.00000E+00	0.25000E+00	0.10000E+03
7	0.25000E+00	0.25000E+00	0.14268E+03
8	0.50000E+00	0.25000E+00	0.17135E+03
9	0.75000E+00	0.25000E+00	0.17561E+03
10	0.10000E+01	0.25000E+00	0.15241E+03
11	0.00000E+00	0.50000E+00	0.10000E+03
12	0.25000E+00	0.50000E+00	0.18453E+03
13	0.50000E+00	0.50000E+00	0.24111E+03
14	0.75000E+00	0.50000E+00	0.24392E+03
15	0.10000E+01	0.50000E+00	0.21538E+03
16	0.00000E+00	0.75000E+00	0.10000E+03
17	0.25000E+00	0.75000E+00	0.29562E+03
18	0.50000E+00	0.75000E+00	0.34509E+03
19	0.75000E+00	0.75000E+00	0.35267E+03
20	0.10000E+01	0.75000E+00	0.30641E+03
21	0.00000E+00	0.10000E+01	0.50000E+03
22	0.25000E+00	0.10000E+01	0.50000E+03
23	0.50000E+00	0.10000E+01	0.50000E+03
24	0.75000E+00	0.10000E+01	0.50000E+03
25	0.10000E+01	0.10000E+01	0.50000E+03

x-coord.	y-coord.	-a11(du/dx)	-a22(du/dy)	Flux Mgntd
0.1057E+00	0.1057E+00	-0.9881E+03	-0.5666E+03	0.1139E+04
0.1057E+00	0.3943E+00	-0.2681E+04	-0.7906E+03	0.2795E+04
0.3943E+00	0.1057E+00	-0.8220E+03	-0.1403E+04	0.1626E+04
0.3943E+00	0.3943E+00	-0.1632E+04	-0.2457E+04	0.2950E+04
0.6057E+00	0.1057E+00	-0.1830E+03	-0.1566E+04	0.1577E+04
0.6057E+00	0.3943E+00	-0.2553E+03	-0.2872E+04	0.2884E+04
0.8943E+00	0.1057E+00	0.8494E+03	-0.1499E+04	0.1723E+04
0.8943E+00	0.3943E+00	0.1152E+04	-0.2709E+04	0.2944E+04
0.1057E+00	0.6057E+00	-0.7097E+04	-0.1718E+04	0.7302E+04
0.1057E+00	0.8943E+00	-0.5095E+04	-0.1272E+05	0.1370E+05
0.3943E+00	0.6057E+00	-0.2009E+04	-0.4743E+04	0.5151E+04
0.3943E+00	0.8943E+00	-0.7522E+03	-0.6432E+04	0.6476E+04
0.6057E+00	0.6057E+00	-0.4085E+03	-0.4237E+04	0.4257E+04
0.6057E+00	0.8943E+00	-0.2876E+03	-0.5918E+04	0.5925E+04
0.8943E+00	0.6057E+00	0.1909E+04	-0.3846E+04	0.4293E+04
0.8943E+00	0.8943E+00	0.1194E+04	-0.6904E+04	0.7006E+04

Example 2: Flow of an incompressible viscous fluid in a slider bearing (see Figure A.4.2) is analyzed using a mesh of six nine-node elements. The mesh used is to illustrate the input data, and it is not representative of the mesh required to obtain an accurate solution. The input data and an edited output from program *FEM2DHT* are given on the next few pages.

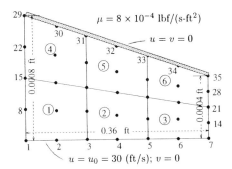

Figure A.4.2: Domain and mesh used for the flow of a viscous incompressible fluid in a slider bearing (Example 2).

Input data and output for Example 2:

```
Example 2: Flow of a viscous fluid in a slider bearing
  1    1    0                                      ITYPE,ISTRS,ITEM
  1    9    2    0                                 IELTYP,NPE,MESH,NPRNT
  6   35                                           NEM, NNM

  5                                                NRECL
  1    7    1    0.0    0.0      0.36  0.0    1.0  NOD1,NODL,NODINC,...
  8   14    1    0.0    5.0E-5   0.36  5.0E-5  1.0
 15   21    1    0.0    1.0E-4   0.36  1.0E-4  1.0
 22   28    1    0.0    3.5E-4   0.36  1.5E-4  1.0
 29   35    1    0.0    8.0E-4   0.36  4.0E-4  1.0
  2                                                NRECEL
  1    3    1    2    9     1    3   17   15   2   10  16   8   9  NEL1,NELL,..
  4    6    1    2    9    15   17   31   29  16   24  30  22  23

 28                                                NSPV
  1    1    1    2    2    1    2    2    3    1    3    2    4    1    4    2    5    1    5    2
  6    1    6    2    7    1    7    2   29    1   29    2   30    1   30    2   31    1   31    2
 32    1   32    2   33    1   33    2   34    1   34    2   35    1   35    2   ISPV(I,J)
 30.0   0.0   30.0   0.0   30.0    0.0  30.0   0.0  30.0   0.0
 30.0   0.0   30.0   0.0    0.0    0.0   0.0   0.0   0.0   0.0
  0.0   0.0    0.0   0.0    0.0    0.0   0.0   0.0      VSPV(I)

  0                                                NSSV
 8.0E-4   8.0E12                                   AMU, PENLTY
 0.0    0.0    0.0                                 FO, FX, FY

Example 2: Flow of a viscous fluid in a slider bearing

            OUTPUT  FROM   PROGRAM   *FEM2DHT*  BY  J. N. REDDY


          A   VISCOUS  INCOMPRESSIBLE FLOW IS ANALYZED

    PARAMETERS OF THE FLUID FLOW PROBLEM:
        Viscosity of the fluid, AMU .............= 0.8000E-03
        Penalty parameter, PENLTY ...............= 0.8000E+13

    CONTINUOUS SOURCE COEFFICIENTS:
        Coefficient, FO  .......................= 0.0000E+00
        Coefficient, FX  .......................= 0.0000E+00
        Coefficient, FY  .......................= 0.0000E+00
```

```
FINITE ELEMENT MESH INFORMATION:
    Element type: 0 = Triangle; >0 = Quad.)..=    1
    Number of nodes per element, NPE ........=    9
    No. of primary deg. of freedom/node, NDF =    2
    Number of elements in the mesh, NEM .....=    6
    Number of nodes in the mesh, NNM ........=   35
    Number of equations to be solved, NEQ ...=   70
    Half bandwidth of the matrix GLK, NHBW ..=   34
    No. of specified PRIMARY variables, NSPV =   28
```

Node	x-coord.	y-coord.	Speci. primary & secondary variables (0, unspecified; >0, specified)			
			Primary DOF		Secondary DOF	
1	0.0000E+00	0.0000E+00	1	2	0	0
2	0.6000E-01	0.0000E+00	1	2	0	0
3	0.1200E+00	0.0000E+00	1	2	0	0
4	0.1800E+00	0.0000E+00	1	2	0	0
5	0.2400E+00	0.0000E+00	1	2	0	0
6	0.3000E+00	0.0000E+00	1	2	0	0
7	0.3600E+00	0.0000E+00	1	2	0	0
8	0.0000E+00	0.5000E-04	0	0	0	0
9	0.6000E-01	0.5000E-04	0	0	0	0
10	0.1200E+00	0.5000E-04	0	0	0	0
11	0.1800E+00	0.5000E-04	0	0	0	0
12	0.2400E+00	0.5000E-04	0	0	0	0
13	0.3000E+00	0.5000E-04	0	0	0	0
14	0.3600E+00	0.5000E-04	0	0	0	0
15	0.0000E+00	0.1000E-03	0	0	0	0
16	0.6000E-01	0.1000E-03	0	0	0	0
17	0.1200E+00	0.1000E-03	0	0	0	0
18	0.1800E+00	0.1000E-03	0	0	0	0
19	0.2400E+00	0.1000E-03	0	0	0	0
20	0.3000E+00	0.1000E-03	0	0	0	0
21	0.3600E+00	0.1000E-03	0	0	0	0
22	0.0000E+00	0.3500E-03	0	0	0	0
23	0.6000E-01	0.3167E-03	0	0	0	0
24	0.1200E+00	0.2833E-03	0	0	0	0
25	0.1800E+00	0.2500E-03	0	0	0	0
26	0.2400E+00	0.2167E-03	0	0	0	0
27	0.3000E+00	0.1833E-03	0	0	0	0
28	0.3600E+00	0.1500E-03	0	0	0	0
29	0.0000E+00	0.8000E-03	1	2	0	0
30	0.6000E-01	0.7333E-03	1	2	0	0
31	0.1200E+00	0.6667E-03	1	2	0	0
32	0.1800E+00	0.6000E-03	1	2	0	0
33	0.2400E+00	0.5333E-03	1	2	0	0
34	0.3000E+00	0.4667E-03	1	2	0	0
35	0.3600E+00	0.4000E-03	1	2	0	0

```
NUMERICAL INTEGRATION DATA:
    Full quadrature (IPDF x IPDF) rule, IPDF =    3
    Reduced quadrature (IPDR x IPDR),    IPDR =    2
    Quadrature rule used in postproc.,  ISTR =    2

S O L U T I O N :
```

Node	x-coord.	y-coord.	Velocity, u	Velocity, v
1	0.00000E+00	0.00000E+00	0.30000E+02	0.00000E+00
2	0.60000E-01	0.00000E+00	0.30000E+02	0.00000E+00
3	0.12000E+00	0.00000E+00	0.30000E+02	0.00000E+00
4	0.18000E+00	0.00000E+00	0.30000E+02	0.00000E+00
5	0.24000E+00	0.00000E+00	0.30000E+02	0.00000E+00

```
 6   0.30000E+00   0.00000E+00   0.30000E+02    0.00000E+00
 7   0.36000E+00   0.00000E+00   0.30000E+02    0.00000E+00
 8   0.00000E+00   0.50000E-04   0.26008E+02   -0.45304E-02
 9   0.60000E-01   0.50000E-04   0.25983E+02    0.22683E-02
10   0.12000E+00   0.50000E-04   0.26085E+02   -0.46442E-02
11   0.18000E+00   0.50000E-04   0.26308E+02    0.21300E-02
12   0.24000E+00   0.50000E-04   0.26982E+02   -0.50555E-02
13   0.30000E+00   0.50000E-04   0.28270E+02    0.14714E-02
14   0.36000E+00   0.50000E-04   0.31533E+02   -0.68447E-02
15   0.00000E+00   0.10000E-03   0.22292E+02    0.10406E-01
16   0.60000E-01   0.10000E-03   0.22259E+02   -0.51970E-02
17   0.12000E+00   0.10000E-03   0.22421E+02    0.10016E-01
18   0.18000E+00   0.10000E-03   0.22847E+02   -0.57272E-02
19   0.24000E+00   0.10000E-03   0.23933E+02    0.86438E-02
20   0.30000E+00   0.10000E-03   0.26237E+02   -0.81796E-02
21   0.36000E+00   0.10000E-03   0.31469E+02    0.26246E-02
22   0.00000E+00   0.35000E-03   0.87401E+01    0.14297E-01
23   0.60000E-01   0.31667E-03   0.10255E+02   -0.80107E-02
24   0.12000E+00   0.28333E-03   0.12079E+02    0.12563E-01
25   0.18000E+00   0.25000E-03   0.14350E+02   -0.99138E-02
26   0.24000E+00   0.21667E-03   0.17197E+02    0.97258E-02
27   0.30000E+00   0.18333E-03   0.20956E+02   -0.13382E-01
28   0.36000E+00   0.15000E-03   0.25982E+02    0.38986E-02
29   0.00000E+00   0.80000E-03   0.00000E+00    0.00000E+00
30   0.60000E-01   0.73333E-03   0.00000E+00    0.00000E+00
31   0.12000E+00   0.66667E-03   0.00000E+00    0.00000E+00
32   0.18000E+00   0.60000E-03   0.00000E+00    0.00000E+00
33   0.24000E+00   0.53333E-03   0.00000E+00    0.00000E+00
34   0.30000E+00   0.46667E-03   0.00000E+00    0.00000E+00
35   0.36000E+00   0.40000E-03   0.00000E+00    0.00000E+00
```

x-coord.	y-coord.	sigma-x	sigma-y	sigma-xy	pressure
0.2536E-01	0.2113E-04	-0.2325E+04	-0.2325E+04	-0.6464E+02	0.2325E+04
0.2536E-01	0.7887E-04	-0.2324E+04	-0.2324E+04	-0.5930E+02	0.2324E+04
0.9464E-01	0.2113E-04	-0.8666E+04	-0.8666E+04	-0.6391E+02	0.8666E+04
0.9464E-01	0.7887E-04	-0.8666E+04	-0.8666E+04	-0.5883E+02	0.8666E+04
0.1454E+00	0.2113E-04	-0.1311E+05	-0.1311E+05	-0.6235E+02	0.1311E+05
0.1454E+00	0.7887E-04	-0.1311E+05	-0.1311E+05	-0.5731E+02	0.1311E+05
0.2146E+00	0.2113E-04	-0.1771E+05	-0.1771E+05	-0.5387E+02	0.1771E+05
0.2146E+00	0.7887E-04	-0.1771E+05	-0.1771E+05	-0.5184E+02	0.1771E+05
0.2654E+00	0.2113E-04	-0.1858E+05	-0.1858E+05	-0.4341E+02	0.1858E+05
0.2654E+00	0.7887E-04	-0.1859E+05	-0.1859E+05	-0.4380E+02	0.1859E+05
0.3346E+00	0.2113E-04	-0.9608E+04	-0.9608E+04	-0.2520E+00	0.9608E+04
0.3346E+00	0.7887E-04	-0.9607E+04	-0.9607E+04	-0.1734E+02	0.9607E+04
0.2536E-01	0.1753E-03	-0.2323E+04	-0.2323E+04	-0.4786E+02	0.2323E+04
0.2536E-01	0.5632E-03	-0.2323E+04	-0.2323E+04	-0.1607E+02	0.2323E+04
0.9464E-01	0.1590E-03	-0.8666E+04	-0.8666E+04	-0.4855E+02	0.8666E+04
0.9464E-01	0.5025E-03	-0.8667E+04	-0.8667E+04	-0.2187E+02	0.8667E+04
0.1454E+00	0.1471E-03	-0.1311E+05	-0.1311E+05	-0.4832E+02	0.1311E+05
0.1454E+00	0.4580E-03	-0.1311E+05	-0.1311E+05	-0.2752E+02	0.1311E+05
0.2146E+00	0.1309E-03	-0.1771E+05	-0.1771E+05	-0.4730E+02	0.1771E+05
0.2146E+00	0.3973E-03	-0.1771E+05	-0.1771E+05	-0.3830E+02	0.1771E+05
0.2654E+00	0.1190E-03	-0.1859E+05	-0.1859E+05	-0.4532E+02	0.1859E+05
0.2654E+00	0.3529E-03	-0.1859E+05	-0.1859E+05	-0.4934E+02	0.1859E+05
0.3346E+00	0.1027E-03	-0.9605E+04	-0.9605E+04	-0.6196E+02	0.9605E+04
0.3346E+00	0.2922E-03	-0.9606E+04	-0.9606E+04	-0.7193E+02	0.9606E+04

A.5 Source Listings of Selective Subroutines

The computer implementation of the finite element formulations presented in Chapters 2 and 3 can be found in Chapter 13 of [1]. A complete listing of the FORTRAN source of FEM2DV2 is also included in [1]. The program *FEM2DHT* consists of the following subroutines:

BOUNRY: Imposes applied boundary conditions on primary and secondary variables.

CONCT: Assembles element coefficient matrices and vectors.

ECHO: Echos the input file.

ELKMFR: Generates element coefficient matrices and vectors for rectangular (quadrilateral) elements.

ELKMFT: Generates element coefficient matrices and vectors for triangular elements.

MSH2DG: Generates finite element meshes for some general domains (see [1] for details).

MSH2DR: Generates finite element meshes for only rectangular domains (see [1] for details).

PSTPRC: Postprocessor

SHPRCT: Evaluates shape (or interpolation) functions and their global derivatives for rectangular (isoparametric) elements.

SHPTRI: Evaluates shape (or interpolation) functions and their global derivatives for triangular (isoparametric) elements.

SOLVER: Solves symmetric, banded system of algebraic equations.

TEMPORAL: Sets up coefficient matrices and column vectors for transient analysis.

To aid the reader in the computer implementation of the finite element formulations presented in Chapters 2 and 3, listings of subroutines ELKMFR, SHPRCT, and TEMPORAL from program *FEM2DHT*, which illustrate the computation of element matrices for quadrilateral elements, are included in the remaining pages of this appendix. The complete FORTRAN source of program *FEM2DHT* or *FEM2DV2* from [1] may be obtained for a small charge from the first author.

Reference

1. J. N. Reddy, *An Introduction to the Finite Element Method*, Second Edition, McGraw-Hill, New York (1993).

Listings of Subroutines *ELKMFR*, *SHPRCT*, and *TEMPORAL*

```
      SUBROUTINE ELKMFR(NPE,NN,ITYPE)
C     _____
C
C
C     Element  calculations based on  linear and quadratic rectangular
C     elements and  isoparametric  formulation are carried out for the
C     heat transfer and penalty model of fluid flow.
C
C     _____
C
      IMPLICIT REAL*8(A-H,O-Z)
      COMMON/STF/ELF(18),ELK(18,18),ELM(18,18),ELXY(9,2),ELU(18),A1,A2
      COMMON/PST/A10,A1X,A1Y,A20,A2X,A2Y,A00,C0,CX,CY,F0,FX,FY,
     1            AMU,PENLTY
      COMMON/SHP/SF(9),GDSF(2,9),SFH(16)
      COMMON/PNT/IPDF,IPDR,NIPF,NIPR
      DIMENSION  GAUSPT(5,5),GAUSWT(5,5)
      COMMON/IO/IN,ITT
C
      DATA GAUSPT/5*0.0D0,  -0.57735027D0,  0.57735027D0,  3*0.0D0,
     2   -0.77459667D0,  0.0D0,  0.77459667D0,  2*0.0D0,  -0.86113631D0,
     3   -0.33998104D0,  0.33998104D0,  0.86113631D0,  0.0D0,  -0.90617984D0,
     4   -0.53846931D0,0.0D0,0.53846931D0,0.90617984D0/
C
      DATA GAUSWT/2.0D0,  4*0.0D0,  2*1.0D0,  3*0.0D0,  0.55555555D0,
     2    0.88888888D0,  0.55555555D0,  2*0.0D0,  0.34785485D0,
     3 2*0.65214515D0,  0.34785485D0,  0.0D0,  0.23692688D0,
     4    0.47862867D0,  0.56888888D0,  0.47862867D0,  0.23692688D0/
C
      NDF = NN/NPE
      NET=NPE
C
C     Initialize the arrays
C
      DO 120 I = 1,NN
      ELF(I)  = 0.0
      DO 120 J = 1,NN
      IF(ITEM.NE.0) THEN
          ELM(I,J)= 0.0
      ENDIF
  120 ELK(I,J)= 0.0
C
C     Do-loops on numerical (Gauss) integration begin here. Subroutine
C     SHPRCT (SHaPe functions for ReCTangular elements) is called here
C
      DO 200 NI = 1,IPDF
      DO 200 NJ = 1,IPDF
      XI  = GAUSPT(NI,IPDF)
      ETA = GAUSPT(NJ,IPDF)
      CALL SHPRCT (NPE,XI,ETA,DET,ELXY,NDF,ITYPE)
      CNST = DET*GAUSWT(NI,IPDF)*GAUSWT(NJ,IPDF)
      X=0.0
      Y=0.0
      DO 140 I=1,NPE
      X=X+ELXY(I,1)*SF(I)
  140 Y=Y+ELXY(I,2)*SF(I)
C
      SOURCE=F0+FX*X+FY*Y
      IF(ITEM.NE.0) THEN
          CT=C0+CX*X+CY*Y
      ENDIF
      IF(ITYPE.LE.0) THEN
          A11=A10+A1X*X+A1Y*Y
          A22=A20+A2X*X+A2Y*Y
      ENDIF
```

```
      II=1
      DO 180 I=1,NET
      JJ=1
      DO 160 J=1,NET
      S00=SF(I)*SF(J)*CNST
      S11=GDSF(1,I)*GDSF(1,J)*CNST
      S22=GDSF(2,I)*GDSF(2,J)*CNST
      S12=GDSF(1,I)*GDSF(2,J)*CNST
      S21=GDSF(2,I)*GDSF(1,J)*CNST
      IF(ITYPE.EQ.0) THEN
C
C     Heat transfer and like problems (i.e. single DOF problems):_____
C
          ELK(I,J) = ELK(I,J) + A11*S11 + A22*S22 + A00*S00
          IF(ITEM.NE.0) THEN
              ELM(I,J) = ELM(I,J) + CT*S00
          ENDIF
      ELSE
C
C     Viscous incompressible fluids:_____
C     Compute coefficients associated with viscous terms (full integ.)
C
          ELK(II,JJ)    = ELK(II,JJ)      + AMU*(2.0*S11 + S22)
          ELK(II+1,JJ)  = ELK(II+1,JJ)    + AMU*S12
          ELK(II,JJ+1)  = ELK(II,JJ+1)    + AMU*S21
          ELK(II+1,JJ+1)= ELK(II+1,JJ+1)  + AMU*(S11 + 2.0*S22)
          IF(ITEM.NE.0) THEN
              ELM(II,JJ)    = ELM(II,JJ)     + CT*S00
              ELM(II+1,JJ+1)= ELM(II+1,JJ+1) + CT*S00
          ENDIF
      ENDIF
  160 JJ = NDF*J+1
C
C     Source of the form fx = F0 + FX*X + FY*Y is assumed
C
      L=(I-1)*NDF+1
      ELF(L) = ELF(L)+CNST*SF(I)*SOURCE
  180 II = NDF*I+1
  200 CONTINUE
C
      IF(ITYPE.GT.0) THEN
C
C     Use reduced integration to evaluate coefficients associated with
C     penalty terms for flows of viscous incompressible fluids.
C
          DO 280 NI=1,IPDR
          DO 280 NJ=1,IPDR
          XI  = GAUSPT(NI,IPDR)
          ETA = GAUSPT(NJ,IPDR)
          CALL SHPRCT (NPE,XI,ETA,DET,ELXY,NDF,ITYPE)
          CNST=DET*GAUSWT(NI,IPDR)*GAUSWT(NJ,IPDR)
C
          II=1
          DO 260 I=1,NPE
          JJ = 1
          DO 240 J=1,NPE
          S11=GDSF(1,I)*GDSF(1,J)*CNST
          S22=GDSF(2,I)*GDSF(2,J)*CNST
          S12=GDSF(1,I)*GDSF(2,J)*CNST
          S21=GDSF(2,I)*GDSF(1,J)*CNST
          ELK(II,JJ)    = ELK(II,JJ)      + PENLTY*S11
          ELK(II+1,JJ)  = ELK(II+1,JJ)    + PENLTY*S21
          ELK(II,JJ+1)  = ELK(II,JJ+1)    + PENLTY*S12
          ELK(II+1,JJ+1)= ELK(II+1,JJ+1)  + PENLTY*S22
  240 JJ=NDF*J+1
  260 II=NDF*I+1
  280 CONTINUE
      ENDIF
```

```
      IF(ITEM.NE.0) THEN
C
C     Compute the coefficient matrices of the final algebraic equations
C     (i.e., after time approximation) in the transient analysis:_____
C
         CALL TEMPORAL(NN)
      ENDIF
      RETURN
      END



      SUBROUTINE SHPRCT(NPE,XI,ETA,DET,ELXY,NDF,ITYPE)
C     _____
C
C
C     The subroutine evaluates the interpolation functions (SF(I)) and
C     their derivatives with respect to global coordinates (GDSF(I,J))
C     for Lagrange linear & quadratic rectangular elements, using  the
C     isoparametric formulation. The subroutine also evaluates Hermite
C     interpolation functions and their  global derivatives  using the
C     subparametric formulation.
C
C     SF(I)........Interpolation function for node I of the element
C     DSF(J,I).....Derivative of SF(I) with respect to XI if J=1 and
C                  and ETA if J=2
C     GDSF(J,I)....Derivative of SF(I) with respect to  X if J=1 and
C                  and  Y  if J=2
C     XNODE(I,J)...J-TH (J=1,2) Coordinate of node I of the element
C     NP(I)........Array of element nodes (used to define SF and DSF)
C     GJ(I,J)......Jacobian matrix
C     GJINV(I,J)...Inverse of the jacobian matrix
C
C     _____
C
      IMPLICIT REAL*8 (A-H,O-Z)
      DIMENSION ELXY(9,2),XNODE(9,2),NP(9),DSF(2,9),GJ(2,2),GJINV(2,2)
      COMMON/SHP/SF(9),GDSF(2,9)
      COMMON/IO/IN,ITT
      DATA XNODE/-1.0D0, 2*1.0D0, -1.0D0, 0.0D0, 1.0D0, 0.0D0, -1.0D0,
     *          0.0D0, 2*-1.0D0, 2*1.0D0, -1.0D0, 0.0D0, 1.0D0, 2*0.0D0/
      DATA NP/1,2,3,4,5,7,6,8,9/
C
      FNC(A,B) = A*B
      IF(NPE.EQ.4) THEN
C
C     LINEAR Lagrange interpolation functions for FOUR-NODE element
C
         DO 10 I = 1, NPE
         XP  = XNODE(I,1)
         YP  = XNODE(I,2)
         XI0 = 1.0+XI*XP
         ETA0=1.0+ETA*YP
         SF(I)    = 0.25*FNC(XI0,ETA0)
         DSF(1,I)= 0.25*FNC(XP,ETA0)
   10    DSF(2,I)= 0.25*FNC(YP,XI0)
      ELSE
         IF(NPE.EQ.8) THEN
C
C     QUADRATIC Lagrange interpolation functions for EIGHT-NODE element
C
            DO 20 I = 1, NPE
            NI   = NP(I)
            XP   = XNODE(NI,1)
            YP   = XNODE(NI,2)
            XI0  = 1.0+XI*XP
            ETA0 = 1.0+ETA*YP
            XI1  = 1.0-XI*XI
            ETA1 = 1.0-ETA*ETA
            IF(I.LE.4) THEN
                SF(NI)    = 0.25*FNC(XI0,ETA0)*(XI*XP+ETA*YP-1.0)
                DSF(1,NI) = 0.25*FNC(ETA0,XP)*(2.0*XI*XP+ETA*YP)
                DSF(2,NI) = 0.25*FNC(XI0,YP)*(2.0*ETA*YP+XI*XP)
            ELSE
```

```
                              SF(NI)      = 0.5*FNC(XI1,ETA0)
                              DSF(1,NI)  = -FNC(XI,ETA0)
                              DSF(2,NI)  = 0.5*FNC(YP,XI1)
                        ELSE
                              SF(NI)      = 0.5*FNC(ETA1,XI0)
                              DSF(1,NI)  = 0.5*FNC(XP,ETA1)
                              IF(I.LE.6) THEN
                        DSF(2,NI)  = -FNC(ETA,XI0)
                              ENDIF
                        ENDIF
                  ENDIF
      20          CONTINUE
            ELSE
C
C
C     QUADRATIC Lagrange interpolation functions for NINE-NODE element
C
                  DO 30 I=1,NPE
                  NI    = NP(I)
                  XP    = XNODE(NI,1)
                  YP    = XNODE(NI,2)
                  XI0   = 1.0+XI*XP
                  ETA0  = 1.0+ETA*YP
                  XI1   = 1.0-XI*XI
                  ETA1  = 1.0-ETA*ETA
                  XI2   = XP*XI
                  ETA2  = YP*ETA
                  IF(I .LE. 4) THEN
                        SF(NI)      = 0.25*FNC(XI0,ETA0)*XI2*ETA2
                        DSF(1,NI)= 0.25*XP*FNC(ETA2,ETA0)*(1.0+2.0*XI2)
                        DSF(2,NI)= 0.25*YP*FNC(XI2,XI0)*(1.0+2.0*ETA2)
                  ELSE
                        IF(I .LE. 6) THEN
                              SF(NI)      = 0.5*FNC(XI1,ETA0)*ETA2
                              DSF(1,NI)  = -XI*FNC(ETA2,ETA0)
                              DSF(2,NI)  = 0.5*FNC(XI1,YP)*(1.0+2.0*ETA2)
                        ELSE
                              IF(I .LE. 8) THEN
                                    SF(NI)      = 0.5*FNC(ETA1,XI0)*XI2
                                    DSF(2,NI)  = -ETA*FNC(XI2,XI0)
                                    DSF(1,NI)  = 0.5*FNC(ETA1,XP)*(1.0+2.0*XI2)
                              ELSE
                                    SF(NI)      = FNC(XI1,ETA1)
                                    DSF(1,NI)  = -2.0*XI*ETA1
                                    DSF(2,NI)  = -2.0*ETA*XI1
                              ENDIF
                        ENDIF
                  ENDIF
      30          CONTINUE
            ENDIF
      ENDIF
C
C
C     Compute the Jacobian matrix [GJ] and its inverse [GJINV]
C
      DO 40 I = 1,2
      DO 40 J = 1,2
      GJ(I,J)   = 0.0
      DO 40 K = 1,NPE
   40 GJ(I,J)   = GJ(I,J) + DSF(I,K)*ELXY(K,J)
C
      DET = GJ(1,1)*GJ(2,2)-GJ(1,2)*GJ(2,1)
      GJINV(1,1) = GJ(2,2)/DET
      GJINV(2,2) = GJ(1,1)/DET
      GJINV(1,2) = -GJ(1,2)/DET
      GJINV(2,1) = -GJ(2,1)/DET
```

```
C
C       Compute the derivatives of the interpolation functions with
C       respect to the global coordinates (x,y): [GDSF]
C
        DO 50 I  = 1,2
        DO 50 J  = 1,NPE
        GDSF(I,J) = 0.0
        DO 50 K  = 1, 2
     50 GDSF(I,J) = GDSF(I,J) + GJINV(I,K)*DSF(K,J)
        RETURN
        END


        SUBROUTINE TEMPORAL(NN)
C
C      _____
C
C       The   subroutine computes   the algebraic equations associated with
C       the   parabolic differential equations by using the alfa-family of
C       approximations. A constant source is assumed.
C
C      _____
C
        IMPLICIT REAL*8(A-H,O-Z)
        COMMON/STF/ELF(18),ELK(18,18),ELM(18,18),ELXY(9,2),ELU(18),A1,A2
C
C       The alfa-family of time approximation for parabolic equations
C
            DO 20 I=1,NN
            SUM=0.0
            DO 10 J=1,NN
            SUM=SUM+(ELM(I,J)-A2*ELK(I,J))*ELU(J)
     10     ELK(I,J)=ELM(I,J)+A1*ELK(I,J)
     20     ELF(I)=(A1+A2)*ELF(I)+SUM
        RETURN
        END
```

Solution of Linear Equations

B.1 Introduction

All of the steady-state iteration methods and most of the time-dependent solution methods described throughout the text lead ultimately to the point where at least one set (matrix) of linear algebraic equations must be solved. In general terms, the system

$$\tilde{\mathbf{K}}\mathbf{U}^* = \tilde{\mathbf{F}} \tag{B.1.1}$$

must be solved where $\tilde{\mathbf{K}}$ is a banded, sparse matrix that may be either symmetric or unsymmetric depending on the characteristics of the partial differential equation describing the physical problem and perhaps the nonlinear solution algorithm. The solution vector (nodal point variables) $\mathbf{U}^*$ may contain several degrees of freedom per node or a single variable, again depending on the specific problem. Other characteristics for $\tilde{\mathbf{K}}$, such as whether or not it is positive definite, and the size of its condition number, are also very dependent on the type of finite element model and the particular solution algorithm. However, these characteristics are extremely important, since they strongly influence the type of matrix solution procedure that can be successfully used for any given problem.

A linear set of equations can be solved by either a direct or iterative method. Direct solvers, like the Gauss elimination method, are often used to solve systems of algebraic equations. They provide the solution after a fixed number of steps and are less sensitive to the conditioning of the matrix. However, the main deficiency of the direct solvers is that they require the coefficient matrix be stored in an ordered format to enhance the matrix band structure and reduce storage requirements. In recent years, the direct solvers have been refined to reduce this deficiency through innovative data management techniques (e.g., frontal solvers, skyline solvers, and others; see Carey and Oden [1]). These improvements enable users to solve moderately large systems of equations efficiently. However, they have been found to be unsuitable for solving very large systems of equations (especially in three-dimensional problems) because they demand out-of-core storage of the equations and hence require large data transfers. In addition, direct methods are difficult to organize for efficient use on multiprocessor, parallel computers and are therefore seeing reduced utilization.

The limitations on CPU time and storage requirements make the use of direct solvers not economical, even on present-day high speed and high memory computers, and using direct solvers for complex problems with more than a quarter million equations is impractical. For large systems, iterative methods are more efficient in that they require less storage and CPU time while giving comparable accuracy in the solution. This is due to the fact that the global matrix formation may be avoided, and the major operation is the matrix-vector multiplication as compared to the matrix reduction in direct methods. Another advantage of the various iterative methods is that the solution algorithm can be effectively performed in parallel on an array of processors. Iterative methods are well suited to many of the linear algebraic equation sets generated by finite element models. However, the methodology is not universal in its application. The matrix problems arising from constrained finite element models, such as incompressible flow or quasi-static electromagnetics, present significant challenges to iterative methods with the major failing being the lack of a suitable preconditioner.

In the following sections a brief outline of some methods for the solution of linear algebraic equations will be given. Further details for this topic, which is outside the major focus of this book, can be found in references such as [2–4].

B.2 Direct Methods

By *direct methods*, we mean those in which simultaneous linear algebraic equations are solved exactly, assuming negligible computational round-off error, by successive elimination of variables and back substitution. The Gauss elimination method is a direct technique [2–4], and frontal [5] and skyline [6] solution methods are examples of direct solution methods that use the Gauss elimination technique efficiently. Direct solution methods involve a fixed number of steps to determine the solution. The direct techniques are useful when the number of equations involved is not too large. The number of operations (i.e., multiplication, division, etc.) for Gauss elimination is of the order $n^3/3 + O(n^2)$, where n denotes the number of operations.

For a comparable direct method, we refer to the *frontal* solution procedure [5]. Because it is faster than most direct solvers, it requires less core space as long as active variables can be kept in the core, and it allows for partial pivoting. An additional advantage is that no stringent node numbering scheme is needed, though a judicious element numbering helps to minimize the front width. Details of the frontal method will not be given here as it is a sophisticated implementation of the simple Gauss elimination method which is adequately studied in numerous texts. Implementation issues for the frontal method can be found in [5].

Due to the fact that the approximation functions are defined only within an element, the coefficient matrix in the finite element method is banded, i.e., $K_{ij} = 0$ for $j > i + m_b$, where m_b is the half bandwidth of the matrix

$[K]$. This greatly reduces the number of operations, if we make note of the fact that elements outside the bandwidth are zero. Of course, the bandwidth size depends on the global node numbering. The skyline technique is one in which bandedness of the finite element equations is exploited by storing the row number m_j of the first nonzero element in column j. The variables $m_i, i = 1, 2, \cdots, n$, define the skyline of the matrix. For additional details see Bathe [6].

B.3 Iterative Methods

B.3.1 General Comments

Among the various iterative methods, the Conjugate Gradient (CG) method [7] is most widely used because it is a finite step method (i.e., apart from round-off errors, the solution is achieved in a fixed number of iterations) and it can be used to determine the inverse. However, the number of iterations required depends on the condition number of the coefficient matrix (the condition number of the coefficient matrix is the ratio of the largest to smallest eigenvalue). The convergence of the conjugate gradient method, and iterative methods in general, can be improved by preconditioning and/or scaling the equations [8–10].

The limitations on storage can be overcome by solving the equations at the element level, i.e., use the Gauss–Seidel iteration idea for the set of variables associated with the element. This approach avoids assembly of element matrices to form the global coefficient matrix. This idea of using the element-by-element data structure of the coefficient matrix was first pointed out by Fox and Stanton [11] and Fried [12–14]. The phrase *element-by-element* refers to a particular data structure for finite element techniques wherein information is stored and maintained at the element level rather than assembled into a global data structure. In this method the matrix-vector multiplications are carried out at the element level and the assembly is carried out on the resultant vector. This idea proves to be very attractive when solving large problems, because the matrix–vector multiplication can be done in parallel on a series of processors. Another advantage of this method is that the resultant savings in storage, compared to direct solvers, allows solution of large problems on small computers. A review of the literature on element-by-element algorithms is presented in [15], and they have been investigated by numerous investigators [16–34].

In summary, for iterative solution methods the advantages of the element-by-element data structure over assembling the global coefficient matrix are

1. the need for formation and storage of a global matrix is eliminated, and therefore the total storage and computational costs are low,

2. the amount of storage is independent of the node numbering and mesh topology and depends on the number and type of elements in the mesh,

and

3. the element-by-element solution algorithms can be vectorized for efficient use on supercomputers.

The major disadvantage of the element-by-element data structure is the limited number of preconditioners that can be formulated from the unassembled matrices. This becomes of critical importance when the linear system is not well-conditioned as in the mixed method, incompressible flow model.

B.3.2 Solution Algorithms

In this section, we review three iterative solvers from [15] that are applicable to nonsymmetric, positive definite equation systems that are typical of isothermal flow algorithms. The three iterative solution schemes used here are the Biorthogonal Conjugate Gradient method [9], the Lanczos ORTHORES [9], and the GMRES [31].

The conjugate gradient method for solving a system of equations can be interpreted as the search for the minimum of the energy E of the system. The energy of the system is a minimum when the residual vector $\mathbf{r} = \tilde{\mathbf{F}} - \tilde{\mathbf{K}}\mathbf{U}^*$ vanishes. The algorithm for the biorthogonal conjugate gradient method (also known as two-term form of the steepest descent method, Lanczos/ORTHOMIN) for unsymmetric systems of equations [9,15] is given in Table B.3.1, and the steps involved in the Lanczos ORTHORES solution algorithm [9,15] are given in Table B.3.2.

Table B.3.1: Steps involved in using biorthogonal conjugate gradient method (Lanczos/ORTHOMIN solver).

Repeat the following steps for each nonlinear iteration:

I. *Initial Calculations:*

(1) Form the element stiffness matrix $\tilde{\mathbf{K}}^e$ and force vector $\tilde{\mathbf{F}}^e$.

(2) Apply essential and/or natural boundary conditions, and modify $\tilde{\mathbf{K}}^e$ and $\tilde{\mathbf{F}}^e$.

(3) Store the element matrices in $\bar{\mathbf{A}}$ (whose dimensions are nem, neleq, neleq).†

---

† nem = number of elements in the finite element mesh, neleq = number of element equations.

(4) Store the inverse of the diagonal terms of the global system in $\mathbf{W}$ ($\mathbf{W}_{ii} = \sum_{e=1}^{\text{nem}} \tilde{K}_{ii}^{-1}$).

(5) Assemble the global force vector.

II. *Preconditioning:*

Form the preconditioned system of equations

$$\bar{\mathbf{K}}\bar{U} = \bar{\mathbf{F}}; \quad \bar{\mathbf{K}} = \mathbf{W}^{-1/2}\tilde{\mathbf{K}}\mathbf{W}^{-1/2}, \quad \bar{U} = \mathbf{W}^{1/2}U^*, \quad \bar{\mathbf{F}} = \mathbf{W}^{-1/2}\tilde{\mathbf{F}}$$

III. *Lanczos ORTHOMIN Algorithm:*

(1) For known initial solution vector $\bar{U}^0$, compute:

$$\mathbf{r}^0 = \bar{\mathbf{F}} - \bar{\mathbf{K}}\bar{U}^0, \quad \mathbf{P}^0 = \mathbf{r}^0, \quad \tilde{\mathbf{r}}^0 = \tilde{\mathbf{P}}^0 = \mathbf{r}^0,$$

$$\alpha_0 = 0, \quad \lambda_0 = \frac{(\mathbf{r}^0, \tilde{\mathbf{r}}^0)}{(\bar{\mathbf{K}}\mathbf{P}^0, \tilde{\mathbf{r}}^0)}, \quad \bar{U}^1 = \bar{U}^0 + \lambda_0 \mathbf{P}^0$$

(2) For each ORTHOMIN iteration $m = 1, 2, 3, \ldots$, compute§:

$$\lambda_m = \frac{(\mathbf{r}^m, \tilde{\mathbf{r}}^m)}{(\bar{\mathbf{K}}\mathbf{P}^m, \tilde{\mathbf{r}}^m)}, \quad \alpha_m = \frac{(\mathbf{r}^m, \tilde{\mathbf{r}}^m)}{(\mathbf{r}^{m-1}, \tilde{\mathbf{r}}^{m-1})},$$

$$\mathbf{P}^m = \mathbf{r}^m + \alpha_m \mathbf{P}^{m-1}, \quad \tilde{\mathbf{P}}^m = \tilde{\mathbf{r}} + \alpha_m \tilde{\mathbf{P}}^{m-1},$$

$$\mathbf{r}^{m+1} = \mathbf{r}^m - \lambda_m \bar{\mathbf{K}}\mathbf{P}^m, \quad \tilde{\mathbf{r}}^{m+1} = \tilde{\mathbf{r}}^m - \lambda_m \bar{\mathbf{K}}^T \tilde{\mathbf{P}}^{m-1},$$

$$\bar{U}^{m+1} = \bar{U}^m + \lambda_m \mathbf{P}^m$$

(3) Convergence criterion: $\|\bar{U}^{m+1}\|/\|\mathbf{r}^0\| \leq 10^{-6}$

(4) If convergence criterion is satisfied $\bar{U}^* = \mathbf{W}^{-1/2}\bar{U}^{m+1}$

The third iterative solver uses the GMRES solution algorithm. For an approximate solution of the form $\bar{U}_0 + \mathbf{z}$, where $\bar{U}_0$ is the initial guess vector and $\mathbf{z}$ is a member of the Krylov space $\mathcal{K}$ of dimension k, the GMRES algorithm determines the vector $\mathbf{z}$ such that $\| \tilde{\mathbf{F}} - \tilde{\mathbf{K}}(\bar{U}_0 + \mathbf{z}) \|$ is minimized, where $\| \cdot \|$ denotes the L_2–norm. The Krylov space is given by $\mathcal{K} = span\{\bar{U}_0, \tilde{\mathbf{K}}\bar{U}_0, \tilde{\mathbf{K}}^2\bar{U}_0, \cdots \tilde{\mathbf{K}}^{k-1}\bar{U}_0\}$. Therefore, when solving large systems of equations, as the value of k increases, the amount of storage required

---

§ $(a, b) = \sum a_i b_i.$

Table B.3.2: Steps involved in using Lanczos/ORTHORES solver.

Repeat the following steps for each nonlinear iteration:

I. *Initial Calculations:*

   (1) Form the element stiffness matrix $\tilde{\mathbf{K}}^e$ and force vector $\tilde{\mathbf{F}}^e$.

   (2) Apply the boundary conditions, and modify $\tilde{\mathbf{K}}^e$ and $\tilde{\mathbf{F}}^e$.

   (3) Store the element matrices in $\bar{\mathbf{A}}$ (whose dimensions are nem, neleq, neleq).

   (4) Store the inverse of the diagonal terms of the global system in $\mathbf{W}$ ($\mathbf{W}_{ii} = \sum_{e=1}^{\text{nem}} \tilde{K}_{ii}^{-1}$).

   (5) Assemble the global force vector.

II. *Preconditioning:*

   Form the preconditioned system of equations

$$\bar{\mathbf{K}}\bar{\mathbf{U}} = \bar{\mathbf{F}}, \quad \bar{\mathbf{K}} = \mathbf{W}^{-1/2}\tilde{\mathbf{K}}\mathbf{W}^{-1/2}, \quad \bar{\mathbf{U}} = \mathbf{W}^{1/2}\tilde{\mathbf{U}}, \quad \bar{\mathbf{F}} = \mathbf{W}^{-1/2}\tilde{\mathbf{F}}$$

III. *Lanczos ORTHORES Algorithm:*

   (1) For known initial solution vector $\bar{\mathbf{U}}^0$, compute: $\mathbf{r}^0 = \bar{\mathbf{F}} - \bar{\mathbf{K}}\bar{\mathbf{U}}^0$

   (2) For each ORTHORES iteration $m = 0, 1, 2, \ldots$, compute ($\tilde{\mathbf{r}} = \mathbf{r}_0, \lambda^0 = 0$):

$$\lambda^{m+1} = \frac{(\mathbf{r}^m, \tilde{\mathbf{r}}^m)}{(\bar{\mathbf{K}}\mathbf{r}^m, \tilde{\mathbf{r}}^m)}$$

$$\beta^{m+1} = \begin{cases} 1 & ; \text{ if } m = 0 \\ \left[1 - \frac{\lambda^{m+1}}{\lambda^m}\frac{(\mathbf{r}^m, \tilde{\mathbf{r}}^m)}{(\mathbf{r}^{m-1}, \tilde{\mathbf{r}}^{m-1})}\frac{1}{\beta_m}\right]^{-1} & ; \text{ if } m \geq 1 \end{cases}$$

$$\mathbf{r}^{m+1} = \beta^{m+1}\left(\mathbf{r}^m - \lambda^{m+1}\bar{\mathbf{K}}\mathbf{r}^m\right) + \left(1 - \beta^{m+1}\right)\mathbf{r}^{m-1}$$

$$\tilde{\mathbf{r}}^{m+1} = \beta^{m+1}\left(\tilde{\mathbf{r}}^m - \lambda^{m+1}\bar{\mathbf{K}}\tilde{\mathbf{r}}^m\right) + \left(1 - \beta^{m+1}\right)\tilde{\mathbf{r}}^{m-1}$$

$$\bar{\mathbf{U}}^{m+1} = \beta^{m+1}\left(\bar{\mathbf{U}}^m + \lambda^{m+1}\mathbf{r}^m\right) + \left(1 - \beta^{m+1}\right)\bar{\mathbf{U}}^{m-1}$$

   (3) Convergence criterion: $\|\bar{\mathbf{U}}^{m+1}\|/\|\mathbf{r}^0\| \leq 10^{-6}$

   (4) If convergence criterion is satisfied $\mathbf{U}^* = \mathbf{W}^{-1/2}\bar{\mathbf{U}}^{m+1}$

also increases. This drawback can be overcome by employing the GMRES algorithm iteratively by using a smaller value for k and restarting the algorithm after every k steps. The restart version of the GMRES algorithm [31,15] is explained in Table B.3.3.

Table B.3.3: Steps involved in using the GMRES solver.

Repeat the following steps for each nonlinear iteration:

I. *Initial Calculations:*

   (1) Form the element stiffness matrix $\tilde{\mathbf{K}}^e$ and force vector $\tilde{\mathbf{F}}^e$.

   (2) Apply the boundary conditions, and modify $\tilde{\mathbf{K}}^e$ and $\tilde{\mathbf{F}}^e$.

   (3) Store the element matrices in $\bar{\mathbf{A}}$ (whose dimensions are nem, neleq, neleq).

   (4) Store the inverse of the diagonal terms of the global system in $\mathbf{W}$ ($W_{ii} = \sum_{e=1}^{\text{nem}} \tilde{K}_{ii}^{-1}$).

   (5) Assemble the global force vector.

II. *Preconditioning:* Form the preconditioned system of equations

   $$\bar{\mathbf{K}}\bar{\mathbf{U}} = \bar{\mathbf{F}}, \quad \bar{\mathbf{K}} = \mathbf{W}^{-1/2}\tilde{\mathbf{K}}\mathbf{W}^{-1/2}, \quad \bar{\mathbf{U}} = \mathbf{W}^{1/2}\tilde{\mathbf{U}}, \quad \bar{\mathbf{F}} = \mathbf{W}^{-1/2}\tilde{\mathbf{F}}$$

III. *GMRES Algorithm:*

   (1) *Start:* Choose $\bar{\mathbf{U}}^0$ and compute

   $$\mathbf{r}^0 = \bar{\mathbf{F}} - \bar{\mathbf{K}}\bar{\mathbf{U}}^0, \text{ and}$$

   $$\mathbf{v}^1 = \bar{\mathbf{U}}^0 / \| \bar{\mathbf{U}}^0 \|$$

   (2) *Iterate:* For $j = 1, 2..., k$ do:

   $$h_{i,j} = (\bar{\mathbf{K}}v_j, v_i), \quad i = 1, 2, ..., j,$$

   $$\hat{v}_{j+1} = \bar{\mathbf{K}}v_j - \sum_{i=1}^{j} h_{i,j}v_i,$$

   $$h_{j+1,j} = \| \hat{v}_{j+1} \|, \text{ and}$$

   $$v_{j+1} = \hat{v}_{j+1} / h_{j+1,j}$$

   (3) *Form the approximate solution:*

   $$\bar{\mathbf{U}}^k = \bar{\mathbf{U}}^0 + \mathbf{V}\mathbf{y}, \text{ where } \mathbf{y} \text{ minimizes } \| \mathbf{e} - \bar{\mathbf{H}}\mathbf{y} \|, \mathbf{y} \in \mathbf{R}^k.$$

(4) *Restart:*

Compute $\mathbf{r}^k = \bar{\mathbf{F}} - \bar{\mathbf{K}}\bar{\mathbf{U}}^k$;

check convergence; if satisfied stop; otherwise, compute

$\bar{\mathbf{U}}^0 := \bar{\mathbf{U}}^k$, $\mathbf{v}_1 := \bar{\mathbf{U}}^k/\|\bar{\mathbf{U}}^k\|$, and go to step 2.

(5) Convergence criterion: $\|\bar{\mathbf{U}}^{m+1}\|/\|\mathbf{r}^0\| \leq 10^{-6}$

(6) If convergence criterion is satisfied $\mathbf{U}^* = \mathbf{W}^{-1/2}\bar{\mathbf{U}}$

where $\mathbf{V}$ is a $N \times k$ matrix whose columns are 1–2 orthonormal basis vectors $\{\mathbf{v}_1, \mathbf{v}_2, \ldots, \mathbf{v}_k\}$, $\mathbf{e} = \{\|\bar{\mathbf{U}}^0\|, 0, \ldots, 0\}$, and $\bar{\mathbf{H}}$ is the upper $k \times k$ Hessenberg matrix whose entries are the scalars $h_{i,j}$. When using the restart version of the GMRES algorithm, the total number of iteration m can be computed from the number of restarts and the dimension of k.

The presence of a penalty matrix in the global coefficient matrix of the penalty model spoils the condition number. This results in slow convergence when using iterative solvers. However, the convergence of the iterative solvers can be improved by preconditioning the system. In [15], the system of equations is transformed using diagonal scaling matrix (Jacobi/diagonal preconditioning). Accordingly, the system of equations

$$\tilde{\mathbf{K}}\mathbf{U}^* = \tilde{\mathbf{F}} \tag{B.3.1}$$

becomes

$$\bar{\mathbf{K}}\bar{\mathbf{U}} = \bar{\mathbf{F}} \tag{B.3.2}$$

$$\bar{\mathbf{K}} = \mathbf{W}^{-1/2}\tilde{\mathbf{K}}\mathbf{W}^{-1/2}; \quad \bar{\mathbf{U}} = \mathbf{W}^{1/2}\mathbf{U}^*; \quad \bar{\mathbf{F}} = \mathbf{W}^{-1/2}\tilde{\mathbf{F}} \tag{B.3.3}$$

where $W_{ii} = \tilde{K}_{ii}^{-1}$ is a diagonal matrix. During the matrix multiplication, the element-by-element data structure is exploited and the multiplications are carried out at the element level and the residuals are then assembled to form the global vector. The diagonal terms are always positive because of the viscous and penalty terms.

References for Additional Reading

1. G. F. Carey and J. T. Oden, *Finite Elements: Computational Aspects*, Prentice Hall, Englewood Cliffs, New Jersey (1984).

2. K. E. Atkinson, *An Introduction to Numerical Analysis*, Wiley, New York (1978).

3. B. Carnahan, H. A. Luther, and J. O. Wilkes, *Applied Numerical Methods*, Wiley, New York (1969).

4. D. K. Fadeev and V. N. Fadeeva, *Computational Methods of Linear Algebra*, Freeman, San Francisco, California (1963).

5. P. Hood, "Frontal Solution Program for Unsymmetric Matrices," *International Journal for Numerical Methods in Engineering*, **10**, 379–399 (1976); also see **10**, 1055 (1976) for a correction.

6. K. J. Bathe, *Finite Element Procedures*, Prentice Hall, Englewood Cliffs, New Jersey (1998).

7. M. R. Hestenes and E. L. Stiefel, "Methods of Conjugate Gradients for Solving Linear Systems," *National Bureau of Standards Journal of Research*, **49**, 409–436 (1952).

8. G. H. Golub and C. F. Van Loan, *Matrix Computations*, Second Edition, The Johns Hopkins University Press, Baltimore, Maryland (1989).

9. K. C. Jea and D. M. Young, "On the Simplification of Generalized Conjugate–Gradient Methods for Nonsymmetrizable Linear Systems," *Linear Algebra Applications*, **52**, 399–417 (1983).

10. D. J. Evans, "Use of Preconditioning in Iterative Methods for Solving Linear Equations with Symmetric Positive Definite Matrices," *Computer Journal*, **4**, 73–78 (1961).

11. R. L. Fox and E. L. Stanton, "Developments in Structural Analysis by Direct Energy Minimization," *AIAA Journal*, **6**, 1036–1042 (1968).

12. I. Fried, "Gradient Methods for Finite Element Eigenproblems," *AIAA Journal*, **7**, 739–741 (1969).

13. I. Fried, "More on Generalized Iterative Methods in Finite Element Analysis," *AIAA Journal*, **7**, 565–567 (1969).

14. I. Fried, "A Gradient Computational Procedure for the Solution of Large Problems Arising from the Finite Element Discretization Method," *International Journal for Numerical Methods in Engineering*, **2**, 477–494 (1970).

15. M. P. Reddy, J. N. Reddy and H. U. Akay, "Penalty Finite Element Analysis of Incompressible Flows Using Element by Element Solution Algorithms," *Computer Methods in Applied Mechanics and Engineering*, **100**, 169-205 (1992).

16. A. J. Wathen, "An Analysis of Some Element-By-Element Techniques," *Computer Methods in Applied Mechanics and Engineering*, **74**, 271–287 (1989).

17. T. J. R. Hughes, I. Levit, and J. Winget, "An Element-By-Element Implicit Algorithm for Heat Conduction," *Journal of the Engineering Mechanics Division, ASCE,* **109**(2), 576–585 (1983).

18. J. M. Winget and T. J. R. Hughes, "Solution Algorithms for Nonlinear Transient Heat Conduction Analysis Employing Element-By-Element Iterative Strategies," *Computer Methods in Applied Mechanics and Engineering,* **52**, 711–815 (1985).

19. G. F. Carey and B. Jiang, "Element-By-Element Linear and Nonlinear Solution Schemes," *Communications in Applied Numerical Methods,* **2**, 145–153 (1986).

20. J. L. Hayes and P. Devloo, "An Element-By-Element Block Iterative Method for Large Non–Linear Problems," in *Innovative Methods for Nonlinear Behavior,* W. K. Liu et al. (eds.), Pineridge Press, Swansea, U.K., 51–62 (1985).

21. E. K. Buratynski, "An Element-By-Element Method for Heat Conduction CAE Including Composite Problems," *International Journal for Numerical Methods in Engineering,* **26**, 199–215 (1988).

22. I. Levit, "Element By Element Solvers of Order N," *Computers & Structures,* **27**(3), 357–360 (1987).

23. A. de La Bourdinnaye, "The Element By Element Method as a Preconditioner for Linear Systems Coming from Finite Element Models," *International Journal of Supercomputer Applications,* **3**, 60–68 (1989).

24. K. V. G. Prakhya, "Some Conjugate Gradient Methods for Symmetric and Nonsymmetric Systems," *Communications in Applied Numerical Methods,* **4**, 531–539 (1988).

25. G. F. Carey, E. Barragy, R. Mclay, and M. Sharma, "Element-by-Element Vector and Parallel Computations," *Communications in Applied Numerical Methods,* **4**, 299–307 (1988).

26. L. J. Hayes and P. Devloo, "A Vectorized Version of a Sparse Matrix-Vector Multiplication," *International Journal for Numerical Methods in Engineering,* **23**, 1043–1056 (1986).

27. F. Shakib, T. J. R. Hughes, and Z. Johan, "A Multi-Element Group Preconditioned GMRES Algorithm for Nonsymmetric Systems Arising in Finite Element Analysis," *Computer Methods in Applied Mechanics and Engineering,* **75**, 415–456 (1989).

28. T. E. Tezduyar and J. Liou, "Element-by-Element and Implicit-Explicit Finite Element Formulations for Computational Fluid Dynamics," *First International Symposium on Domain Decomposition Methods for Partial Differential Equations,* R. Glowinski, G. H. Golub, G. A. Meurant, and J. Periaux (eds.), SIAM, Philadelphia, Pennsylvania, 281–300 (1988).

29. T. E. Tezduyar and J. Liou, "Grouped Element-By-Element Iteration Schemes for Incompressible Flow Computations," *Computational Physics Communications*, **53**, 441–453 (1989).

30. J. Liou and T. E. Tezduyar, "A Clustered Element-By-Element Iteration Method for Finite Element Computations," University of Minnesota Supercomputer Institute Research Report, 90/116 (1990).

31. Y. Saad and M. H. Schultz, "GMRES: A Generalized Minimal Residual Algorithm for Solving Nonsymmetric Linear Systems," *SIAM Journal of Scientific and Statistical Computations*, **7**(3), 856–869 (1986).

32. J. A. Mitchell and J. N. Reddy, "A Multilevel Hierarchical Preconditioner for Thin Elastic Solids," *International Journal for Numerical Methods in Engineering*, **43**, 1383–1400 (1998).

33. J. A. Mitchell and J. N. Reddy, "A High Performance Iterative Solution Procedure for the Analysis of Structural Problems," *Journal of High Performance Computing*, **5**(1), 3–13 (1999).

34. J. A. Mitchell and J. N. Reddy, "A Hierarchical Iterative Procedure for the Analysis of Composite Laminates," *Computer Methods in Applied Mechanics and Engineering*, **181** 237–260 (2000).

Fixed Point Methods
and Contraction Mappings

C.1 Fixed Point Theorem

The discussion in this section will focus on the use of fixed point iteration schemes for the solution of the system of nonlinear equations of the form

$$\mathcal{R}(\mathbf{x}) = \mathbf{0} \tag{C.1.1}$$

or its fixed point form

$$\mathbf{x} = \mathcal{G}(\mathbf{x}) \equiv \mathbf{x} - \mathcal{R}(\mathbf{x}) \tag{C.1.2}$$

where $\mathbf{x}$ represents a vector of unknowns. Solutions of Eq. (C.1.2) are called *fixed points* of the mapping $\mathcal{G}$, because they are unchanged by the operation of $\mathcal{G}$. The form of (C.1.2) immediately suggests the following iteration scheme [1]:

$$\mathbf{x}^{n+1} = \mathcal{G}(\mathbf{x}^n) \tag{C.1.3}$$

where $n = 0, 1, \cdots$. Equation (C.1.3) is the method of successive substitutions known as the Picard iteration method. The convergence of the method in (C.1.3) is guaranteed by the *Banach Fixed Point Theorem*, which gives the sufficient conditions for the existence and uniqueness of solutions (see Reddy [1], pp. 202–208). Before we present the theorem, it is necessary to define a contraction mapping.

An operator $\mathcal{G}$ is termed a *contraction mapping* if there exists a positive number λ, $0 < \lambda < 1$, such that

$$\| \mathcal{G}(\mathbf{x}) - \mathcal{G}(\mathbf{y}) \| \leq \lambda \| \mathbf{x} - \mathbf{y} \| \tag{C.1.4}$$

holds for all $\mathbf{x}, \mathbf{y}$ in the closed ball $\mathcal{N} = \{\mathbf{x} : \| \mathbf{x} - \mathbf{x_0} \| \leq r\}$ of a normed space $\mathcal{V}$, with center at $\mathbf{x_0}$ and radius r. Here λ denotes the contraction factor. The notation $\| \cdot \|$ is an appropriate norm for the operator in $\mathcal{N}$ (see [1]). It is clear from Eq. (C.1.4) that a contraction mapping "contracts distances": the distance between the images $\mathcal{G}(\mathbf{x})$ and $\mathcal{G}(\mathbf{y})$ is smaller by a scale factor of λ than the distance between the elements $\mathbf{x}$ and $\mathbf{y}$.

Theorem: Let $\mathcal{G}(\mathbf{x})$ be a contraction mapping on a closed subset $\mathcal{N}$ of a closed normed space (or a Banach space) $\mathcal{V}$, with a contraction factor λ and let $\mathbf{x}_0$ be such that

$$\frac{1}{1-\lambda} \parallel \mathcal{G}(\mathbf{x}_0) - \mathbf{x}_0 \parallel = r_0 \leq r$$

Then

1. The sequence defined by $\mathbf{x}^{n+1} = \mathcal{G}(\mathbf{x}^n)$ converges to a point $\mathbf{x}^*$ in $\mathcal{N}$.

2. $\mathbf{x}^*$ is a fixed point of the operator $\mathcal{G}$.

3. $\mathbf{x}^*$ is the unique fixed point of $\mathcal{G}$ in $\mathcal{N}$.

The proof of this theorem and a detailed background on nonlinear solution methods are available in [1,2].

Based on the above theorem, a further result can be obtained if a few restrictions are placed on $\mathcal{G}$. Assume that (C.1.3) has a solution $\mathbf{x}^*$ and that $\mathcal{G}$ has a continuous Jacobian that satisfies

$$\parallel \mathcal{G}'(\mathbf{x}) \parallel_\infty \leq \lambda < 1 \qquad\qquad (C.1.5)$$

where $\parallel \cdot \parallel_\infty$ denotes the *sup–norm* ("sup" stands for supremum). Then it can be shown that for any initial estimate $\mathbf{x}^0$ satisfying

$$\parallel \mathbf{x}^0 - \mathbf{x}^* \parallel_\infty \leq r \qquad\qquad (C.1.6)$$

the iterative scheme in Eq. (C.1.3) will converge to the unique solution $\mathbf{x}^*$. The conditions (C.1.5) and (C.1.6) basically state that the norm of the Jacobian of the iteration operator must be less than unity and the initial estimate for the solution must be within the ball of contraction, if the fixed point scheme is to converge. The quantity r is also referred to as the radius of convergence of the iteration scheme.

Generally, it is quite difficult to prove that a given operator $\mathcal{G}$ is a contraction mapping, especially for the systems generated by finite element applications. It is somewhat easier to use (C.1.5) and estimate the norm of $\mathcal{G}'$ to test for a convergent process. However, the real benefit from the ideas of fixed point operators is derived from the iteration scheme in (C.1.3) and its generalization,

$$\mathbf{x}^{n+1} = \mathbf{x}^n - \alpha \mathbf{A}(\mathbf{x}^n)\mathcal{F}(\mathbf{x}^n) \qquad\qquad (C.1.7)$$

where α is a constant and $\mathbf{A}$ is a nonsingular matrix function of $\mathbf{x}$. In the following, a few of the basic iteration schemes will be defined based on the definition of $\mathbf{A}$.

C.2 Chord Method

The simplest choice for $\mathbf{A}$ is a constant which defines the parallel chord method. For a one degree-of-freedom problem, this becomes

$$x^{n+1} = x^n - af(x^n) \qquad \text{(C.2.1)}$$

which is shown graphically in Figure C.2.1. In the multi-dimensional case, the line with slope $\frac{1}{a}$ is replaced by a hyperplane. The behavior of this method depends strongly on the form of $\mathbf{A}$. One possibility is to set $\mathbf{A} = \alpha\mathbf{I}$ ($\mathbf{I}$ is the identity matrix) in which case each component of $\mathbf{x}$ is corrected in proportion to α. If $\mathbf{A} = \alpha^T\mathbf{I}$ where α is a vector, then each component of $\mathbf{x}$ can be corrected individually. The problem remains, however, as to how to select α.

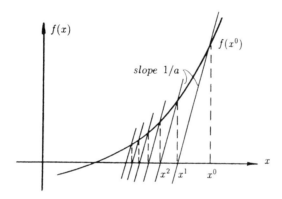

Figure C.2.1: Illustration of the chord method for one degree-of-freedom problem.

C.3 Newton's Method

Extending the idea that $\mathbf{A}$ is in some sense an inverse "slope", it is appropriate to relate $\mathbf{A}$ to the derivative of $\mathcal{F}$. Therefore, let

$$\mathbf{A} = \left[\frac{\partial \mathcal{F}}{\partial \mathbf{x}}\right]^{-1} = \mathbf{J}^{-1} \qquad \text{(C.3.1)}$$

in which case Eq. (C.1.7) becomes

$$\mathbf{x}^{n+1} = \mathbf{x}^n - \alpha\mathbf{J}^{-1}(\mathbf{x}^0)\mathcal{F}(\mathbf{x}^n) \qquad \text{(C.3.2)}$$

where the Jacobian is evaluated at $\mathbf{x}^0$. For the one degree-of-freedom case, this becomes ($\alpha = 1$)

$$x^{n+1} = x^n - \frac{1}{f'(x^0)}f(x^n) \qquad \text{(C.3.3)}$$

which is illustrated in Figure C.3.1. This is recognized as a modified Newton's method since the Jacobian is not updated at each iteration.

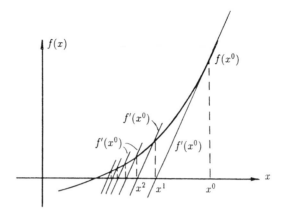

Figure C.3.1: Illustration of modified Newton's method.

C.4 The Newton–Raphson Method

The modified Newton's scheme suggests a rational manner in which to select **A**. However, its convergence is little better than the chord method due to the use of a constant Jacobian. Allowing the Jacobian to change at each iteration produces the standard Newton (or Newton–Raphson) procedure

$$\mathbf{x}^{n+1} = \mathbf{x}^n - \alpha \mathbf{J}^{-1}(\mathbf{x}^n)\mathcal{F}(\mathbf{x}^n) \tag{C.4.1}$$

This is shown graphically in Figure C.4.1 for a one-component system. Note that if $\alpha \neq 1$ a damped Newton scheme is produced.

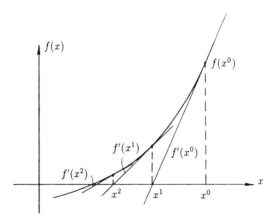

Figure C.4.1: Illustration of Newton's method.

C.5 Descent Methods

Descent methods reinterpret the general algorithm in (C.1.7) as

$$\mathbf{x}^{n+1} = \mathbf{x}^n - \alpha \mathbf{D}^n \qquad (C.5.1)$$

where $\mathbf{D}$ is a vector that determines the direction of the correction and α controls the magnitude of the correction. In addition, rather than directly finding the solution to (C.1.1), descent methods work by trying to minimize the function

$$R(\mathbf{x}) = \mathcal{F}^T(\mathbf{x})\mathcal{F}(\mathbf{x}) \qquad (C.5.2)$$

The simplest descent methods are the univariant or relaxation schemes in which $\mathbf{D}$ is selected so that only one component of $\mathbf{x}$ is corrected at each iteration. A better approach involves the gradients of R, such as the method of steepest descent. In this case

$$\mathbf{D}^n = \mathbf{g}^n = \nabla R(\mathbf{x}^n) \qquad (C.5.3)$$

The method of steepest descent has a linear rate of convergence.

Finally, the conjugate gradient method is a sequential method in which the correction vectors are computed from the recursion relation

$$\mathbf{D}^n = \nabla R(\mathbf{x}^n) + \beta^{n-1}\mathbf{D}^{n-1} \qquad (C.5.4)$$

Here the new correction vector is a combination of the gradient at $\mathbf{x}^n$ plus the old correction direction. The parameter β is defined to insure that the new and old correction vectors are properly orthogonal (conjugate).

Full details for all of the above algorithms and the general concepts behind nonlinear solution methods can be found in References 3–5.

References for Additional Reading

1. J. N. Reddy, *Applied Functional Analysis and Variational Methods in Engineering*, McGraw-Hill, New York (1986); reprinted by Krieger Publishers, Melbourne, Florida (1991).

2. M. R. Hestenes, *Optimization Theory: The Finite Dimensional Case*, John Wiley & Sons, New York (1975).

3. E. B. Becker, G. F. Carey, and J. T. Oden, *Finite Elements, An Introduction*, Vol. I, Prentice Hall, Englewood Cliffs, New Jersey (1981).

4. J. M. Ortega and W. C. Rheinboldt, *Iterative Solution of Nonlinear Equations in Several Variables*, Academic Press, New York (1970).

5. C. Johnson, *Numerical Solution of Partial Differential Equations by the Finite Element Method*, Cambridge University Press, Cambridge, U.K. (1987).

Subject Index

A

Absorbtivity, 23

Acceleration tensor, 10,11,162, 264

Acceleration vector, 98,187

Activation energy, 14

Adams–Bashforth predictor, 97,187

Adams–Bashforth–Moulton family, 100

Adaptive time step, 94,107

ALE method, 205,380

Ampere's law, 361

Amplification factor, 103

Amplification matrix, 103

Anisotropic thermal conductivity, 108,131

Approximation functions, 31
    also see, interpolation functions

Area coordinates, 52,85,118

Arrhenius form, 14,93,289

Assembly, 33,43–45

Axisymmetric problems, 27,49,197, 242

B

Backward difference scheme, 48,95, 96,103,186

Backward facing step, 233

Banach Fixed Point Theorem, 455

Banach space, 456

Bar element, 116

Best approximation, 38

Bilinear form, 37,51,156

Bingham fluid, 301,310,333

Bingham model, 301,310

Body force, 5,256,311

Boundary conditions
    contact, 119,124
    convection, 19,51,79,90
    convective heat transfer, 256,257

Dirichlet, 15
    displacement, 361
    electromagnetics, 371
    essential, 19
    natural, 19,321
    Neumann, 15
    porous flow, 18
    radiation, 19,91,142
    thermal, 19,33,79,298
    traction, 361

Boundary flux, 64

Boussinesq approximation, 5,256

Boussinesq hypothesis, 215,273

Brazing, 136

Brick elements, 83

Brinkman formulation, 18

Brinkman model, 11,18,287

Brinkman viscosity, 270

Bulk nodes, 125

Buoyancy, 5,12,273,336,400

C

Capacitance coefficient, 12,111

Capacitance matrix, 96,102

Capacitance method, 110

Carreau model, 300

Cauchy stress tensor, 5,311,359

Cauchy–Green tensor, 312,317

Cavity flow, 226,229,278,336,338

Change of phase, 21,269,276

Chemical reactions, 128,275

Chemically reactive flows
    finite element formulation, 274
    governing equations, 12–15

Chord method, 178,457

Codeformational derivative,
    see upper-convected derivative

Coefficient matrix, 39,82,155

Coefficient of thermal expansion, 6,256